Grassland: a global resource

Grassland:
a global resource

edited by:

D.A. McGilloway

Subject headings:
Agriculture
Environment
Land use

ISBN 907699871X

First published, 2005

Wageningen Academic Publishers
The Netherlands, 2005

The XX International Grassland Congress took place in Ireland and the UK in June-July 2005. The main congress took place in Dublin from 26 June to 1 July and was followed by post congress satellite workshops in Aberystwyth, Belfast, Cork, Glasgow and Oxford. The meeting was hosted by the Irish Grassland Association and the British Grassland Society.

Organising Committee of the XX International Grassland Congress

Mr J. Flanagan, President, Teagasc
Dr. F.P. O'Mara, Secretary, University College Dublin
Prof. R.J. Wilkins, British Grassland Society
Dr. P McFeely, Irish Grassland Association
Mr. N.P. McGill, Department of Agriculture and Food, Ireland
Prof. M.P Boland, University College Dublin
Prof. J.H. Roche, University College Dublin
Dr. M. Camlin, Department of Agriculture and Rural Development of Northern Ireland
Mrs. J.M. Crichton (deceased), British Grassland Society
Dr. T.F. Nolan, Teagasc.

Scientific Committee of the XX International Grassland Congress

Prof R.J. Wilkins, Chairperson
Mr. D.A. Davies
Dr. P. French
Prof. M.B. Jones
Dr. A.S. Laidlaw
Mrs. C.A. Marriott
Dr. D.A. McGilloway
Dr. P. O'Kiely

Reviewers for invited papers

M.T. Abberton	A. Kinsella	D. Scholefield
M. Askew	A.S. Laidlaw	L. Shalloo
I.R. Calder	L. 't Mannetje	P. Smith
M. Camlin	C.S. Mayne	R. Smith
D. Christian	R.J. Merry	J. Stevens
D.L. Easson	J.A. Milne	M. Walace
M. Gill	M.A. O'Donovan	R.J. Wilkins
A. Hameleers	N.W. Offer	B. Winter
N.M. Holden	R. Pakeman	I.A. Wright
R. Jones	B. Rees	

Supporters of the XX International Grassland Congress, Dublin

Department of Agriculture and Food, Ireland
Teagasc, Ireland
Bord Bia, Ireland
Allied Irish Banks, Plc
FBD Trust, Ireland
Environmental Protection Agency, Ireland
University College Dublin, Ireland
USDA-Agricultural Research Service
Food and Agriculture Organization of the United Nations
Department for Environment Food and Rural Affairs, UK
USDA-National Resources Conservation Service
USDA-Cooperative State Research, Education, and Extension Service
US Bureau of Land Management
US Environmental Protection Agency
Alltech (Ireland) Ltd
ACCBank, Ireland
Irish Farmers Journal
Greenvale Animal Feeds
Biotechnology and Biological Sciences Research Council, UK
Irish Farmers Association
British Grassland Society
Irish Grassland Association
American Society of Agronomy
Crop Science Society of America
Soil Science Society of America
Glanbia Plc
K & S Kali, Germany
Stapledon Memorial Trust, UK
Grassland Fertilizers Ltd, Ireland
Grazing Lands Conservation Initiative, USA

Organisations that assisted the attendance of delegates from developing countries

Development Cooperation Ireland
Swiss Agency for Development and Cooperation
CTA Technical Centre for Agricultural and Rural Cooperation ACP-EU
Joint FAO/IAEA Division of Nuclear Techniques in Food and Agriculture
Australian Centre for International Agricultural Research
Syngenta Foundation for Sustainable Agriculture, Switzerland
Japanese Society of Grassland Science
Ministry for Foreign Affairs of Finland

Foreword

This book contains a compilation of four plenary papers and twenty nine invited papers presented at the main congress of the XX International Grassland Congress held in University College Dublin, Ireland from 26 June to 1 July 2005. It is complemented by six other books arising from the congress as listed on the back cover: the book of offered papers from the main congress and five books containing the proceedings of five satellite workshops held immediately after the main congress at locations in Ireland and the UK (Aberystwyth, Belfast, Cork, Glasgow and Oxford). The workshops were designed to facilitate more in-depth presentations and discussions on more specialised topics of worldwide significance.

Grasslands: a Global Resource draws together contributions from leading researchers, educators, policy makers and farmers from around the world, to espouse current knowledge and understanding of this complex ecosystem, the ways in which it can be enhanced and utilised and where the research challenges are for the future. The following themes unite the book:
- Efficient production from grassland.
- Grassland and the environment.
- Delivering the benefits from grassland.

This volume offers the reader insights into likely future demands from grasslands, and potential methodologies to addressing these demands. Appropriately the implications for grasslands-based livestock production systems are considered against a rising demand for meat and milk in developing countries where it is estimated that by 2020, these developing countries will consume approximately 72 million metric tons (mmt) more meat and 152 mmt more milk compared to 2002/3, dwarfing developed-country increases of 9 and 18 mmt for meat and milk respectively. Meeting this demand will challenge all concerned with managing this global resource. In contrast to the life and death issues of securing a reliable food supply, the problems of the more developed affluent world are also considered e.g. extensification, biodiversity and how these objectives might be reconciled with maintaining a viable grassland based ruminant production industry. Alternative uses of grasslands e.g. as biofactories, fibre, biomass and energy production are also considered.

In theme two, the emphasis shifts from production, to a consideration of the environmental benefits and challenges associated with grasslands. These include the contribution of grazing land to carbon sequestration, the implications of the growing demand for scarce water resources and effects on biodiversity.

The final section deals with perhaps the most practical aspects of global grassland biology. How do we communicate knowledge and understanding of this complex system to those individuals most in need of the advice?

These then are the big canvasses on which the papers in this volume are based.

David McGilloway

Table of contents

Foreword   7

Section 1: Efficient production from grassland   11

Grassland in Ireland and the UK   13
M. Rath and S. Peel

Rising demand for meat and milk in developing countries: implications for grasslands-based livestock
production   29
C.L. Delgado

Improving the quality of products from grassland   41
N.D. Scollan, R.J. Dewhurst, A.P. Moloney and J.J. Murphy

Grass and forage improvement: temperate forages   57
C.J. Pollock, M.T. Abberton and M.O. Humphreys

Grass and forage plant improvement in the tropics and sub-tropics   69
L. Jank, C.B. do Valle and R.M.S. Resende

Foraging behaviour and herbage intake in the favourable tropics/sub-tropics   81
S.C. Da Silva and P.C. de F. Carvalho

Interactions between foraging behaviour of herbivores and grassland resources in the eastern Eurasian steppes   97
D. Wang, G. Han and Y. Bai

Strategies to mitigate seasonality of production in grassland-based systems   111
C. Porqueddu, S. Maltoni and J.G. McIvor

Overcoming seasonality of production: opportunities offered by forage conservation technologies   123
P. O'Kiely and A.G. Kaiser

Evolution of integrated crop-livestock production systems   137
M.H.Entz, W.D.Bellotti, J.M. Powell, S.V. Angadi, W. Chen, K.H. Ominski and B. Boelt

Adoption of tropical legume technology around the world: analysis of success   149
H.M. Shelton, S. Franzel and M. Peters

Grasses as biofactories: scoping out the opportunities   167
N. Roberts, K. Richardson, G. Bryan, C. Voisey, W. McNabb, T. Conner, M. Christey and R. Johnson

The potential of grassland and associated forages to produce fibre, biomass, energy or other feedstocks
for non-food and other sectors: new uses for a global resource   179
M.F. Askew

Section 2: Grassland & the environment   191

Grasslands for production and the environment   193
D.R. Kemp and D.L. Michalk

Soil microbial community: understanding the belowground network for sustainable grassland management   209
Y.G. Zhu, W.D. Kong, B.D. Chen, Z.B. Nan and P. Christie

Soil quality assessment and management   219
M.G. Kibblewhite

Water resources, agriculture and pasture: implications of growing demand and increasing scarcity   227
M.W. Rosegrant, R.A. Valmonte-Santos, S.A. Cline, C. Ringler and W. Li

Grassland productivity and water quality: a 21st Century issue   239
D.M. Nash and P.M. Haygarth

Global atmospheric change and its effect on managed grassland systems   251
A. Lüscher, J. Fuhrer and P.C.D. Newton

Grazing land contributions to carbon sequestration   265
R.F. Follett and G.E. Schuman

Methane and nitrous oxide emissions from grazed grasslands   279
H. Clark, C. Pinares-Patiño and C. deKlein

Relationships between biodiversity and production in grasslands at local and regional scales   295
A. Hector and M. Loreau

Enhancing grassland biodiversity and its consequences for grassland management and utilisation   305
J. Isselstein

Section 3: Delivering the benefits from grassland   **321**

Grassland and forage to improve livelihoods and reduce poverty   323
S.G. Reynolds, C. Batello, S. Baas and S. Mack

Participatory approach to common use grazing management in dry area developing countries   339
J.A. Tiedeman, A. Larbi, F. Ghassali and N. Battikha

Adoption of *Brachiaria* grasses in Mexico and Central America: a successful story   343
F. Holmann, P.J. Argel and C.E. Lascano

Improved livelihoods from grasslands; the case of Napier grass in smallholder dairy farms in Kenya   347
D.M. Mwangi, D. Romney, S. Staal, I. Baltenweck and S.W. Mwendia

Role of information and information providers in technology transfer   351
D.J. Undersander

Participatory research for smallholder livestock systems - applying common sense to complex problems   359
P.M. Horne and W.W. Stür

The contribution of participatory research: on-farm research   375
P.F. Fennessy, N.J. Daniels, S.A. Chadwick and P.A. Speck

Computer-based forage management tools: historical, current, and future applications   389
D.H. Hannaway, C.P.Q. Daly, D.F Chapman, B.B. Baker and A.S. Cooper

Paying for our keep: grasslands decision support in more-developed countries   403
A.D. Moore

Decision support for grassland systems in developing countries   415
P.K. Thornton

Keyword index   427

Author index   429

Section 1

Efficient production from grassland

Grassland in Ireland and the UK

M. Rath[1] and S. Peel[2]
[1]*University College Dublin, Department of Animal Science, Belfield, Dublin 4, Ireland,
Email: myles.rath@ucd.ie*
[2]*Rural Development Service, Department for Environment Food and Rural Affairs, Otley
Road, Lawnswood, Leeds LS16 5QT, UK*

Key points

1. Grassland is the dominant land use option in Ireland and the UK, and is characterised by a long growing season.
2. Dynamic, interactive systems of grassland management have been developed which combine high grass dry matter intakes with good sward quality. In the better grassland areas milk yields in excess of 7000 kg/cow are attainable with low levels of concentrate supplementation.
3. In the times to come, measures to protect the environment will constrain stocking rates, and fertiliser and manure use on intensive grassland enterprises.
4. A high proportion of beef and sheep farms participate in voluntary, EU-funded agri-environmental schemes that promote less intensive production systems and high standards of environmental protection.
5. Access for the public to, and conservation by farmers of, the countryside have become increasingly important in the last 20 years. In the future, grasslands will have to meet a variety of demands and be truly multifunctional.

Keywords: intensive, dairy, pollution, biodiversity, multifunctional

Introduction: background and context

Irish agriculture is overwhelmingly grass based, and concerned with the conversion of grass to milk, beef or sheep-meat. Agriculture in the UK is similar except that cereal production also assumes importance. Details on land use, livestock numbers and farm structure and type are given in Tables 1 and 2. Permanent grassland is by far the largest land use option for agricultural land in Ireland, accounting for almost 80% of the land area. This is almost twice the proportion in the EU-15 as a whole. Grassland also dominates in the UK, though cereal production accounts for 18% of the land area. This compares to 7% and 26% for Ireland and the EU-15 respectively. While grassland and rough grazing are the major land use options, it is worth noting that many dairy farms in the UK (and Ireland to a lesser extent) also use maize silage or whole-crop cereal silage in addition to grass silage.

The vast majority of farms in Ireland are owned and operated by farmers. Historically most farmland in the UK was tenanted, but today 66% of farms are owned by the occupying farmer. A large proportion of farm households in both countries also have a source of off-farm income. Dairy herds, usually concentrated on the more productive land, account for 19% and 11% of farms in Ireland and the UK respectively. However the most common farm type in Ireland is the beef cattle farm, while in the UK it is the sheep farm. Many of the beef cattle and sheep farms in both countries are in areas classified under EU regulations as Less Favoured Areas (LFA). There are relatively high numbers of beef cows in both countries. Sheep production is also an important enterprise in both Ireland and the UK.

Table 1 Land use and livestock numbers

	Ireland	UK
Total agricultural area (000 ha)	4,370	18,449
Grassland	3,466	6,884
Rough grazing	468	5,565
Forage maize	16	119
Arable silage (mainly wholecrop cereals)[1]	29	38
Other forages (roots, green crops)	5	55
Breeding livestock (000)		
Dairy cows	1,156	2,192
Beef cows	1,187	1,700
Ewes	3,615	14,926

Data for 2003 from: Ireland - Central Statistics Office; UK - Defra.
[1](Wilkinson & Toivonen, 2003)

Table 2 Structure of agricultural holdings in Ireland and the UK (Charlier, 2003)

	Ireland	UK
Size of farm (ha)	31	68
Proportion of farms by type (%)		
Specialist dairy	19	11
Specialist cattle – rearing and fattening	51	14
Sheep and other grazing livestock	20	38
Specialist cereals, general cropping or horticulture	4	22
Proportion of organic farms	1	1

The role of grassland in Ireland and the UK is greatly influenced by the Common Agricultural Policy (CAP) of the European Union. Following entry in 1973 the rapid expansion of both milk and meat production continued, and sheep numbers rose dramatically, encouraged by CAP headage payments. However, the growth of intervention stocks of many agricultural products created difficulties, and measures to limit production began in 1984 with milk quotas, which are still in place today. Milk quotas reversed the growth in dairy cow numbers. Under the McSharry reforms in 1992 there was a shift from support for market prices, to headage payments for both sheep and beef. The growth in animal numbers contributed to overgrazing of grassland and resulted in damage to some environmentally sensitive areas. In 2003 a radical reform of the CAP was agreed, resulting in complete decoupling of income support from production in both Ireland and the UK. The many headage payments for livestock (and area payments for crops) have been discontinued. Income support for farmers, now known as the 'Single Farm Payment', is not dependent on the production of any specific crop or livestock. The introduction of decoupling may lead to profound changes in grassland usage within the EU in the years ahead.

Through the 1970s and 1980s, evidence began accumulating of the impact of the intensification and specialisation of agriculture on both surface water and groundwater. Agriculture was also found to be impacting on soil and air quality, and on the landscape and biodiversity. A number of statutory and voluntary schemes have been introduced to promote less intensive and more environmentally friendly systems of animal production. The most recent EU reform of the CAP includes a clause that makes all direct payments conditional on cross-compliance by farmers with a range of food safety, animal welfare and environmental measures. In Ireland farmers must follow the Code of Good Farming Practice and in the UK the land must be kept in Good Agricultural and Environmental Condition.

Two other major EU Directives, the Water Framework Directive and the Nitrate Directive, also have major implications for livestock production and for grassland management. It is likely that on intensive grassland farms, mainly in the dairy sector, farmers may have to reduce stocking rates, or export animal manures to neighbouring farms. In addition to requiring farmers to avoid air and water pollution the public, especially in the UK, are also seeking access to a countryside that exhibits attractive landscapes, ecological balance and biodiversity. These pressures will all impinge on the use of grassland in the future, and the mix of animal enterprises and management practices may therefore change significantly in years ahead.

Effects of location and climate on grass production

Ireland and the UK are located on the northwest edge of Europe. Dublin, at 53°N latitude, is further north than Calgary (Canada) or Irkutsk (Siberia). The most southern point of the UK lies further north than the 49°N parallel, which forms much of the border between Canada and the United States. Ireland and the UK benefit greatly from the moderating influence of the Atlantic, and particularly with the warming effects of the Gulf Stream. The maritime climate is mild and moist which is good for growing grass and is especially important in giving a long growing season. Likewise the length of the grazing season in the most favourable areas in Ireland and the UK is a major advantage relative to many parts of Europe. However, one commonly overlooked disadvantage arising from the location and climate in both countries is that the intensity of radiation during the summer may be sub-optimal for grass growth and, more particularly, for some other high output forage crops. A comparison of the Irish climate with some selected regions is given in Table 3 (Keane & Sheridan, 2004).

Taking into account the altitude and distance from the sea, the January temperatures illustrate the mildness of the winters in Ireland and in many areas of England and Wales. However, the July temperatures show why Ireland is disadvantaged in relation to forage maize production compared to the other sites. The July rainfall values are also noteworthy – rainfall in Ireland and the UK is similar to the other sites, but tends to fall in prolonged, relatively light showers rather than in heavy 'downpours' which are more common on continental Europe. The favourable climatic conditions of Christchurch are also apparent. (Christchurch is located on the South Island in New Zealand which has much less favourable growing conditions than the North Island where most of the dairy cows in New Zealand are located.)

Regional and annual variation within Ireland and the UK

While the climate in Ireland and the UK is, in general, very suitable for grass production, it is also suitable for growing a range of other crops including cereals and potatoes. This is particularly so in the eastern half of Britain and in more limited areas in the eastern and southern parts of Ireland. Such crops compete with grassland as land use options for some of the best land, and tend to dominate in the low rainfall areas, especially in eastern England (Hopkins, 2000). Grassland is the predominant land use option in all other areas, especially in the hills and in areas where wetter soils make the growing and harvesting of arable crops more difficult. In these areas the less intensive use of grassland is the norm, with sheep production and suckler cows more common than milk production.

Table 3 Temperature, precipitation and sunshine at selected met stations, 1961-90 (Keane & Sheridan, 2004)

Country		Ireland	England	France	Netherlands	N. Zealand	Poland
Station		Birr	Lyneham	Nantes	De Bilt	Christchurch	Poznan
Altitude (m)		73	147	27	3	37	84
Mean	Jan	4.8	3.4	5.2	2.2	17.2	-2.0
Temperature	April	7.9	7.7	10.3	8.0	12.2	7.6
(°C)	July	15.0	16.0	19.0	16.8	5.8	18.0
	Oct	10.2	10.4	12.7	10.5	11.8	8.8
Precipitation	Jan	76	64	87	69	46	30
(mm)	April	53	45	50	53	53	36
	July	59	55	46	76	68	69
	Oct	84	64	79	75	44	39
Sunshine	Jan	50	53	72	47	215	40
(hr)	April	139	153	187	153	143	152
	July	131	205	267	187	126	218
	Oct	83	101	141	103	187	102

Hopkins (2000), outlined the variation in grass growing days in the UK based on temperature, adjusted for drought and altitude. The variation was from less than 200 days to in excess of 300 days. In Ireland a large area of the country has a growing season of between 270 and 300 days (Burke *et al.*, 2004). This is an area devoted to intensive grassland enterprises, mainly dairying. Peel & Matkin (1982) described 7 climatic zones in relation to grass productivity in England and Wales. They used a calculated 'drought factor' and noted that even in areas with high concentrations of dairy cows, that summer rainfall was significantly less than the potential evapotranspiration, so that soil moisture deficits were not uncommon.

Collins *et al.* (2004), reviewed climate and soil management in Ireland, whilst the relationship between soil type and grassland productivity have been summarised by Ryan (1972). Soils described as either dry and light, or dry and loamy, predominate in the main agricultural areas. With the exception of some restrictions on summer growth due to low rainfall these soils have few limitations for grass production. Soils described as wet and heavy, or wet and peaty have serious restrictions on both the production, and utilisation of grass for periods of the year. Intensive dairying is mainly based on dry loamy land, with some also on wet heavy land in the southern part of the country. Two thirds of the dairy cows in Ireland are located in Munster (1 of 4 provinces). Beef production in Ireland is distributed across all areas, with some based on intensive grassland located on good land across the country.

Shalloo *et al.* (2004a) compared the profitability of a typical dairy enterprise on free draining, or on badly drained soil in southern Ireland using the most suitable technology on both sites. Very large differences in annual profitability (*circa* €28,000) were observed, such that, even with relatively high milk prices, milk production was hardly viable on the badly drained site. The difference in profitability was due to a variety of factors arising mainly from the longer

Grassland: a global resource

overwintering period, but also from interruption to grazing due to soil conditions following rainfall. Additional concentrate feeding and silage making costs accounted for almost half the difference in profitability. Capital charges arising from higher infrastructure costs plus land charges accounted for most of the rest (L. Shalloo, *pers. comm.*). If the expected drop in EU milk price occurs, milk production in the less favoured areas (even in Ireland) can only be viewed as a transition phase, probably to non-intensive part-time farming, combined with off-farm employment. This is a situation that is likely to be repeated in many areas across Europe.

Variation from year to year is of great significance for the management of intensive, grass-based animal production enterprises. In spring-calving dairy herds, poor grass growth in the critical early lactation period due to a 'slow' spring can have serious consequences for the rest of the lactation. Likewise an unexpected period of high growth can lead to a rapid deterioration in sward quality. This unpredictable variation in grass growth has provided the impetus in recent years for the development of dynamic interactive systems of grassland management to replace more static, date-based guidelines. Burke *et al.* (2004) presented Irish data on the variation in expected growth between regions, and between years for a 10-year period. In the January to April (winter-spring) period and also in the May to August (summer) period, the difference between the best and the worst years was in excess of 1000 kg DM/ha. In the September to December (autumn-winter) period the variation was little more than half the variation in the other two periods. Since the overall growth in winter-spring was much less than in summer, the relative variation in growth rates in the spring is much higher.

Again, as in the UK, soil moisture deficits are a continuing feature affecting grass growth in parts of Ireland. Burke *et al.* (2004) present data for a 20-year period from 1956 to 1975. The average losses in growth were relatively low, but in the eastern regions the average losses were in excess of 10% of annual growth. In the 5 driest years, a loss of at least 20% of annual output occurred in most areas, while the losses in the low rainfall areas were in excess of 30% of annual dry matter output.

The evidence that the climate is changing seems to be quite strong - four of the five warmest years recorded in central England since records began in 1772 have occurred since 1990. Rainfall in the last 30 years has been higher in winter but lower in summer compared to the 150-year historical data (Defra, 2004). It is expected that weather will also become more variable with more extremes occurring. In future, therefore, it is likely that soil moisture deficits during the summer months will become more serious in the main agricultural areas in Ireland and the UK, and grass growth during the summer months will be adversely affected. Temperature and rainfall are expected to increase in winter. Grass growth over the winter months will increase but the wetter soils will make grazing difficult. If these changes transpire, it is likely that forage maize for silage will become more important and that grass may be devoted almost entirely to grazing, with a small amount of surplus grass made into silage in the early part of the grazing season. It is also possible that there will be greater interest in *Medicago sativa* (lucerne) and other drought-resistant forages.

Forage species, fertilizers and conservation: the basic framework

Forage species

The forage area in Ireland and the UK is overwhelmingly devoted to permanent grassland, along with a substantial area that can be described as rough grazing. A relatively small proportion of the total grassland area forms part of a regular rotation with arable crops – amounting to about 10% in England and Wales, with substantially less in Ireland. However, at least half of all enclosed grassland has been sown since the mid 20th century – often reseeded directly from grass to grass. Hopkins (2000) estimated that 70% of swards over 20 years old in the UK could be classified as first or second grade *Lolium perenne* L. (perennial ryegrass) pastures. By far the most widely sown grass species is *L. perenne*, accounting for some 80% of agricultural grass seed sold in the UK (Defra, 2003a). Most of the remainder is either *Lolium multiflorum* Lam. (Italian ryegrass) or *Lolium hybridum* (hybrid ryegrass) for use in 2 - 4 year grass leys. The only other significant sown species is *Phleum pratense* L. (Timothy), which is often included in long-term mixtures as a minor component. The position in Ireland is similar but with *L. perenne* in an even more dominant position.

Of the legumes, *Trifolium repens* (white clover) is the only widely used species, and is a component of most long-term mixtures. In recent times, there has been increased research interest in *Trifolium pratense* (red clover), but this has only made a major impact on organic farms. *Lotus* spp. (Trefoils) have been shown to be suitable legumes on more difficult soils, but uptake has been very small. *M. sativa* is scarcely used.

The context for the choice of forage species is that Ireland and the UK are very suitable for ryegrass species, and ryegrass is suitable for both grazing and silage. Fertiliser nitrogen has been relatively inexpensive, and was heavily promoted from the 1960's to the 1980's. Robust and productive legume varieties have been bred, and legume-based systems have been extensively researched, but they have only been widely adopted on organic farms.

Fertilisers

Average fertiliser applications on grassland in Ireland, and particularly in the UK, are high by European standards, though they are much lower than the recommended economically optimum rates for intensive grassland enterprises. Data for enclosed grassland in Britain (Defra & SEERAD, 2004) shows that the proportion of grassland receiving the different elements is 73% for nitrogen, 60% for phosphate, 59% for potash and 6% for sulphate with average rates of application of 89, 20, 25 and 44 kg/ha respectively (Figure 1). Areas cut for silage receive much higher rates (133 kg/ha of N) than areas for grazing. Also much higher rates are applied on dairy farms than on beef and sheep farms.

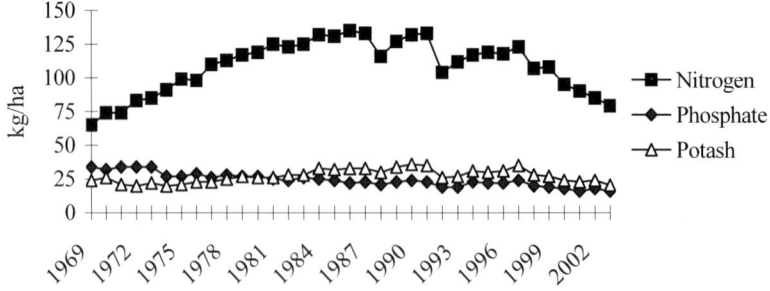

Figure 1 Overall fertiliser use on grassland in England and Wales

In recent years application rates in the UK are falling. The annual dataset for England and Wales (Figure 1) goes back to 1969, when overall rates were 65 kg/ha N, 34 kg/ha phosphate and 22 kg/ha potash. Nitrogen use peaked in the mid 1980's at 130 kg/ha. The recent trend reflects a number of factors including lower farming profits, recommendations for lower levels of N application, a reduction in the amount of grass ensiled and an increased participation by farmers in agri-environmental schemes.

Forage conservation

Silage is the dominant means of grass conservation. Wilkinson & Toivonen (2003) summarised the silage making practices in the UK and Ireland. Grass silage accounts for 83% and 70% of the forage conserved in Ireland and the UK respectively, with only 16% and 18% as hay. Most of the grass silage in Ireland, and in the western part of the UK is low dry matter silage, either direct cut or subjected to only a brief wilt. Additives are widely used - bacterial inoculants and enzyme products now more popular than acids and salts. A large silage cut taken in late May or early June is the norm on dairy and cattle farms. A smaller second cut is also common. Machinery contractors with high capacity systems carry out a high proportion of silage making - 60% in the UK, a higher proportion in Ireland. Big-bale silage has been widely adopted on smaller farms, and also on larger farms for supplementary cuts, and as an aid to the management of sward quality - 35% of grass silage in the UK is made as big bale silage.

The area of forage conserved in the UK is falling - from 2.7 million ha in 1994 to 2.2 million ha in 2000. This may reflect the fall in cattle numbers, and perhaps also recognition that grass silage is expensive compared with grazing. As yield of dry matter for individual cuts of grass silage are low relative to the yield of forage maize or whole crop cereal, costs of ensiling are high. When dry matter yields of grass silage are low, the total cost per ton of dry matter for grass silage are similar to the cost of cereals purchased at harvest and stored on farm.

There has been a substantial increase in the use of whole crop cereals for silage in the UK and in the use of *Zea mays* (maize) for silage in both Ireland and the UK. The major limitations of grass silage, especially low dry matter silage, arise from its low intake characteristics. The higher intakes achieved with *Z. mays* silage make it a valuable alternative in the feeding of high yielding dairy cows, and in the intensive fattening of beef cattle. Keady (2003) reviewed the use of *Z. mays* silage in Northern Ireland. He concluded that *Z. mays* silage had a role to play even under the rather unfavourable climatic conditions in Northern Ireland using the complete cover plastic mulch system, which improved both dry matter yields and feeding value.

Intensive grassland production systems

Targets for milk output

The development of efficient grassland-based systems of milk production, which rely heavily on grazed grass, has been the subject of much research in Ireland over the last decade. This work has mainly focused on assessing the suitability of such feeding systems for Holstein-Friesian dairy cows with high genetic potential for milk production. The suitability of different strains of Holstein-Friesian, of alternative dairy breeds and, more recently, of crossbred animals has also been assessed. The emphasis in Ireland has been on spring calving dairy cows (Buckley *et al*., 2000; Kennedy *et al*., 2002; Horan *et al*., 2005) while work in Northern Ireland has focused on autumn-calving herds (Ferris *et al*., 2002).

The challenge of achieving high milk yields per cow from grazing systems has been reviewed by Mayne *et al.* (2000), and by Peyraud *et al.* (2004), who stated that the gap between the potential (or expected) milk yield and the actual milk yield achieved on grass increases progressively as the potential milk yield increases. The gap is substantial at potential milk yields of 40 kg/day, which is well within the capability of modern dairy cows. The limitation of grass as a feed is probably the major reason why the average milk yield per cow in both Ireland and New Zealand is low by international standards. In addition, seasonal-calving dairy herds must maintain a 365-day calving interval, combined with low levels of involuntary culling, which may be a much more challenging target than achieving high milk yields *per se.*

The milk output achieved by Kennedy *et al.* (2002) in southern Ireland is a benchmark in Western Europe for high genetic merit Holstein-Friesian dairy cows in spring-calving systems, under favourable grassland conditions. Average annual milk yields per cow were 7389 and 8461 kg for low concentrate (377 kg) and high concentrate (1540 kg) groups respectively (with some groups producing in excess of 9000 kg per annum). These yields correspond to levels of *milk from forage*[1] of approximately 6500 and 5000 kg. (*Milk from forage* will normally be reduced as a result of additional concentrate feeding, due to the phenomenon of substitution of the additional concentrates for the basal forage intake). Likewise, Ferris *et al.* (2002) achieved benchmark levels of milk output from autumn-calving dairy cows under slightly less favourable conditions for grass production and utilisation in Northern Ireland. Animals received 928 kg concentrate dry matter (DM) in a system based on top quality grass silage combined with generous allowances of high quality grass for grazing. Cows consumed 1895 kg grass silage DM, and 3119 kg grass DM, and achieved an annual milk yield of 7868 kg per cow corresponding to a *milk from forage* value of almost 5500 kg. Results from commercial dairy farms in the UK show lower levels of *milk from forage – circa* 2700 kg on average, but with important regional variations (Simpson, 2004). Selected groups of dairy farmers were achieving a very efficient combination of milk yields of 7700 kg along with *milk from forage* values of more than 4000 kg per cow.

Feeding systems which achieve high milk yields per cow from animals receiving low levels of concentrates, require high grass DM intakes by the grazing animal and very high levels of technical and biological efficiency in the utilisation of grass (Kennedy *et al.*, 2003). However, such high levels of technical and biological efficiency are not necessarily synonymous with optimum economic efficiency. Analysis of the data from Kennedy *et al.* (2002) by Shalloo *et al.* (2004b), found that high levels of concentrate feeding were economically justified in certain circumstances with very high genetic merit animals, even though *milk from forage* and grass utilisation were reduced somewhat. Also, when rigid quantitative limits on milk output were imposed, as in the operation of the EU milk quota regime in Ireland, dairy cow genotypes which produced somewhat lower levels of milk output from low levels of concentrate input were economically very efficient. The data of Ferris *et al.* (2002) also showed that maximising milk output from forage was not necessarily the most economic in all circumstances. Nevertheless, achieving high levels of *milk from forage*, especially from grazed grass, on a per cow basis as well as on a per unit area basis, will probably remain a key objective in the profitable production of milk in temperate grassland regions for the foreseeable future.

---

[1] *Milk from forage* is calculated by first estimating the milk yield equivalent of the concentrates fed per cow using nutrient requirement tables. This is then deducted from the total yield per cow to obtain *milk from forage.*

While very high milk yields have been achieved with Holstein-Friesian cows, there have been problems with reproductive performance and with maintaining body condition. Limitations on grass DM intake may contribute to these problems, but it is important to note that feeding high levels of concentrate did not overcome poor reproductive performance. However, there have been promising results with some strains of Holstein-Friesian cows, which have slightly lower milk yields, and also with alternative breeds and crossbred dairy cows (B. Horan, *pers. comm.*). In the future the focus may be on animal genotypes and feeding systems that combine slightly lower milk yields with good reproductive performance. However, even in such systems, a target for milk from forage in excess of 5000 kg per cow should be realistic, arising from a total milk yield per cow of between 6000 and 7000 kg combined with a moderate levels of supplementary concentrate feeding.

Evolution of grassland management to optimise milk production

Until the establishment of milk quotas within the EU in 1984, the emphasis was on increasing output per unit area, mainly through increasing stocking rate. This focus will become important again if milk quotas are removed. However, since 1984 the emphasis has switched to reducing the costs of production - large differences in costs per unit dry matter has led to attempts to replace concentrates by forages, and especially by grazed grass. The cost of ground, pelleted concentrates is relatively high, and in Ireland and parts of the UK total mixed ration systems are not common, due to small herd size. Grass silage is also a relatively expensive feed source. Optimising the use of grazed grass has therefore been a high priority.

Post-grazing residual sward surface height and herbage allowance
Mayne *et al.* (1987, 1988) and Stakelum (1993), emphasised the importance of post-grazing residual sward surface height (PGRSSH) and its relationship to dry matter intake and milk yield, and to sward morphology and digestibility in subsequent grazings. Previously grassland management systems tended to be inflexible, with fixed proportions devoted to grazing and to conservation at various times during the year. This resulted in a serious deterioration in sward quality when grass growth was higher than normal during the first half of the grazing season. Guidelines for PGRSSH, and the need to adjust grazing plans for changing sward conditions are important components of current recommendations for optimum grassland management.

The effects of daily herbage allowance (DHA) on milk yield and milk composition have been addressed by Maher *et al.* (1999). O'Donovan *et al.* (1998, 2000) established that the inclusion of DHA along with PGRSSH considerably improved the management of grass DM intake at farm level. Selection of the appropriate level of DHA is important in the development of grassland management plans for high yielding dairy cows. Guidelines for DHA, in addition to PGSSH guidelines, are now considered to be critical components of current recommendations for the optimum management of grazing animals.

Feed budgeting and average pasture cover
Clark and Jans (1995) refer to the concepts of feed profiling, feed budgeting and grazing plans, and to the development of decision support models for pasture management in New Zealand. Stakelum, (1996) refers to annual, intermediate and short term feed budgeting, and to the concept of average pasture cover. This refers to the amount of grass dry matter per hectare, averaged across all paddocks within the grazing area. This was further developed by O'Donovan *et al.* (1997, 1998), who also addressed the problem of estimating herbage mass at farm level. Targets were developed for average pasture cover, expressed on either a per

hectare or a per cow basis. The latter is probably more functional across a range of grassland situations. Likewise it may be easier to communicate the concept to farmers, if pasture covers were expressed as the number of cow grazing days ahead of the herd rather, than as kg DM/ha.

Average pasture cover is important because it allows short term feed budgets to be constructed based on the feed requirements of the animals when it is combined with expected grass growth for the period ahead. Any deviation from the target cover signals that the overall strategic plan for the grassland area requires tactical adjustments to the short term and/or intermediate term feed budgets. This is important because grass growth is variable during spring, mainly due to variations in temperature, while soil moisture deficits lead to variation in growth in summer in regions with low rainfall.

Feed budgeting is one of the most important concepts to have been introduced into grassland management for dairying in Ireland and the UK in the last decade. It has been critically important in enabling farmers to exploit the benefits of early turnout to pasture in spring (Dillon & Crosse, 1994; Sayers & Mayne, 2001), and of extended grazing in the late autumn – early winter period. Progressive dairy farmers have, in general, adopted dynamic interactive systems of grassland management that involve the setting of targets for various milk output and pasture parameters. Continuous monitoring of these parameters, and adjustment of the grassland and feeding programme is required. Comprehensive guidelines for dynamic interactive systems of grassland management are now available (O'Donovan et al., 1998; Mayne, 2000; MDC, 2003; Teagasc, 2004b).

Other issues
The suitability of various cultivars of perennial ryegrass has been investigated (Gilliland et al., 2002; Gowen et al., 2002; Wilkins & Humphreys, 2003). The impact of sward factors on dry matter intake and milk output, and the impact of the quantity and the composition of supplementary feeding for the grazing animal are under consideration (McGilloway & O'Riordan, 1999; Mayne & Laidlaw, 1999; Mayne et al., 2000b; Peyraud et al., 2004). The preceeding discussion of developments in intensive grassland management has focused almost entirely on dairying, but similar concepts have been developed or are being developed in relation to beef and sheep production (Mayne et al., 2000a; Steen, 1998; Teagasc 2004a).

Environmental aspects of intensive grass and forage systems

Statutory measures to reduce nitrate pollution of surface waters and groundwater began to be introduced in the 1990s. Nitrate Vulnerable Zones (NVZs) were designated in all areas where the existing or predicted concentration of nitrate N was greater than 50 mg/l. Nitrate Vulnerable Zones currently cover about 55% of England and a small proportion of the rest of the UK. The position in Ireland is under review at the time of writing, but it is likely that the entire country will be designated as a NVZ. In these zones farmers are required to keep records of fertiliser and manure applications. Fertiliser N must not be applied in excess of crop or grassland requirements, and a limit is set on the maximum loading of organic N in animal excreta. For grassland this was initially set at 250 kg/ha, but is now expected to be reduced to 170 kg/ha. This means that on intensive grassland farms (mainly the dairy sector), farmers may have to reduce stocking rates, or export animal manures to neighbouring farms.
Other environmental problems addressed by the EU Water Framework Directive include pollution of surface waters by phosphorus and soil runoff, and atmospheric pollution by ammonia, nitrous oxide and other gaseous emissions including methane. In Ireland, where

the Kyoto targets will be very difficult to meet, the national greenhouse gas abatement strategy requires a significant contribution from agriculture, which may increase the pressure to reduce the total number of ruminant animals in the country.

Extensive grassland: a multifunctional resource

Farmers have traditionally grazed sheep and beef (including suckler cows), on non-intensive grassland and rough grazing. However, increasing public interest in access to the countryside, may in time impact on such grassland farmers. As well as an intensive network of public footpaths, there is now a 'right to roam' on most open grazing land in Britain, and it is recognised that in many rural parts of Ireland and the UK, that tourism now has a higher economic value than agriculture. Whilst it can be difficult to reconcile public access with farming, it does present farmers with commercial opportunities – with farmers paid from public funds to care for and maintain extensive grassland as part of a multifunctional resource.

Landscape, environmental features and farm types

In most of Scotland and Wales and in parts of Ireland and northern England, the landscape is dominated by hills and rugged terrain that is typically between 300m and 1000m altitude. It has a combination of difficulties including steep slopes, rocky outcrops, and acid soils, some of which are permanently waterlogged. Most of it is unenclosed moorland characterised by heather and other dwarf shrubs, and has traditionally served the dual function of rough grazing for sheep or beef cattle, and hunting of wild deer and birds, particularly the red grouse. In nature conservation terms it is valued highly. The UK contains a substantial proportion of the world's resource of this habitat. Heather and other shrubs can only tolerate limited grazing – about 40% of each year's growth. Increased sheep numbers have contributed to a major decline in heather cover, particularly in England and Wales, and have led to its replacement by grasses such as *Nardus stricta*.

In the lowlands, particularly in the UK, the landscape has a characteristic 'patchwork quilt' appearance of fields enclosed by walls or hedgerows. Some of these are of great antiquity, and many more were constructed in the 18th and 19th centuries following government land reforms. Until the widespread use of inorganic fertilisers in the 1960s, the hay meadows, pastures and grazing marshes within this landscape were highly biodiverse. Since then such habitats have become rare.

The best of the historic environment and the richest examples of wildlife habitats, through the hills, uplands and lowlands, are now protected by European and national legislation. This includes the widespread network of 'Natura 2000' sites in the UK designated in response to the EU Birds and Habitats Directives. In addition to these high profile sites, many environmental features such as hedgerows and semi-natural grassland are also now protected through the use of cross-compliance. This was further extended on 1 January 2005 - from this date the Single Farm Payment can be partially or wholly withheld if such features are damaged.

The differentiation of the hill and upland sectors from the lowland sector is more distinct in the UK than in Ireland, and the descriptions that follow are more typical of the UK situation than the Irish situation.

Hill farms

Most of the land area is moorland rough grazing. Breeding sheep are the main or only enterprise, where hardy breeds such as the Scottish Blackface or Welsh Mountain produce on average only one lamb per ewe per year. They graze the moorland for most of the year and are brought down to lower ground to mate in the autumn and to lamb in the spring. The ewes may be mated with a ram of a larger and more prolific breed to produce female lambs, which can be sold to upland or lowland farmers. These females will then be mated with a ram such as a Suffolk or Texel to produce prime lamb for meat. This tradition of sheep moving from hills to uplands to lowlands is known as the 'stratification' of the sheep industry.

Hill farms often have very little 'in-bye' land from which to cut grass for hay or silage. They are unable to support large numbers of cattle since these usually have a much higher requirement for winter-feed. Farms that have more 'in-bye' may have suckler cows, usually of a hardy breed such as Galloway or Welsh Black. Although cattle have been less profitable than sheep for several decades, cattle may in future be required for maintaining diverse moorland and grassland. In both hill sheep and cattle production there is a need for low-cost, 'easy-care' systems, which nevertheless have good animal welfare and are compatible with protection of the environment.

Upland farms

The convention in the UK is that those farms in hilly areas where the majority or all of the land area is enclosed grassland are known as upland farms. Some of these were created from moorland from the 1940s to 1980s. Its productivity was improved by the use of lime, fertilisers, cultivation and/or reseeding. The native grasses, typically *Festuca rubra* (Red fescue), *F. ovina* (Sheep's fescue) and *Agrostis capillaries* (Bent grass), were replaced with *L. perenne* and *T. repens*. More recently many of these swards have partially or wholly reverted to the native grass species as inputs have been reduced. These farms often have both sheep and cattle enterprises, and farmers may have invested heavily in silage pits and winter housing or hard-standing areas. This enabled the farmer to achieve higher stocking rates, have greater flexibility to breed animals out of season and fatten livestock.

Lowland farms

Grassland farms in the lowlands are often intensively managed for dairying or other enterprises. However, there are also extensive grassland-based enterprises that may occur for two main reasons. They may be farms with an area of existing permanent grassland that is difficult to manage intensively because of steepness, wetness, obstructions or accessibility. In many cases the farmer is eligible for agri-environment payments to maintain or restore this grassland. A second group may have income from another business or off-farm employment. Extensive grassland on lowland farms is often integrated with more intensive grassland, and/or arable crops. It can complement the other land such as providing summer grazing. The extensive livestock 'enterprises' found on lowland farms are hugely varied. In Ireland extensive systems of beef and sheep production are widely practiced on high quality soils in lowland areas. The main reason is probably historical in that the creameries that processed manufacturing milk were not distributed around the whole country but were concentrated in the southern region.

Support for extensive grassland farming

All hill land, most uplands and some lowlands with heavy wet soils, are classified within the EU as Less Favoured Areas (LFA). Forty four per cent of agricultural land in the UK, and

53% in Ireland, is classified as LFA. Until 2004 all beef and sheep, managed extensively or intensively, received support payable per head, with the payment being higher within the LFA. In future all headage for beef and sheep payments are consolidated into the Single Farm Payment, which is decoupled from production. This decoupling may lead to a substantial reduction in beef and sheep numbers, especially in the LFA. A recent survey of farmers in Ireland anticipates a reduction in sheep numbers, with an increase in the area devoted to forestry.

Since the mid 1980s a small but increasing proportion of government support for agriculture has been through voluntary Agri-Environment (AE) schemes. The first of a number of schemes in the UK was the 'Environmentally Sensitive Areas Scheme', which was introduced in 1986. These schemes offer annual and capital payments to farmers for restoring or recreating plant, bird or other wildlife habitats by reducing or ceasing fertiliser inputs to grassland. They were superseded in England in 2005 by the more ambitious 'Environmental Stewardship Scheme', which also includes soil and water protection as an objective. The aim is for the majority of all land in England to be in at least the lower tier of this scheme within the next few years.

In Ireland, the 'Rural Environment Protection Scheme' (REPS) was introduced in 1994 and initially covered 33% of farmland. The emphasis was on limiting the input of both organic and inorganic nitrogen on grassland, the development of nutrient management plans and the avoidance of pollution. In the third version of REPS (recently introduced) there is greater emphasis on broader environmental objectives with farmers expected to be managers of the natural heritage. These schemes are popular with farmers, and also have the support of the public.

Conclusions

Conditions in the more favourable areas in Ireland and the UK are very suitable for grass production over a long growing season. Intensive systems of milk production, and to a lesser extent beef and sheep production, using high inputs of fertiliser nitrogen, have been developed and are widely used by farmers. Guidelines for dynamic interactive systems of grassland management that rely heavily on grazed grass have been developed. Very high levels of biological and economic efficiency in both seasonal and non-seasonal systems of milk production can be achieved.

Changes in the EU income support system for beef and sheep farmers agreed in 2003, may lead to significant reductions in the numbers of beef and sheep in Ireland and the UK. Intensive grass-based systems of animal production create risks of pollution to surface- and ground- water. A high proportion of beef and sheep farms participate in voluntary, EU-funded agri-environmental schemes. All EU direct income support payments are dependant on cross compliance with a range of food safety, animal welfare and environmental measures. Grassland farmers in particular, are regarded as custodians of the countryside. However the public, especially in the UK increasingly require access to the countryside which must not be polluted and which demonstrates ecological balance and attractive landscapes.

References

Buckley, F., P. Dillon, S. Crosse, F. Flynn & M. Rath (2000). The performance of Holstein Friesian dairy cows of high and medium genetic merit for milk production on grass-based feeding systems. *Livestock Production Science*, 64, 107-119.

Burke, J.I., A.J. Brereton, P. O'Kiely & R.P. Schulte (2004). Weather and crop production. In: T. Keane & J.F. Collins (eds.) Climate, Weather and Irish Agriculture, AgMET, Dublin, 161-210.

Charlier, H (2003). Structure of agricultural holdings in the EU. *Eurostat, Statistics in focus, Agriculture and fisheries, Theme 5*, 7p. (http://europa.eu.int/comm/eurostat/Public/datashop/print-)

Clarke, D.A. & F. Jans (1995). High forage use in sustainable dairy systems. In: M. Journet, E. Grenet, M-H. Farce, M. Theriez & C. Demarquilly (eds.) Recent developments in the nutrition of herbivores. INRA, 497-526.

Collins, J.F., F.J. Larney & M.A. Morgan (2004). Climate and soil management. In: T. Keane & J.F. Collins (eds.) Climate, Weather and Irish Agriculture, AgMET, Dublin, 119-160.

Defra (2003a). Seed traders annual return. DEFRA, 2p.
(http://www.defra.gov.uk/corporate/regulat/forms/plantvar/star2003.pdf).

Defra (2003b). The environment in your pocket. DEFRA, 8p.
(http://www.defra.gov.uk/environment/statistics/eiyp/index.htm).

Defra and SEERAD (2004). The British Survey of Fertiliser Practice 2003. DEFRA.
http://defraweb/environ/pollute/bsfp/2002/bsfp2002.pdf).

Dillon, P. & S. Crosse (1994). Summer milk production – the role of grazed grass. *Irish Grassland and Animal Production Association Journal*, 28, 23-35.

Ferris, C.P., D.C. Patterson & J. Murphy (2002). Grassland-based systems of milk production for autumn calving dairy cows. Agricultural Research Institute of Northern Ireland, 75th Annual Report (2001-2002), 44-57.

Gilliland, T.J., P.D. Barrett & R.E. Agnew (2002). Variety diversity in key characters determining grazing value in ryegrass (*Lolium perenne L.*). *Irish Grassland and Animal Production Association Journal*, 36, 26-33.

Gowen, N., M. O'Donovan, I. Casey & G. Stakelum (2002). Improving cow performance at grass, what do grass cultivars offer? *Irish Grassland and Animal Production Association Journal*, 36, 33-44.

Hopkins, A. (2000). Introduction. In: A. Hopkins (ed.) Grass, its production and utilisation, 3rd Edition, Blackwell Science Ltd., 1-12.

Horan, B., P. Dillon, P.Faverdin, L.Delaby, F. Buckley & M. Rath (2005). The interaction of strain of Holstein-Friesian cows and pasture-based feed systems on milk yield, body weight and body condition score. *Journal of Dairy Science*, 88,1231-1243.

Keady, T.W.J. (2003). Maize silage in the diet of beef and dairy cattle – the influence of maturity at harvest and grass silage feed value, and feeding value relative to whole crop wheat. Agricultural Research Institute of Northern Ireland, 76th Annual Report (2002-2003), 43-54.

Keane, T. & T. Sheridan (2004). Climate of Ireland. In: T. Keane & J.F. Collins (eds.) Climate, Weather and Irish Agriculture, AgMET, Dublin, 27-62.

Kennedy, J., P. Dillon, P. Faverdin, L. Delaby, F. Buckley & M. Rath (2002). The Influence of cow genetic merit for milk production on response to level of concentrate supplementation in a grass based system. *Animal Science*, 75, 433-446.

Kennedy, J., P. Dillon, P. Faverdin, L. Delaby, G. Stakelum & M. Rath (2003). Effect of genetic merit and concentrate supplementation on grass intake and milk production with Holstein Friesian dairy cows. *Journal of Dairy Science*, 86, 610-621.

Maher, J., G. Stakelum, F. Buckley & P. Dillon (1999). The effect of level of daily grass allowance on the performance of Spring-calving dairy cows. *Irish Grassland and Animal Production Association Journal*, 33, 36-47.

Mayne, C.S. (2000). Getting more milk from grass. In: Proceedings of research seminar on ruminant production – 26 May 2000. Occasional Publication No. 29, Agricultural Research Institute of Northern Ireland, 77-89.

Mayne, C.S. & A.S. Laidlaw (1999). Managing swards to achieve high grass intakes with dairy cows. Agricultural Research Institute of Northern Ireland, 72nd Annual Report (1998-1999), 31-39.

Mayne, C.S., R.D. Newberry & S.C.F. Woodcock (1988). The effects of a flexible grazing management strategy and leader/follower grazing on the milk production of grazing dairy cows and on sward characteristics. *Grass and Forage Science*, 43, 137-150.

Mayne, C.S., R.D. Newberry, S.C.F. Woodcock & R.J. Wilkins (1987). Effect of grazing severity on grass utilisation and milk production of rotationally grazed dairy cows. *Grass and Forage Science*, 42, 59-72.

Mayne, C.S., A.J. Rook, J.L. Peyraud, J.W. Cone, K. Martinsson & A. Gonzales (2004). Improving the sustainability of milk production systems in Europe through increasing reliance on grazed pasture. In A. Luscher (ed.) Land use systems in grassland dominated regions. *Grassland Science in Europe*, 9, 584-586.

Mayne, C.S., R.W.J. Steen & J.E. Vipond (2000a). Grazing management for profit. In: A.J. Rook & P.D. Penning (eds.) Grazing Management, BGS Occasional Symposium No. 34, 201-210.

Mayne, C.S., I.A. Wright & G.E.F. Fisher (2000b). Grassland management under grazing and animal response. In: A. Hopkins (ed.) Grass, its production and utilisation, 3rd Edition, Blackwell Science Ltd., 247-291.

McGilloway, D.A. & E. O'Riordan (1999). The potential for grassland based ruminant production systems beyond 2000. *Irish Grassland and Animal Production Association Journal*, 33, 3-16.

MDC (2003). MDC Grass +, Grassland management improvement programme. Milk Development Council, Cirencester GL7 6JN.

O'Donovan, M., P. Dillon & G. Stakelum (1998). Grassland management – the effect on herd performance. *Irish Grassland and Animal Production Association Journal,* 32, 31-40.

O'Donovan, M., P. Dillon, G. Stakelum & S. Crosse (1997). Using pasture measurements to monitor performance on intensive dairy farms. *Irish Grassland and Animal Production Association Journal*, 31, 40-49.

O'Donovan, M., P. Dillon, G. Stakelum, N. Gowen & M. Rath (2000). The feed budgets and herd performance of six spring-calving dairy herds over a two year period (1997-1998). In: A.J. Rook & P.D. Penning (eds.) Grazing Management, British Grassland Society Occasional Symposium No. 34, 247-248.

Peel, S. & E.A. Matkin, (1982). The productivity of grassland farms in seven climatic zones in England and Wales. *Grass and Forage Science*, 37,299-310.

Peyraud, J.L., R. Delagarde & L. Delaby (2001). Relationships between milk production, grass dry matter intake and grass digestion. *Irish Grassland and Animal Production Association Journal,* 35, 27-48.

Peyraud, J.L., R. MOsquera-Losada & L. Delaby (2004). Challenges and tools to develop efficient dairy systems based on grazing: how to meet animal performance and grazing management. In: A. Luscher (ed.) Land use systems in grassland dominated regions. *Grassland Science in Europe*, 9, 373-384.

Ryan, M. (1972). Productivity of grassland is closely related to soil type. *Farm and Food Research,* 3, 28-31

Sayers, H.J. & C.S. Mayne (2001). Effect of early turnout to grass in spring on dairy cow performance. *Grass and Forage Science*, 56, 259-267.

Shalloo, L., P. Dillon, J. O'Loughlin, M. Rath & M. Wallace (2004a). Comparison of a pasture-based system of milk production on a high rainfall, heavy clay soil with that on a lower rainfall, free draining soil. *Grass and Forage Science,* 59, 157-168

Shalloo, L., J. Kennedy, M. Wallace, M. Rath & P. Dillon (2004b). Modelling the influence of cow genetic potential for milk production and concentrate supplementation on profitability of pasture based systems under different EU quota scenarios. *Journal of Agricultural Science*, (in press).

Simpson, R. (2004). Personal Communication from Kingshay Dairy Manager Reports July 2004. <http://www.kingshay.com>

Stakelum, G. (1993). Achieving high performance from dairy cows on grazed pastures. *Irish Grassland and Animal Production Association Journal,* 27, 9-18.

Stakelum, G. (1996). Practical grazing management for dairy cows. *Irish Grassland and Animal Production Association Journal*, 30, 33-45.

Steen, R. J. (1998). A comparison of high-forage and high-concentrate diets for beef cattle. Agricultural Research Institute of Northern Ireland, 71st Annual Report (1997-1998), 30-41.

Teagasc/IFI (2004a). End of Project Reports – Beef- Various Reports, Numbers 3962, 4276,4281,4489, 4582. <http://www.teagasc.ie/research/reports/beef/index.htm>

Teagasc/IFI (2004b). End of Project Reports - Dairy-4351- Measurement of grassland management practice on commercial dairy farms. <http://www.teagasc.ie/research/reports/beef/index.htm>

Wilkins, P.W. & M.O. Humphreys (2003). Progress in breeding perennial forage grasses for temperate agriculture. *Journal of Agricultural Science*, 140, 120-150.

Wilkinson J.M. & M.I. Toivonen (2003). World silage: a survey of forage conservation round the world. Chalcombe Publications, Lincoln.

Rising demand for meat and milk in developing countries: implications for grasslands-based livestock production[2]

C.L. Delgado

The International Livestock Research Institute (ILRI)-International Food Policy Research Institute (IFPRI) Joint Program on Livestock Market Opportunities, c/o IFPRI, 2033 K St., N.W. Washington, D.C. 20006, USA
Email: c.delgado@cgiar.org

Key points

1. Meat and milk consumption in developing countries has grown three times as fast as in developed countries over the past 30 years.
2. By 2020, developing countries will consume 72 million metric tons (mmt) more meat and 152 mmt more milk compared to 2002/3, dwarfing developed-country increases of 9 mmt for meat and 18 mmt for milk.
3. Ruminant livestock will account for 27% of the increase in global meat consumption between 2003 and 2020, up from 23% over the previous two decades.
4. The inflation-adjusted prices of livestock and feed grains are expected to fall only marginally by 2020, compared to precipitous declines in the past 30 years.
5. Production growth of monogastric livestock in Asia and South America will continue, but at a slower rate as environmental, health, and feed cost problems become more acute.

Keywords: developing countries, ruminants, price projections

Introduction

From the beginning of the 1970s to the mid-1990s, consumption of meat in developing countries increased by 70 million metric tons (mmt), almost triple the increase in developed countries, and consumption of milk by 105 mmt of liquid milk equivalents (LME), more than twice the increase that occurred in developed countries. The market value of that increase in meat and milk consumption totalled approximately $155 billion (1990 US$), more than twice the market value of increased cereal consumption under the better known 'Green Revolution' in *Triticum* spp. (wheat), *Oryza sativa* spp. (rice) and *Zea mays* (maize). The population growth, urbanisation, and income growth that fuelled the increase in meat and milk consumption are expected to continue well into the new millennium, creating a veritable 'livestock revolution'. As these events unfold, the diet for many people will change, some for the better, but others for the worse - especially if food contamination is not controlled. Farm income could rise dramatically, but whether resource-poor smallholders and landless agricultural workers who need it most will share that gain is still undetermined. The environmental, nutritional and public health impact of rapidly rising livestock production in close proximity to population centers also needs attention (Delgado *et al.,* 1999).

---

[2]This paper draws upon Delgado (2003), which it updates from 1997 to 2003, and for which it also disaggregates the grasslands-relevant livestock commodities, while attempting to address the particular market issues facing grasslands livestock producers. Grateful acknowledgement is made to Mark Rosegrant of IFPRI for the disaggregated July 2002 IMPACT projections model results and of Nancy Morgan of FAO Commodities Division (FAO/ESC) for providing very useful datasheets on updated meat statistics to 2003 on her global list-serve pertaining to meat issues.

The livestock revolution

The livestock revolution is propelled by demand. People in developing countries are increasing their consumption from the low levels of the past have a long way to go before coming near developed country averages. In developing countries people consumed an annual average in 2002/03 of 29 kg/capita meat and 45 kg/capita milk, roughly one-third the meat and one-fifth the milk consumed by people in developed countries. Average meat consumption per capita per annum increased by 263% in developing countries since the early 1970s, but only by 22% in developed countries (Table 1). Nevertheless, the caloric contribution per capita of meat, milk and eggs in developing countries at the start of the new millennium was still only a quarter that of the same absolute figure for developed countries, and at 10% accounted for only half the share of calories from animal sources observed in the developed countries (Delgado, 2003).

Table 1 Annual per capita human food consumption (kg/capita) 1973 and 2003

Commodity	Developed countries		Developing countries	
	1973	2003	1973	2003
Beef	26	23	4	6
Mutton and goat	3	2	1	2
Pork	26	30	4	12
Poultry	11	27	2	8
Four major meats[1]	67	82	11	29
Milk and products excluding butter[2]	188	202	29	45

[1]Four major meats = beef, pork, mutton and goat, and poultry, and may vary slightly from the sum of the individual entries due to rounding; [2]data for milk pertain to 2002.
Source: Values for 1973 are three-year moving averages based on the year shown, from Delgado *et al.* (1999); data for meats are preliminary estimates for 2003 from worksheets obtained from the FAO Commodities Division. Data for milk products are in Liquid Milk Equivalents and are the most recent year available at the time of writing from FAO 2005.

Throughout this paper, 'food' will be used to distinguish direct food consumption by humans from uses of animal products as feed, fuel, cosmetics, or coverings.

Per capita consumption is rising fastest in regions where urbanisation and rapid income growth result in people adding variety to their diets. Across countries, per capita consumption is determined by average capita income and whether or not one is living in a city. City life hastens cultural change, increases the frequency of eating outside the home, and increases the choice of foods available locally, all of which are positively associated with meat consumption. Aggregate consumption grows fastest where rapid population growth augments income and urban growth. Since the early 1980s, total meat and milk consumption grew at 6 and 4% per year respectively throughout the developing world. In East and Southeast Asia - where income grew at 4-8% per year between the early 1980s and 1998, population at 2-3% per year, and urbanisation at 4-6% per year -meat consumption grew between 4 and 8% per year (Cranfield *et al.*, 1998; Delgado & Courbois, 1998; Rae, 1998).

The Livestock Revolution has been most evident in East Asia, as illustrated by the per capita figures for China in Table 2. The significance of the per capita figures is more striking when they are multiplied by population, since four-fifths of the latter is in developing countries and their share is growing.

Table 2 Per capita meat and milk consumption (kg) by region, 1982/84 and 2002/03

Region	Meat (kg)		Milk (kg)	
	1983	2003	1983	2002
China	16	54	3	13
India	4	5	46	63
Other South Asia	6	10	47	64
Southeast Asia	11	22	10	13
Latin America	40	62	93	105
WANA*	20	24	86	74
Sub-Saharan Africa	10	14	32	29
Developing world	14	29	35	45
Developed world	74	82	195	202
United States	107	125	237	262
World	30	40	76	79

*WANA = Western Asia and North Africa data
Source: Values for 1983 are three-year moving averages based on the year shown from Delgado *et al.*, 1999; data for meats are preliminary estimates as food use for 2003 from worksheets obtained from the FAO Commodities Division; data for milk products in 2002 are in liquid milk equivalents for food use excluding butter and are the most recent year from FAO (2005).

Using current FAO estimates of Chinese consumption in 1982/84 and 2002/03, the share of the world's meat consumed in developing countries rose from 37 to 57%, and their share of the world's milk rose from 34 to 45%. Conversely, both per capita and aggregate milk and meat consumption stagnated in the developed world, where saturation levels of consumption have been reached and population growth is small.

China and Brazil account for a major proportion of the meat component within the Livestock Revolution. However, the doubling of aggregate milk consumption as food in India between the early 1980s and the early 2000s suggests that the Livestock Revolution extends beyond just meat and China and Brazil. At 66 mmt of LME in 2002 (a figure from FAO (2005) - considered low by many Indian dairy analysts), Indian milk consumption amounted to 13.5% of the world's total and 30% of milk consumption in all developing countries. The high milk consumption of Latin America in 2002, at 105 kg/capita, is half way between the developing world as a whole (45 kg/capita) and the developed countries (202 kg/capita), because of the very high level (75%) of urbanisation in Latin America (Table 2).

The share of the developing countries in world use of cereals for feed went from 21% in 1982/84 to 36% in 1996/98. This salient fact has inspired many observers to consider if the rise of production of grain-fed monogastric livestock products for the urban middle class would jack up the price of cereals to the poor in both rural and urban areas of developing countries. A further consideration is whether the trends portrayed above could continue far into the future, without resource scarcities or import constraints raising prices to the point that the growth in consumption would peter out (Delgado *et al.*, 1999).

Whether these trends will continue was explored in 1998 with IFPRI's International Model for Policy Analysis of Agricultural Commodities and Trade (IMPACT), a global food model first reported in Rosegrant *et al.* (1995). Results were put into the context of growing concern about livestock issues in Delgado *et al.* (1999) and updates were reported in 2003.

Rising consumption of meat and milk to 2020

For the 1996/98 to 2020 period, IMPACT predicts developing country aggregate consumption growth rates of meat and milk to be 3.0 and 2.9% per year, respectively, compared to 0.8 and 0.6%, respectively, in developed countries. Aggregate meat consumption in developing countries is projected to grow by 72 mmt between 2003 and 2020, whereas the corresponding figure for developed countries is 9 mmt (Table 3). Similarly, additional milk consumption in the developed countries of 18 mmt of LME will be dwarfed by the additional consumption in developing countries of 152 mmt. Poultry consumption in developing countries is projected to grow at 3.9% per annum through 2020, followed by beef at 2.9% and pork at 2.4%. In the developed countries, poultry consumption is projected to grow at 1.5% per annum through 2020, with other meats growing at 0.5% or less (Table 3).

Table 3 Projected food consumption trends of various livestock products to the year 2020

Region	Projected growth of consumption 1997-2020 (% per annum)	Total consumption (million mt)			% of world total	Per capita consumption (kg)
		1997	2003	2020	2020	2020
Developed world						
Beef	0.5	30	30	34	40	25
Mutton	0.8	3	3	5	26	3
Pork	0.4	36	39	39	33	29
Poultry	1.5	28	35	39	36	29
Meat	0.8	98	108	117	35	87
Milk	0.6	251	268	286	43	210
Developing world						
Beef	2.9	27	32	52	61	9
Mutton	2.3	8	10	14	74	2
Pork	2.4	47	60	81	67	13
Poultry	3.9	29	41	70	64	11
Meat	3.0	111	145	217	65	36
Milk	2.9	194	223	375	57	62

Sources: Values for 1997 are three-year moving averages based on the year shown from Delgado *et al.* (2003b); data for meats are preliminary estimates for food use for 2003 from worksheets obtained from the FAO commodities division; 'mutton' refers to meat from all sheep and goats; data for milk products pertain to 2002 and are in liquid milk equivalents and are the most recent year from FAO (2005). The 2020 projections are from the July 2002 version of Mark Rosegrant's IMPACT model (Rosegrant *et al.*, 2001; Delgado *et al.*, 2003b).

In the developing countries, 27% of the additions to meat consumption from the early 1980s to the late 1990s were from ruminant animals; in the developed countries, the comparable figure was 0%. From 1996/98 to 2003, 16% of additions to meat consumption in developing countries were from ruminant animals, whereas absolute consumption of ruminant meat actually declined in the developed countries. The IMPACT projections suggest that over one-quarter of global additions to meat consumption from the late 1990s to 2020 will involve meats from ruminants, with positive growth in consumption occurring in both the developed an developing world (Table 4).

Table 4 Increase in total annual meat[1] consumption[2] 1982 to 2020, actual and predicted

	Actual 1983 to 1997 (million mt)	Actual 1997 to 2003 (million mt)	Projected 2003 to 2020 (million mt)
Developed countries			
Ruminants	0	-2	+6
Monogastrics	+11	+10	+4
Developing countries			
Ruminants	+17	+5	+14
Monogastrics	+45	+26	+50
World all sources (mmt)	+73	+39	+74
Share of ruminants (%)	23	8	27

[1]Meat = beef, pork, mutton and goat, and poultry; [2]consumption = direct use as food, uncooked weight bone-in. Sources: Increases in total annual meat consumption between 1983 and 1997 are based on differences between annual three-year annual averages based on the year shown, calculated from FAOStat (FAO, various years). The figures for 2003 are derived from preliminary worksheets obtained form the FAO commodities division. The 2020 projections are from the July 2002 version of Mark Rosegrant's IMPACT model (Rosegrant *et al.*, 2001; Delgado *et al.*, 2003b).

As the growth rates in Table 5 suggest, high growth in consumption of livestock-source foods is spread throughout the developing world and not limited to China, India and Brazil, although the sheer size and vigour of those countries will mean that they will continue to increase their dominance of world markets for livestock products. Experience for individual commodities will vary widely among different parts of the developing world, with China leading the way on meat with a near-doubling of the total quantity consumed; the increments are primarily poultry and pork. India and the other South Asian countries will drive a large increase in total milk consumption.

Impact on world relative prices of beef and lamb versus pork and chicken

Since so much of the expansion in meat consumption, and thus production, comes from monogastric livestock such as pigs and poultry, effective demand for concentrate feeds in developing countries will continue to increase. Projections from IMPACT suggest a worldwide expansion of an additional 295 mmt of cereals used as feed per year by 2020, compared to the 1996/98 annual average. This can be compared to an average annual US *Zea mays* (corn) crop of about 200 mmt in the 1990s. Developing countries accounted for 36% of cereals feed use in 1996/98, but are projected to account for 46% in 2020. On a human per capita basis, cereals feed use in 2020 in developed countries is projected to be 375 kg, compared to 72 kg in developing countries.

Table 5 Projected food consumption[1] trends of meat[2] and milk[3], 1997-2020

Region	Projected annual growth (% per year) 1997-2020		Total consumption (million mt) 2020		Per capita consumption (kg) 2020	
	Meat	Milk	Meat	Milk	Meat	Milk
China	3.1	3.8	107	24	73	16
India	3.5	3.5	10	133	8	105
Other East Asia	3.2	2.5	5	2	54	29
Other South Asia	3.5	3.1	7	42	13	82
Southeast Asia	3.4	3.0	19	12	30	19
Latin America	2.5	1.9	46	85	70	130
of which Brazil	2.4	1.8	20	30	94	145
WANA[4]	2.7	2.3	13	42	26	82
Sub-Saharan Africa	3.2	3.3	11	35	12	37
Developing world	3.0	2.9	217	375	36	62
Developed world	0.8	0.6	117	286	86	210
World	2.1	1.7	334	660	45	89

[1]Consumption = direct use as food, uncooked weight bone-in; [2]Meat = beef, pork, mutton and goat, and poultry; [3]milk = milk and milk products in liquid milk equivalents excluding butter; [4]WANA = Western Asia and North Africa data. [Metric tons and kilograms are three-year moving averages based on the year shown].
Sources: Total and per capita meat consumption for 1997 are annual averages of 1996 to 1998 values, calculated from FAO 2005. The 2020 projections are from the July 2002 version of Mark Rosegrant's IMPACT model (Rosegrant *et al.*, 2001; and Delgado *et al.*, 2003b).

With these large projected increases in animal food product consumption and cereals use as feed, it is interesting to review inflation-adjusted prices of livestock and feed commodities to 2020. Real prices for these items fell sharply from the early 1970s to the early 1990s, in most cases stabilised in the mid 1990s, and fell again thereafter. Real beef prices fell by a factor of three from 1970/72 to 1996/98. Real *Zea mays* prices did not fall over the 1990s, reflecting perhaps high demand for feed under the Livestock Revolution (Delgado *et al.*, 2003b). The stability of feed grain prices however was not matched by stable prices for pork and poultry over the 1990s.

Hog producers have had a miserable time since 1991 on the price front, reflecting first rapid growth in supply and later on demand problems (Table 6). Poultry producers also started to suffer after 1997, as a combination of effects of the Asian economic crisis, the Russian financial crisis of 1998, avian influenza, and robust growth in supply from Latin America and China all began to have an impact. Beef on the other hand has seen real price growth since 1997, and overall sheep meat producers have held their own.

Looking to the future, IMPACT projects expected changes in real prices to 2020 relative to 1996/98. The overall picture for 2020 is a noticeable real decline for *Triticum* and *Oryza* spp. (8 and 11%), a similar decline for milk (8%), more modest decreases for meats (3%) and stability or slight increases for feed grains (+11 and –4% for *Zea mays* and *Glycine max* (soybeans), respectively). The results lend support to the view that the main effect of the Livestock Revolution on agricultural prices is to stem the fall in feed grain prices, such that *Zea mays* and *Glycine max* will increase in value over time compared to *Triticum* and *Oryza* spp., whose real prices will fall.

Table 6 Actual and predicted total real price changes for meats 1991 to 2020

	(total % change)			
	Monogastrics		Ruminants	
	Chicken	Pork	Beef	Lamb
1991/93 to 1996/98 (actual)	+11	-30	-41	+10
1996/98 to 2001/03 (actual)	-44	-36	+7	-5
1996/98 to 2020 (predicted)	-2	-3	-3	-3

Source: Actual price changes are computed from worksheets of nominal US dollar prices for benchmark world series obtained form the FAO commodities division and deflated using the US Department of Commerce Seasonally Adjusted Quarterly US GDP deflator. Percentage differences were measured between the midpoints of the annual averages shown. The commodities were represented as follows: chicken - Brazilian free-on-board (f.o.b.) export series for broilers; pork - US frozen pork export unit values; beef - Australian manufacture cow beef charges-interest-freight (c.i.f.) prices to the US; lamb - New Zealand frozen whole carcass sales in London wholesale markets. Projected price changes are from the July 2002 version of Rosegrant's IMPACT model as reported in Delgado *et al.* (2003b).

Cheap feed grains facilitated the rapid expansion of monogastric production in the 1980's and early 1990's, but that situation may have been temporary. In Southwest China, for example, the price of pork increased 25% from June 2003 to June 2004, whereas industrial prices were flat and grain prices rose 32% (Fuller *et al.*, 2001). Basically anyone who could produce meat without much grain did very well, and those that could not, did not. The suddenness of the onset of the demand-led Livestock Revolution in the early 1980s led to a rapid market response of investment in short-cycle animals under controlled conditions. Over time, ruminant production for milk and meat may be beginning to catch up in China, South Asia and Africa.

In summary, the Livestock Revolution will cushion if not prevent the further fall in real global livestock prices, and also ensure that the costs of production of monogastric livestock (about two-thirds of which are concentrate feed costs) will remain stable if feed conversion ratios do not decline further. Technical progress in lowering feed conversion ratios for monogastric livestock in developing countries has been spectacular over the last 30 years due to both scientific advances and catch-up with the industrial world. However, biological limits to further scientific advances may be coming nearer for poultry at least, and the rapidly expanding industrial livestock sectors of Asia are already approaching the productivity levels of the industrial countries (Delgado *et al.*, 2003a).

Furthermore, based on the comparison of the long-term price projections (over 23 years from 1996/98 to 2020) to the medium-term actuals (5 years from 1996/98 to 2003) in Table 6, readers could be pardoned for wondering if the long-term projections for monogastric prices were too optimistic from a producer standpoint, even though the ones for ruminants seem about right. The price and cost conditions that favoured rapid responses by the monogastric sectors over the past two decades may be less evident in the next twenty years. Furthermore, other events not factored explicitly into the projections suggest that limits to the continued rapid expansion of monogastric livestock production compared to ruminants are in sight (Delgado *et al.*, 2003a).

Factors other than relative prices that will favour grasslands production over monogastrics

On the supply side, there is increased awareness in developing countries of the environmental and public health issues raised by increasing monogastric animal densities. Major attention is now being devoted to observed nutrient loading of nitrogen and phosphorous in the soils of many emerging environmental 'hot spots' in zones of high production of pigs and poultry, especially in the East coast provinces of China, large parts of Southeast Asia, Central America, Southern Brazil, Northern Europe and the Middle Atlantic region of North America (de Haan *et al.*, 1997).

Property rights systems that do not internalise externalities (where private costs do not adequately reflect true social costs) are responsible for most problems of this kind. Recent research shows that larger livestock (primarily monogastric) farms in India, the Philippines, Thailand and Brazil tend to create larger nutrient surpluses per unit of land than do small farmers, implying a strong probability that they pollute more. Separate calculations by the same authors show that larger farms also tend to spend less per unit output on mitigating the negative effects of pollution by livestock waste than do small farms in the same areas (Delgado *et al.,* 2003a).

Growing concentrations of animals and people in the major cities of developing countries also notably increased the incidence of zoonotic diseases such as infections from *Salmonella, E. coli,* and Avian Flu - diseases that can only be controlled through enforcement of zoning and health regulations. Greater intensification of livestock production, especially monogastrics, has caused a build-up of pesticides and antibiotics in the food chain in both the developed and developing world.

There is mounting evidence from around the world that governments are moving to have producers internalise (i.e. pay the cost of) negative environmental externalities that they are creating. Animal disease outbreaks and issues with chemical residues in the context of expanding world trade are also leading many governments to enforcing previous lax regulations with respect to the pig and poultry sectors (Delgado *et al.*, 2003a).

On the demand-side, evidence has yet to be generated that large numbers of people in developing countries are prepared to pay a premium for red meat that is both more tender and leaner. However, history elsewhere suggests that this trend will be observable in important markets such as the major urban markets of the faster growing countries, e.g. China. Consumers also tend to place a premium on diversity of diet as they become wealthier, which may help explain why China went from 0 to 5% of world beef production in less than 20 years. Demand for improved qualities of livestock products from grasslands is likely to be added to continued demand for higher quantities of meat and milk in developing countries over the next two decades, at a time that traditional competition from monogastric livestock will become more constrained by cost and side-effect factors than it was in the past.

Conclusions: opportunities for poverty alleviation through ruminant livestock systems

The principal conclusion of the IMPACT projections is to confirm the view that the Livestock Revolution in developing countries will continue at least through the arbitrary horizon of 2020 and will increasingly drive world markets for meat, milk and feed grains. The main trade implication predicted by IMPACT is that developing countries will increase their already large net imports of cereals to an annual amount in 2020 of approximately the same magnitude as the annual US corn crop (193 mmt). About half (92 mmt) of these net imports will be *Zea mays* and cereals other than *Triticum* and *Oryza* spp.; most of the coarse grains will probably go to feeding, as may some of the *Triticum*. Meat and milk production increases in developing countries will largely match the big consumption increases, and meat exports from Latin America to Asia will soar.

The projections suggest relatively little change in 2020 inflation-adjusted prices relative to real price levels in the base years 1996/98. This is principally because of the propping-up effect of net import demand from developing countries; feed grain prices will remain at about 1996/98 average levels. Meat prices as a whole will fall in the range of 3%, whereas the milk price is projected to fall 8%. These falls would be substantially higher without the livestock revolution. Experience since the late 1990s may prompt one to wonder whether the real prices of pigs will not decline even more over time relative to other meat animals, and what the net impacts of this will be on feed grains and other markets. Part of the answer will depend on imponderables such as biotechnology (for both feed and swine) and consumer acceptance of GMO products in major markets.

Even as price and cost trends over the past three decades favored production response to the demand surges through large-scale farming, particularly of monogastric livestock, price trends over the next two decades are likely to be more encouraging for producers of ruminant livestock. Even without projecting changes in relative output price trends, projections suggest that while meat from ruminants accounted for 23% of global increases in meat consumption from 1982/84, the share from 2003 to 2020 is likely to be 27%. Furthermore, the absolute increases in consumption of ruminant meat in developing countries will be substantially more than twice as large in developing countries than in developed.

The rapid rise in aggregate consumption of meat and milk is propelled by (literally) billions of people diversifying primarily starch-based diets into a small amount of milk and meat, and this fundamental structural shift is not likely to be policy-changeable. Nor should it be, at least not for human nutrition reasons, as average per capita consumption in developing countries is still in most cases far below what is desirable. In any event, whether it is a good thing is not the issue; it is a phenomenon that will occur. What is much less certain is who will produce the extra meat and milk, what share of the meat will be from ruminants, and what proportion will be through large-scale or industrial methods.

Increased consumption of meat and milk will offer a major opportunity to improve the incomes per capita of resource-poor farmers and food processors in developing countries. Considerable evidence from in-depth field studies of rural households in Africa and Asia shows that the rural poor and landless traditionally receive a higher share of their income from livestock than do better-off rural people (von Braun & Pandya-Lorch 1991; Delgado *et al.,* 1999). The exception tends to be in Latin America, where relative rural wealth correlates more clearly with cattle holdings. In Africa, there is an overwhelming correlation between poverty and ruminant livestock keeping (Thornton *et al.*, 2002). Since Africa also contains most of the world's under-used grazing land, a bright future for products of ruminant livestock is helpful in thinking about how to best assist one of the world's poorest regions.

In most of the developing world, a goat, a pig, some chickens, or a milking cow can provide a key income supplement for the landless and otherwise asset-poor. Ruminants in particular offer the poor of the world a path for improving their livelihoods. Smallholder dairy operations are thriving in many areas around the world, and are growing in importance as a livelihood source for the poor in East Africa and South Asia. There are virtually no economies of scale in production in this activity, although the organisation of post-harvest support systems is critical to the expanded participation of smallholders (Rangnekar & Thorpe, 2002). There are large economies of scale in processing livestock-origin food products and perhaps in input supply, but far less in production itself if market-distortions are removed.

The prospects for using monogastric livestock production as an engine for poverty alleviation are still being debated. Clearly the magnitude of production increases suggest that every effort should be devoted to keeping small-scale and resource-poor producers in developing countries engaged in that sector. Policy reforms that help create a more level playing field by forcing the internalisation of the unpaid costs of pollution (which are larger per unit output for larger operations) will help, as will institutional innovations such as contract farming that help overcome economies of scale in input supply. However, the best scope for using the demand surge from the livestock revolution to help the poor directly is likely to be in the ruminant sectors in many cases.

Grassland and ruminant livestock sciences can serve the cause of poverty alleviation in developing countries in two main ways, depending on whether the zone in question is favorable (or not) to increased intensification of mixed crop-livestock systems. On a global basis, livestock use 3.4 billion hectares of grazing land, in addition to the production of about a quarter of the land under crops (FAO, 2005).

Access by resource-poor people to common lands for grazing or fodder production in large parts of Africa, the Middle East, and Central Asia is shrinking over time, and feed is a growing constraint for many livestock keepers in low potential areas worldwide. Crop farming increasingly impinges on pastureland, and herd transhumance that was traditionally used to optimise pastures under climatic changes has become less viable. Pastoralists have become increasingly dependent on marginal lands, leading to overgrazing and land degradation. Insecure land rights and regular periods of severe drought often exacerbate these processes. In such lower potential zones, there is above all a need for institutional innovations to protect common property resources as a major asset of resource-poor people. Science on both the livestock and pasture side can then help maintain and possibly enhance the productivity of a fragile resource base.

Where there is both market and agronomic potential for further sustainable intensification of farming systems, ruminant livestock play a key role in increasing the overall productivity of small farms. The demand surge for milk and meat, especially within 100 km of cities, provides a growing outlet for products, and helps ensure impact for the combined efforts of crop and livestock scientists whose objective is to increase the productivity and sustainability of the crop-livestock system as a whole. The scope for doing this is much better in a world where both meat and grain prices are rising, which gives an added advantage to ruminant livestock production on pastures, forages and crop by-products. The overall message is that the livestock revolution is changing the context for success in our work: although we may picture a sea of grass, no producer or consumer is an island.

References

Cranfield, J.A.L., T.W. Hertel, J.S. Eales & P.V. Preckel (1998). Changes in the structure of global food demand. *American Journal of Agricultural Economics*, 80, 1042-1050.

de Haan, C., H. Steinfeld & H. Blackburn (1997). Livestock and the environment: finding a balance. Report of a study coordinated by the Food and Agriculture Organization of the United Nations, the United States Agency for International Development, and the World Bank. Brussels: European Commission Directorate-General for Development. <http://www.virtualcentre.org>

Delgado, C. (2003). Rising consumption of meat and milk in developing countries has created a new food revolution. *Journal of Nutrition (133) 11: Supplement II on Animal Source Foods,* 3907s-3910s

Delgado, C. & C. Courbois (1998). Trade-offs among fish, meat, and milk demand in developing countries from the 1970s to the 1990s. In: A. Eide & T. Vassdal (eds.) Proceedings of the IXth Biennial Conference of the International Institute of Fisheries Economics and Trade (IIFET 98), July 7-11, (University of Tromso, Norwegian School of Fisheries, Tromsö, Norway, 755-764.

Delgado, C., C. Narrod & M. Tiongco (2003a). Policy, technical and environmental determinants and implications of the scaling-up of livestock production in four fast-growing developing countries. Final Research Report of Phase II, submitted to FAO/LEAD, Rome, July 24, 305+xipp. <http://www.virtualcenter.org>

Delgado, C., M. Rosegrant, H. Steinfeld, S. Ehui & C. Courbois (1999). Livestock to 2020: the next food revolution. Food, Agriculture, and the Environment Discussion Paper 28. International Food Policy Research Institute, Washington, DC, 72 + *viii*pp.

Delgado, C., M. Rosegrant & M. Wada (2003b). Meating and milking global demand: stakes for small-scale farmers in developing countries. In: A.G. Brown (ed.) The livestock revolution: a pathway from poverty? Record of a conference conducted by the ATSE Crawford Fund, Parliament House, Canberra, August 13. A festschrift in honour of Derek E. Tribe. The ATSE Crawford Fund, Parkville, Vic., Australia, 1-12

Food and Agriculture Organization of the United Nations. FAOStat statistical database. FAO. <http://faostat.fao.org/default.htm>

Fuller, F., D. Hu, J. Huang & D. Hayes (2001). Livestock production and feed use by rural households in China: a survey report. Staff Report 01 SR-96, Center for Agricultural and Rural Development, Iowa State University, Ames, IA, 51pp.

Rae, A.N. (1998). The effects of expenditure growth and urbanization on food consumption in East Asia: a note on animal products. *Agricultural Economics*, 18, 291-299.

Rangnekar D. & W. Thorpe (2002). (eds.) Smallholder dairy production and marketing - opportunities and constraints. Proceedings of a South-South workshop held at NDDB, Anand, India, March 13–16 (2001). NDDB (National Dairy Development Board), Anand, India, and ILRI (International Livestock Research Institute), Nairobi, Kenya, 538pp.

Rosegrant, M.W., M. Agcaoili-Sombilla & N. Perez (1995). Global food projections to 2020: implications for investment. 2020 Vision Discussion Paper No. 5. International Food Policy Research Institute, Washington, DC, 54pp.

Rosegrant, M.W., M. Praisner, S. Meijer & J. Witcover (2001). Global food projections to 2020: emerging trends and alternative futures. Occasional Paper. International Food Policy Research Institute, Washington, DC, 206pp.

Thornton, P., R. Kruska, N. Henninger, P. Kristjanson, R. Reid, F. Atieno, N. Odero & T. Ndegwa (2002). Mapping poverty and livestock in the developing world. A report commissioned by the UK department for International Development, on behalf of the Inter-Agency Group of Donors Supporting Research on Livestock Production and Health in the Developing World. International Livestock Research Institute, Nairobi, Kenya, 124pp.

United Nations (1998). World population prospects, 1998 revision. Comprehensive tables, United Nations, New York. <http://esa.un.org/unpp>

von Braun, J. & R. Pandya-Lorch (1991). Income sources of malnourished people in rural areas: a synthesis of case studies and implications for policy. In: J. von Braun & R. Pandya-Lorch (eds.) Income sources of malnourished people in rural areas: microlevel information and policy implications. Working Papers on Commercialization of Agriculture and Nutrition, No. 5. International Food Policy Research Institute, Washington, D.C, 206 + *iii*pp.

Improving the quality of products from grassland

N.D. Scollan[1], R.J. Dewhurst[1], A.P. Moloney[2] and J.J. Murphy[3]
[1]*Institute of Grassland and Environmental Research, Plas Gogerddan, Aberystwyth, Wales, SY23 3EB, United Kingdom*
Email: nigel.scollan@bbsrc.ac.uk
[2]*Teagasc, Grange Research Centre, Dunsany, Co. Meath, Ireland*
[3]*Teagasc, Moorepark Research Centre, Fermoy, Co. Cork, Ireland*

Key points

1. Consumers are increasingly aware of the links between diet and health, and place increasing emphasis on nutritional quality as a component of product quality.
2. Meat and milk products are rich sources of nutrients such as omega-3 (*n*-3) fatty acids and conjugated linoleic acid, which offer health benefits to consumers.
3. Green plants are the primary source of *n*-3 fatty acids in the food chain.
4. Grassland production systems have the potential to enhance the content of beneficial fatty acids, improve stability (from higher antioxidant content) and alter sensory attributes of meat and milk.
5. Grassland offers considerable scope to help create product differentiation in increasingly competitive markets.

Keywords: consumer, health, grass, fatty acids, quality

Introduction

The impetus to enhance the quality of animal products has increased in recent years as industry seeks to meet the rapidly changing requirements of consumers who require food that is safe, healthy, traceable, of consistent eating quality, diverse and convenient. In addition to consumer related issues, increased globalisation, reduced commodity prices, reform of Common Agricultural Policy in Europe, increased interest on animal welfare and environmental issues, have combined to reduce profit margins, particularly for the primary producer. Improving the quality of milk and meat is considered to maintain or increase consumption and help to *add-value* across the food chain.

Consumers are increasingly aware of the relationships between diet and health, particularly in relation to cancer and atherosclerosis. Knowledge of these relationships has increased interest in identifying food components that may improve health, and hence "nutritional" quality is becoming a more important dimension of product quality. Consumers often have a negative perception of food products derived from animals relative to plants. However, there is increasing evidence that the consumption of milk and meat and associated products may confer additional health benefits, including protection against cancer. Table 1 presents an overview of components in milk and meat that are considered to provide health benefits in relation to factors regulating their concentration. The primary factors relate to the crop and animal but it must be recognised that further modification may occur during manufacturing of the raw material or by fortification, for example addition of vitamins during processing to increase nutritional value. It is evident that the main components influenced by crop genetics and management are the lipids. This paper will focus on recent research on manipulating the fatty acid composition of milk and meat (mostly beef) and the implications for important quality characteristics such as colour shelf life and sensory attributes.

Table 1 Factors influencing milk and meat components shown to have health benefits for man

	Crop		Animal				Manufac-turing	Fortifi-cation
	genetics	management	feeding	genetics	management	rumen		
Protein			✓	✓				
Bioactive peptides			(✓)	(✓)			✓	
Sphinolipids								
Stanols/sterols								✓
Ca	(?)		✓	✓				
Vitamins								
A			✓					✓
B$_6$, B$_{12}$					✓			✓
C								✓
D			✓					✓
E	✓	✓	✓	species				✓
Immunoglobulins					✓		✓	
Melatonin					✓			
Fatty acids								
n-3 PUFA	✓	✓	✓		✓			✓
CLA/TVA	✓	✓	✓	✓	✓		✓	
Iso-C15:0			✓		✓			
Butyric acid			✓					
Lactose			limited	limited			✓	
Oligosaccharides				✓	✓			
Nucleosides					✓	(?)		
Fe	(?)	(?)	✓					
Zn	(?)	(?)						
Se	(?)	✓	✓					

Relationships between dietary fat and human health

The links between dietary fat and incidence of non-communicable diseases such as coronary heart disease (CHD) and stroke are well established. However, the qualitative composition of fats in the diet has a significant role in modifying the risk, and this has resulted in specific guidelines for intake of different fat types. It is suggested that the contribution of fat and saturated fatty acids (SFA) to dietary energy intake should not exceed 0.35 and 0.1 of total intake, respectively. The ratio of polyunsaturated to saturated fatty acids (P:S ratio) should be around 0.4, and the ratio of n-6 to n-3 polyunsaturated fatty acids (PUFA) should be less than 4 (Department of Health, 1994; Leaf et al., 2003; Simopoulos, 2001). Ruminant fat typically contains a high proportion of SFA (as a consequence of microbial biohydrogenation within the rumen) and monounsaturated fatty acids (MUFA) and small amounts of PUFA (see below). Linoleic and α-linolenic acids are the main PUFA while oleic acid (18:1n-9) is the most prominent MUFA, with the remainder of the MUFA occurring mainly as cis and trans isomers of 18:1. The PUFA and MUFA are generally regarded as beneficial for human health and there is even recent evidence of beneficial effects of 18:1 trans-11 (Corl et al., 2003), though other work suggests negative effects (Clifton et al., 2004). The predominant SFA are 14:0, 16:0 and 18:0. There are concerns about the effects of SFA of plasma cholesterol, though 18:0 is regarded as neutral in this regard (Yu et al., 1995) and 16:0 is not hypercholesterolemic if the diet contains high levels of linoleic acid (18:2n-6; Clandinin et al., 2000). Myristic acid (14:0) is regarded as more potent than palmitic acid (16:0) in raising plasma lipids (Zock et al., 1994). Meat and milk products from ruminants are also the main dietary sources of conjugated linoleic acid (CLA; Ritzenthaler *et al.*, 2001), which have being identified as processing a range of health promoting biological properties including anticarcinogenic activity of the dominant CLA in milk and meat, the *cis*-9, *trans*-11 isomer. Research has focused much attention on methods of enhancing the nutritional value of milk and meat by decreasing their content of SFA and increasing *n*-3 PUFA and CLA and assessing the implications for other aspects of product quality.

Milk and meat quality and characteristics that influence processability

Milk fat is an important source of dietary nutrients and energy (Chen *et al.*, 2004). The diet of the animal is the primary determinant of the fatty acid composition of milk and it is evident that milk fatty acid composition can influence technological qualities of milk and milk products including 'softness', shelf life and flavour. The effects of fatty acids on softness (and hence for example spreadability of butter) is due to the different melting points of the fatty acids in milk. As unsaturation increases, melting point decreases. Hence dairy products from cows fed diets rich in unsaturated fatty acids are softer and less viscous and those rich in PUFA are very fluid at higher temperatures, which affect physical structure (Chen *et al.*, 2004). The effects of fatty acids on shelf life are related to the tendency of unsaturated fatty acids to oxidise, leading to rancidity as storage or display time increases. Hence, milk containing higher levels of PUFA is prone to oxidation and addition of antioxidants such as vitamin E is important to limit the development of oxidation in dairy products. Oxidised milk and milk products are characterised by metallic, cardboardy or stale flavours and a paler colour (Timmons *et al.*, 2001). For meat, colour is an important purchase decision by consumers, while tenderness, juiciness and flavour are important characteristics of the eating experience and all may be influenced by the diet of the animal. Total fat rather than individual fatty acids is considered to influence tenderness and juiciness (Wood *et al.*, 2003). The colour of adipose tissue largely reflects the concentrations of β-carotene and lutein. The colour of muscle largely reflects the concentration (and oxidation state) of myoglobin. Thus, when beef is initially cut, the myoglobin oxidises, giving rise to a bright red colour. On exposure to air, its colour changes slowly to brown due to conversion of myoglobin to metmyoglobin. The rate of loss of the more desirable red colour is related to degree of unsaturation of lipids and presence of antioxidants, supplied through the animal's diet or added at processing. Disruption of the tissue during processing provides an additional challenge to the colour and lipid stability of the resulting meat product. As for milk, fatty acids in meat influence flavour. This is related to the production of volatile, odourous, lipid oxidation products during cooking linking with Maillard reaction products to form other volatiles which contribute to aroma and flavour (Wood *et al.*, 2003).

The fatty acid composition of milk and beef

Lean beef has an intramuscular fat content of around 5% with on average 0.45 - 0.48, 0.35 – 0.45 and up to 0.05 of total fatty acids as SFA, MUFA and PUFA, respectively (Moloney *et al.*, 2001). Of the total SFA, 0.3 are represented by stearic acid (18:0). The P:S ratio for beef is typically low at around 0.1 (Scollan *et al.*, 2001), except for double muscled animals which are very lean (<1% intramuscular fat) where P:S ratios are typically 0.5-0.7 (Raes *et al.*, 2001). The *n*-6:*n*-3 ratio for beef is beneficially low, typically less than 3. This reflects the considerable amounts of beneficial *n*-3 PUFA in beef, particularly α-linolenic (18:3*n-3*) and the long chain PUFA, eicosapentaenoic acid (EPA; 20:5*n-3*) and, docosahexaenoic acid (DHA; 22:6*n-3*). For many people, meat is a significant source of *n*-3 PUFA (British Nutrition Foundation, 1999). Milk fat generally contains 0.6 - 0.7, 0.25 - 0.35 and up to 0.05 as SFA, MUFA and PUFA, respectively (Jensen, 2002). Linoleic and α-linolenic acids are the main PUFA - typically 0.02 and 0.005 of milk fatty acids. Oleic acid is on average 0.65 of the MUFA (0.2 of total fatty acids), with the remainder of the MUFA mainly *cis* and *trans* isomers of 18:1.

The following sections will focus on manipulating the beneficial fatty acids in milk and beef by enhancing 18:3*n*-3, CLA and trans-vaccenic acid (18:1 *trans*-11; TVA). The main approach is by changing dietary ingredients which are known sources of long chain PUFA, such as 18:3*n*-3, 20:5*n*-3 and 22:6*n*-3. There are four main sources of fatty acids in ruminant diets: (1) fresh and ensiled forages, (2) oils and oilseeds, (3) fish oil and marine algae and (4) fat supplements. Green plants are the primary source of *n*-3 fatty acids, and forages such as grass and clover contain a high proportion (0.5-0.75) of total fatty acids as α-linolenic acid (18:3*n*-3). Hence for animals fed on high proportions of forage in the diet, this represents an important source of fatty acids.

Increasing the polyunsaturated fatty acid content of milk

Levels of PUFA in milk are usually very low, though this is a direct consequence of rumen biohydrogenation, and it is feasible to obtain much higher levels if biohydrogenation can be avoided. Feeding high levels of encapsulated sunflower oil led to milk with 35 g/100g of fatty acids as 18:2*n*-6 (Chilliard *et al.*, 2000), whilst feeding very high levels of a product in which linseed oil was protected with formaldehyde-treated proteins led to milk with 20 g/100g of milk fat as 18:3*n*-3 (McDonald & Scott, 1977). Petit *et al.*, (2002) obtained milk with 14 g/100g of milk fat as 18:3*n*-3 by duodenal infusion of linseed oil. A number of protected oilseed products have been fed to dairy cows at a range of levels, resulting in a series of levels of 18:3*n*-3. For example, Goodridge *et al.* (2001) obtained milk with 6.4 g/100g of milk fat as 18:3*n*-3 using a protected linseed product, whilst a poorly-protected product fed by Petit *et al.*, (2002) resulted in only 2.0 g/100g as 18:3*n*-3. Processing oilseeds is generally far less effective at increasing 18:2*n*-6 and 18:3*n*-3 in milk than feeding rumen protected lipid supplements (Kennelly, 1996). The efficiency of transfer of long-chain omega-3 fatty acids from fish oil (Offer *et al.*, 1999) or algal isolates (Offer *et al.*, 2001) is also low.

The highest recorded levels of CLA (5.1% of milk fatty acids) and TVA (17.0% of milk fatty acids) in cows' milk were observed by Gulati *et al.* (2003), who fed 1.1 kg/day of a mixture of soyabean oil and fish oil mixture (70/30) in which the rumen protection technology was weak. This milk also contained an exceptionally low level of 18:0 (2.8% of milk fatty acids). More typical levels of fish oil supplementation (250 g/day) led to CLA and TVA levels of 1.55 and 7.50% of milk fatty acids respectively (Offer *et al.*, 1999).

A number of studies have identified the general relationship between levels of 18:3*n*-3 in herbage and levels of 18:3*n*-3 and CLA in milk. Thomson & Van Der Poel (2000) showed this effect in relation to changes in concentrations of fatty acids in grasses over the grazing season, whilst Chouinard *et al.* (1998) showed effects of the stage of growth at which grass was cut for silage-making. However, Loyola *et al.* (2002) showed differences in the CLA content of milk from cows grazing different ryegrass cultivars, despite their similar fatty acid profiles.

Some of the most marked effects of forages on milk PUFA result from feeding fresh forage as opposed to conserved hay or silage; Tables 2 and 3 provide a summary of reported effects of fresh herbage on 18:3*n*-3 and CLA, respectively. Many of these effects reflect the loss of PUFA during field wilting (see below). The highest levels of 18:3*n*-3 and CLA in milk from pasture-fed cows were 2.31 and 2.21% of milk fatty acids respectively. The level of 18:3*n*-3 achieved with pasture feeding was only around one-tenth of that achieved under extreme conditions, and one-third of levels achieved with a more normal level of protected fat supplement. The level of CLA in milk that was achieved by pasture feeding was comparable to levels achieved with safe levels of fish oil (200-300 g/day).

Table 2 Effect of the forage component of diets on the α-linolenic acid (18:n-3) content of milk fat (g/100g total fatty acids)

| | Diets based on: | |
	Fresh forage	Conserved forage
Timmen & Patton (1988)	0.84 (pasture)	0.36 (grass/wheat silage)
Aii et al. (1988)	1.97 (grass)	1.46 (grass hay)
Aii et al. (1988)	1.34 (grass)	1.13 (grass hay)
Hebeisen et al. (1993)	2.31 (grass)	0.45 (conserved grass)
Kelly et al. (1998)	0.95 (grass-white clover)	0.25 (maize and legume silages)
Dhiman et al. (1999)	2.02 (grass-white clover)	0.81 (lucerne hay; grass-white clover)
White et al. (2001)	0.73 (grass-white clover)	0.37 (maize and lucerne silages)
Schroeder et al. (2003)	0.57 (winter oat pasture)	0.07 (maize silage)
Whiting et al. (2004)	1.13 (lucerne)	0.83 (lucerne silage)

Table 3 Effect of the forage component of diets on the conjugated linoleic acids[1] content of milk fat (g/100g total fatty acids)

| | Diets based on: | |
	Fresh forage	Conserved forage
Timmen & Patton (1988)	1.34 (pasture)	0.27 (grass and wheat silages)
Precht & Molkentin (1997)	0.76 (grass)	0.38 (maize and grass silages)
Precht & Molkentin (1997)	1.05 (grass)	0.55 (grass silage; green maize)
Kelly et al. (1998)	1.09 (grass-white clover)	0.54 (maize and legume silages)
Dhiman et al. (1999)	2.21 (grass-white clover)	0.89 (lucerne hay; grass-white clover)
White et al. (2001)	0.66 (grass-white clover)	0.36 (maize and lucerne silages)
Schroeder et al. (2003)	1.12 (winter oat pasture)	0.41 (maize silage)

[1]generally cis-9, trans-11 18:2

A number of applied studies have investigated effects of restricting pasture, and shown either no change or small decreases in levels of 18:3n-3 and CLA in milk when pasture allocation is reduced (Stanton et al., 1997; Stockdale et al., 2003). Loor et al. (2003) showed 40-70% increases in levels of 18:3n-3, CLA and TVA in milk when cows were grazed for 7-8 hour periods in addition to being fed conserved forages in a total mixed ration (TMR). Alpine pasture leads to production of milk (cheese) with enhanced levels of CLA and n-3 PUFA (Innocente et al., 2002; Hauswirth et al., 2004). Collomb et al. (2002) investigated these effects further and identified relationships between milk fatty acids and the species contained within herb-rich pastures. A number of these associations merit further attention in terms of potential mechanisms for increasing milk PUFA.

Recent studies with *Trifolium pratense* L. (red clover) silage (Dewhurst et al., 2003; Lee et al., 2003) have identified a substantial reduction in the extent of rumen biohydrogenation of

18:3n-3 when feeding *T. pratense* silage as opposed to grass silage. This effect relates, in part at least, to the effect of polyphenol oxidase (PPO), which is activated when *T. pratense* tissue is damaged, reducing the extent of lipolysis (Lee *et al.*, 2004).

There is limited evidence of potential for production of 20:5n-3 by fatty acid chain elongation from (forage) 18:3n-3 in the mammary gland (Hebeisen *et al.*, 1993). Increasing supply of the precursor, 18:3n-3 does not always increase 20:5n-3 (Petit *et al.*, 2002; Whiting *et al.*, 2004).

Increasing the polyunsaturated fatty acid content of beef

Increasing the P:S ratio, reducing the n-6:n-3 ratio and increasing CLA are important targets for research aimed at improving the nutritional value of beef. In general, nutritional manipulation does not increase the P:S ratio in the meat above the normal range (0.06-0.15), reflecting the high degree of biohydrogenation of dietary PUFA in the rumen. Providing equivalent amounts of 18:3n-3 as linseed in the diet (L) or linseed oil (LO) into the small intestine increased the percentage and amount (mg/100 g muscle) of 18:3n-3 in total lipid to 1.0, 1.5 and 8.7 and 15.8, 26.3 and 176.5 for the control, L and LO, respectively (Gatellier *et al.*, 2004). The LO treatment resulted in a high P:S ratio (0.495 relative to the recommended target of > 0.4) and low n-6:n-3 ratio (1.04 relative to the recommended target of < 2-3). This demonstrates the high potential to deposit 18:3n-3 in muscle lipids. Similarly, using ruminally protected lipids (rich in both 18:2n-6 and 18:3n-3) resulted in meat characterised by higher P:S ratio (0.22), lower n-6:n-3 ratio (1.8) and 3.0% 18:3n-3 (Scollan *et al.*, 2004).

Feeding pasture relative to concentrates, rich in 18:3n-3 and 18:2n-6, respectively, results in higher concentrations of n-3 PUFA in muscle lipids (French *et al.*, 2001; Nuernberg *et al.*, 2004). Grass relative to concentrate feeding not only increased 18:3n-3 in muscle phospholipid but also 20:5n-3, 22:5n-3 and 22:6n-3 (Warren *et al.*, 2002). Concentrates rich in 18:2n-6 lead to higher concentrations of 18:2n-6 and associated longer chain derivatives (20:4n-6). French *et al.* (2001) demonstrated that an increase in the proportions of grass in the diet decreased SFA concentration, increased P:S and n-3 PUFA concentration and decreased n-6:n-3 PUFA. These beneficial responses with grass have been related to time at pasture (Table 4; Noci *et al.*, 2003). Feeding mixtures of grass and clover (both white and red clover) relative to grass alone increased the deposition of both n-6 and n-3 PUFA in muscle of finishing beef steers, contributing to increases in the P:S ratio (Scollan *et al.*, 2002).

Table 4 Nutritionally important fatty acids of *longissimus thoracis* muscle in heifers fed on grass for differing times (Noci *et al.*, 2003)

% of muscle fatty acids	Days at grass				s.e.d.	P[1]
	0	40	99	158		
Sum SFA	45.4	45.8	45.5	43.2	0.77	**L,Q
TVA	1.35	1.93	2.27	3.01	0.18	L
CLA cis-9, trans-11	0.50	0.50	0.57	0.71	0.06	***L
Sum n-6 PUFA	3.25	3.20	2.97	3.31	0.23	NS
Sum n-3 PUFA	1.79	2.06	1.91	2.43	0.17	**L
n-6:n-3 ratio	2.00	1.79	1.56	1.32	0.10	***L
P:S ratio	0.12	0.14	0.12	0.15	0.009	*

[1]L and Q are significant linear and quadratic effects of days at grass, respectively; SFA = saturated fatty acids. *P < 0.05; **P < 0.01; ***P < 0.001.

The main CLA isomer in beef is CLA *cis*-9, *trans*-11 and it is mainly associated with the neutral lipid fraction (tyically 92% of total CLA in muscle lipid) and hence is positively correlated with fatness. As with milk, CLA in tissue is influenced by (1) CLA and TVA produced in the rumen, and (2) conversion of TVA to CLA in the tissue. Moloney *et al.* (2001) have reviewed the concentrations of CLA in beef across a range of production systems and found values ranging from 1.2-12.5 mg/g fat. Supplementing diets with sunflower oil or seeds, linseed or soya bean oil increases CLA and TVA in beef (Mir *et al.*, 2003). Grass relative to concentrate feeding, results in high CLA, and a positive association between CLA content and duration at pasture before slaughter has been reported (Table 4).

Relationship between fatty acid composition of milk and processability

The fatty acid composition of milk has a large impact on quality and processability characteristics. Feeding formaldehyde treated oilseeds to help reduce ruminal biohydrogenation of dietary lipids, resulted in milk fat containing up to 350 g/kg of 18:2*n*-6 and 220 g/kg of 18:3*n*-3 (Cook *et al.*, 1972; Bitman *et al.*, 1975; Wrenn *et al.*, 1976). This milk fat was highly susceptible to oxidation and had an unsatisfactory melting profile with the fat becoming liquid at about 20 °C. Diets that increased the proportion of oleic acid (18:1*n*-9) in milk also generally decrease the proportion of 16:0, resulting in 'softer' milk fat with a more desirable melting profile. The results in Figure 1 illustrate the effect of feeding approximately 700 g/day of rapeseed oil, as processed full fat rapeseeds, on the fatty acid composition and the resulting melting profile of the milk fat. The short to medium chain fatty acids and 16:0 were reduced proportionally by 0.40 and 0.24, respectively, while 18:1*n*-9 was increased by 0.47. These fatty acid changes resulted in a solid fat at 10°C of 341 g/kg compared to 401 g/kg in the milk fat from unsupplemented cows.

Comparing milk fat from cows fed control diets indoors and at pasture, showed that milk fat produced at pasture was proportionally lower in 16:0 and higher in 18:1*n*-9 by 0.19 and 0.55, with solid fat contents at 10°C of 520 and 401 g/kg, respectively (Murphy, 2000). Similar seasonal variation in the softness/spreadability of butter was observed in New Zealand (MacGibbon & McLennan, 1987) and related to variation in milk PUFA (Thomson & Van der Poel, 2000). These data highlight the enhanced fatty acid composition of milk fat produced at pasture and the opportunity to manipulate composition in a beneficial way by relatively simple supplementation with whole oilseeds containing high proportions of 18-carbon fatty acids. 'Spreadable' butter is being produced commercially in Northern Ireland using this strategy and factors that influence the consistency of the fat produced are being investigated (Magowan *et al.*, 2003, 2004). Milk having a high level of PUFA is more susceptible to autoxidation than conventional milk. Production of oxidised flavour at 8 days post sampling was positively correlated with levels of 18:2*n*-6 (r=0.49), 18:3*n*-3 (r= 0.55) and total PUFA (r= 0.50) in milk fat (Timmons *et al.*, 2001). Milk for cows fed on *T. pratense* compared to grass silage contained more 18:2*n*-6 and 18:3*n*-3 PUFA which resulted in increased oxidative deterioration of milk (Al-Mabruk *et al.*, 2004). The latter could be corrected by feeding supplemental vitamin E. The taste of milk from diets containing *T. pratense* was negatively affected (Bertilison & Murphy, 2003).

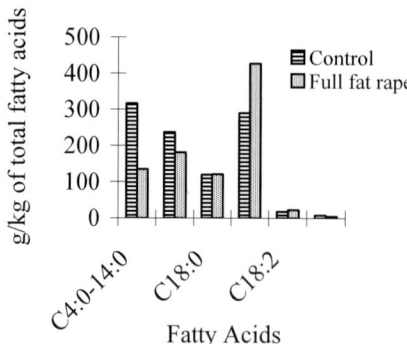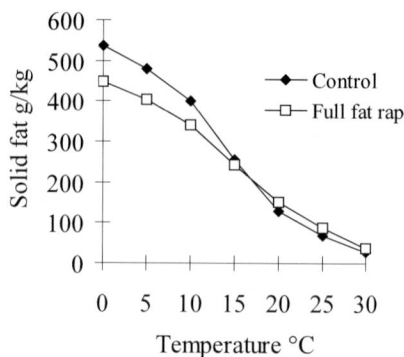

Figure 1 The effect on milk fatty acid composition and resulting melting profile of the milk fat of feeding approximately 700 g/day of rapeseed oil, in the form of processed full fat rapeseeds, to dairy cows at pasture (Murphy, 2000)

Relationship between fatty acid composition of meat and processability

Differences between the quality characteristics of meat from grass-fed or non-grass (usually grain)-fed cattle or sheep have been reviewed (Muir *et al*., 1998; Priolo *et al*., 2001; Geay *et al*., 2001). In general there were no consistent effects of these feeds on meat colour, pH, drip loss, water holding capacity, cooking loss, tenderness or juiciness. Grazing consistently results in an increase in yellowness of fat (Muir *et al*., 1998), reflecting the higher concentrations of β-carotene and lutein in grass compared to alternative feedstuffs.

Even though pasture feeding results in higher concentrations of more oxidisable *n*-3 PUFA in muscle lipids, the meat is more resistant to lipid oxidation than grain fed-beef (O'Sullivan *et al*., 2003). This reflects the higher deposition of plant derived anti-oxidants, in particular vitamin E, in meat from pasture-fed cattle, but also increased activity of some anti-oxidant enzymes (Gatellier *et al*., 2004). This observation generally holds for fresh or aged meat (Yang *et al*., 2002) and meat that has undergone long-term frozen storage (Farouk & Wieliczko, 2003), but the opposite may occur when this meat is minced (Realini *et al*., 2004). The authors suggest that mincing disrupts cellular integrity and exposes more of the polyunsaturated fatty acids to oxidation - providing a greater test of anti-oxidative protection of the *n*-3 PUFA in grass-fed beef. Similarly, Yang *et al*., (2002) reported that at similar vitamin E concentrations, pasture-fed beef was less stable than concentrate beef, again highlighting the influence of the fatty acid composition of pasture-fed beef. The appropriate ratio of vitamin E (and other antioxidants) to *n*-3 PUFA in meat to ensure lipid and colour stability during processing remains to be determined. With regard to general functional properties, Farouk & Wieliczko (2003), concluded that there is not much difference between beef finished on pasture or grain even after long-term frozen storage.

In a recent review examining the relationships between fatty acids and meat quality, it was concluded that only when concentrations of linolenic acid approach 3% of lipids are there any adverse effects on lipid stability, colour stability or flavour. In support of this conclusion, when the concentration of linolenic acid in beef was increased from 0.7% to 1.2% of total lipids by feeding linseeds (Vatansever *et al*., 2000), or to 1.9% by using a ruminally-protected lipid supplement (Scollan *et al.*, 2003; Enser *et al.*, 2001), there was little effect on lipid stability or flavour characteristics. When linolenic acid was increased to 2.8% of total fatty acids, higher sensory scores for 'abnormal' and 'rancid' were recorded (Scollan *et al.*, 2004).

Since grazing *per se* does not increase the proportion of linolenic acid to the extent achieved by the use of protected lipid supplements, little difference between fatty acid-derived flavours might be expected between grass-fed and grain-fed beef. Muir *et al.*, (1998) concluded that on the basis of similar weight or fat cover, the differences resulting from these feeding regimes were not sufficient to generate differences in flavour 'factors' which were large enough to be detected by the sensory panellists. Clearly much of the perception of differences between grass and grain-fed beef flavour is confounded by other factors, in particular fatness, which contributes to flavour. Moreover, when preferences for a particular product are expressed, they may be influenced by prior experience (Sanudo *et al.*, 1998). Non-lipid compounds and their metabolites, derived from pasture constituents may also contribute directly to flavour. Thus, compared with a grain diet, grass feeding increased the concentration of diterpenoids (which derive from chlorophyll breakdown in the rumen) (Melton, 1990) and 3-methylindole (skatole) (Young *et al.*, 1997).

Little has been published on differences in the quality of beef from cattle fed different forages. Scollan *et al.*, (2002) found little difference in the proportions of 18:2 or 18:3 in total lipids of beef from cattle fed grass or mixtures of grass and red or white clover, and no difference in colour stability, lipid stability or flavour. Wilting grass prior to ensiling did not change the quality characteristics of beef compared to beef from cattle fed unwilted grass silage from a similar sward (Moloney *et al.*, 2004). In contrast, extensive rather than restricted fermentation in the silo resulted in beef that was more colour stable and visually more desirable (O'Sullivan *et al.*, 2004). This reflected differences in vitamin E concentrations indicating that conservation strategies should consider loss of antioxidants as well as PUFA.

The effects of different forages on the quality of lamb have been recently reviewed (Duckett & Kuber, 2001). These authors concluded that compared to grazed grass, the intensity of flavour in lamb is increased with grazing of *Trifolium repens* L. (white clover), *Medicago sativa* (lucerne), and certain crop aftermaths, but that these differences can be decreased by grazing grass for 2 to 3 weeks before slaughter. In contrast, Vipond *et al.* (1995) failed to demonstrate an effect of clover on the flavour of lamb. They suggested that differences between the experiments might be due to the proportion of clover in the diet, in that many earlier studies used pure swards in their comparisons, which are unlikely to be used in commercial practise. There also seems to be seasonal effects on flavour intensity with maximum intensity occurring in meat from lambs grazing in cooler months (Duckett & Kuber, 2001). Young *et al.* (1994) reported that feeding lambs on a range of different pasture species resulted in different sensory attributes of the meat.

Breeding and managing forage for fatty acids

The transfer of forage linolenic acid to milk or meat is dependent on two important processes, (1) increasing the level of 18:3*n*-3 in the feed and (2) reducing the extent of biohydrogenation. The former is dependent on maximising levels in the fresh herbage, and for silage reducing losses during field wilting. A number of authors (Dewhurst *et al.*, 2001; Elgersma *et al.*, 2003a; Boufaied *et al.*, 2003) have recently commented on genetic variation in herbage fatty acid levels. Quantitative Trait Loci (QTL) have been identified in a ryegrass mapping family, which will facilitate rapid selection for high-lipid grasses. As with most herbage quality traits, there are substantial environmental effects and interactions that make exploitation more difficult. The most notable effects are the number and timing of cuts and season (Saito et al., 1969; Bauchart et al., 1984; Dewhurst et al., 2001; Boufaied et al., 2003; Elgersma et al., 2003a,b). Management which inhibits the initiation of flowering (e.g. two early cuts in the work of Bauchart et al., 1984 and 9-cuts per annum in the work of Dewhurst et al., 2002) increases fatty acid levels.

Oxidative loss of PUFA during field wilting represents a major loss to the food chain, with substantial losses of 18:3n-3 during hay-making (Aii et al., 1988), and modest losses during wilting prior to ensiling (Dewhurst & King 1998; Boufaied et al., 2003). These losses are associated with the lipoxygenase system - a plant defence mechanism which is initiated in damaged tissues. Plant lipases release free 18:3n-3 and 18:2n-6 from damaged membranes (Thomas, 1986), and these are rapidly converted to hydroperoxy PUFA by the action of lipoxygenases (Feussner & Wasternack, 2002). The hydroperoxy PUFA are further catabolised to yield a range of volatile anti-microbial and anti-fungal compounds, such as leaf aldehydes and alcohols (Fall et al., 1999). These compounds develop rapidly and provide the smell of freshly cut herbage. 'Stay-green' grasses have provided one approach to reducing wilting losses. 'Stay-green' grasses, which lack one of the enzymes involved in chlorophyll breakdown (Thomas & Smart, 1993) and so retains thylakoid membrane structure later in senescence, have been investigated. Stay-green material showed substantially reduced losses of fatty acids when artificially senesced by excision and incubation on moist filter paper in darkness (Harwood et al., 1982). Dewhurst et al. (2002) found a small reduction in losses of fatty acids during wilting, though the effect may have been restricted by the rapid drying conditions.

Rumen biohydrogenation

The extent of biohydrogenation of dietary PUFA is very high, averaging approximately 86 and 92% for 18:2n-6 and 18:3n-3, respectively. However, this process does give rise to a number of metabolically important intermediates including 18:1trans isomers and CLA, both of which determinant the levels of CLA in milk and meat. Whilst forages usually represent the main source of 18:3n-3 in the diet, this tends to be counteracted by increasing biohydrogenation with increasing forage proportion in the diet (Kalscheur et al., 1997; Kucuk et al., 2001), reflecting the predominant role of the fibrolytic bacterium Butyrivibrio fibrisolvens in this process (Demeyer & Doreau, 1999).

As discussed, high levels of 18:3n-3 may be achieved in milk and meat by using ruminally protected lipids, hence reducing the extent of biohydrogenation. Establishing natural methods of modifying the extent of biohydrogenation and generating important biohydrogenation intermediates is an important target for on-going research. For forages, promising results have appeared for T. pratense expressing high PPO activity. Studies also suggest that forages expressing high PPO activity may reduce lipid losses in the rumen (Lee et al., 2004). In vivo studies have demonstrated a lower degree of biohydrogenation on T. pratense compared with grass, and beneficially higher flows of linolenic acid (18:3n-3) to the duodenum (Lee et al., 2003).

Use of existing knowledge now

Increased health consciousness among consumers has led to a growing preference for healthier, more nutritious and more functional food products. It is evident that milk and meat contain components which offer beneficial effects for human health, and which are improved by grass feeding. This provides a basis on which to develop a range of novel 'functional foods'. The term functional foods is a generic term used to describe foods or food components that have beneficial effects on human health above that expected on the basis of nutritive value (Milner, 1999). Such products are targeted at disease prevention and are aimed at healthy people. Milk and meat and associated products offer exciting opportunities in this rapidly developing area, and the functional food components in milk and meat help to illustrate the important role of these products in the human diet. Indeed in some countries, omega-3 enriched milk and dairy products are on the market. The meat industry has been the slowest section of the food industry to embrace the functional trend and incorporate functional ingredients into their products, but this is changing. Grassland production systems have the potential to enhance the content of beneficial fatty acids in meat and milk, improve stability (from higher antioxidant content) and alter sensory attributes. These attributes offer considerable scope to help create product differentiation in increasingly competitive markets.

Future research

Exploitation of forage based systems as a route to the production of meat and milk and associated products with enhanced product quality characteristics offers outstanding opportunities. However, such systems do present significant challenges and research must address a number of issues to help ameliorate particular problems. Some of these are discussed below:

1. Variability in both quality and quantity of forage represents a major challenge for these systems, which can lead to variability in product quality. Traditionally producers have compensated for variations in nutrients supply from forage by the strategic use of concentrates. Season, drought and flowering all have large effects on forage characteristics. Extensive forage systems that consistently deliver (all-year-round) high quality milk and meat are an important target. More attention will need to be given to the appropriate choice of animal genotype to fit these systems.
2. Development of future forage genotypes should emphasise characteristics that facilitate consistency of supply and effects of quality, including fatty acids, shelf life/stability and sensory attributes of meat and milk products. Progress should continue in improving intake and digestibility characteristics of forages, but efforts, for example to increase the lipid content (and proportion of 18:3n-3) through genetic selections based on QTL offer scope for reasonably fast progress. Recent transgenic research has reported the production of longer chain n-6 and n-3 PUFA in higher plants (Arabidopsis thaliana; Qi et al., 2004). There is considerable scope for similar approaches in grasses as an alternative and more sustainable route to enhancing delivery of longer chain PUFA into the food chain than using fish oils.
3. Biohydrogenation of dietary PUFA by ruminal micro-organisms is a major limitation to our ability to beneficially enhance the fatty acid composition of ruminant products. Understanding the major microbial species involved in biohydrogenation and how they are influenced by diet will permit an increased understanding of methods of modifying this process.

4. Grass feeding may have some beneficial effects on improving lipid and colour stability of raw meat and milk, but this will merit further attention when these materials are processed further. In this respect, the relationships between vitamin E (and other antioxidants) and *n*-3 PUFA in meat and milk to ensure adequate stability during processing require attention.
5. The relationships between pasture species, and in particular herb-rich pastures and fatty acid composition and sensory attributes of meat and milk is interesting, and further studies here may reveal useful mechanisms for increasing the quality of ruminant products. Greater integration of research across the various levels of the food chain and increased cooperation with industry will aid the development of foods with higher quality and safety together with clear health benefits for consumers.

References

Aii, T., S. Takahashi, M. Kurihara & S. Kume (1988). The effects of Italian ryegrass hay, haylage and fresh Italian ryegrass on the fatty acid composition of cows' milk. *Japanese Journal of Zootechnical Science*, 59, 718-724.

Al-Mabuk, R.M., N.F.G. Beck & R.J. Dewhurst (2004). Effects of silage species and supplemental vitamin E on the oxidative stability of milk. *Journal of Dairy Science*, 87, 406-412.

Bauchart D., R. Verite & B. Remond (1984). Long-chain fatty acid digestion in lactating cows fed fresh grass from spring to autumn. *Canadian Journal of Animal Science*, 64 (Suppl.), 330-331.

Bertilison, J. & M. Murphy (2003). Effects of feeding clover silages on feed intake, milk production and digestion in dairy cows. *Grass and Forage Science*, 58, 309-322.

Bitman, J., T.R. Wrenn, D.L. Wood, G.C. Mustakes, E.C. Baker & W.J. Wolf (1975). Effects of feeding formaldehyde treated, full fat soyabean flours on milk fat polyunsaturated fatty acids. *Journal of American Oil Chemists' Society*, 52, 415-418.

Boufaïed, H., P.Y. Chouinard, G.F. Tremblay, H.V. Petit, R. Michaud & G. Bélanger (2003). Fatty acids in forages. I. factors affecting concentrations. *Canadian Journal of Animal Science*, 83, 501-511.

British Nutrition Foundation (1999). *Meat in the Diet*. British Nutrition Foundation, London

Chen, S., G. Bobe, S. Zimmerman, E.G. Hammond, C.M. Luhman, T.D. Boylston, A.E. Freeman & D.C. Beitz (2004). Physical and sensory properties of dairy products from cows with various milk fatty acid compositions. *Journal of Agricultural and Food Chemistry*, 52, 3422-3428.

Chilliard, Y., A. Ferlay, R.M. Mansbridge & M. Doreau (2000). Ruminant milk fat plasticity: nutritional control of saturated, polyunsaturated, *trans* and conjugated fatty acids. *Annales de Zootechnie*, 49, 181-205

Chouinard, P.Y., L. Corneau, M.L. Kelly, J.M. Griinari & D.E. Bauman (1998). Effect of dietary manipulation on milk conjugated linoleic acid concentrations. *Journal of Animal Science*, 76 (Suppl.1), 233 (abstract).

Clandinin, M.T., Cook, S.L., Konrad, S.D. and French, M.A. (2000). The effect of palmitic acid on lipoprotein cholesterol levels. *International Journal of Food Sciences and Nutrition*, 51, S61-S71.

Clifton, P.M., J.B. Keogh & M. Noakes (2004). Trans fatty acids in adipose tissue and the food supply are associated with myocardial infarction. *Journal of Nutrition*, 134, 874-879.

Collomb, M., U. Bütikofer, R. Sieber, B. Jeangros & J-O. Bosset (2002). Correlation between fatty acids in cows' milk fat produced in the lowlands, mountains and highlands of Switzerland and botanical composition of the fodder. *International Dairy Journal*, 12, 661-666.

Cook, L. J., T.W. Scott & Y.S. Pan (1972). Formaldehyde-treated casein-safflower oil supplement for dairy cows, II. Effect on the fatty acid composition of plasma and milk lipids. *Journal of Dairy Research*, 39, 211-218.

Corl, B.A., D.M. Barbano, D.E. Bauman & C. Ip (2003). *Cis*-9, *trans*-11 CLA derived endogenously from trans-11 18:1 reduces cancer risk in rats. *Journal of Nutrition*, 133, 2893-2900.

Demeyer, D. & M. Doreau (1999). Targets and procedures for altering ruminant meat and milk lipids. *Proceedings of the Nutrition Society*, 58, 593-607.

Department of Health (1994). Report on health and social subjects No. 46. Nutritional aspects of cardiovascular disease. HMSO, London.

Dewhurst, R.J., W.J. Fisher, J.K.S. Tweed & R.J. Wilkins (2003). Comparison of grass and legume silages for milk production. 1. Production responses with different levels of concentrate. *Journal of Dairy Science*, 86, 2598-2611.

Dewhurst, R.J. & P.J. King (1998). Effects of extended wilting, shading and chemical additives on the fatty acids in laboratory grass silages. *Grass and Forage Science*, 53, 219-224.

Dewhurst, R.J., J.M. Moorby, N.D. Scollan, J.K.S. Tweed & M.O. Humphreys (2002). Effects of a stay-green trait on the concentrations and stability of fatty acids in perennial ryegrass. *Grass and Forage Science*, 57, 360-366.

Dewhurst, R.J., N.D. Scollan, S.J. Youell, J.K.S. Tweed & M.O. Humphreys (2001). Influence of species, cutting date and cutting interval on the fatty acid composition of grasses. *Grass and Forage Science*, 56, 68-74.

Dhiman, T.R., G.R. Anand, L.D. Satter & M.W. Pariza (1999). Conjugated linoleic acid content of milk from cows fed different diets. *Journal of Dairy Science*, 82, 2146-2156.

Duckett, S.K. & P.S. Kuber (2001). Genetic and nutritional effects on lamb flavour. *Journal of Animal Science,* 79 (E. suppl.), E249-E259.

Elgersma, A., G. Ellen, H. van der Horst, B.G. Muuse, H. Boer & S. Tammga (2003a). Comparison of the fatty acid composition of fresh and ensiled perennial ryegrass (*Lolium perenne* L.), affected by cultivar and regrowth interval. *Animal Feed Science and Technology*, 108, 191-205.

Elgersma, A., G. Ellen, H. van der Horst, B.G. Muuse, H. Boer & S. Tammga (2003b). Influence of cultivar and cutting date on the fatty acid composition of perennial ryegrass (*Lolium perenne* L.). *Grass and Forage Science*, 58, 323-331.

Enser, M., N. Scollan, S. Gulati, I. Richardson, G. Nute & J. Wood (2001). The effects of ruminally-protected dietary lipid on the lipid composition and quality of beef muscle. *Proceedings of the 47th International Congress of Meat Science of Technology*, Krakow, Poland, 1, 12-13.

Fall, R., T. Karl, A. Hansel, A, Jordan & W. Lindinger (1999). Volatile organic compounds emitted after leaf wounding: On-line analysis by proton-transfer-reaction mass spectrometry. *Journal of Geophysical Research-Atmospheres*, 104, 15963-15974.

Farouk, M.M. & K.J. Wieliczko (2003). Effect of diet and fat content on the functional properties of thawed beef. *Meat Science*, 64, 451-458.

Feussner, I. & C. Wasternack (2002). The lipoxygenase pathway. *Annual Review of Plant Biology*, 53, 275-297.

French, P., E.G. O'Riordan, F.J. Monahan, P.J. Caffrey, M.T. Mooney, D.J. Troy & A.P. Moloney (2001). The eating quality of meat from steers fed grass and/or concentrates. *Meat Science* 57, 379-386.

Gatellier, P., Y. Mercier & M. Renerre (2004). Effect of diet finishing mode (pasture or mixed diet) on antioxidant status of Charolais bovine meat. *Meat Science,* 67, 385-394.

Geay, Y., D. Bauchart, J-F. Hocquette & J. Culioli (2001). Effect of nutritional factors on biochemical, structural and metabolic characteristics of muscles in ruminants, consequences on dietetic value and sensorial qualities of meat. *Reproduction Nutrition and Development,* 41, 1-26

Goodridge J., J.R. Ingalls & G.H. Crow (2001). Transfer of omega-3 linolenic acid and linoleic acid in milk fat from flaxseed or linola protected with formaldehyde. *Canadian Journal of Animal Science,* 81, 525-532.

Gulati, S.K., S. McGrath, P.C. Wynn & T.W. Scott (2003). Preliminary results on the relative incorporation of docosahexaenoic and eicoapentaenoic acids into cows milk from two types of rumen protected fish oil. *International Dairy Journal*, 13, 339-343.

Harwood, J.L., A.W.H.M. Jones & H. Thomas (1982). Leaf senescence in a non-yellowing mutant of *Festuca pratensis*. III. Total acyl lipids of leaf tissues during senescence. *Planta*, 156, 152-157.

Hauswirth, C.B., M.R.L. Scheeder & J.H. Beer (2004). High ω-3 fatty acid content in alpine cheese. The basis for an alpine paradox. *Circulation*, 109, 103-107.

Hebeisen, D.F., F. Hoeflin, H.P. Reusch, E. Junker & B.H. Lauterburg (1993). Increased concentrations of omega-3 fatty acids in milk and platelet rich plasma of grass-fed cows. *International Journal of Vitaminology and Nutrition Research*, 63, 229-233.

Innocente, N., D. Praturlon & C. Corradini (2002). Fatty acid profile of cheese produced with milk from cows grazing on mountain pastures. *Italian Journal of Food Science*, 14, 217-224.

Jensen, R.G. (2002). The composition of bovine milk lipids: January 1995 to December 2000. *Journal of Dairy Science*, 85, 295-350.

Kalscheur, K.F., B.B. Teeter, L.S. Piperova & R.A. Erdman (1997). Effect of forage concentration and buffer addition on duodenal flow of trans-$C_{18:1}$ fatty acids and milk fat production in dairy cows. *Journal of Dairy Science*, 80, 2104-2114.

Kelly, M.L., E.S. Kolver, D.E. Bauman, M.E. Van Amburgh & L.D. Muller (1998). Effect of intake of pasture on concentrations of conjugated linoleic acid in milk of lactating cows. *Journal of Dairy Science*, 81, 1630-1636.

Kennelly, J.J. (1996). The fatty acid composition of milk fat as influenced by feeding oilseeds. *Animal Feed Science and Technology*, 60, 137-152.

Kucuk, O., B.W. Hess, P.A. Ludden & D.C. Rule (2001). Effect of forage:concentrate ratio on ruminal digestion and duodenal flow of fatty acids in ewes. *Journal of Animal Science*, 79, 2233-2240.

Leaf, A., Y.F. Xiao, J.X. Kang & G.E. Billamn (2003). Prevention of sudden cardiac death by *n*-3 polyunsaturated fatty acids. *Pharmacology and Therapeutics*, 98, 355-377.

Lee, M.R.F., L.J. Harris, R.J. Dewhurst, R.J. Merry & N.D. Scollan (2003). The effect of clover silages on long chain fatty acid rumen transformations and digestion in beef steers. *Animal Science*, 76, 491-501.

Lee, M.R.F., A.L. Winters, N.D. Scollan, R.J. Dewhurst, M.K. Theodorou & F.R. Minchen (2004). Plant-mediated lipolysis and proteolysis in red clover with different polyphenol oxidase activities. *Journal of the Science of Food and Agriculture*, 84, 1639-1645.

Loor, J.J., F.D. Soriano, X. Lin, J.H. Herbein & C.E. Polan (2003). Grazing allowance after the morning or afternoon milking for lactating cows fed a total mixed ration (TMR) enhances trans11-18:1 and cis9,trans11-18;2 (rumenic acid) in milk fat to different extents. *Animal Feed Science and Technology*, 109, 105-119.

Loyola, V.R., J.J. Murphy, M. O'Donovan, R. Devery, M.D.S. Oliveira and C. Stanton (2002). Conjugated linoleic acid (CLA) content of milk from cows on different ryegrass cultivars. *Journal of Dairy Science*, 85 (Suppl.1), (abstract), p313.

MacGibbon, A.K.H. & W.D. McLennan (1987). Hardness of New Zealand patted butter: Seasonal and regional variations. *New Zealand Journal of Dairy Science and Technology*, 22, 143-156.

Magowan, E., A.M. Fearon, D.C. Patterson & J.A.M. Beattie (2003). Effect of buffer feeding on milk fat composition from dairy cows offered a high lipid ration at pasture. *Proceedings British Society of Animal Science*, ISBN 0906562 41 4, p205.

Magowan, E., A.M. Fearon, D.C. Patterson, D.J. Kilpatrick & J.A.M. Beattie (2004). The effect of dietary lipid content and composition on the milk fat iodine value of dairy cows. *Proceedings BSAS meeting 2004*, ISBN 0906562 45 7, p50.

McDonald, I.W. & T.W. Scott (1977). Foods of ruminant origin with elevated content of polyunsaturated fatty acids. *World Review of Nutrition and Dietetics*, 26, 144-207.

Melton, S.L. (1990). Effects of feeds on flavour of red meat, a review. *Journal of Animal Science*, 68, 4421-4435.

Milner, J.A. (1999). Functional foods and health promotion. *Journal of Nutrition*, 129, 1395S-1397S.

Mir, P.S., M. Ivan, M.L. Ha, B. Pink, E. Okine, L. Goonewardene, T.A. McAllister, R. Weselake & Z. Mir (2003). Dietary manipulation to increase conjugated linoleic acids and other desirable fatty acids in beef: a review. *Canadian Journal of Animal Science*, 83, 673-685.

Moloney, A.P., M.T. Mooney, J.P. Kerry & D.J. Troy (2001). Producing tender and flavoursome beef with enhanced nutritional characteristics. *Proceedings of the Nutrition Society*, 60, 221-229.

Moloney, A.P., B. Murray, D.J. Troy, G.E. Nute & R.I. Richardson (2004). The effects of fish oil inclusion in the concentrate and method of silage preservation on the colour and sensory characteristics of beef. *Proceedings Agricultural Research Forum*, Tullamore, Ireland, p13.

Muir, P.D., J.M. Deaker & M.D. Brown (1998). Effects of forage-and grain-based feeding systems on beef quality: A review. *New Zealand Journal of Agricultural Research*, 41, 623-635.

Murphy, J.J. (2000). Synthesis of milk fat and opportunities for nutritional manipulation. *Milk Composition – Occasional Publication No. 25*, British Society of Animal Science, 201-222.

Noci, F., A.P. Moloney, P. French & F.J. Monahan (2003). Influence of duration of grazing on the fatty acid profile of *M. longissimus dorsi* from beef heifers. *Proceedings of Bristish Society of Animal Science*, Winter meeting, York, p23.

Nuernberg, K., D. Dannenberger, G. Nuernberg, K. Ender, J. Voigt, N.D. Scollan, J.D. Wood, G.R. Nute & R.I. Richardson (2004). Effect of a grass-based and a concentrate feeding system on meat quality characteristics and fatty acid composition of *longissimus* muscle in different cattle breeds. *Livestock Production Science*, (in press).

Offer, N.W., M. Marsden, J. Dixon, B.K. Speake & F.E. Thacker (1999). Effect of dietary fat supplements on levels of n-3 poly-unsaturated fatty acids, trans acids and conjugated linoleic acid in bovine milk. *Animal Science*, 69, 613-625.

Offer, N.W., M. Marsden, & R.H. Phipps (2001). Effect of oil supplementation of a diet containing a high concentration of starch on levels of trans fatty acids and conjugated linoleic acids in bovine milk. *Animal Science*, 73, 533-540.

O'Sullivan, A., K. Galvin, A.P. Moloney, D.J. Troy, K. O'Sullivan & J.P. Kerry (2003). Effect of pre-slaughter ration of forages and or concentrates on the composition and quality of retail packaged beef. *Meat Science*, 63, 279-286.

O'Sullivan, A., K. O'Sullivan, K. Galvin, A.P. Moloney, D.J. Troy & J.P. Kerry (2004). Influence of concentrate composition and forage type on retail packaged beef quality. *Journal of Animal Science*, 82, 2384-2391.

Petit, H.V., R.J. Dewhurst, N.D. Scollan, J.G. Proulx, M. Khalid, W. Haresign, H. Twagiramungu & G.E. Mann (2002). Milk production and composition, ovarian function, and prostaglandin secretion of dairy cows fed omega-3 fats. *Journal of Dairy Science*, 85, 889-899.

Precht, D. & J. Molkentin (1997). Effect of feeding on conjugated *cis*Δ9,*trans*Δ11,-octadecadienoic acid and other isomers of linoleic acid in bovine milk fats. *Nahrung*, 41, 330-335.

Priolo, A., D. Micol & J. Agabriel (2001). Effects of grass feeding systems on ruminant meat colour and flavour: a review. *Animal Research*, 50, 185-200.

Qi, B.X., T. Fraser, S. Mugford, G. Dobson, O. Sayanova, J. Butler, J.A. Napier, A.K. Stobart & C.M. Lazarus (2004). Production of very long chain polyunsaturated omega-3 and omega-6 fatty acids in plants. *Nature Biotechnology*, 22 (6), 739-745.

Raes, K., S. De Smet & D. Demeyer (2001). Effect of double-muscling in Belgian Blue young bulls on the intramuscular fatty acid composition with emphasis on conjugated linoleic acid and polyunsaturated fatty acids. *Animal Science*, 73, 253-260.

Realini, C.E., S.K. Duckett, G.W. Brito, M. Dalla Rizza & D. De Mattos (2004). Effect of pasture vs. concentrate feeding with or without antioxidants on carcass characteristics, fatty acid composition and quality of Uruguayan beef. *Meat Science*, 66, 567-577.

Ritzenthaler, K.L., M.K. McGuire, R. Falen, T.D. Schultz, N. Dasgupta & M.A. McGuire (2001). Estimation of conjugated linoleic acid intake by written dietary assessment methodologies underestimates actual intake evaluated by food duplicate methodology. *Journal of Nutrition*, 131, 1548-1554.

Saito, T., S. Takadama, H. Kasuga & T. Nakanishi (1969). Effects on fatty acid composition of lipids in cows milk by grass and legume fed (VI) Differences of effects of district and seasons, on fatty acid composition of lipids in grass and legume. *Japanese Journal of Dairy Science*, 18, 183-189. (*In Japanese, with English Summary and Tables*)

Sanudo, C., G.R. Nute, M.M. Campo, G. Maria, A. Baker, I. Sierra, M. Enser & J.D. Wood (1998). Assessment of commercial lamb meat quality by British and Spanish taste panels. *Meat Science*, 48, 91-99.

Schroeder, G.F., J.E. Delahoy, I. Vidaurreta, F. Bargo, G.A. Gagliostro & L.D. Muller (2003). Milk fatty acid composition of cows fed a total mixed ration or pasture plus concentrates replacing corn with fat. *Journal of Dairy Science*, 86, 337-3248.

Scollan, N.D., N.J. Choi, E. Kurt, A.V. Fisher, M. Enser & J.D. Wood (2001). Manipulating the fatty acid composition of muscle and adipose tissue in beef cattle. *British Journal of Nutrition*, 85, 115-124.

Scollan, N.D., A. Cooper, P. Evans, M. Enser, R.I. Richardson, G.R. Nute, A.V. Fisher & J.D. Wood (2002). Effect of forage legumes on the fatty acid composition of beef and other aspects of meat quality. *Proceedings of the 48th International Congress of Meat Science and Technology*, Rome, Italy, 1, 356-357.

Scollan, N.D., M. Enser, S. Gulati, I. Richardson & J.D. Wood (2003). Effects of including a ruminally protected lipid supplement in the diet on the fatty acid composition of beef muscle in Charolais steers. *British Journal of Nutrition*, 90, 709-716.

Scollan, N.D., M. Enser, R.I. Richardson, S. Gulati, K.G. Hallett, G.R. Nute & J.D. Wood (2004). The effects of ruminally protected dietary lipid on the fatty acid composition and quality of beef muscle. *Proceedings of the 50th International Congress of Meat Science and Technology*, Helsinki, Finland, p116.

Simopoulos, A.P. (2001). *n*-3 fatty acids and human health: defining strategies for public policy. *Lipids*, 36; S83-S89.

Stanton, C., F. Lawless, G. Kjellmer, D. Harrington, R. Devery, J.F. Connolly & J. Murphy (1997). Dietary influences on bovine milk *cis*-9, *trans*-11-conjugated linoleic acid content. *Journal of Food Science*, 62, 1083-1086.

Stockdale, C.R., G.P. Walker, W.J. Wales, D.E. Dalley, A. Birkett, Z. Shen & P.T. Doyle (2003). Influence of pasture and concentrates in the diet of grazing dairy cows on the fatty acid composition of milk. *Journal of Dairy Research*, 70, 267-276.

Thomas, H. (1986). The role of polyunsaturated fatty acids in senescence. *Journal of Plant Physiology*, 123, 97-105.

Thomas, H. & C.M. Smart (1993). Crops that stay green. *Annals of Applied Biology*, 123, 193-219.

Thomson, N.A. & W. Van Der Poel (2000). Seasonal variation of the fatty acid composition of milk fat from Friesian cows grazing pasture. *Proceedings of the New Zealand Society of Animal Production*, 60, 314-317.

Timmen, H. & S. Patton (1988). Milk fat globules: fatty acid composition, size and in vivo regulation of fat liquidity. *Lipids*, 23, 685-689.

Timmons, J.S., W.P. Weiss, D.L. Palmquist & W.J. Harper (2001). Relationships among roasted soybeans, milk componnets, and spontaneous oxidized flavor of milk. *Journal of Dairy Science*, 84, 2440-2449.

Vatansever, L., E. Kurt, M. Enser, G.R. Nute, N.D. Scollan, J.D. Wood & R.I. Richardson (2000). Shelf life and eating quality of beef from cattle of different breeds given diets differing in n-3 polyunsaturated fatty acid composition. *Animal Science*, 71, 471-482.

Vipond, J.E., S. Marie & E.A. Hunter (1995). Effects of clover and milk in the diet of grazed lambs on meat quality. *Animal Science*, 60, 231-238.

Warren, H.E., N.D. Scollan, K. Hallett, M. Enser, R.I. Richardson, G.R. Nute & J.D. Wood (2002). The effects of breed and diet on the lipid composition and quality of bovine muscle. *Proceedings of the 48th Congress of Meat Science and Technology*, Rome, Italy, 1, 370-371.

White, S.L., J.A. Bertrand, M.P. Wade, S.P. Washburn, J.T. Green & T.C. Jenkins (2001). Comparison of fatty acid content of milk from Jersey and Holstein cows consuming pasture of total mixed ration. *Journal of Dairy Science*, 84, 2295-2301.

Whiting, C.M., T. Mutsvangwa, J.P. Walton, J.P. Cant & B.W. McBride (2004). Effects of feeding either fresh alfalfa or alfalfa silage on milk fatty acid content in Holstein dairy cows. *Animal Feed Science and Technology*, 113, 27-37.

Wood, J.D., R.I. Richardson, G.R. Nute, A.V. Fisher, M.M. Campo, E. Kasapidou, P.R. Sheard & M. Enser (2003). Effects of fatty acids on meat quality: a review. *Meat Science*, 66, 21-32.

Wrenn, T. R., J.R. Weyant, D.J. Wood, J. Bitman, R.M. Rawlings & K.E. Lyon (1976). Increasing polyunsaturation of milk fats by feeding formaldehyde protected sunflower-soyabean supplement. *Journal of Dairy Science,* 59, 627-635.

Yang, A., M.C. Lanari, M. Brewster & R.K. Tume (2002). Lipid stability and meat colour of beef from pasture and grain-fed cattle with or without vitamin E supplement. *Meat Science,* 60, 41-50.

Young, O.A., J.L. Berdagué, C. Viallon, S. Rousset-Akrim & M. Theriez (1997). Fat-borne volatiles and sheep meat odour. *Meat Science,* 45, 183-200.

Young, O.A., G.J. Cruickshank, K.S. MacLean & P.D. Muir (1994). Quality of meat from lambs grazed on seven pasture species in Hawkes Bay. *New Zealand Journal of Agricultural Research,* 37,177-186.

Yu, S., J. Derr, T.D. Etherton & P.M. Kris-Etherton (1995). Plasma cholesterol-predictive equations demonstrate that stearic acid is neutral and monounsaturated fatty acids are hypocholesterolemic. *American Journal of Clinical Nutrition*, 61, 1129-1139.

Zock, P.L., J.H.M. De Vries & M.B. Katan (1994). Impact of myristic acid versus palmitic acid on serum-lipid and lipoprotein levels in healthy women and men. *Arteriosclerosis and Thrombosis*, 14, 567-575.

Grass and forage improvement: temperate forages

C.J. Pollock, M.T. Abberton and M.O. Humphreys
Institute of Grassland and Environmental Research, Plas Gogerddan, Aberystwyth SY23 3EB, Wales, UK
Email: chris.pollock@bbsrc.ac.uk

Key points

1. Plant breeding has contributed significantly to the development of effective grassland production systems.
2. New technologies offer enhanced precision in breeding and access to wider genetic variation.
3. The requirement for more sustainable production systems will require genetic improvements in complex traits where the use of new technology will be vital.

Keywords: selection, genetic mapping, molecular markers, grass breeding, clover

Introduction

The grazing ruminant animal has been a key element of temperate agriculture since the domestication of sheep and oxen in around 5000 BC (Humphreys, 2003). Ruminant animals provided meat, milk and motive power, and were able to live on plants that humans could not use directly. As agricultural systems developed, their role in recycling nutrients was also recognised, animal production forming a key element in rotational agriculture in the middle ages. Less attention was paid to the forages that ruminants ate, although the value of grasses and clovers in short-term leys was established by the 12th century, and it was well recognised that some permanent pastures were, by nature, more productive than others.

Following the development of practical genetics-based plant breeding in the early 20th century, forage improvement became more formalised. However, the complexities of working with perennial outbreeders in mixed systems that were very sensitive to management meant that the history of true genetic-based breeding of temperate forages is relatively short. The perennial ryegrass variety S 24 for example, was produced in the 1930s based on collected ecotypes from UK pastures and unrestricted crossing between selected plants. This generated a variety that was resilient, persistent and productive (in terms of the norm for UK pastures at the time), but one that was very heterogeneous and where the potential for genuine breeding advances was not fully realised. Since then, for both grasses and legumes, much more attention has been paid to precision crossing and to the incorporation of specific traits from individual genotypes, often collected outside the UK. The results of this process have been a continued improvement in yield and quality with some gains in disease resistance and grass-clover compatibility.

Improved pasture supports the bulk of livestock production in the UK, with grazed forages providing the cheapest feed for ruminant animals, as well as bringing environmental benefits. As we enter the 21st century, both the technology and the targets for temperate forage breeding are changing rapidly. There is increased pressure in many developed countries for sustainable farming systems that yield both a viable economic return and a range of environmental and social goods. This is changing the nature of plant breeding. It is our view that success in breeding for targets such as nutrient use efficiency, palatability and improvements in livestock product quality cannot be done by traditional selection methods, since the traits are complex, expensive to measure and subject to considerable genotype x environment (GxE) interactions. For this reason we first consider what forage breeders can achieve with the application of such new technology, and then discuss what targets are likely to be important and how the outputs of such programmes will be used.

Table 1 Breeding advances in forage crops

Species	Annual yield (dry matter)[1]	Digestibility[2]	Intake[2]	Crude protein[2]
		Average genetic gain (%)		
Lolium perenne	3.8	1.0	14.7	6.5
Dactylis golomerata	3.3	2.4		
Trifolium repens	4.9			
Medicago sativa	4.1	3.3		2.7

[1]per decade[-1] (after Humphreys, 1999); [2]per generation cycle[-1] (after Casler, 2001).

The impact of modern breeding technologies

Over the last twenty-five years, genetic knowledge has expanded extremely rapidly, driven by advances in molecular biology. The sequencing of entire higher plant genomes has generated a huge reference resource, which has proved particularly valuable where close synteny exists between sequenced organisms and those used in agriculture. Among crop plants, *Oryza sativa* (rice) has been sequenced and the degree of alignment between the rice genome and other gramineae is very high. This allows traits to be characterised in non-sequenced organisms and equivalent areas of the sequenced genome in rice to be identified, helping to identify at the sequence level the basis of useable variation for breeding programmes (Armstead *et al.*, 2004). Sequencing work is also under way for the model legumes *Medicago truncatula* and *Lotus japonicus* (Young *et al.*, 2003). This body of information will increasingly influence the way in which other forage legumes will be bred in the future.

Marker assisted selection

Of the emerging genetic technologies, marker assisted selection (MAS) is the most likely to have an immediate impact on plant breeding. It depends on the detection of DNA variation among individuals using a variety of techniques (Henry, 2001; Forster *et al.*, 2004). Successful breeding using MAS depends on understanding the genetic architecture of relevant traits. Traditionally, major gene and polygenic variation has been analysed in different ways, but the techniques of Quantitative Trait Locus (QTL) analysis now allow a more integrated approach in dissecting complex traits and assessing gene effects. Quantitative Trait Locus analysis can be used to determine the location or locations of variation for a complex trait directly onto the genetic map of a particular species. At the same time, the density of markers on these maps has increased dramatically associated with the use of the DNA markers. Heterozygous molecular markers that are tightly linked to traits of interest can be used to speed up selection in breeding programmes. As a result of the International *Lolium* Genome Initiative (ILGI; Jones *et al.*, 2002) involving IGER and groups in Australia, Japan and France, a reference linkage map of *L. perenne* has been produced. It comprises 240 loci covering 811cM on seven linkage groups. The map contains 124 co-dominant markers, of which 109 are heterologous anchor restriction fragment length polymorphisms (RFLP) probes from from *Triticum* spp. (wheat), *Hordeum* spp.

(barley), *Avena* spp. (oats) and *O. sativa*, allowing comparative relationships to be investigated between *L. perenne* and other *Poaceae*. The linkage groups of *L. perenne* are numbered to correspond to the homoeologous groups of the *Triticeae* cereals. The genetic maps of *L. perenne* and the *Triticeae* cereals are highly conserved in terms of synteny and colinearity (Jones *et al.,* 2002). There is also general agreement over the syntenic relationships between *L. perenne, Festuca pratensis* (meadow fescue), oat and rice, and those between the *Triticeae* and these species.

An 'IGER' linkage map of *L. perenne* based on an F2 mapping population was produced initially for the genetic analysis of water-soluble carbohydrate (WSC) accumulation. This map identified 7 linkage groups covering a total of 515cM and was compared with the ILGI map produced from a BC1-type population (Armstead *et al.,* 2002). The maps could be aligned using 38 common loci, and a marker order for all mapped loci in either population was identified in an integrated map. Using this map, QTL associated with WSC accumulation in the leaves and leaf sheaths of ryegrass have been identified (Humphreys & Turner, 2001; Turner *et al.,* 2001). Interestingly, some QTL overlie or fall close to the location of genes with known function. The genomic region associated with a major flowering time QTL in ryegrass shows a high degree of synteny with the location of genes in rice, having homology with the *Arabidopsis* flowering time gene *CONSTANS* (*CO*) (Donnison *et al.,* 2002; Armstead *et al.,* 2004). Allelic variation in the *CO*-like genes correlate with differences in heading date and may account for up to 70% of the total variation.

In forage legumes, the majority of work has been carried out in *Medicago sativa* (lucerne), primarily in the USA, but with some effort in Europe. In both *M. sativa* and other forage legumes, current effort is directed mainly towards mapping, with little directed MAS except under experimental conditions. The first *Trifolium repens* map has been developed using self-compatible inbred lines in a collaboration between IGER and The Plant Biotechnology Centre, Victoria (Abberton *et al.,* 2000b). This utilises amplified fragment length polymorphisms (AFLPs) and microsatellites. Previous work had used random ampligfication of polymorphic DNA (RAPDs) to assess the extent of genetic diversity between inbred lines and compare this with the degree of heterosis in crosses between them (Joyce *et al.,* 1999). Quantitative Trait Locus for agronomically important characters in single plants (including seed yield), have been located in a meta-analysis across sites and years. This work is now being extended to studies of plot performance. New mapping families are being used to investigate the genetic basis of stolon morphology and, in collaboration between IGER and Teagasc, Ireland, to locate and clone genes involved in resistance to stem nematode. In parallel with the mapping studies, development of EST collections is being undertaken in a joint programme between PBC and AgResearch, New Zealand (Forster *et al.,* 2001). A *Trifolium pratense* L. (red clover) linkage map has also been published (Isobe *et al.,* 2003).

The use of mapping populations to generate the resources to facilitate MAS poses challenges. There is a need for accurate, rapid and ideally non-invasive phenotyping so that the trait variation can be characterised as well as the marker profiles. High throughput chemometry is being developed for metabolomic studies, and there are real opportunities to use these advances in phenotyping. Near infra-red reflectance spectrometry (NIRS) is an example of one of these unbiased high throughput systems that is already in use as a routine analytical tool (Wilman *et al.,* 2000), and mass spectrometry-based approaches are also in increasing use. For developmental traits, imaging methodologies are being developed, although the challenges are greater. Hand in hand with this type of phenotyping goes the need for good informatics and statistical approaches to ensure that reliable information is derived, analysed and accurately collated with DNA-based information from the same plants. There is also a growing need for robust process models for land use systems that will allow breeders to estimate the likely significance of potential improvements. Such models, for example of N flows in grassland, do exist, and some of them are scale dependent (Bhogal *et al.,* 2001), but they have rarely been used to identify and validate breeding targets.

Introgression

Introgression has been used as a breeding tool for many years, but the development and employment of molecular markers improves both speed and precision. At IGER, the main targets for introgression have been the introduction of biotic and abiotic stress tolerance from *Festuca* spp. into *Lolium multiflorum* or *L. perenne* (Humphreys *et al.,* 1998, 2003). Introgression into *L. perenne* is necessary following the initial generation of the hybrids in order to ensure that the high nutritive quality of *L. perenne* spp. is retained together with the improved stress tolerance derived from *Festuca*. A number of mapping families have been produced to facilitate marker-assisted introgression. Having identified plants that express introgressed *Festuca*-derived traits, they can also be analysed using *in situ* hybridisation. The close relationship between *L. perenne* and *F pratensis* has allowed work on introgression mapping and alignment of the genetic map with a physical map (Armstead *et al.,* 2001; King *et al.,* 2001). Genomic in situ DNA hybridisation can be used to determine the location of alien *Festuca* genes, and genetic markers ascribed to the targeted sequence (Humphreys *et al.,* 1998). *L. perenne* lines carrying different *Festuca* genes that convey a range of adaptations to abiotic stresses are currently being developed alongside 'breeder's toolkits' to aid their commercial exploitation.

Introgression studies have also been carried out on a range of forage legumes, with effort concentrated on *M. sativa* and *T. repens*. Some of these studies are beginning to employ marker-assisted introgression using both molecular and cytological markers. Interspecific hybrids have been produced between *T. pratense* and some of its relatives, but with little impact to date. Significant programmes on *T repens* are in place at IGER (using molecular markers) and Ag Research New Zealand (using cytological markers). Both groups have focused on two introgressions, with *T. nigrescens* and with *T. ambiguum. T. nigrescens* is an annual diploid species that is sexually compatible with *T repens,* and is believed to be one of its ancestral genomes. In associated research, fluorescent *in situ* hybridisation of nucleolar organiser regions has been used to explore relationships between *T. repens* and related species including *T. nigrescens* (Ansari *et al.,* 1999). Bulked segregant AFLP approaches (BSA-AFLP) have been used to identify markers for high seed yield potential in second and third generation backcross hybrids, with *T. repens* as the recurrent parent (Abberton *et al.,* 2000a). Hybrids between *T. repens* and the rhizomatous very persistent *T. ambiguum* (Caucasian clover) require ovule culture or embryo rescue, which adds complexity. However, backcross hybrids show considerably enhanced drought tolerance in relation to the *T. repens* parent, together with somewhat reduced protein content, giving the potential to reduce nitrogenous pollution in livestock feeding.

Genetic manipulation

Genetic manipulation (GM) may be an alternative to conventional breeding or an additional element in the programme, and can aid breeding by accessing or creating novel sources of specific plant variation. Genetic manipulation also provides tools to understand basic plant biology and help to identify genes controlling the physiological and biochemical processes underlying traits of agronomic importance. Transformation systems are now well established for all major forage grasses and legumes. Precision genetic manipulation requires accurate targeting of transgene expression, either to different cell types or to specific intercellular compartments. Some progress has been made using cell specific promoters in grasses and legumes (but many more cell specific promoters are needed), and for compartment targeting using 5' and 3' signal sequences in grasses associated with specific wall degrading enzymes (Morris & Langdon, 2001). Plastid transformation gives higher expression of prokaryotic genes, and may facilitate 'containment' of transgenes within the transformed plants. However, there are marked species-dependent differences in the level of paternal inheritance of plastid DNA. In addition, there are considerable regulatory challenges to be met before GM grasses and legumes are approved for widespread cultivation in areas where compatible wild relatives abound. Currently GM is used predominantly in forage breeding for 'proof of principle' studies, where transgenic material up or down-regulated in an individual gene product, provides a clear approach to establishing gene function in a way that natural variation rarely permits.

Genetic resources

In all cases, variation for traits of interest is being incorporated into potential new varieties. Capturing, curating and maintaining genetic resources that allow access to such variation is central to breeding programmes worldwide. In the case of forages, most of this will be in direct relatives of the agricultural species. Approaches to defining core collections on the basis of allele variation rather than phenotype variation will simplify curation (van Hintum *et al.*, 2001), and the use of molecular markers will speed the process of screening collections for favourable alleles. However, the commitment to collection, characterisation and maintenance of genetic resources is considerable and funding is not easy to obtain. There is a real need to ensure, preferably on an international scale, that funding for this activity is maintained, but also that every opportunity to streamline the process and minimise costs without losing vital material is considered.

Current targets for precision breeding in forages

It has long been known that there are big differences between species in both the suitability of forages for animal production and in their capacity to grow and survive in particular areas. Traditionally, improvement programmes have concentrated on those species that are suited to the locale and that deliver the best return in terms of animal products. Within that, agronomic characters such as yield, flowering date, persistency and compatibility have been priority targets, with less emphasis on improving intake, digestibility, resource use efficiency and end-product quality. There are sound reasons for this approach. Primary agronomic characters can be measured easily, are important in grassland management and show tight genetic control. However, the application of the technological approaches outlined previously allows more complex traits to be manipulated and there is increasing evidence of the relevance of such an approach.

One of the best examples of precision breeding for quality traits in grasses is the ability to manipulate the efficiency of rumen function through altering WSC content. Elevated soluble sugars in grazed herbage provide an accessible carbon source in the early stages of colonisation of fresh feed by rumen microorganisms (Miller *et al.,* 2001; Lee *et al.,* 2003), thus reducing the breakdown of peptides and deamination of amino acids. Table 2 indicates the benefits that accrue in terms of conversion efficiency and reduced environmental deposition of N. Quantitative Trait Locus analysis has indicated a number of regions of the *L. perenne* genome involved in WSC accumulation. Currently MAS is being used to identify new genotypes with the high sugar trait (Humphreys *et al.*, 2003).

Table 2 Effects of water-soluble sugar content (WSC) of grass on nitrogen partitioning into milk and urine (after Miller *et al.*, 2001)

	WSC (%)	CP (%)	N intake (g/day)	N output (g/day)	
				Milk	Urine
High sugar ryegrass	20.1	9.2	268	82	71
Normal ryegrass	12.9	10.6	278	69	100

This approach is being applied to a number of quality and efficiency traits that are known to be important in grasses and legumes - including nitrogen and phosphorous use efficiency, mineral composition, and developmental and morphological traits that improve compatibility. In principle, if defined mapping populations exhibiting variation in an important trait can be established, then QTL analysis and MAS can be used, with a concomitant gain in speed and simplicity of selection. Even very complex derivative traits like palatability and grazing preference are susceptible to this approach, although these would require a considerable experimental investment to derive the original mapping populations.

Marker-assisted introgression is also being used to manipulate important traits. Using a range of *Lolium-Festuca* introgressions, variation in drought, heat and cold tolerance, resistance to crown rust and improved protein stability during digestion has been incorporated into a *L. perenne* background and is being assessed. Additionally the stay-green character, which has considerable relevance to turf grasses, has been transferred into *Lolium* and used to generate commercial varieties. Hybrids between *T. repens* and *T. ambiguum* have been used at IGER to produce drought resistant material with both stolons and rhizomes (Abberton *et al.*, 2000a) and in New Zealand primarily for virus resistance (Woodfield & Brummer, 2001). Molecular markers have been used to select for the rhizomatous habit in plants of backcross families (Abberton *et al.*, 2000b) and this material is currently being assessed for cold tolerance at several sites in Northern Europe.

There is increasing interest in improving grass-clover compatibility for mixed swards, and compatibility is now starting to be considered as much from the nutritional point of view as agronomic. The development of marker-based approaches will make this more feasible. It is known, for example that root structures are important determinants of compatibility (Collins *et al.*, 2003), and these are very difficult to select for without good markers. In the future, greater knowledge of rhizosphere processes, including N transfer from clover to grass is also likely to be a key area, particularly since compatability issues are likely to be affected by climate change.

Other targets

There are other compositional traits that are known to affect the quality of livestock products. These include lipid composition; particularly the high levels in forages of polyunsaturated fatty acids and the omega-3 fatty acids, which have positive health benefits and also promote the colour and keeping quality of beef (Wood *et al.,* 2003; Scollan *et al.,* 2005). There are also other approaches to reducing proteolysis within the rumen. Plant-mediated proteolysis is demonstrable in the rumen (Kingston-Smith *et al.,* 2003), opening up the possibility of producing forages with reduced protease contents. There is also evidence that protein protection mediated by polyphenol oxidase is under genetic control (Evans *et al.,* 2002).

Development in *T. repens* of near-isogenic lines (NILS) has allowed study of the expression of recessive genes. For example, one of the lines is deficient in N fixation, permitting the protein content of clover to be manipulated via contrasting levels of fertilizer application. This has been used as 'proof of principle' that the lower protein content of legumes during ensiling can result in silage with similar retained protein content, but with significantly lower environmental losses. This is now a target for current precision breeding approaches. The NILS have functioned as a 'GM analogue' in terms of trait identification and characterisation. The availability within the NILS of 'control' lines that are very closely related to the non-fixing lines has been of great value (Abberton *et al.,* 1998).

There is also physiological evidence for a genetic component affecting tolerance of heavy metals, opening up opportunities for developing grass varieties suitable for managing degraded land. However, the level of investment in basic and strategic science that is required to underpin variety development makes exploitation of the full range of possible targets somewhat unlikely. Reform of the European support system for agriculture will alter the ways in which farmers receive payment from the state, and this in turn is likely to affect their requirements for new forage varieties. However, it is not currently clear what sorts of agricultural systems will prevail and thus what the new targets will be. From an environmental standpoint, the benefits of rotational agriculture, spring cereal growing and hay-making are abundantly clear (Chamberlain *et al.,* 2000). If agricultural support mechanisms end up promoting such practices, then this will stimulate the use of *T. pratense,* cereal/legume mixtures and grass varieties optimised for short-term leys, all of which are relatively minor components of current farming systems. Nevertheless, the generic technologies of genome analysis, marker development and gene identification will continue to influence forage breeding regardless of the specific breeding targets.

Opportunities, challenges and constraints

Within grassland agriculture in developed countries, managing the balance between the natural and the farmed environment is likely to assume increased importance over the next 10-15 years. There are contrasting pressures, with an increased demand for animal products on one hand and evidence of the need for reduction of the impacts of agricultural systems on the other. The Environment Agency in the UK has identified pollution from agriculture as a major issue, with farming being responsible for 27% of major pollution incidents (Anon, 2002). Grassland farming contributes to pollution via losses of nitrate and phosphate into ground water, ammonia and methane into the atmosphere, and by the direct effect of slurry and silage effluent leaking into watercourses. Intensive livestock agriculture involving silage production also has a negative impact on biodiversity, with the declining presence of seed-eating birds in grasslands being significantly correlated with the move away from hay production in the 1960s and 70s (Vickery *et al.,* 2001). Against this must be set the vital role of the grazing animal in preserving key habitats. Unimproved semi-natural pasture requires careful grazing if species diversity and habitat are to be maintained. Such management requires a viable livestock agriculture to achieve this objective, together with a support system that provides incentives for farmers to deliver environmental goods. To achieve this more broadly based set of objectives, will require a much closer linkage between forage breeding and the development of novel management systems.

It is already known that modelling the processes of nitrogen movement within grassland can identify the optimal times for nitrogen application, leading to significant reductions in both fertiliser application and nitrate leaching without a concomitant reduction in forage yield or animal production (Laws *et al.*, 2000). Linking approaches like this to the genetic improvement of nutrient use efficiency offers further potential gains at the system level. The beneficial effect that *T. repens* has on soil quality in heavy and compacted soils is also a target for genetic improvement, but must be pursued in concert with management systems that are directed towards preserving the improvement attributable to *T. repens*. Current and proposed environmental legislation provides a strong impetus for integrated research of this kind within the EU, and there are welcome signs that research funders are recognising the value of this approach. Increasingly, breeding advances will have to be delineated at the system level, which has consequences for variety assessment. A further approach to improving nitrogen use efficiency at the sward level by increasing protein protection in the rumen involves the presence in a mixed sward of a tannin- containing species such as *Lotus corniculatus* (birdsfoot trefoil). Utilisation of this species can also result in a reduction of the parasite burden in sheep (Marley *et al.*, 2003) although it is not clear whether this effect is mediated through stimulation of the immune response or via a direct effect of tannins.

The Curry Report into UK agriculture argued that the removal of direct production subsidies will broaden the base of UK land use, and it will be important to ensure that these changes feed through into flexible forage improvement programmes. For niche producers, where high on-farm value and product quality are paramount, pasture quality becomes of comparable importance in terms of maximising the contribution of a cheap, traceable and 'natural' component of the diet. As well as the direct quality traits indicated above, more complex traits that promote the stability of production over time, particularly in mixed swards, will become important. Again, the challenge here is to define the trait under realistic conditions involving animals, and then to use DNA markers to access and utilise the available genetic variation. A more multifunctional land use base might also generate novel opportunities for forage species. Already sports turf accounts for 30% of UK grass seed sales and the IGER amenity breeding programme has focussed on 'non-agricultural' traits such as prostrate habit and wear-resistance (Thorogood, 2003).

There are other opportunities in the development of alternative fibre and biomass crops (J. Valentine, *pers. comm.*). Work is also in progress to assess the available genetic variation in grasses for environmental targets such as C-sequestration in the soil, support for soil biodiversity, bioremediation and flood mitigation (J. Macduff, *pers. comm.*). Candidate traits for C-sequestration include shoot and root turnover rates, rooting depth, lignification and tissue density. Candidate traits impacting on soil microbial biodiversity at the single plant level include C and N root exudation and rhizosphere acidification, as well as those relevant to C-sequestration. Bioremediation targets include (i) heavy metal (Zn, Pb, Fe) uptake/tolerance; (ii) absorption of atmospheric NOx; and (iii) buffer zone efficacy with respect to capture of soil nitrate. Flood mitigation involves work on aspects of rooting and canopy architecture that confer increased infiltration rates into the soil profile as well as those conferring increased transpiration rates, such as leaf water conductance, and at a field scale canopy conductance. These novel traits will need to be allied to specified and detailed management systems to deliver benefits.

The challenge involved in getting broader targets incorporated into forage improvement programmes is principally one of cost. It is estimated that the annual wholesale value of forage seed sold to agriculture in the UK is approximately £8M (Caligari, 2002). In contrast, grassland supports approximately 40% of agricultural output, which is estimated to be worth approximately £5billion. Thus the genetic improvement of grassland species has broad social benefits because of its wider impact on agriculture. However the market is capable of supporting only a relatively small research base and needs to ensure a supply of new varieties for existing uses before being able to invest speculatively. For basic and enabling research, collaboration between researchers is recognised as essential, and has led to a number of the advances described above. The technologies involved in basic crop research are likely to become increasingly generic, in turn offering better opportunities for co-ordination. Unfortunately, the need to protect intellectual property and the differing sectoral requirements of different regions mean that downstream improvement programmes remain quite fragmented. This severely restricts the level of effort available to create new niches for forage varieties.

As the targets for forage breeding broaden, and as the need to carry out system studies involving animals increases, the validity of the current variety evaluation systems used for registration become increasingly questionable. Simple determination of the distinctiveness, uniformity and stability (DUS) of varieties remains straightforward and desirable, but the kind of performance measures needed to generate a recommended list add considerably to the costs of variety production. It can be argued that the use of cutting regimes to simulate grazing is not representative of a significant amount of livestock agriculture. The ideal would be to produce data that allowed users to assess the value of the variety within the production systems for which it is intended, and this would have the advantage of continuing to provide robust, experimentally obtained and consistent data. However, it is difficult to see how this ideal can be realised within the current cost structure of grassland farming. The industry may have to rely on data produced directly by breeders and users although it would be beneficial to have some independent monitoring of such information.

Conclusions

The precision breeding and improvement of temperate forage varieties will continue, powered by developments in marker-assisted selection, genome mapping and introgression. Genetic manipulation remains a potentially useful approach to accessing novel genetic variability, but there are operational and public perception issues that need to be resolved first. Changes in temperate agriculture will emphasise a systems approach to both breeding and the development of precision management. There are new opportunities for forages outside the farming sector, but it is not clear whether enabling research in this area would attract enough support to create new markets. Maintenance of current programmes will increasingly come to rely on co-operative approaches to the generation of basic knowledge and to the development of new breeding technologies.

Acknowledgements

Financial support for the IGER forage improvement programme is received from The Biotechnology and Biological Sciences Research Council, The Department for Environment, Food and Rural Affairs, and Germinal Holdings Limited.

References

Abberton, M.T., T.P.T. Michaelson-Yeates & J.H. Macduff (1998). Characterization of novel inbred lines of white clover (*Trifolium repens* L.) in flowing solution culture. I. Dynamics of plant growth and nodule development in flowing solution culture. *Euphytica,* 103, 35-43.

Abberton, M.T., A.H. Marshall, T.P.T. Michaelson-Yeates, T.A. Williams, W. Thornley, W. Prewer, C. White & I. Rhodes (2000a). An integrated approach to introgression breeding for stress tolerance in white clover. In: G. Parente & J. Frame (eds.) Crop development for the cool and wet regions of europe: achievements and future Prospects. *Proceedings of the Final COST Action 814 Conference,* Pordenone, Italy, May 10-13 (EUR 19683) Office for Official Publications of the European Communities. 311-314.

Abberton, M.T., T.P.T. Michaelson-Yeates & I.A.H. Marshal (2000b). Molecular marker analysis in white clover. In: N.A. Provorov, I.A. Tichonovich & F. Veronesi (eds.) New approaches and techniques in breeding sustainable fodder crops and amenity grasses. All-Russia Institute for Agricultural Microbiology, St Petersburg, 192-195.

Anonymous (2002). Agriculture and natural resources: benefits, costs and potential solutions. The Environment Agency, U.K. 100pp.

Ansari, H., N.W. Ellison, S.M. Reader, E.D. Badaeva, B. Friebes, T.E. Miller & W.M. Williams (1999). Molecular cytogenetic organisation of 5S and 18S-26S rDNA loci in white clover (*Trifolium repens* L.) and related species. *Annals of Botany,* 83, 199-206.

Armstead, I.P., A. Bollard, W.G. Morgan, J.A. Harper, I.P. King & R.N. Jones, (2001). Genetic and physical analysis of a single *Festuca pratensis* chromosome segment substitution in *Lolium perenne. Chromosoma,* 110, 52-57.

Armstead, I.P., L.B. Turner, I.P. King, A.J. Cairns & M.O. Humphreys (2002). Comparison and integration of genetic maps generated from F2 and BC1-type mapping populations in perennial ryegrass (*Lolium perenne* L.). *Plant Breeding,* 121, 501-507.

Armstead, I.P., L.B. Turner, M. Farrell, L. Skot, P. Gomez, T. Montoya, I.S. Donnison, I.P. King & M.O. Humphreys (2004). Synteny between a major heading-date QTL in perennial ryegrass (*Lolium perenne* L.) and the Hd3 heading-date locus in rice. *Theoretical and Applied Genetics,* 108, 822-828.

Bhogal, A., M.A. Shepherd, D.J. Hatch, L. Brown & S.C. Jarvis (2001). Evaluation of two N cycle models for the prediction of N mineralization from grassland soils in the UK. *Soil Use and Management,* 17, 163-172.

Caligari, P.D.S. (2002). The Role of future public research investment in the genetic improvement of UK grown crops. Report to the Department for Environment, Food & Rural Affairs by BioHybrids International Ltd. and ADAS Consulting Ltd. 222pp.

Casler, M.D. (2001). Breeding forage crops for increased nutritional value. *Advances in Agronomy,* 71, 51-107.

Chamberlain, D.E., R.J. Fuller, R.J.H. Bunce, J.C. Duckworth & M. Shrubb (2000). Changes in the abundance of farmland birds in relation to the timing of agricultural intensification in England and Wales. *Journal of Applied Ecology,* 37, 771-788.

Collins, R.P., M. Fothergill, J.H. Macduff & S. Puzio (2003). Morphological compatibility of white clover and perennial ryegrass cultivars grown under two nitrate levels in flowing solution culture. *Annals of Botany,* 92, 247-258.

Donnison, I.S., P. Cisneros, T. Montoya, I.P. Armstead, B.J. Thomas, A.M. Thomas, R.N. Jones & P. Morris (2002). The floral transition in model and forage grasses. *Flowering Newsletter,* 32, 42-48.

Evans, M., S. Thoma, M. Sullivan & R. Hatfield (2002). Isolation and characterisation of polyphenol oxidase genes from alfalfa. *Annual Meeting of the American Society of Plant Biology,* August 2003, Denver, USA, p161.

Forster, J.W., E.S. Jones, R. Kolliker, M.C. Drayton, J.L. Dumsday, M.P. Dupal, K.M. Guthridge, N.L. Mahoney, E. van Zijl deJong & K.F. Smith (2001). Development and implementation of molecular markers for forage crop improvements. In: G. Spangenberg (ed.) Molecular breeding of forage crops. Kluwer Academic Publishers, Dordrecht, 101-134.

Forster, J.W., E.S. Jones, J. Batley & K.F. Smith (2004). Molecular marker-based genetic analysis of pasture and turf grasses. Developments in plant breeding. In: A. Hopkins, Z-Y Wang, R. Mian, M. Sledge & R.E. Barker (eds.) Molecular breeding of forage and turf. *Proceedings 3rd International Symposium on Molecular Breeding of Forage and Turf,* Kluwer Academic Publishers, Dordrecht, 197-238

Henry, R.J. (2001). Plant genotyping - the DNA fingerprinting of plants. CABI Publishing, Wallingford, UK, 325pp.

van Hintum, Th.J.L., N.R. Sackville-Hamilton, J.M.M. Engels & R. van Treuren (2001). Accession management strategies: splitting and lumping. In: J.M.M. Engels, V.R. Rao, A.H.D. Brown & M.T. Jackson (eds.) Managing plant genetic resources, CABI Publishing, Wallingford, UK, 113-120,

Humphreys, M.O. (1999). The contribution of conventional plant breeding to forage crop improvement. In: J.G. Buchanan-Smith, L.D. Bailey & P. McCaughey (eds.) *Proceedings of the XVIII International Grassland Congress*, 1997, Vol 3, pp71- 78. Canadian Forage Council, Canadian Society of Agronomy, Canadian Society of Animal Science.

Humphreys, M.O. & L.B. Turner (2001). Molecular markers for improving nutritive value in perennial ryegrass plant Breeding: sustaining the future. *Abstracts 16th Eucarpia Congress*, Edinburgh, 10-14 September 2001 (Abstract P1.17, p8).

Humphreys, M.O., L.B. Turner, L. Skot, M.W. Humphreys, I.P. King, I. Armstead & P.W. Wilkins (2003). The use of genetic markers in grass breeding. *Czech Journal of Genetics and Plant Breeding*, 39, 112-119.

Humphreys, M.W., I. Pašakinskienė, A.R. James & H. Thomas (1998). Physically mapping quantitative traits for stress-resistance in the forage grasses. *Journal of Experimental Botany*. 49, 1611-1618.

Humphreys, M.W. (2003). The exploitation of chromosome recombination between *Lolium* and *Festuca* to assemble, locate, and identify gene combinations that may underpin future sustainable grassland production. In: Z. Zwierzykowski, M. Surma & P Kachlicki (eds.) *Proceedings PAGEN International Workshop, application of novel cytogenetic and molecular techniques in genetics and breeding of the grasses*, 21-33.

Isobe, S., I. Klimenko, S. Ivashuta, M. Gau & N.N. Kozlov (2003). First RFLP linkage map of red clover (*Trifolium pratense* L.) based on cDNA probes and its transferability to other red clover germplasm. *Theoretical and Applied Genetics*, 108, 105-112.

Joyce, T.A., M.T. Abberton, T.P.T. Michaelson-Yeates, & W.J. Forster (1999). Relationships between genetic distance measured by RAPD-PCR and heterosis in inbred lines of white clover (*Trifolium repens*). *Euphytica*, 107, 159-165.

Jones, E.S., M.L. Mahoney, M.D. Hayward, I.P. Armstead, J.G. Jones, M.O. Humphreys, I.P. King, T. Kishida, T. Yamada, F. Balfourier, G. Charmet & J.W. Forster (2002). An enhanced molecular marker-based genetic map of perennial ryegrass (*Lolium perenne* L.) reveals comparative relationships with other *Poaceae* genomes. *Genome*, 45, 282-295.

King, I., H.M. Thomas, I.P. Armstead, L.A. Roberts, W.G. Morgan & M. Kearsey (2001). Introgression mapping elucidates the relationship between chiasma frequency and genetic distance and how genetic distance relates to physical distance. *Journal of Experimental Botany*, 52 (Supplement), C2.13.

Kingston-Smith, A.H., A.L. Bollard, I.P. Armstead & M.K. Theodorou (2003). Proteolysis and cell death in clover leaves is induced by grazing. *Protoplasma*, 220, 119-129.

Laws, J.A., B.F. Pain, S.C. Jarvis & D. Scholefield (2000). Comparison of grassland management systems for beef cattle using self-contained farmlets: effects of contrasting nitrogen inputs and management strategies on nitrogen budgets, and herbage and animal production. *Agriculture, Ecosystems and Environment*, 80, 243-254.

Lee, M.R.F., R.J. Merry, D.R. Davies, J.M. Moorby, M.O. Humphreys, M.K. Theodorou & N.D. Scollan (2003). Effect of increasing availability of water-soluble carbohydrates on *in vitro* rumen fermentation. *Animal Feed Science and Technology*, 104, 59-70.

Marley, C.L., R. Cook, J. Barrett, R. Keatinge, N.H. Lampkin & S.D. McBride (2003). The effect of dietary forage on the development and survival of helminth parasites in ovine faeces. *Veterinary Parasitology*, 118, 93-107.

Miller, L.A., J.M. Moorby, D.R. Davies, M.O. Humphreys, N.D. Scollan, J.C. Macrae & M.K. Theodorou (2001). Increased concentration of water-soluble carbohydrate in perennial ryegrass (*Lolium perenne* L.). Milk production from late-lactation dairy cows. *Grass and Forage Science*, 56, 383-394.

Morris, P., & T. Langdon (2001). Manipulation of the phenolic acid content and digestibility of forage grass cell walls by targeted expression of a ferulic acid esterase gene. *US Provisional Patent # 249608*.

Scollan, N.D., R.J. Dewhurst, A. Moloney & J. Murphy (2005). Improving the quality of products from grassland. In: D.A. McGilloway (ed.) Grassland – a global resource. *Proceedings of the XX International Grassland Congress*, Dublin, Ireland, Wageningen Academic Publishers (in press).

Thorogood, D. (2003). Perennial ryegrass. In: M.D. Casler & R.R. Duncan (eds.) Turfgrass biology, genetics and breeding. John Wiley & Sons Ltd., 75-105.

Turner, L.B., M.O. Humphreys, A.J. Cairns & C.J. Pollock (2001). Comparison of growth and carbohydrate accumulation in seedlings of two varieties of *Lolium perenne*. *Journal of Plant Physiology*, 158, 891-897.

Vickery, J.A., J.R.B. Tallowin, R.E. Feber, E.J. Asteraki, P.W. Atkinson, R.J. Fuller & V.K. Brown (2001). The management of lowland neutral grasslands in Britain: effects of agricultural practices on birds and their food resources. *Journal of Applied Ecology*, 38, 647-664.

Wilman, D., M. Field, S.J. Lister & D. I. Givens (2000). The use of near infrared spectroscopy to investigate the composition of silages and the rate and extent of cell-wall degradation. *Animal Feed Science and Technology*, 88, 139-151.

Wood, J.D., R.I. Richardson, G.R. Nute, A.V. Fisher, M.M. Campo, E. Kasapidou, P.R. Sheard & M. Enser (2003). Effects of fatty acids on meat quality: a review. *Meat Science*, 66, 21-32.

Woodfield, D.R. & E.C. Brummer (2001). Integrating molecular techniques to maximise the genetic potential of forage legumes. In: G. Spangenberg (ed.) Molecular breeding of forage crops, Kluwer Academic Publishers, Dordrecht, 51-65.

Young, N.D., J. Mudge, & N. Ellis (2003). Legume genomes: more than peas in a pod. *Current Opinion in Plant Biology*, 6, 199-204.

Grass and forage plant improvement in the tropics and sub-tropics

L. Jank, C.B. do Valle and R.M.S. Resende
Embrapa Gado de Corte, CxP. 154, 79002-970, Campo Grande, MS, Brazil
Email: liana@cnpgc.embrapa.br

Key points

1. The majority of tropical and subtropical forage grass genera and/or species have not yet been collected, or need further collection to be representative of their natural distribution.
2. New biotechnological techniques will only result in the release of superior forage cultivars if supported by strong breeding programs.
3. More funding and investment in the formation of strong public research teams in forage conservation and improvement are needed to guarantee the sustainability of tropical and subtropical pasture-based livestock systems in the future.
4. The creation of a permanent international working group on tropical and subtropical forages is essential to assist the International Plant Genetic Resources Institute (IPGRI) in prioritising collection, conservation, evaluation and adoption in the tropical/subtropical world for the benefit of mankind.

Keywords: apomixis, genetic improvement, molecular breeding, selection

Introduction

Approximately half the world bovine meat production comes from tropical and subtropical countries (Table 1). These countries have had a 200% increase in beef and veal production in the last 40 years, compared with world production which exhibited an increase of 91% (FAOSTAT data, 2004). According to Pinstrup-Andersen & Pandya-Lorch (1997), world population is projected to rise to 8 billion by 2020, resulting in an increase of 110% in demands for food, cereals and meat, mostly in developing countries. These numbers pose a challenge and an opportunity for R&D, and innovation in production of quality animal products and encompasses animal nutrition, animal health, and particularly pasture and forage science.

Table 1 Comparison of beef and veal production in the tropics and subtropics to that of the world in the last 40 years (FAOSTAT data, 2004)

Beef and Veal Production (Mt)	Year		Percent increase
	1963	2003	
Tropics and subtropics[1]	10,352,277	30,873,627	200
World	30,855,743	58,922,239	91

[1]Estimate since some countries contain both subtropical and temperate areas.

Animal production in tropical and subtropical areas of the world is largely dependent on either native or planted pastures; therefore the demand for high quality, productive and adapted forages is high. Brazil, for instance, has the number one commercial beef cattle herd in the world (~189 million heads in 2003), and the evolution of improved pastures has occupied extensive areas of native grasslands, especially in the savannahs of central Brazil. Native pastures decreased from 103 to 78 million ha from 1970 to 1996, while cultivated pastures increased from 30 to 100 million ha in this period (IBGE Censo Agropecuário, 1995-1996). Poor management and overgrazing due to an increase in the demand for animal products has resulted in large expanses of degraded pastures. The long dry season in many tropical and subtropical countries has also contributed to pasture degradation, and due to a lack of available adapted forages the sustainability of pasture systems in these regions is limited.

Species and genetic resources

The main institutions responsible for the development of tropical pastures in the last 30 years were the International Centre for Tropical Agriculture (CIAT) and the Commonwealth Scientific and Industrial Research Organisation (CSIRO), which played a fundamental role in collecting, maintaining, exchanging and evaluating forage germplasm, especially legumes (Miles, 2001; Valle, 2001).

Forage germplasm collection in tropical and subtropical regions were a priority in the 1970's and 1980's, and CIAT and CSIRO together with the International Livestock Research Institute (ILRI) and the International Plant Genetic Resources Institute (IPGRI) with the collaboration of many national Institutes gathered vital forage genetic resources (Schultze-Kraft et al., 1993; Hanson & Maass, 1999). It is estimated that the three main international forage genebanks in the world hold about 30,000 accessions of the main forage species (Hanson & Maass, 1999), with 2.5 to 8.5 times more legumes than grasses. The bank with the most legume accessions is CIAT (15,981) and with the most grasses is CSIRO (2,666). The most representative genera in these banks are *Brachiaria* and *Panicum* in the grasses, and *Stylosanthes*, *Desmodium* and *Centrosema* in the legumes.

In recent times, a shortage of funding and staff have jeopardised the conservation of this germplasm for future generations, which is causing serious concern among pasture scientists worldwide (Valle, 2001). In addition, the diversity may considerably less than the numbers suggest since estimates of 30% duplication of accessions in these genebanks has been reported (Hanson & Maass, 1999). Furthermore, tropical grass seeds are known to lose their viability - even under ideal storage. Budget restrictions in the late 1990's, cut back CIAT's and CSIRO's staff and funds for adequately maintaining and renewing these collections, thus 15,000 accessions of the least important genera in the Australian Tropical Forage Genetic Resource Centre (ATFGRC) were sent to CIAT and ILRI for storage.

Apomictic forage grasses display wide variation in nature, however, little of this diversity has been sampled (Savidan, 2000). The first species intensively collected was *Panicum maximum* in East Africa by the Institute de Recherche pour le Developpment[3] (Combes & Pernès, 1970), and later by Hojito & Horibata (1982). Thus the diversity of this species is well represented in *ex situ* collections (Savidan et al., 1989). *Brachiaria* was also extensively collected by CIAT in East Africa (Keller-Grein et al., 1996), but important species (*B. mutica* for example) and sexual pools of *B. brizantha* and *B. humidicola* are still lacking, thus the existing collection cannot be considered representative of the natural distribution. The genus *Paspalum* spp. of South American origin, has also been extensively collected by the Brazilian Agricultural Research Corporation (Embrapa) but the complete collection of ca. 1600 accessions has not been fully characterized to determine if this genebank is representative or not.

---

[3]Former ORSTOM (Office de la Recherche Scientifique et Technique d'Outre-Mer).

Genera such as *Pennisetum, Hyparrhenia, Melinis, Setaria, Andropogon, Hemarthria, Chloris*, and *Cenchrus* should be included in forthcoming collecting efforts to ensure enough diversity for future improvement efforts, and sexual forms within apomictic *Hyparrhenia* and *Melinis* must be sought (Jank *et al.*, in press). It is essential that germplasm development from collection, plot evaluation, regional trials and animal performance trials in pastures through to its use by farmers be considered of high priority for the adoption of forage-based technologies (Peters & Lascano, 2003).

The creation of an IPGRI-sponsored, permanent international working group on forages, as recommended since 1979, and emphasised by Schultze-Kraft *et al.*, (1993) and Valle (2001), is strongly urged by the scientific community. This group could be a forum for discussions and communication among forage germplasm specialists, and assist IPGRI in taking action to guarantee forage collection, conservation, evaluation and adoption in the tropical/subtropical world for the benefit of mankind.

Breeding targets and priorities

Breeding is the key for the future development of superior forages, as selection programmes based on large germplasm progress and superior accessions are released. The tropical world has still to profit from the genetic manipulation of tropical forages through breeding. Most cultivars of tropical grasses available commercially are wild ecotypes selected from natural diversity (Hacker & Jank, 1998). Others on the verge of release may still add to several production systems before hybrids come along. However, breeding is only justified after the germplasm of a genus or species has been explored and specific problems identified (Cameron, 1983). According to Miles (2001), it is premature to begin hybridisation programs of a species prior to its wide use in production systems, or before large germplasm resources are available and have been evaluated, or without knowledge of the biology and genetics of the species.

Breeding of forages has similar targets to those of crops, e.g. increased productivity and quality, resistance to pests and diseases, efficient use of fertilisers, and adaptation to edaphic and climatic stresses. However, there are additional requisites since forages have no intrinsic value unless converted into animal products (meat, milk, hide, calves), which implies indirect evaluation to verify worth. Through the years, selection has involved the choice of the most vigorous plants, and field-oriented selection has contributed greatly to genetic advance (Busey, 1989).

The main objective of forage breeding programs thus far has been improvement of forage yield and quality (Hacker, 1986; Oram, 1986; Vogel *et al.*, 1989; Burton *et al.*, 1993; Hacker & Jank, 1998; Jank *et al.*, 2001). Whereas direct selection has resulted in the selection and release of cultivars 80 to 130% more productive in terms of leaf yield than the commercial standard, bred cultivars have also led to increases of 26 and 47% in digestibility and productivity respectively (Burton, 1989; Burton *et al.*, 1993; Jank *et al.*, in press). Improvement for yield and quality had direct benefits to the farmers by improving animal performance. Thus the adoption of released cultivars have resulted in increases in 6% milk production, and 25 to 28% liveweight gain per area (Lowe *et al.*, 1991; Euclides *et al.*, 1993).

Leaf yield, leaf percentage, seed production and aggressive regrowth are the objectives of *P. maximum* breeding in Brazil (Jank *et al.*, 2001, 2004; Muir & Jank, 2004; Resende *et al.*, 2004). The *Brachiaria* breeding programs at CIAT and Embrapa, intend to select hybrids which combine the qualities of three species: the high forage quality and determined flowering cycle of *B. ruziziensis*, the yield and resistance to spittlebug of *B. brizantha* and the vigour and adaptation to acid, infertile soils of *B. decumbens* (Miles & Valle, 1996; Miles, 1999; Valle *et al.*, 2000; Peters & Lascano, 2003). At CIAT the *Brachiaria* breeding program also involves *Rhizoctonia* resistance, tolerance to drought, and increased seed production (Peters & Lascano, 2003).

Ease of management through grazing and control of excessive stemmy herbage during periods of active growth were the objectives of the CIAT *A. gayanus* improvement project. Although synthetic lines were developed, they were not commercialised and the programme was discontinued. In Brazil, *A. gayanus* cv. Baeti was bred for quick establishment and stand uniformity (Batista & Godoy, 1995). After the release of this cultivar the programme was also discontinued.

Seed production is also a factor considered in breeding programs (Hacker, 1991a,b; Hacker *et al.*, 1993; Diz & Schank, 1993). Poor seed production results in high seed prices and consequently little adoption. This was the case with many good legume cultivars released, e.g. the *Brachiaria* hybrid cv. Mulato (Miles, 1999). Despite the use of vegetative propagation in certain systems and in some countries, seed producing cultivars are easier to establish, faster to be adopted and much more widespread. In this sense, one of the objectives of the *P. purpureum* breeding programmes in Florida and Brazil is selection of seed producing cultivars, which are obtained by crossing *P. purpureum* and *P. glaucum* (L.) R. Br. The progenies from this cross are triploid and sterile, but once doubled by colchicine become hexaploid and fertile. The improvement of this species also involves the incorporation of apomixis and maintenance of perenniality. The species reproduces sexually and can be crossed with *P. squamulatum* Fresen which reproduces apomictically (Pereira *et al.*, 2001). The hybrid vigour is fixed in apomictics and the hybrid may be released as a cultivar for pastures or for cut and carry.

Other traits bred and selected for are leafiness, establishment, stand uniformity in sexual reproducing plants, late flowering, spring productivity and winter survival, early growth and regrowth, adaptation, mineral composition and disease resistance (Table 2).

Contrary to the scenario with cool-season forages, there has been very little or no consideration given to the impact on the environment as a consequence of livestock production, either in terms of water requirements or diffuse pollution. For this reason, environmental impact is not a priority in the breeding programs of most tropical and subtropical forage grasses and legumes. However, the principle of 'Best Production Practices', which comprises environmental and social concerns, are a market demand and are being incorporated into beef supply chains in Brazil and other major beef/milk producing and exporter countries in the tropics.

Table 2 Main tropical and subtropical genera and summarised objectives in the breeding of tropical forage species[1]

Species	Breeding target	References
Andropogon gayanus	Establishment, uniformity,late flowering, vigour	Batista & Godoy, 1995; Grof (*pers. comm.*).
Brachiaria spp.	Spittlebug resistance, nutritive value, adaptation to acid soils, seed yield, Rhizoctonia resistance	Miles & Valle, (1996); Valle *et al.*, (2000); Miles, (1999); Peters & Lascano, (2003)
Cenchrus ciliaris	Digestibility, overall performance, spring productivity, winter survival	Pengelly *et al.*, (1992); Hacker *et al.*, (1995); Hussey & Bashaw, (1996)
Chloris gayana	Early growth and regrowth, mineral composition	Nakagawa *et al.*, (1993); Jones *et al.*, (1995)
Cynodon spp.	Yield and digestibility	Burton & Monson, (1988); Burton *et al.*, (1993)
Digitaria spp.	Digestibility, seed yield and quality, leafiness, spring yield	Hacker, (1986); Terblanche *et al.*, (1996); Hacker *et al.*, (1993; 1995)
Hemarthria altissima	Yield, winterhardiness	Oakes, (1979); Quesenberry *et al.*, (1978)
Panicum maximum	Leaf yield, adaptation, leafiness, regrowth, seed production	Machado *et al.*, (1988); Sato *et al.*, (1993); Jank *et al.*, (2001, 2004); Muir & Jank, (2004); Resende *et al.*, (2004)
Paspalum spp.	Yield, nutritive value, ergot resistance	Burton, (1989); Schrauf *et al.*, (2003); Batista & Godoy, (2000); Venuto *et al.*, (2003)
Pennisetum spp.	Seed production, growth habit, spittlebug resistance, resistance to 'kikuyu yellows' (virus?), digestibility	Pereira *et al.*, (2001); Wouv *et al.*, (1999); Diz & Schank, (1993); Luckett *et al.*, (1996)
Setaria sphacelata	Yield, seed production, winter greenness	Hacker, (1991a; 1991b); Oram, (1986; 1990); Jank *et al.*, (2002)

[1]Some programs are the result of direct germplasm selection, and some have been discontinued.

Tropical forage breeding has been reviewed by many authors (Cameron, 1983; Miles, 2001; Pereira *et al.*, 2001; Valle, 2001, Miles *et al.*, 2004) who identify the most important factors limiting progress are as follows:

- Access to germplasm representing natural variation, especially of exotic grasses.
- Large number of candidate species/genera of legumes with little to no information about the biology, genetic variation and agronomy.
- Important species with a complex reproductive structure (polyploidy, apomixis), non-domesticated (no seed retention, anti-quality compounds etc.) and dependence on existing breeding methods not necessarily efficient for the particular program.
- Complex criteria of merit and deficient screening techniques. Unlike cereals, forages have no specific point in time when yield can be fully evaluated, due to conversion into animal

products over their growth curve. Merit depends on co-ordinated effort of teamwork, over a wide range of ecosystems to ascertain G x E interaction, and involves time-consuming and costly evaluations of pastures under grazing.

- Little knowledge of genetic control of agronomic traits to be improved.
- Lack of forage breeding courses or tropical forage breeders, thus no academic expertise being developed; few genera/species being tackled.
- Funding and personnel limitations. Complete reliance on public research institutes for the activity with little input from the private sector, the direct beneficiary of the activity.

Research needs and prospects for using new technologies

Tropical and subtropical forage grasses still need to undergo domestication, the process which requires continuous cycles of cultivation and selection to fit human needs. Common breeding and screening techniques to fit the fundamentals of forages also need to be developed to shorten the process of cultivar release.

Conventional breeding has contributed substantially to the genetic improvement of tropical and subtropical forages. With the development of DNA-based molecular markers (RFLP, RAPD, AFLP, SSR) and other biotechnological tools, new and numerous possibilities of utilisation are being pursued.

The main application of molecular markers in tropical forages thus far, has been in the evaluation of genetic divergence among accessions in germplasm banks and commercial cultivars, as in *Cynodon* spp. (Assefa *et al.*, 1999; Karaca *et al.*, 2002), *Paspalum notatum* (Daurelio *et al.*, 2004), *C. gayana* (Perez *et al.*, 1999; Ubi *et al.*, 2003), *Pennisetum purpureum* (Smith *et al.*, 1993). These studies are useful to direct germplasm collection and utilisation.

Another very important application of molecular markers is fingerprinting of genetic materials for the legal rights associated with the development and release of protected cultivars.

Genomic maps are being produced, and markers linked to economic important traits are being sought (Jessup *et al.*, 2000 in *Cenchrus*), as the apomixis locus in many apomictic species permit early identification of reproductive modes of hybrid populations (Pessino *et al.*, 1998; Casa *et al.*, 2002; Goel *et al.*, 2003). The greatest potential of these techniques are: a) the early identification of hybrids to reduce the time involved in the process of selection and release of cultivars; b) the determination of heredity mode in polyploid species (such as disomic or tetrasomic); c) the identification of apomictic or sexual hybrids in apomictic species of grasses; d) aid in the selection of resistance to pests and diseases; and e) aid in the selection of characters of difficult measurement (quantitative traits), such as factors associated to nutritional quality and digestibility, and adaptation to soil characteristics.

The use of quantitaive trait loci (QTL's) have not as yet proved useful, since low heritability of production related characters (associated with a high number of loci) in forages, may limit progress in subsequent generations (Bernardo, 2001).

The use of transgenesis in tropical forages should impact on many production systems, as the mechanisms of floral initiation, the biological clock, apical dominance, root development and other mechanisms being studied in various grain-crop grasses become better understood. Grasses in general, present a high degree of synteny in their genomes, thus specific information generated for *Zea mays* (maize), *Oryza* spp. (rice) or *Triticum* spp. (wheat) has potential application to forage grasses (Morgan *et al*., 2002). Transformation procedures have been developed for some forage grasses but need to be improved. The difficulty in transgenesis lies in identifying candidate genes to introgress, since most important traits are quantitative (Morgan *et al*., 2002).

The main impact of transgenesis will be in characters associated with nutritional quality (Hancock & Ulyatt, 2001; Spangenberg *et al*., 2001). The development of transgenic cultivars resistant to herbicides should impact on commercial seed production, reducing the cost of production and increasing purity and quality of the seeds. Also, cultivars resistant to pests and diseases are necessary in tropical conditions, and the incorporation of resistance genes is a shorter and more efficient method.

The genomic approach in studies with tropical forages depends on building infrastructure for the sequencing of the genome, and in obtaining the expressed sequence tags (EST) and segregating populations for the characters under study. This infrastructure is being built for grasses of the *Brachiaria* genus at Embrapa and CIAT. The main focus for these studies is to elucidate the complex molecular base of environmental stress tolerance, mainly aluminium tolerance. Another objectives of these studies is to develop QTL's that co-segregate with aluminium tolerance in the populations, to enhance the efficiency of breeding programs in the future. The data of EST's will be used for association studies with the available QTL's from genomes of different grasses, e.g. *Oryza* spp. (Ishitani *et al*., 2004), *Triticum* spp. and *Zea mays*.

However, it is important to say that these techniques will not result in the release of superior forages unless there are strong breeding programs to support them.

User participation in breeding

Livestock is an important component in many smallholder farming systems throughout the tropics (Pengelly *et al*., 2003). The demand for livestock products has increased in the past 10 years, and will increase even further in the future, which will have major impacts on household, farm and regional economics. New adapted forages to address these increasing demands are necessary, and the participation of farmers is essential in the evaluation and adoption of new cultivars (Stür *et al*., 2002). Smallholder farmers have different needs for forages, which need to be adapted to their utilisation systems as well as to labour availability. Selection in these systems should therefore include on-farm evaluation and participatory research, which is more effective and less time-consuming. The adoption of forage thus selected is also more effective and quicker to expand.

The main limitation to livestock productivity in many smallholder systems is the lack of adequate quality pastures, especially over the 5-7 months dry season (Gobius *et al*., 2001). Other limitations include the low efficiency of many systems, difficulty in accessing and adopting technology and poor farm management techniques. The reasons for the poor adoption are lack of tradition in using forages, the long-term or indirect benefits of using forages as compared to crops, and unavailability of planting material because of lack of demand for seeds (Kumwenda & Ngwira, 2003; Peters & Lascano, 2003; Roothaert *et al*., 2003). In some countries, these problems are being bypassed by Government incentives such as in NE Thailand (Gobius *et al*., 2001), where promotion of pasture use and farmer-awareness have demonstrated that livestock play a critical role in the sustainability and intensification of agricultural productivity (Kumwenda & Ngwira, 2003).

In SE Asia, forage research only began in 1992, with the selection of environmentally adapted germplasm through projects from CIAT, CSIRO, AusAid and the Asian Development Bank (Roothaerdt *et al.*, 2003). Many projects are being developed in Africa and Southeast Asia to improve family welfare by implementing small-scale mixed farming systems with promising results (Ayele, 2003; Gobius *et al.*, 2001; Roethardt *et al.*, 2003; Stür *et al.*, 2002).

Funding issues and training in plant breeding

It has been estimated, that the value of a breeding program is at least US$ 100,000 per year. Considering that the development of a forage cultivar takes around 10 years, because cultivars are perennial and so need to be evaluated over extended periods and under grazing, a new grass cultivar may represent an investment of over one million dollars (Vogel, 1989).

Until 2001, no private companies had been involved in tropical forage plant breeding (Miles, 2001), and most tropical forage releases have been funded exclusively by Government investments, with nil participation of private companies. Even in turf grasses, public plant breeding resulted in the release of the most widely used warm-season turf grasses (Busey, 1989) although private efforts have been notable in development and marketing.

In many cases, breeders have to seek external financing for continuing their research programmes. Thus many small projects, which integrate large breeding programs are being financed by both public and private organisations. Other sources of funding include seed companies, which invest in public plant breeding in return for the exclusiveness in commercialising the released cultivars. This is the case of UNIPASTO (Association for Promotion of Breeding Research in Tropical Forages) in Brazil, by which 40 seed companies finance breeding programmes developed by Embrapa, and the Seed Company Papalotla in Mexico, which finances the *Brachiaria* breeding program at CIAT.

The main revenue to breeding projects world-wide is from the royalties paid to the breeders from plant breeding rights and from seed sales (Dale, 2004). In many institutions, e.g. the University of Florida, royalties revert to the breeding programs, which permit their continuation. As more plant variety protection legislation is adopted in more countries, there will be a tendency for multinational seed companies to enter the tropical seed market (Miles, 2001). However difficulties in obtaining financing, and the limited numbers of researchers world-wide working with tropical forage plant breeding, have not permitted the expansion of many plant breeding programs, and at times, have resulted in the discontinuation of many programs. This has occurred especially so in Australia, with the retirement of active plant breeders.

Most Universities around the world have maize and other crop breeders in genetics and plant improvement departments, but only a very few, particularly in the USA actually have forage breeders. This results in biases towards crops of alogamous or autogamous reproduction, and forages or apomixis become the theme of only a few lectures. Much is yet to be done. Studies concerning apomixis and the apomictic grasses are too numerous to be omitted from genetic courses, especially since at present times many attempts are being made to transfer apomixis to important cultivated crops such as *Oryza sativa*, *Z. mays* and *Triticum* spp.

Concluding remarks

In the past, much was invested in tropical forage germplasm collection, conservation, distribution and evaluation. Many cultivars have been released, and a smaller proportion adopted in the various countries. A number of important advances in pasture/livestock research made in the last decade are expected to have a major impact on the productivity, persistence and sustainability of pasture-based livestock systems in tropical and subtropical areas.

However, the downsizing of international pasture research programmes, have left germplasm conservation under extreme pressure. Therefore, it is of major importance that more progressive and large national programmes covering a range of agroecosystems (e.g. Embrapa in Brazil), assume the leading role in this area of research. Furthermore, action must be undertaken to create an international working group on forages to guarantee investments in forage research and adoption worldwide. The continuation of sustainable pasture based livestock systems in the tropics and subtropics depends on research to achieve further progress.

References

Assefa, S., C.M. Taliaferro, M.P. Anderson, B.G. de los Reyes & R.M. Edwards (1999). Diversity among *Cynodon* accessions and taxa based on DNA amplification fingerprinting. *Genome*, 42, 465-474.

Ayele, Z. (2003). Community-based forage development program: the experiences of FARM Africa Goat Project in Ethiopia. *Tropical Grasslands*, 37, 257-261.

Batista, L.A.R. & R. Godoy (1995). BAETÍ - Embrapa 23, uma nova cultivar do capim andropogon (*Andropogon gayanus* Kunth.) (BAETÍ - Embrapa 23, a new cultivar of andropogon grass (*Andropogon gayanus* Kunth.)). *Revista Brasileira de Zootecnia*, 24, 204-213.

Batista, L.A.R. & R. Godoy (2000). Caracterização preliminar e seleção de germoplasma do gênero *Paspalum* para produção de forragem. (Preliminary characterization and selection of germplasm of the genus *Paspalum* for hay production). *Revista Brasileira de Zootecnia*, 29, 23-32.

Bernardo, R. (2001). What if we knew all the genes for a quantitative trait in hybrid crops? *Crop Science*, 41, 1-4.

Burton, G.W. (1989). Registration of 'Tifton 9' Pensacola bahiagrass. *Crop Science*, 29, 1326.

Burton, G.W., R.N. Gates & G.M. Hill (1993). Registration of 'Tifton 85' bermudagrass. *Crop Science*, 33, 644.

Burton, G.W. & W.G. Monson (1988). Registration of 'Tifton 78' bermudagrass. *Crop Science*, 28, 187-188.

Busey, P. (1989). Progress and benefits to humanity from breeding warm-season grasses for turf. In: D.A. Sleper, K.H. Asay & J.F. Pedersen (eds.) Contributions from breeding forage and turf grasses, CSSA Special Publication no. 15, Crop Science Society of America, Wisconsin, 49-70.

Cameron, D.F. (1983). To breed or not to breed. In: J.G. McIvor & R.A. Bray (eds.) Genetic resources of forage plants, CSIRO, Melbourne, 237-337.

Casa, A.M., S.E. Mitchell, C.R. Lopes & J.F.M. Valls (2002). RAPID analysis reveals genetic variability among sexual and apomictic *Paspalum dilatatum* Poiret biotypes. *Journal of Heredity*, 93, 300-302.

Combes, D. & J. Pernès (1970). Variations dans le nombres chromosomiques du *Panicum maximum* Jacq. en relation avec le mode de reproduction (Variation in the number of chromosomes in *Panicum maximum* Jacq. in relation to mode of reproduction). *Comptes Rendues Academie des Science*, Paris, Sér. D, 270, 782-785.

Dale, P.J. (2004). Public-good plant breeding: what should come next? *Journal of Commercial Biotechnology*, 10, 199-208.

Daurelio, L.D., F. Espinoza, C.L. Quarin & S.C. Pessino (2004). Genetic diversity in sexual diploid and apomictic populations of *Paspalum notatum* situated in sympatry or allopatry. *Plant Systematics and Evolution*, 244, 189-199.

Diz, D.A. & S.C. Schank (1993). Breeding hexaploid pennisetums for improved seed production. In: *Proceedings of the XVII International Grassland Congress*, Palmerston North, NZ, 1663-1664.

Euclides, V.P.B., M.C.M. Macedo, A. Vieira & M.P. de Oliveira (1993). Evaluation of *Panicum maximum* cultivars under grazing. In: *Proceedings of the XVII International Grassland Congress*, Palmerston North, NZ, 1999.

FAOSTAT data (2004). <http://apps.fao.org/faostat/collections?version=ext&hasbulk=0>

Gobius, N.R., C. Phaikaew, P. Pholsen, O. Rodchompoo & W. Susena (2001). Seed yield and its components of *Brachiaria decumbens* cv. Basilisk, *Digitaria milanjiana* cv. Jarra and *Andropogon gayanus* cv. Kent in north-east Thailand under different rates of nitrogen application. *Tropical Grasslands*, 35, 26-33.

Goel, S., Z.B. Chen, J.A. Conner, Y. Akiyama, W.W. Hanna & P. Ozias-Akins (2003). Delineation by fluorescence *in situ* hybridization of a single hemizygous chromosomal region associated with aposporous embryo sac formation in *Pennisetum squamulatum* and *Cenchrus ciliaris*. *Genetics*, 163, 1069-1082.

Hacker, J.B. (1986). Selecting for nutritive value in *Digitaria milanjiana*. 1. Breeding of contrasting full-sib clones differing in leaf digestibility. *Australian Journal of Experimental Agriculture*, 26, 543-549.

Hacker, J.B. (1991a). Evaluation of bred populations and cultivars of *Setaria sphacelata*. *Tropical Grasslands*, 25, 245-252.

Hacker, J.B. (1991b). Seed production potential in bred populations and cultivars of *Setaria sphacelata*. *Tropical Grasslands*, 25, 253-261.

Hacker, J.B. & L. Jank (1998). Breeding tropical and subtropical forage plants. In: J.H. Cherney & D.J.R. Cherney (eds.) Grass for dairy cattle, CABI Wallingford, 49-71.

Hacker, J.B., R.J. Williams & J.N. Coote (1995). Productivity in late winter and spring of four cultivars and 21 accessions of *Cenchrus ciliaris* and *Digitaria eriantha* cv. premier. *Tropical Grasslands*, 29, 28-33.

Hacker, J.B., G.P.M. Wilson & L. Ramirez (1993). Breeding and evaluation of *Digitaria eriantha* for improving seed yield and seed production. *Euphytica*, 68, 193-204.

Hancock, K.R. & M.J. Ulyatt (2001). Opportunities in molecular biology: enhancement of the nutritional value of forages. In: *Proceedings of the XIX International Grassland Congress*, Piracicaba, Brazil, 627-632.

Hanson, J. & B.L. Maass (1999). Conservation of tropical forage genetic resources. In: *Proceedings of the XVII International Grassland Congress*, Palmerston North, NZ, 31-36.

Hojito, S. & T. Horibata (1982). Plant exploration, collection and introduction from Africa. In: Nekken Shiryo, TARC, Japan, 58, 1-120.

Hussey, M.A. & E.C. Bashaw (1996). Performance of buffelgrass germplasm with improved winter survival. *Agronomy Journal*, 88, 944-946.

IBGE Censo Agropecuario (1995-1996). <http://www.ibge.gov.br>

Ishitani, M., I. Rao, P. Wenzl, S. Beebe, J. Tohme (2004). Integration of genomics approach with traditional breeding towards improving abiotic stress adaptation: drought and aluminium toxicity as case studies. *Field Crops Research*, 90, 35-45.

Jank, L., C.B. do Valle & P. de F. Carvalho. New grasses and legumes: advances and perspectives for the tropical zones of Latin America. In: S. Reynolds (ed.) Grasslands – Future Perspectives, FAO. (in press).

Jank, L., C.B. do Valle, J. de Carvalho & S. Calixto (2001). Evaluation of guineagrass (*Panicum maximum* Jacq.) hybrids in Brazil. In: *Proceedings of the XIX International Grassland Congress*, Piracicaba, Brazil, 498-499.

Jank, L., K.H. Quesenberry, A.R.S. Blount & P. Misley (2002). Selection in *Setaria sphacelata* for winter survival. *New Zealand Journal of Agricultural Research*, 45, 273-281.

Jank, L, R.M.S. Resende, S. Calixto, M.M. Gontijo Neto, V.A. Laura, M.C.M. Macedo & C.B. do Valle (2004). Preliminary performance of *Panicum maximum* accessions and hybrids in Brazil. In: In: *Proceedings of the XX International Grassland Congress*, Dublin, offered paper (in press).

Jessup, R.W., B.L. Burson, A.H. Paterson & M.A. Hussey (2000). Breeding apomictic forage grasses: molecular strategies. *Proceedings of the 55th Southern Pasture and Forage Crop Conference*, Raleigh, NC, June 12-14, 1-4.

Jones, R.J., D.S. Loch, R.P. Lefeuvre (1995). Differences in mineral concentration among diploid and tetraploid cultivars of rhodesgrass (*Chloris gayana*). *Australian Journal of Experimental Agriculture*, 35, 1123-1129.

Karaca, M., S. Saha, A. Zipf, J.N. Jenkins & D.J. Lang (2002). Genetic diversity among forage bermudagrass (*Cynodon* spp.): evidence from chloroplast and nuclear DNA fingerprinting. (Cell Biology & Molecular Genetics). *Crop Science*, 42, 2118-2127.

Keller-Grein, G., B.L. Maass & J. Hanson (1996). Natural variation in *Brachiaria* and existing germplasm collections. In: J.W. Miles, B.L. Maass & C.B. do Valle (eds.) *Brachiaria*: biology, agronomy, and improvement, CIAT Publication No 259, 16-42.

Kumwenda, M. & A. Ngwira (2003). Forage demand and constraints to adoption of forage technologies by livestock keepers in Malawi. *Tropical Grasslands*, 37, 274-278.

Lowe, K.F., R.J. Moss, R.T. Cowan, D.J. Minson & J.B. Hacker (1991). Selecting for nutritive value in *Digitaria milanjiana* 4. Milk production from an elite genotype compared with *Digitaria eriantha* ssp. pentzii (pangola grass). *Australian Journal of Experimental Agriculture*, 31, 603-608.

Luckett, D., A. Kaiser & A. Virgona (1996). Innovative breeding of high digestibility kikuyu cultivars to increase milk production. Final Report to the Dairy Research and Development Corporation, Project DAN 063. DRDC, Melbourne,

Maass, B.L. & B.C. Pengelly (2001). Tropical forage genetic resources - will any be left for future generations? In: *Proceedings of the XIX International Grassland Congress*, Piracicaba, Brazil, 541-542.

Machado, H., R. Roche, A. Tamayo & E. Segui (1988). Selection of sexual plants and possibility of cross breeding *Panicum maximum* in Cuba. *Pastos y Forrajes*, 11, 31-36.

Miles, J.W. (1999). Nuevos híbridos de *Brachiaria* (New hybrids of *Brachiaria*). *Pasturas Tropicales*, 21, 78-80.

Miles, J.W. (2001). Achievements and perspectives in the breeding of tropical grasses and legumes. In: (editors) *Proceedings of the XIX International Grassland Congress*, Piracicaba, Brazil, 509-515.

Miles, J.W. & C.B. do Valle (1996). Manipulation of apomixis in *Brachiaria* breeding. In: J.W. Miles, B.L. Maass & C.B. do Valle (eds.) *Brachiaria*: biology, agronomy, and improvement, CIAT Publication No 259, 164-177.

Miles, J.W., C.B. do Valle, I. Rao & V.P.B. Euclides (2004). Brachiariagrasses. In: L.E. Sollenberger, L. Moser & B. Burson (eds.) Warm-season (C4) grasses, *Agronomy monograph*, ASA-CSSA-SSSA, Madison, 45, 745-783.

Morgan, P.W., S.A. Finlayson, K.L. Childs, J.E. Mullet & W.L. Rooney (2002). Opportunities to improve adaptability and yield in grasses: lessons from *Sorghum*. (Review & Interpretation). *Crop Science*, 42, 1780-1791.

Muir, J. & L. Jank (2004). Guineagrass. In: L.E. Sollenberger, L. Moser & B. Burson (eds.) Warm-season (C4) grasses, *Agronomy monograph*, ASA-CSSA-SSSA, Madison, 45, 589-621.

Nakagawa, H., N. Shimizu & H. Sato (1993). A new registered cultivar 'Hatsunatu' rhodesgrass (*Chloris gayana* Kunth). *Bulletin of the Kiushu National Agricultural Experiment Station*, 27, 399-416.

Oakes, A.J. (1979). Winterhardiness in limpograss, *Hemarthria altissima*. In: *Proceedings of the 39th Soil and Crop Science Society of Florida*, 86-88.

Oram, R.N. (1986). *Setaria sphacelata* (Schumach.) Moss var. *sericea* (Stapf) Clayton (setaria) cv. Solander (Reg. No. A-8a-5). *Journal of the Australian Institute of Agricultural Science*, 52, 180-181.

Oram, R.N. (1990). (Ed.) Register of Australian Herbage Plant Cultivars. CSIRO, Melbourne.

Pengelly, B.C., J.B. Hacker & D.A. Eagles (1992). The classification of a collection of buffel grasses and related species. *Tropical Grasslands*, 26, 1-6.

Pengelly, B.C., A. Whitbread, P.R. Mazaiwana & N. Mukombe (2003). Tropical forage research for the future – better use of research resources to deliver adoption and benefits to farmers. *Tropical Grasslands*, 37, 207-216.

Pereira, A.V., C.B. do Valle, R. de P. Ferreira & J.W. Miles (2001). Melhoramento de forrageiras tropicais (Breeding of tropical forages). In: L.L. Nass, A.C.C. Valois, I.S. de Melo & M.C. Inglis-Valadares (eds.) Recursos Genéticos & Melhoramento - Plantas. Fundação MT. Rondonópolis, 549-601.

Perez, H., S. Bravo, V. Ongaro, A. Castagnaro, L.G. Seffino & E. Taleisnik (1999). *Chloris gayana* cultivars: RAPD polymorphism and field performance under salinity. *Grass and Forage Science*, 54, 289-296.

Pessino, S., C. Evans, J.P.A. Ortiz, I. Armstead, C.B. do Valle & M.D. Hayward (1998). A genetic map of the apospory-region in *Brachiaria* hybrids: identification of two markers closely associated with the trait. *Hereditas*, 128, 153-158.

Peters, M. & C.E. Lascano (2003). Forage technology adoption: linking on-station research with participatory methods. *Tropical Grasslands*, 37, 197-203.

Pinstrup-Andersen, P. & R. Pandya-Lorch (1997). A vision of the future world food production and implications for the environment and grasslands. In: *Proceedings XVIII International Grassland Congress*, Calgary, 11-16.

Quesenberry, K.H., L.S. Dunavin, E.M. Hodges, G.B. Killinger, A.E. Kretchmer, Jr., W.R. Ocumpaugh, R.D. Rousch, O.C. Ruelke, O.C. Smith, S.C. Schank, G.H. Snyder & R.L. Stanley (1978). Redalta, Greenalta, and Bigalta limpograss, promising forages for Florida. Institute of Food and Agricultural Sciences Experiment Station Bulletin No. 802, 18pp.

Resende, R.M.S., L. Jank, C.B. do Valle & A.L.V. Bonato (2004). Biometrical analysis and selection of tetraploid progenies of *Panicum maximum* using mixed model method. *Pesquisa Agropecuária Brasileira*, 39, 335-341.

Roethaert, R., P. Horne & W. Stür (2003). Integrating forage technologies on smallholder farms in the upland tropics. *Tropical Grasslands*, 37, 295-303.

Sato, H., N. Shimizu, H. Nakagawa, K. Nakajima & M. Ochi (1993). A new registered cultivar 'Natsukaze' guineagrass (*Panicum maximum* Jacq.). *Bulletin of the Kyushu National Agricultural Experiment Station*, 27, 417-437.

Savidan, Y.H. (2000). Apomixis: genetics and breeding. *Plant Breeding Reviews*, 18, 13-86.

Savidan, Y.H., L. Jank, J.C.G. Costa & C.B. do Valle (1989). Breeding *Panicum maximum* in Brazil: 1. Genetic resources, modes of reproduction and breeding procedures. *Euphytica*, 41, 107-112.

Schrauf, G.E., M.A. Blanco, P.S. Cornaglia, V.A. Deregibus, M. Madia, M.G. Pacheco, J. Padilla, A.M. García & C. Quarin (2003). Ergot resistance in plants of *Paspalum dilatatum* incorporated by hybridisation with *Paspalum urvillei*. *Tropical Grasslands*, 37, 182-186.

Schultze-Kraft, R., W.M. Williams & J.M. Keoghan (1993). Searching for new germplasm for the year 2000 and beyond. In: *Proceedings of the XVII International Grassland Congress*, Palmerston North, NZ, 181-188.

Smith, R.L., M.E. Schweder, M.K.U. Chowdhury, J.C. Seib & S.C. Schank (1993). Development and application of RFLP and RAPD DNA markers in genetic improvement of *Pennisetum* for biomass and forage production. *Biomass & Bioenergy*, 5, 51-62.

Spangenberg, G., R. Kalla, A. Lidgett, T. Sawbridge, E.K. Ong & U. John (2001). Transgenesis and genomics in molecular breeding of forage plants. In: *Proceedings of the XIX International Grassland Congress*, Piracicaba, Brazil, 615-625.

Stür, W.W., P.M. Horne, F.A. Gabunada, P. Phengsavanh & P.C. Kerridge (2002). Forage options for smallholder crop-animal systems in Southeast Asia: working with farmers to find solutions. *Agricultural Systems*, 71, 75-98.

Terblanche, C.J., A. Smith, M.F. Smith & R. Greyling (1996). The improvement of *Digitaria eriantha* (Smuts finger grass). In: *Proceedings abstracts of the Grassland Society of South Africa Congress*, 31, 45pp.

Ubi, B.E., R. Kolliker, M. Fujimori & T. Komatsu (2003). Genetic diversity in diploid cultivars of rhodesgrass determined on the basis of amplified fragment length polymorphism markers. *Crop Science*, 43, 1516-1522.

Valle, C.B. do. (2001). Genetic resources for tropical areas: achievements and perspectives. In: *Proceedings of the XIX International Grassland Congress*, Piracicaba, Brazil, 477-482.

Valle, C.B. do., M.C.M. Macedo & S. Calixto (2000). Avaliação agronômica de híbridos de *Brachiaria* (Agronomic evaluation of *Brachiaria* hybrids). In: Reunião Anual da Sociedade Brasileira de Zootecnia, 37, CD Rom, ID No. 383.

Venuto, B.C., B.L. Burson, M.A. Hussey, D.D. Redfearn, W.E. Wyatt & L.P. Brown (2003). Forage yield, nutritive value, and grazing tolerance of dallisgrass biotypes. *Crop Science*, 43, 295-301.

Vogel, K.P., H.J. Gorz & F.A. Haskins (1989). Breeding grasses for the future. In: D.A. Sleper, K.H. Asay & J.F. Pedersen (eds.) Contributions from breeding forage and turf grasses, CSSA Special Publication no. 15. Crop Science Society of America, Wisconsin, 105-122.

Wouv, M. van de., J. Hanson & S. Luethi (1999). Morphological and agronomic characterisation of a collection of napier grass (*Pennisetum purpureum*) and *P. purpureum* x *P. glaucum*. *Tropical Grasslands*, 33, 150-158.

Foraging behaviour and herbage intake in the favourable tropics/sub-tropics

S.C. Da Silva[1] and P.C. de F. Carvalho[2]
[1]Universidade de São Paulo, E.S.A. "Luiz de Queiroz", Departamento de Zootecnia, Av. Pádua Dias, 11 – 13418-900 Piracicaba, SP, Brazil
Email: scdsilva@esalq.usp.br
[2]Universidade Federal do Rio Grande do Sul, Faculdade de Agronomia, Av. Bento Gonçalves, 7712 – 91501-970 Porto Alegre, RS, Brazil

Key points

1. Herbage intake by animals grazing tropical/sub-tropical pastures is directly related to bite mass, as it is for those grazing temperate pastures.
2. Where these swards have low proportions of stem and dead material (controlled swards), herbage intake follows a similar pattern to that of temperate pasture species, but leaf characteristics, such as lamina length play an important role and influence the short-term rate of intake.
3. Sward structural characteristics and behavioural factors are relatively more important than nutritional factors in terms of herbage intake regulation. The feeding value of the herbage produced is potentially adequate to sustain high levels of beef cattle performance under controlled sward conditions, but relatively limited for dairy cows, since nutrient concentration in the forage is not optimal for high levels of daily milk yield.
4. Sward targets for attaining production objectives are now a feasible management practice on tropical pastures and should be evaluated further.

Keywords: tropical grasslands, animal performance, grazing management

Introduction

Grasslands cover about 3 billion hectares, approximately a fifth of the world's land area of 14.9 billion hectares (Hadley, 1993), with a large proportion concentrated in the tropical and sub-tropical region. Here climatic conditions and the availability of plant and animal resources favour the production of environmentally safe, healthy and low cost animal products, via efficient, competitive and economically viable pastoral systems. However, animal production from pastures in the tropics has consistently been related to low levels of performance, 'explained' by the classical argument that ruminant ingestion and digestion processes are negatively affected by several chemical and physical constraints associated with the available herbage. Recent evidence illustrates these potential limits, but demonstrates that the problem can be minimised by appropriate harvesting of herbage through efficient and well-planned grazing strategies in an environment extremely favourable to the growth of forage species, particularly vigorous C_4 grasses. In fact, there is now enough evidence to support the argument that animal performance can be considerably improved (Andrade, 2003; Martinichen, 2003), and that there exists the opportunity for the development of sustainable-high productivity pastoral systems.

In a scenario characterised by the competing requirements of plants and animals, knowledge about the plant-animal interface is essential. Foraging behaviour is certainly one of the key components for the integration of plant and animal responses in pastures and will be the main objective of this review paper.

The tropical and sub-tropical pastoral environments

The tropical and sub-tropical region is situated between the latitudes 35° N and 35° S. The humid tropical climate is characterised by high annual rainfall (> 1000 mm) evenly distributed throughout the year, or with dry periods of 2 to 6 months during the autumn/winter period. Mean annual temperatures vary from 25 to 27°C, with little seasonal variation (< 3°C) and a daily range no bigger than 7 to 14°C. During the coldest month of the year, mean temperatures are higher than 18°C. The sub-tropical climate has a similar total annual rainfall, but dry periods may be concentrated in winter as well as in summer. Annual temperature varies widely, defining clear contrasts between warm and cold seasons. Average summer and winter temperatures vary from 24 to 27°C and 5 to 13°C, respectively, with the possibility of some frosts during the coldest month of the year (Rocha, 1991).

In general, grasslands are subject to distinct moist (usually summer) and dry (usually winter) seasons (Hardy *et al.*, 1997), resulting in large fluctuations in quantity and/or quality of the available herbage. Pastures vary from cultivated to non-cultivated and can provide conditions for high levels of animal productivity (25,000 to 30,000 kg milk/ha per year and 1,000 to 1,600 kg LWG/ha per year – Corsi *et al.*, 2001), even though average figures for Brazil are low (around 800 to 1000 kg milk/ha per year and 60 to 100 kg LWG/ha per year – Nascimento Jr *et al.*, 2003). Cultivated pastures are mainly composed of *Brachiaria* and *Panicum* species - C4 grasses with high herbage dry matter (DM) yield potential. These swards are very different from most temperate swards with a wide range of plant morphology and structure, varying from prostrate/semi-prostrate to tall-tufted, erect growing plants. Non-cultivated or native pastures are formed by a mixture of plant species varying from herbaceous (grass, legumes and composites) to shrubs and trees varying from predominantly grass stands to mixed grass legume pastures, and open woodland savannas. Examples of non-cultivated pastures are the savannas in the *Cerrado* region (areas of *campo limpo* and *campo sujo* in the central-west part of Brazil), the wetlands of *Pantanal* in central Brazil and the *campo nativo* in Rio Grande do Sul (Brazil), extending to Uruguay and northeast of Argentina (Campos biome - Carvalho, 2002).

Foraging behaviour and herbage intake by the grazing animal

General principles

Much of the available evidence on foraging behaviour and herbage intake comes from studies on temperate forage species with domesticated animals under controlled experimental conditions (Gordon & Lascano, 1993; Hodgson *et al.*, 1994; Ungar, 1996; Hodgson *et al.*, 1997; Illius, 1997; Prache & Peyraud, 2001; Sollenberger & Burns, 2001). In general, these reviews relate to small-scale studies aimed at providing a comprehensive understanding of the mechanics of the grazing process, with particular emphasis on bite dimensions, bite mass and short-term rate of intake. This involves the cropping, mastication and swallowing of herbage, with constraints related to grazing (e.g. time spent feeding, choice of feeding station, activity budget). Digestive processes (e.g. ease of digestion and turnover of digesta in the rumen) are generally dealt with in longer term studies (Fryxell, 1991).

The early work of Allden & Whittaker (1970) defined herbage intake in terms of the components of ingestive behaviour (intake per bite, rate of biting and grazing time). This work provided the basis for understanding the influence of sward structure on bite mass, the reciprocal relationship between intake per bite and rate of biting, and the effect of these on daily herbage intake. With the development of techniques for close control and manipulation of both sward and animals, a detailed knowledge base evolved of the plant-animal interface (Hodgson *et al.*, 1994). However, extrapolation to practical grazing conditions has been limited (Taylor, 1993), largely due to spatial and temporal heterogeneity, and so animals need to make choices relating to landscape, vegetation community, patch and bite characteristics (foraging strategy) as well as to aspects related to digestion and time available for grazing (Gordon & Lascano, 1993). Although foraging behaviour is highly scale dependent (time, size and space) (Fryxell *et al.*, 2001), it has been argued that the basic parameters of foraging strategy are essentially the same across a range of grazing conditions, suggesting that the understanding developed under controlled conditions can be applied in less controlled environments (Hodgson *et al.*, 1997).

Bite mass is the most important determinant of daily herbage intake (Hodgson *et al.*, 1994; Illius, 1997) and is primarily influenced by bite depth, since bite area is relatively less sensitive to variation in sward height or herbage mass. Variation in herbage bulk density may contribute independently to bite mass (Hodgson *et al.*, 1994), although variation in sward height generates a wider range in intake per bite (Mitchell *et al.*, 1991). Bite rate is generally negatively related to intake per bite, a consequence of the larger number of manipulative jaw movements (prehension and chewing) with increasing bite mass. Despite this negative association between bite rate and bite mass, the short-term rate of intake still tends to increase asymptotically with bite mass or sward height.

According to Poppi *et al.* (1987), the ability of the grazing animal to harvest herbage appears to be the most important factor limiting intake when intake is responding steadily with increase in bite mass, i.e. in the middle section of the asymptotic curve. This ability is influenced by sward structure and grazing behaviour (diet selection, grazing time, bite mass and rate of biting), characteristics strongly influenced by grazing management practices. Approaching the asymptote, nutritional factors such as herbage digestibility, the time feed stays in the rumen, and concentration of metabolic products appear to be important in controlling intake, although there is increasing evidence that non-nutritional factors may also operate at this level (Carvalho, 1997). In this context, grazing time is considered the link between the short-term rate of intake and daily forage intake of the grazing animal, since it may be constrained to some extent by rumination (related to diet characteristics) and idling time (related to animal factors such as nutritional status) (Hodgson *et al.*, 1997). These general relationships appear to 'hold up' quite well for temperate pasture conditions, where plant size and the stem component of herbage growth are both small in comparison to tropical and sub-tropical forage species. In the tropics and sub-tropics sustainability is an important issue, but the requirement to meet human food needs should not be forgotten, thus an understanding of foraging behaviour and herbage intake should embrace productivity goals.

Cultivated pastures in the favourable tropics/sub-tropics

The pioneer work of Stobbs (1973a,b) and Chacon & Stobbs (1976) with cultivated tropical pastures indicated that the major animal factor influencing herbage intake was bite mass. However, in contrast to temperate pastures where variation in sward height strongly influences intake per bite, the major sward characteristics influencing bite mass and short-term rate of intake of tropical pastures were leaf mass, leaf-to-stem ratio and sward bulk density. These findings suggested that tropical and temperate swards are quite different and that tropical pastures would impose physical (behavioural) as well as chemical (nutritive value) constraints that would severely limit herbage intake and performance of grazing animals (Hardy *et al.*, 1997). Little progress was made on the grazing ecology of tropical and sub-tropical pastures during the 1980's and 1990's, and it became generally accepted that tropical pastures only produced stemmy, low density, poor quality herbage, suitable only for low levels of animal performance (Hardy *et al.*, 1997; Sollenberger & Burns, 2001).

This seems a rather simplistic generalisation of a complex problem, since the experimental treatments and protocol in Stobbs (1973b) were based on only two forage species (*Setaria anceps* cv Kazungula and *Chloris gayana* cv Pioneer). There was no sward structure control - simply a description of sward components for fixed regrowth periods. The general results presented by Stobbs (1973b) indicate quite clearly that during the time plants were vegetative and/or with little contribution of stem to herbage mass (2 and 4-week regrowth periods), bite mass was influenced by sward height/mass. This trend changed when swards were reproductive and/or stem and dead material altered sward structure significantly (6 and 8-week regrowth periods). Chacon & Stobbs (1976) produced a similar data set, with bite mass being highly correlated with sward height and/or mass at the early stages of the grazing process, when animals consumed mainly leaf. This changed towards the end of the grazing period, when stem and dead components were increasing in the grazing strata.

In all those situations, data sets were analysed with no distinction made between vegetative or reproductive state (Stobbs, 1973b), and/or within phases of the same grazing period (Chacon & Stobbs, 1976). It was concluded that leaf mass, sward bulk density and leaf-to-stem ratio provided a better expression of forage supply than grazing pressure i.e. herbage mass, and better explained variation in bite mass in 'tropical' pastures. However, it can be argued that the process should be studied in two ways: situations where swards are leafy, with low proportions of stem and dead material (e.g. most of the evidence available for temperate pasture conditions), and situations where swards are stemmy, with high proportions of dead material (e.g. reproductive and/or uncontrolled temperate and tropical swards).

Griffiths *et al.* (2003a,b), working with *Lolium perenne* (perennial ryegrass) and studying multiple patch choices offered to cows under controlled conditions, demonstrated that animals, in order to optimise their rate of herbage intake, preferred to graze on tall patches unless tallness reflected greater maturity (e.g. high stem content and/or bite effort). In this situation, preference switched to short immature patches. A similar response was reported by Prache *et al.* (1998) for grazing sheep. This indicates that for temperate pastures the distinction between leafy and stemmy sward conditions is also relevant, and suggests that when considered on an equivalent basis, temperate and tropical pasture species may similarly influence bite mass and herbage intake of grazing animals.

More recent experimental evidence with tropical species seems to support this argument. Sarmento (2003), working with *Brachiaria brizantha* cv. Marandu (a popular semi-prostrate forage species in Brazil) under tight sward state control (sward height maintained at 10, 20, 30 and 40 cm), reported causative relationships between short-term rate of herbage intake and sward height (Figures 1a,b,c &d). However, due to the long leaf laminae on the tallest swards (Table 1), increasing time per bite (Figure 1e) resulted in a reduction in rate of intake (Figure 1f) regardless of the greater bite mass, implying that even when there was no limitation in herbage allowance (Table 1) or chemical composition (85% leaf, 12.5% CP, 61.7% NDF, 65% IVOMD - Andrade, 2003), behavioural factors restricted herbage intake.

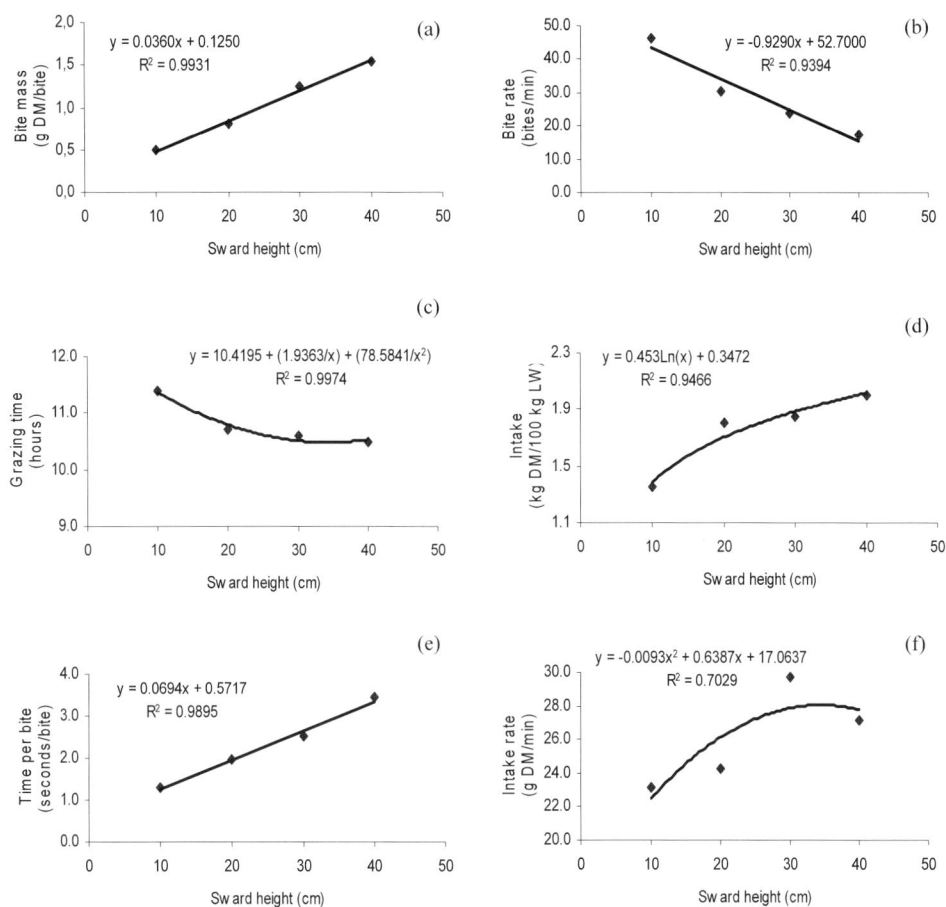

Figure 1 Grazing behaviour and herbage intake of beef cattle heifers grazing *Brachiaria brizantha* cv. Marandu pastures maintained at 10, 20, 30 and 40 cm sward surface height by continuous stocking and variable stocking rate (adapted from Sarmento, 2003)

Table 1 Sward structural characteristics of *Brachiaria brizantha* cv Marandu swards maintained at 10, 20, 30 and 40 cm by continuous stocking and variable stocking rate – summer period (December to March)[1]

Characteristic	Sward surface height (cm)				
	10	20	30	40	SEM[2]
Herbage mass (kg DM/ha)	4630	8210	11920	14420	512
Leaf-to-stem ratio	0.72	0.84	0.78	0.69	0.08
Total bulk density (kg DM/ha per cm)	460	410	400	360	31
Leaf bulk density[3] (kg DM/ha per cm)	150	130	120	90	10
Leaf lamina length[4] (cm)	10.3	14.9	19.1	20.6	0.29

[1]Molan (2004); [2]Standard error of the mean; [3]Top 50% of sward height; [4]Sbrissia (2004);

Animal performance was more related to intake (Table 2) than to the nutritive value of the herbage, since there was no difference in chemical composition of the ingested herbage and animals grazed a relatively constant proportion of sward height (top 33% - Gonçalves, 2002). Under these circumstances, herbage mass increased with increments in sward height, but herbage bulk density decreased (Table 1). However, variation in leaf bulk density (top 50% of sward height) was relatively small (150 to 90 kg DM/ha cm - Molan, 2004), suggesting that variations in bite mass were mainly due to variations in depth of the leaf lamina and/or grazing strata and characteristics of individual leaves (e.g. lamina length, mid rib diameter and shearing force).

Table 2 Daily herbage intake[1] and performance[2] of beef cattle heifers grazing *Brachiaria brizantha* cv Marandu swards maintained at 10, 20, 30 and 40 cm by continuous stocking and variable stocking rate – summer period (December to March)

Variable	Sward surface height (cm)				
	10	20	30	40	SEM[3]
Herbage intake (kg DM/100 kg LW per day)	1.3	1.8	1.8	2.0	0.07
Live weight gain (kg/animal per day)	0.19	0.51	0.75	0.93	0.10

[1]Sarmento (2003); [2]Andrade (2003); [3]Standard error of the mean

Similar evidence has been generated from grazing experiments with *Panicum maximum* cv Mombaça, a tall tussock-forming tropical species. Silva (2004) studied the short-term rate of herbage intake of dairy heifers grazing Mombaça grass pastures, and found that bite mass decreased and bite rate increased with decreasing pre-grazing sward height (Figure 2a,b). Despite the greater bite mass in taller swards, time per bite increased (leaf lamina length around 70 cm) and intake rate decreased (maximum around 100 cm sward height) (Figure 2c,d). Further, as pre-grazing sward height decreased, the number of feeding stations increased and the number of steps between feeding stations decreased (Figure 2e,f), showing the importance of controlling and manipulating sward structure (Table 3).

Table 3 Sward structural characteristics of *Panicum maximum* cv Mombaça pastures intermittently stocked and grazed down from 60, 80, 100, 120 and 140 cm pre-grazing sward surface height by dairy heifers[1]

Characteristic	Sward surface height (cm)					
	60	80	100	120	140	SEM[2]
Herbage mass (kg DM/ha)	7570	9130	11060	13130	17250	1080
Leaf-to-stem ratio	0.48	1.07	1.47	1.06	1.37	0.39
Total bulk density (kg DM/ha per cm)	126	114	110	109	123	13
Leaf bulk density (kg DM/ha per cm)	21	36	43	44	54	5
Leaf lamina length (cm)	22.2	48.8	62.7	65.5	66.5	3.8

[1]Silva (2004); [2]Standard error of the mean

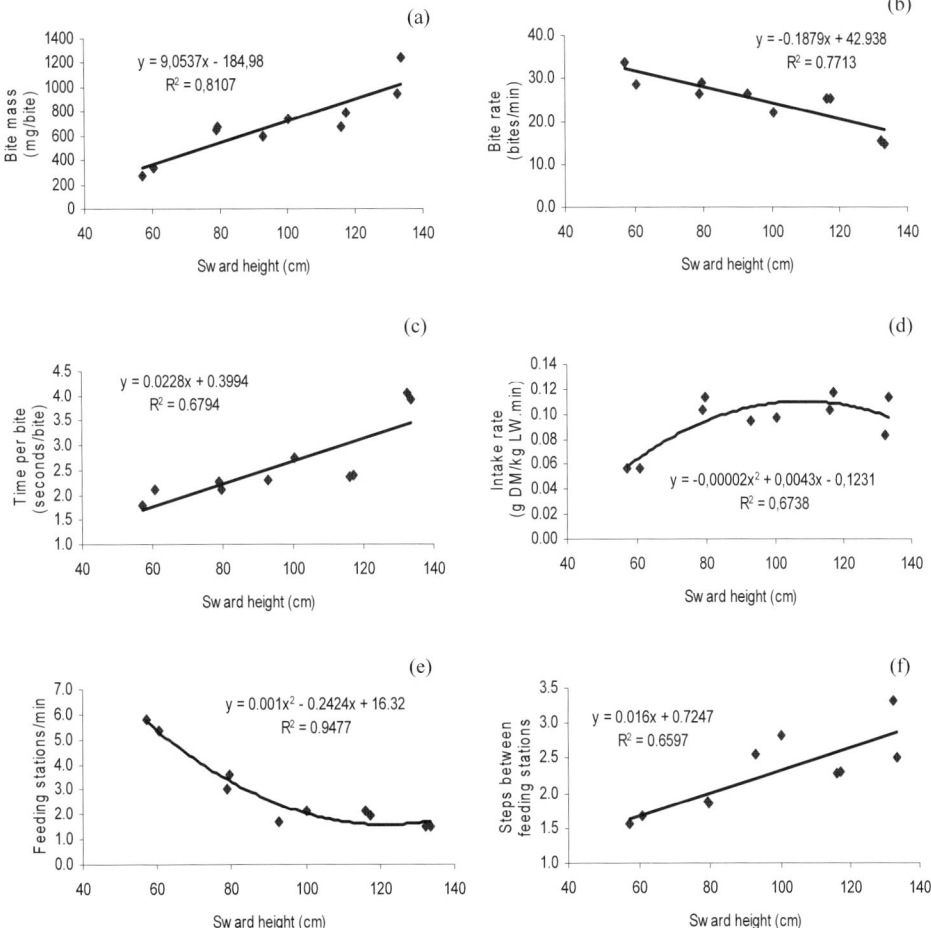

Figure 2 Grazing behaviour and short-term rate of herbage intake of dairy heifers during the grazing down process of intermittently defoliated *Panicum maximum* cv Mombaça pastures (Adapted from Silva, 2004)

The results reported for herbage and leaf bulk densities in these two studies (Tables 1 and 3) are within the range of values for temperate pasture species (Sollenberger & Burns, 2001), and are higher than those from Stobbs (1973b) (14-98 and 12-43 kg DM/ha per cm respectively). This suggests that there may be more similarities than differences between tropical/sub-tropical and temperate pasture species than otherwise implied, the main difference being stem elongation during vegetative development and its contribution to herbage bulk density.

Plant developmental responses have also been found to be closely related to manipulation of sward conditions, indicating the potential use of sward targets as a means of managing animal production on tropical and sub-tropical pastures (Hodgson & Da Silva, 2002). Carnevalli (2003), studying the dynamics of regrowth in Mombaça grass pastures rotationally grazed by dairy cows at either 95 or 100% canopy light interception (LI), found that leaf accumulation was the predominant growth process during the early stages of regrowth, until the canopy intercepted 95% of the incident light (leaf lamina length around 70-80 cm). From this point onwards, leaf accumulation decreased and stem and senescent material accumulation increased (Figure 3), indicating that regrowth in tropical/sub-tropical grass species is a two-stage process, and that control of stem development could be effectively achieved by monitoring and management of sward structure, particularly via grazing frequency. Overall, annual production (26900 kg DM/ha – Carnevalli, 2003) and nutritive value of the herbage produced (11.5% CP and 60% IVOMD – Bueno, 2003) were higher for the 95% LI pre-grazing condition, with losses due to grazing being higher for the 100% LI treatments (25.9% - Carnevalli, 2003). There was a high correlation between light interception and sward height (0.84), with 95% LI being consistently reached at 90 cm (leaf horizon) throughout the year, regardless of the physiological state of plants and post-grazing residues used (30 and 50 cm).

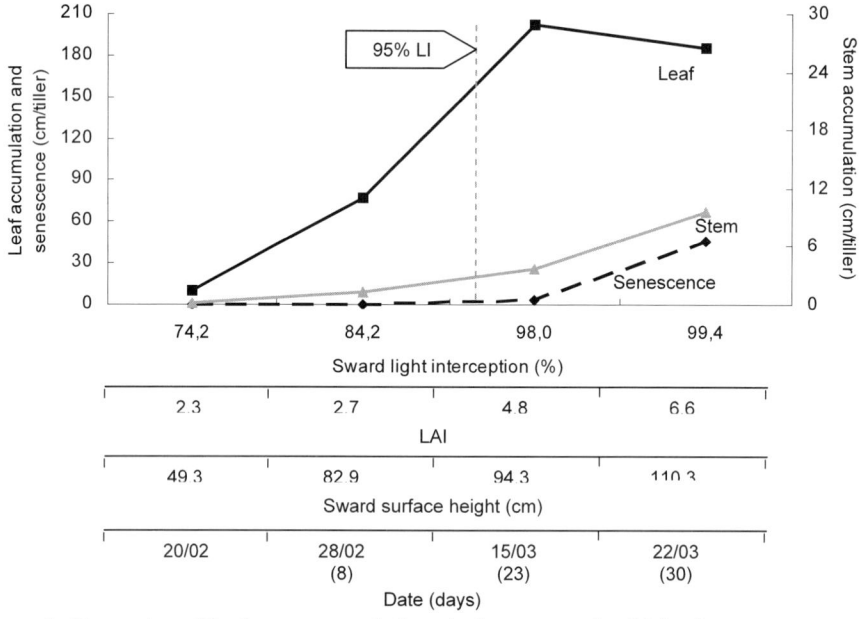

Figure 3 Dynamics of herbage accumulation during regrowth of Mombaça grass grazed at 100% sward light interception and post-grazing residue of 50 cm (from Carnevalli, 2003)

Apparently, the same pre-grazing sward condition that would optimise herbage production and nutritive value would also maximise herbage intake rate for this forage species (Figure 2d). Additionally, such a tall tufted-growing forage species would have high herbage and leaf bulk densities (Table 3), conditions that would likely favour animal performance. The results of Martinichen (2003) support this assumption, since dairy cows grazing Mombaça grass exclusively (and offered the same herbage allowance from sward structures similar to the recommended targets above) produced 13.5 to 15 kg milk/cow per day. In this case there was no difference in total herbage intake but ingestion of leaf varied, indicating that sward structure may operate independently in defining leaf accessibility.

These findings reveal a considerable potential for animal production and highlight the need for reviewing the generally held concept that tropical forages are of low feeding value. They also suggest that non-nutritional factors like leaf size (e.g. lamina length) might operate even in conditions where feed quantity is not limited (e.g. herbage mass, leaf mass, leaf bulk density), and interfere with herbage intake by increasing the need for manipulative jaw movements, and as a consequence, decrease the rate of intake of the grazing animal.

Efforts to describe sward vertical structure are usually made by quantifying each morphological component in 'volumetric layers' on a DM basis (e.g. mg DM/cm^3 or kg DM/ha per cm). The challenge at this scale is how to characterise the sward structure that animals actually respond to (e.g. size, shape and spatial distribution of individual components like leaf and stem). The wide range of herbage mass in *Brachiaria brizantha* (4630 to 14420 kg DM/ha) accounted for differences in leaf bulk density of only 60 kg DM/ha per cm (0.6 mg/cm^3) (Molan, 2004), indicating that the cropping process could not be understood or predicted from sward structure being described only on a DM basis. In short and tall managed *Brachiaria brizantha* and *Pennisetum americanum* (L.) Leeke (pearl millet) pastures, leaf lamina length can be increased two-fold (Sbrissia, 2004), or even five-fold (Castro, 2002). Thus, if the grazing layer does not change significantly in terms of leaf bulk density, the number of grazed leaves will decrease with increasing sward height, and animals will be forced to graze leaves individually when swards are very tall. This was observed with lambs grazing *Pennisetum americanum* (Castro, 2002), hoggets grazing *Panicum maximum* cv Tanzânia (Carvalho *et al.*, 2001) and Holstein heifers grazing *Panicum maximum* cv. Mombaça (Silva, 2004). Despite the importance of leaf lamina length to the intake process, the sward condition from which animals would start grazing leaves individually for different tropical grass species remains to be determined.

The basis of the functional response in grazing herbivores (i.e. the relationship between intake rate and herbage availability) is the response of bite mass to herbage mass, with the asymptotic part of the curve being explained by saturation of the grazing process (Gordon & Lascano, 1993). When focus is concentrated at the leaf lamina level, it becomes clear that not only bite dimensions and dynamics of jaw movements contribute to the functional response, but also the structure of the leaf lamina itself plays an important role. In tropical grasses, as sward height increases, leaf lamina length increases linearly, while increase in leaf mass per unit leaf lamina length is quadratic (Castro, 2002; Silva, 2004). This indicates that (i) the ascending section of the functional response curve is not strictly a function of increasing bite volume; (ii) a progressive decline can be expected in the rate of mass acquisition per unit of bite volume from tall swards, when leaf mass per unit of leaf lamina length reaches a plateau; and (iii) for tropical/sub-tropical species, it is important to describe leaf lamina structure more precisely in order to develop our understanding of the grazing process.

Non-cultivated pastures in the favourable tropics/sub-tropics

Animals grazing non-cultivated pastures face a much more diverse and variable foraging environment. The magnitude of this spatial and temporal variability and the interaction between them usually create a large mosaic of patches that, in turn, vary in time and space (O'Reagain & Schwartz, 1995). Animals deal with this condition by continuously sampling the environment and learning about it (Provenza & Launchbaugh, 1999). In this process they also remember preferred patches (Dumont & Gordon, 2003), and integrate information about their internal state with that of the changing environment (Laca, 2000). In cultivated pastures, heterogeneity is significantly lower and spatial scale seldom extends beyond patch level.

Consequently, cultivated, as opposed to non-cultivated pastures, are better suited to testing hypotheses concerning the grazing process.

Despite the large area occupied by natural grasslands (around 78 million ha) and its importance to animal production in Brazil, there is a paucity of research about the ecosystem and its dynamics. However, studies on *campo nativo* in southern Brazil (Campos biome) are starting to provide knowledge on the ecology and management of native pastures. Vegetation in these areas is highly complex and characterised by the co-existence of more than 30 forage species/m^2 (Boldrini, 1993), differing in structure, quality and metabolic pathway (C_3 and C_4). Herbage intake rate is a function of vertical as well as horizontal heterogeneity in varying scales, integrating biotic and abiotic factors that determine large differences between potential and actual bites and/or in spatial distribution of these bites. In such an environment, tufted grasses increase their abundance according to grazing pressure and can comprise more than 50% of ground cover at low grazing intensities, and less than 5% at high grazing intensities. Tufted grasses may represent a forage resource depending on the availability of preferred species (Soares, 2002), and so the concept of 'forage' itself may be quite variable, and represents an additional challenge to the characterisation of the grazing environment.

The spatial heterogeneity of non-cultivated pastures is known to impact on the grazing process. In a double strata sward structure (typical of certain zones of the Campos biome) grazing time decreased by 66.7 minutes per cm increase in inter-tussock vegetation height (Pinto, *pers. comm.*).. In these circumstances, total pasture herbage mass increased with decreasing grazing pressure, a result of a more pronounced increase in herbage mass of the tussock in relation to the inter-tussock vegetation. This indicates that the usually high correlation found between herbage mass and time spent grazing on a patch in cultivated pastures, is not necessarily applicable to non-cultivated pastures. This is one reason for the lack of accuracy in predicting patch residence time from herbage mass data only (Santos *et al.*, 2003).

Inter-tussock vegetation cover decreases from 100 to 67% when grazing pressure increases from 8.8 to 25 kg LW/kg DM per day, although the actual grazing intensity at the inter-tussock vegetation remains similar (Goret, 2005). This indicates that grazing behaviour depends on the dynamics of the inter-tussock/tussock vegetation cover, and responds predominantly to inter-tussock vegetation characteristics in a similar manner to the model proposed by Gordon (2000) for complex communities dominated by *Nardus stricta*. Consequently, it seems important to establish causative relationships between inter-tussock characteristics and animal responses. Thus Gonçalves (unpublished data) created controlled inter-tussock areas with sward height varying from 3 to 18 cm, and verified that grazing behaviour responses were similar to the classical responses reported for animals grazing cultivated pastures. Under those circumstances bite rate decreased, as did the number of feeding stations visited when inter-tussock sward height increased. Response of short-term rate of intake was quadratic, with a maximum at 13 cm (11.6 g DM/min – heifer beef cattle). At 18 cm, it was reduced to 6.6 g DM/min, a consequence of an increase in reproductive structures and their influence on the grazing process. Although tussock vegetation cover has a minor direct influence on the herbage intake of grazing animals, it represents a dry matter reserve that can be used depending on the type and class of the animals, inter-tussock characteristics, time of the year etc. Moreover tussock vegetation, are areas of high plant diversity (Goret, 2005), and some species only flower when protected by the tussocks.

In the Pantanal, a biome situated in central-west Brazil and characterised by periodical floods, Santos *et al.* (2003) evaluated the grazing behaviour of beef cows from bite- to home range scale studies. She concluded that animals tended to maximise herbage intake rate by selecting patches containing preferred species with the highest crude protein and the lowest neutral detergent fibre content. Similarly Boggiano (1995) demonstrated that in Campos, steers show preference for plants that can be easily accessed or prehended. The main characteristics associated with preference were: (1) leaves with low resistance to fracture, (2) herbaceous texture, (3) leaves with medium width, and (4) leaf laminae with a flat ('open') cross-section, as opposed to those that are folded or rolled around the mid-rib. *Paspalum notatum* and *Paspalum plicatulum* (typical inter-tussock species) were preferred against *Aristida jubata* (typical tall tussock species). However, the possibility of diet selection, even in heterogeneous native pastures, depends on herbage availability, being increased with increasing herbage allowance (Boggiano, 1995; Piaggio, 1994).

Sheep and cattle consistently select a better quality diet than that of the herbage available (Montossi *et al.*, 1998). Generous allowances however, are not necessarily related to high rates of herbage intake. Piaggio (1994) showed a decrease in herbage intake rate from 590 to 370 g OM/steer per hour with increasing herbage allowance, probably due to the significant increase in pasture heterogeneity as a consequence of the too generous allowances used. This illustrates the difficulty in determining available herbage in such a diverse natural pasture, and makes predictions from models developed for simple swards difficult. Further, it highlights the need for specific, planned efforts to generate knowledge and understanding relating to the ecosystem, and its potential development for sustainable pastoral systems.

Implications and perspectives

The tropical/sub-tropical environment is unique, requiring creative and site-specific solutions to overcome production constraints in order to realise its potential. The range of plant species, their varying size, morphology and physiology highlights the need to review some of the concepts and general views relating to animal performance from these pastures. Recent experimental work on pasture ecophysiology and grazing ecology in Brazil has been conceived under the conviction that control, monitoring and manipulation of sward state is an important feature for the establishment of meaningful cause and effect relationships to understand plant, animal and plant-animal responses to grazing management. This is very different from the traditional and simplistic view of production, in which control of the grazing process is made by means of fixed stocking rates, herbage allowances, grazing intervals and grazing method (Carvalho, 1997; Nascimento Jr *et al.*, 2003), and allows for significant variation in sward state and/or structure.

Although limited, evidence generated under these conditions clearly demonstrates that tropical/sub-tropical forage species can support high stocking rates during spring and summer (seasonality of production), and produce herbage of sufficiently high quality to ensure satisfactory animal performance throughout the year. Low quality herbage during the winter period (Hardy *et al.*, 1997) can be a result of inefficient harvest during previous favourable growing seasons and is not necessarily an intrinsic characteristic of the herbage produced. Additionally, non-nutritional factors have a greater relative importance than nutritional factors regulating herbage intake of grazing animals.

In general, pastures need to be grazed more frequently than traditionally in order to generate sward control and optimise herbage production and nutritive value (quantitative restriction). At the lower end of the herbage abundance scale (controlled swards) the response of short-term rate of intake to variation in sward conditions follow a similar pattern to that of temperate pasture species, but with one significant difference - long leaf laminae can result in a reduction in herbage intake rate. However, when the main objective is animal performance and high herbage allowances are used (upper end of the herbage abundance), utilisation is low and stemmy and dead material accumulate. In this situation, herbage intake/performance responses are related to sward leaf mass and leaf-to-stem ratio.

In this context, aspects relating to plant size, morphology, structure and distribution play an important role in determining time per bite, bite effort, the search for grazing patches and stations, and grazing time, indicating that careful control, monitoring and description of sward conditions should be essential features of experiments with tropical/sub-tropical forage species. Further, foraging behaviour variables should be considered as a family of interactive factors governing herbage intake and performance of animals grazing either tropical/subtropical or temperate pasture species. Hence a new window of opportunity is created for improving animal production from the available pasture-forming species, augmenting the benefits from breeding and the introduction of new, more productive and resilient forage species.

Animal production systems for tropical/sub-tropical pastures have an additional and significant constraint, i.e. the pronounced seasonality of herbage production. This generates variation in the feed supply-demand balance of the system between and within seasons of the year, which must be managed if sward control is to be achieved effectively. During periods of feed surplus, increasing stocking rate or conservation cuts is the most feasible action to take. Conversely, during periods of feed deficit, reduction in stocking rate and use of supplementary feeds are necessary. Either way, manipulation of stocking rate is a means of achieving sward control. Management practices like these have direct and indirect impacts on sward control, structure and animal performance that need to be known in order to allow for the correct planning and decision making process on a farm scale. In summary, management of tropical/sub-tropical grasslands should be based on the careful control and planning of sward state, with the objective of providing conditions for the construction of ecologically sound pastoral environments to optimise the harvest of nutrients through grazing and favour animal production.

Acknowledgements

Thanks are due to Professors John Hodgson, Carlos G.S. Pedreira, Domicio do Nascimento Jr and Dr Valéria P.B. Euclides for their helpful comments on the manuscript.

References

Allden, W.G. & I.A. Whittaker (1970). The determinants of herbage intake by sheep: the interrelationship of factors influencing herbage intake and availability. *Australian Journal of Agricultural Research*, 21, 755-766.
Andrade, F.M.E. (2003). Produção de forragem e valor alimentício do capim-Marandu submetido a regimes de lotação contínua por bovinos de corte (Herbage production and feeding value of Marandu grass pastures submitted to continuous stocking regimes by beef cattle). M.Sc. thesis, Universidade de São Paulo Brazil, Escola Superior de Agricultura "Luiz de Queiroz", 125pp. <http://www.teses.usp.br>
Boldrini, I. (1993). Dinâmica de vegetação de uma pastagem natural sob diferentes níveis de oferta de forragem e tipos de solo, Depressão Central RS (Vegetation dynamics of a native pasture submitted to levels of herbage on offer). Ph.D. thesis, Universidade Federal do Rio Grande do Sul Brazil, 262pp.
Boggiano, P.R.O. (1995). Relações entre estrutura da vegetação e pastejo seletivo de bovinos em campo natural (Relationship between sward structure and selective grazing of cattle in natural pastures). M.Sc. thesis, Universidade Federal do Rio Grande do Sul, 159pp.

Bueno, A.A.O. (2003). Características estruturais do dossel forrageiro, valor nutritivo e produção de forragem em pastos de capim-Mombaça submetidos a regimes de lotação intermitente (Canopy structural characteristics, nutritive value and herbage production of Mombaça grass pastures submitted to intermittent defoliation regimes). M.Sc. thesis, Universidade de São Paulo Brazil: Escola Superior de Agricultura "Luiz de Queiroz", 124pp. <http://www.teses.usp.br>.

Carnevalli, R.A. (2003). Dinâmica da rebrotação de pastos de capim-Mombaça submetidos a regimes de desfolhação intermitente (Regrowth dynamics of Mombaça grass pastures submitted to intermittent defoliation regimes). Ph.D. thesis, Universidade de São Paulo, Brazil: Escola Superior de Agricultura "Luiz de Queiroz", 136pp. <http://www.teses.usp.br>

Carvalho, P.C.F. (1997). A estrutura da pastagem e o comportamento ingestivo de ruminantes em pastejo (Sward structure and ingestive behaviour of grazing ruminants). In: C.C. Jobim, G.T. Santos & U. Cecato (eds.). *Proceedings of the Simpósio sobre Avaliação de Pastagens com Animais*, Maringá, PR, Brazil, 25-52.

Carvalho, P.C.F. (2002). Country Pasture/Forage Resources Profile: Brazil, FAO, Rome. <http://www.fao.org/waicent/faoinfo/agricult/agp/agpc/doc/counprof/brazil.htm>

Carvalho, P.C.F., G.K. Marçal, H.M.N. Ribeiro Filho, C.H.E.C. Poli, J.K. Trindade, J.O.R. Oliveira, C. Nabinger & A. Moraes (2001). Pastagens altas podem limitar o consumo dos animais (Tall swards can limit animal intake). In: W.R.S. Mattos, V.P. De Faria, S.C. Da Silva, L.G. Nussio & J.C. Moura (eds.) *Proceedings of the 38th Annual Meeting of Sociedade Brasileira de Zootecnia*, Piracicaba, SP, Brazil, 265-266.

Castro, C.R.C. (2002). Relações planta-animal em pastagem de milheto (*Pennisetum americanum* (L.) Leeke.) manejada em diferentes alturas com ovinos (Plant-animal relations in pearl millet pastures (*Pennisetum americanum* (L.) Leeke.) managed in different sward heights by sheep). M.Sc. thesis, Universidade Federal do Rio Grande do Sul Brazil: Faculdade de Agronomia, 185pp.

Chacon, E. & T.H. Stobbs (1976). Influence of progressive defoliation of a grass sward on the eating behaviour of cattle. *Australian Journal of Agricultural Research*, 27, 709-727.

Corsi, M., G.B. Martha Jr, D. Nascimento Jr & M.A.A. Balsalobre (2001). Impact of grazing management on productivity of tropical grasslands. In: J.A. Gomide, W.R.S. Mattos & S.C. Da Silva (eds.) *Proceedings of the 19th International Grassland Congress*. São Pedro, SP, Brazil, 801-806.

Dumont, B. & I.J. Gordon (2003). Diet selection and intake within sites and across landscapes. In: L't Mannetje, L. Ramirez-Avilés, C. Sandoval Castro & J.C. Ku Vera (eds.) *Proceedings of the VI International Symposium on the Nutrition of Herbivores*, Merida, Mexico, 175-194.

Fryxell, J.M. (1991). Forage quality and aggregation by large herbivores. *American Naturalist*, 138, 478-498.

Fryxell, J.M., C.B.D. Fortin & J. Wilmshurst (2001). On the scale dependence of foraging in terrestrial herbivores. In: J.A. Gomide, W.R.S. Mattos & S.C. Da Silva (eds.) *Proceedings of the 19th International Grassland Congress,* São Pedro, Brazil, 271-275.

Gonçalves, A.C. (2002). Características morfogênicas e padrões de desfolhação em pastos de capim-Marandu submetidos a regimes de lotação contínua (Morphogenetic characteristics and defoliation patterns in Marandu grass pastures submitted to continuous stocking regimes). M.Sc. thesis, Universidade de São Paulo Brazil: Escola Superior de Agricultura "Luiz de Queiroz", 124pp. <http://www.teses.usp.br>.

Gordon, I.J. (2000). Plant-animal interactions in complex communities: from mechanism to modelling. In: J. Hodgson, G. Lemaire, A. Moraes, P.C.F. Carvalho & C. Nabinger (eds.) Grassland ecophysiology and grazing ecology. CAB International, Wallingford, Oxon, UK, 191-207.

Gordon, I.J. & C. Lascano (1993). Foraging strategies of ruminant livestock on intensively managed grasslands: potentials and constraints. In: M.J. Baker, J.R. Crush & L.R. Humphreys (eds.) *Proceedings of the 17th International Grassland Congress,* Hamilton, New Zealand, 681-689.

Goret, T. (2005). Influence de l'intensité de pâturage et de la richesse du sol sur la biodiversité des prairies naturelles du Campos au sud du Brésil. (Effect of grazing intensity and soil richness upon the biodiversity of native pastures from Campos, southern Brazil). Université Catholique de Louvain, Belgium, 151p.

Griffiths, W.M., J. Hodgson & G.C. Arnold (2003a). The influence of sward canopy structure on foraging decisions by grazing cattle. I. Patch selection. *Grass and Forage Science*, 58, 112-124.

Griffiths, W.M., J. Hodgson & G.C. Arnold (2003b). The influence of sward canopy structure on foraging decisions by grazing cattle. II. Regulation of bite depth. *Grass and Forage Science*, 58, 125-137.

Hadley, M. (1993). Grasslands for sustainable ecosystems. In: M.J. Baker, J.R. Crush & L.R. Humphreys (eds.) *Proceedings of the 17th International Grassland Congress*, Hamilton, New Zealand, 21-27.

Hardy, M.B., H.H. Meissner & P.J. O'Reagain (1997). Forage intake and free-ranging ruminants: a tropical perspective. In: J.G. Buchanan-Smith, L.D. Bailey & P. McCaughey (eds.) *Proceedings of the 18th International Grassland Congress*, Winnipeg, Canada, 45-52.

Hodgson, J., D.A. Clark & R.J. Mitchell (1994). Foraging behaviour in grazing animals and its impact on plant communities. In: G.C. Fahey (ed.) Forage quality, evaluation and utilization. *American Society of Agronomy, Crop Science Society of America, Soil Science of America*, 796-827.

Hodgson, J., G.P. Cosgrove & S.J.R. Woodward (1997). Research on foraging behaviour: progress and priorities. In: J.G. Buchanan-Smith, L.D. Bailey & P. McCaughey (eds.) *Proceedings of the 18th International Grassland Congress*, Winnipeg, Canada, 109-118.

Hodgson, J. & S.C. Da Silva (2002). Options in tropical pasture management. In: A.M.V. Batista, S.B.P. Barbosa, M.V.F. Santos & L.M.C. Ferreira (eds.) *Proceedings of the 39th Annual Meeting of Sociedade Brasileira de Zootecnia*, Recife, Brazil, 180-202.

Illius, A.W. (1997). Advances and retreats in specifying the constraints on intake in grazing ruminants. In: J.G. Buchanan-Smith, L.D. Bailey & P. McCaughey (eds.) *Proceedings of the 18th International Grassland Congress*, Winnipeg, Canada, 39-44.

Laca, E.A. (2000). Modelling spatial aspects of plant-animal interactions. In: J. Hodgson, G. Lemaire, A. Moraes, P.C.F. Carvalho & C. Nabinger (eds.) Grassland ecophysiology and grazing ecology. CAB International, Wallingford, Oxon, UK, 209-231.

Martinichen, D. (2003). Efeito da estrutura do capim Mombaça sobre a produção de vacas leiteiras (Effect of the structure of Mombaça grass on the production of dairy cows). M.Sc. thesis, Universidade Federal do Paraná Brazil: Setor de Ciências Agrárias, 87pp.

Mitchell, R.J., J. Hodgson & D.A. Clark (1991). The effect of varying leafy sward height and bulk density on the ingestive behaviour of young deer and sheep. In: A. Tarry (ed.) *Proceedings of the New Zealand Society of Animal Production*, 51, 159-166.

Molan, L.K. (2004). Estrutura do dossel, interceptação luminosa e acúmulo de forragem em pastos de capim-Marandu submetidos a alturas de pastejo por meio de lotação contínua (Sward structure, light interception and herbage accumulation of Marandu grass swards submitted to grazing intensities by continuous stocking regimes). M.Sc. thesis, Universidade de São Paulo Brazil: Escola Superior de Agricultura "Luiz de Queiroz", 159pp. <http://www.teses.usp.br>

Montossi, F., E.J. Berreta, G. Pigurina, I. Santamarina, M. Bemhaja, R. San Julián, D.F. Risso & J. Mieres (1998). Estúdios de selectividad de ovinos y vacunos en diferentes comunidades vegetales de la region de basalto (Study on the selective grazing of sheep and cattle on different vegetation communities in the basaltic region). In: E. Berreta (ed.) Seminário de actualizacion em tecnologias para basalto. Instituto Nacional de Investigacion Agropecuaria, Tacuarembo, Uruguay, 257-286.

Nascimento Jr, D., R.A. Barbosa, K.R.A. Marcelino, A.F.Garcez Neto, G.S. Difante & B.A. Lopes (2003). A produção animal em pastagens no Brasil: uso do conhecimento técnico e resultados (Animal production from pastures in Brazil: use of technical knowledge and results). In: A.M. Peixoto, J.C. Moura, S. C Da Silva & V.P. De Faria (eds.) *Proceedings of the 20th Simpósio sobre Manejo da Pastagem*, Piracicaba, SP, Brazil, 1-82.

O'Reagain, P.J. & J. Schwartz (1995). Dietary selection and foraging strategies of animals on rangeland. Coping with spatial and temporal variability. In: M. Journet, E. Grenet, M-H. Farce, M. Theriez & C. Demarquilly (eds.) Recent developments in the nutrition of herbivores - *Proceedings of the 4th International Symposium on the nutrition of herbivores*, Clermont-Ferrand, 419-424.

Piaggio, L.M. (1994). Parâmetros determinantes do consumo e seletividade de novilhos em pastejo de campo nativo melhorado (Key variables determining intake and selectivity of steers grazing improved native pastures). M.Sc. thesis, Universidade Federal do Rio Grande do Sul, 438pp.

Poppi, D.P., T.P. Hughes & P.J. L'Huillier (1987). Intake of pasture by grazing ruminants. In: A.M. Nicol (ed.) Feeding livestock on pasture. *New Zealand Society of Animal Production*, Occasional Publication N° 10, 55-63.

Prache, S., I. Gordon & A.J. Rook (1998). Foraging behaviour and diet selection in domestic herbivores. *Annales de Zootechnie*, 47, 335-345.

Prache, S. & J.L. Peyraud (2001). Foraging behaviour and intake in temperate cultivated grasslands. In: J.A. Gomide, W.R.S. Mattos & S.C. Da Silva (eds.) *Proceedings of the 19th International Grassland Congress*, São Pedro, Brazil, 309-319.

Provenza, F.D. & K.L. Launchbaugh (1999). Foraging on the edge of chaos. In: K.L. Launchbaugh, J.C. Mosley & K.D. Sanders (eds.) Grazing behaviour of livestock and wildlife. University of Idaho, USA, 1-12.

Rocha, G.L. (1991). Clima e áreas de pastejo (Climate and grazing areas). In: J.C. Moura (ed.) Ecossistemas de pastagens – Aspectos Dinâmicos. Sociedade Brasileira de Zootecnia, Fundação de Estudos Agrários Luiz de Queiroz – FEALQ, Piracicaba, SP, Brazil, 1-42.

Santos, S.A., C. Costa, G. Silva e Souza, S.M.A. Crispim, R.A. Pearson & R. Gutierrez (2003). Foraging strategy of cattle in Pantanal rangeland, Nhecolândia sub-region, Brazil. In: N. Allsopp, A R. Palmer, S.J. Milton, G.I.H. Kerley, K.P. Kirkman, R. Hurt & C. Brown (eds.) Rangelands in the new millennium - *Proceedings of the 7th International Rangeland Congress, Durban, South Africa,* 23-30.

Sarmento, D.O.L. (2003). Comportamento ingestivo de bovinos em pastos de capim-Marandu submetidos a regimes de lotação contínua (Ingestive behaviour of beef cattle grazing *Brachiaria brizantha* cv. Marandu pastures submitted to continuous stocking regimes). M.Sc. thesis, Universidade de São Paulo Brazil: Escola Superior de Agricultura "Luiz de Queiroz", 76pp. <http://www.teses.usp.br>

Sbrissia, A.F. (2004). Morfogênese, dinâmica do perfilhamento e do acúmulo de forragem em pastos de capim-Marandu sob lotação contínua (Morphogenesis, tillering and herbage accumulation dynamics in Marandu grass swards under continuous stocking). Ph.D. thesis, Universidade de São Paulo Brazil: Escola Superior de Agricultura "Luiz de Queiroz", 171pp. <http://www.teses.usp.br>

Silva, A.L.P. (2004). Estrutura do dossel e o comportamento ingestivo de novilhas leiteiras em pastos de capim Mombaça (Sward structure and ingestive behaviour of dairy heifers in Mombaça grass pastures). Ph.D. thesis, Universidade Federal do Paraná Brazil: Setor de Ciências Agrárias, 104pp.

Soares, A.B. (2002). Efeito da alteração da oferta de matéria seca de uma pastagem natural sobre a produção animal e a dinâmica da vegetação (Effect of variation in dry matter allowance of a native pasture on animal production and vegetation dynamics). Ph.D. thesis, Universidade Federal do Paraná Brazil: Setor de Ciências Agrárias, 180pp.

Sollenberger, L.E. & J.C. Burns (2001). Canopy characteristics, ingestive behaviour and herbage intake in cultivated tropical grasslands. In: J.A. Gomide, W.R.S. Mattos & S.C. Da Silva (eds.) *Proceedings of the 19th International Grassland Congress*, São Pedro, SP, Brazil, 321-327.

Stobbs, T.H. (1973a). The effect of plant structure on the intake of tropical pastures. I. Variation in the bite size of grazing cattle. *Australian Journal of Agricultural Research*, 24, 809-819.

Stobbs, T.H. (1973b). The effect of plant structure on the intake of tropical pastures. II. Differences in sward structure, nutritive value, and bite size of animals grazing *Setaria anceps* and *Chloris gayana* at various stages of growth. *Australian Journal of Agricultural Research*, 24, 821-829.

Taylor, J.A. (1993). Chairperson's summary paper. Session 19: Foraging strategy. In: M.J. Baker, J.R. Crush & L.R. Humphreys (eds.) *Proceedings of the 17th International Grassland Congress*, Hamilton, New Zealand, 739-740.

Ungar, E.D. (1996). Ingestive behaviour. In: J. Hodgson & A.W. Illius (eds.) The ecology and management of grazing systems. CAB International, Wallingford, UK, 185-218.

Interactions between foraging behaviour of herbivores and grassland resources in the eastern Eurasian steppes

Wang Deli[1], Han Guodong[2] and Bai Yuguang[3]
[1]*Institute of Grassland Science, Key Laboratory of Vegetation Ecology, Northeast Normal University, Changchun 130024 P.R. China*
Email: wangd@nenu.edu.cn
[2]*Department of Grassland Science, Inner Mongolia Agricultural University, Hohhot 010018 P.R. China,* [3]*Department of Plant Sciences, University of Saskatchewan, Saskatoon, S7N 5A8 Canada*

Key points

1. In rangeland areas such as the eastern Eurasian steppes (Mongolia and China), foraging behaviour is influenced by plant or vegetation properties with high heterogeneity.
2. Until recently foraging theory has not accounted for the foraging process or ingestive behaviour. Existing theories on foraging behaviour need to evolve and begin to coalesce, and combine with observations or manipulative experiments.
3. Plant and patch properties such as diversity and height influence animal foraging behaviour (related to foraging process or diet selection) in heterogeneous steppes.
4. Stocking rate is the most important management factor for grazing or vegetation management, and determining the optimal stocking rate in steppes depends upon variable annual forage production, vegetation regrowth and animal production targets.

Keywords: animal-plant interactions, stocking rate, heterogeneity

Introduction

Rangeland areas comprise 25% of the world's grassland (Hodgson, 1990), and are found in prairie, pampas, veld and campos, savanna, and steppe. This paper will focus on the latter, located within the Eurasian continent ranging from Ukraine to Kazakhstan, Russia, Mongolia and China (35-57°N). Although steppe landscape is greatly modified by human activity, it is still the climax vegetation existing over a range of climates from continental to arid and semiarid climate (annual precipitation = 200-450 mm) (Zhu, 1993). The eastern Eurasian steppes are located on the Mongolian Plateau and Songliao Plains (in Mongolia and China) with three zonal types including 'meadow steppe', 'typical steppe' and 'desert steppe' (Wu, 1990). The dominant species influenced by climate or grazing are *Stipa* and *Leymus* genus (*Stipa baicalensis, S. grandis, S. breviflora, S. kelemenzii* and *Leymus chinensis*), and forbs such as *Artemisia frigida* and *Filifolium sibiricum*. Each type of steppe plays an important role in animal production, providing over 90% of the food source for maintaining basic requirements of herbivores. Critical problems facing farmers include how to avoid or reduce overgrazing and how to enhance forage availability or grazing efficiency within the highly heterogeneous steppes.

The ecological relationship or interaction between plant and animal is universal, and of fundamental importance (Howe & Westley, 1988). Herbivores interact with plants to maintain their requirements for growth and reproduction, but the availability of plants also regulates the dynamics and production of animal populations. Herbivores play a role in controlling the function of whole ecosystems, and show an asymmetry in the interaction of plant and herbivore (Crawley, 1983). The plant-animal interface is a fundamental interaction between trophic levels, and is the central feature of natural and artificial grassland ecosystems (Ungar, 1996). A better understanding on how animals can effectively exploit plants, and how plants alter their fitness and adaptive strategies in response to animal foraging will provide a baseline for the sustainable utilisation of grassland resources for animal grazing. Theoretical models on the interaction of plant and herbivore lay a foundation for interpreting mechanisms of co-evolution between plant and animals in nature (Belovsky *et al.*, 1999; Loeuille *et al.*, 2002).

Two important research aspects for interactions between plants and animals in grasslands are: i) grazing intensity, i.e. how herbivores exploit plants, and ii), herbivore behaviour, i.e. individual animal performance. These are considered the most direct ways that grazing animals interact with a plant community (Newman *et al.*, 1995). Previous studies concentrated on the effects of herbivore grazing intensity (stocking rate, carrying capacity) (Heady & Child, 1994; Roe, 1997). Current interest lies in determining the relationship between herbivore behaviour and vegetation property. However, it is necessary to consider grazing intensity and animal behaviour synchronously within the context of plant and herbivore interactions. Herbivores interact with foraging plants at multiple levels in grasslands. Grazing intensity embodies the circumstance at a higher level (herbivore group or population effect), whilst at a lower level foraging behaviour is often 'individual-dependent'. Thus, the interaction between animal and plant can be summarised by relationships of grazing intensity and plant regrowth, and of foraging behaviour and spatial vegetation pattern (heterogeneous characteristics of vegetation) within grazing systems.

In this paper, experimental results on foraging behaviour especially diet selection, and the relationship between stocking rate and vegetation regrowth on 'natural steppes' in the eastern Eurasian steppes are presented. Existing theories on foraging behaviour of herbivores are reviewed.

The foraging process and its theoretical basis

Foraging process

The foraging process can be divided into two phases: decision-making (searching), and intake or ingestion (cutting, chewing, swallowing and digesting) (Ungar, 1996; Manning & Dawkins, 1998). Decision-making exhibits variable time scales (i.e. second to second). The ingestion phase, during which animals obtain energy or nutrients, follows each decision-making event. Most studies on foraging process focus on the former phase (Bazely, 1990; Howery *et al.*, 2000). Distinguishing the two phases by time sequence is difficult because animals may determine their foraging direction during ingestion. It is possible that different mechanisms drive the two phases. An animal's learning and experience (cognition), and biological innate potentials play major roles in decision-making and ingestion, respectively (Bailey *et al.*, 1996). This hypothesis, however, remains to be tested. For the whole foraging process, components such as decision, currency and constraint should be considered to determine mechanisms related to foraging behaviour.

Theories on foraging process

Rules-of-thumb

Simple Rules-of-Thumb (RT) have been used to describe foraging and behaviour. For example, Iwasa *et al.* (1981) used the number, time, and 'quitting time' of animal foraging when herbivores were facing different food distributions on patches. Rules of thumb can provide a primary understanding of foraging decision or diet selection of herbivores in heterogeneous grasslands, and are an option where the time or effort to obtain information is prohibitive (Ward, 1992; Bailey *et al.*, 1996). The RT hypothesis has several limitations, for example, it is difficult to explain foraging 'optimal' solution using RT for homogeneous environments with no food differences. In addition, it is unclear whether RT can play a role in foraging selection when animals have basic perception and spatial memory. There is also a lack of experimental evidence to verify this hypothesis.

Marginal value theorem

Marginal value theorem (MVT) was developed by Charnov (1976), and has been used to describe foraging strategy when animals exploit 'patchy' resources. The food quality of a patch, residence time within the patch, moving time between patches (departure), and foraging energy are useful parameters explaining the foraging decisions of animals. However, several experiments have shown that measured values such as residence time and intake rate, is often lower than that predicted by MVT, even though the behaviour of small herbivores such as insects fits MVT (Roguet *et al.*, 1998; Prache & Peyraud, 2001). The disadvantage of MVT is that it was hypothesised that the forager could not accumulate information on patches, whereas there is experimental evidence to suggest that both cattle and sheep can accumulate experiences of patch characteristics (Edwards *et al.*, 1996; Bailey *et al.*, 2000).

Optimal foraging theory

'Optimal foraging theory' (OFT) is useful in situations when a forager makes decisions about current resource consumption based on tradeoffs in resource attributes (Gerber *et al.*, 2004). It provides a functional approach for examining grazing behaviour (Bailey *et al.*, 1996) and can quantitatively account for the foraging decisions of animals (Stephens & Krebs, 1986). Maximisation of energy intake rate and minimisation of the time necessary to obtain nourishment are two measures of foraging success that remain in standard use. Foraging success is assumed commensurate with animal fitness (Perry & Pianka, 1997). However, it needs to be developed with more manipulative experiments in both the laboratory and the field (Perry & Pianka, 1997). Optimal foraging theory may be an over-simplified representation of the reality (Prache & Peyraud, 2001). For example, maximisation of reproductive fitness has been simplified into, maximisation of various surrogate currencies such as rate of nutrient intake and energy, because measuring fitness is difficult or impossible in most cases (Lemon, 1991). Nutrient balance seems to be more important than energy intake for herbivores, even in poor food environments where animals have to forage on plant species with a low frequency distribution, and reducing their mean energy intake efficiency, while OFT only emphasizes net energy (nutrient total) or fitness.

Theory of minimal total discomfort

The theory of minimal total discomfort (MTD) was proposed by Forbes (1999, 2001), and is based on the physiological state of animals. A total 'discomfort signal' (factor) integrated from metabolisable energy (ME), crude protein (CP) and neutral detergent fibre (NDF) was used as a parameter estimating animal's foraging behaviour. Animals tend to reach the state of MTD during feeding or the foraging process, and so alter intake or food selection so as to minimise total discomfort (Figure 1). This model suggests that food choice is a physiological requirement, and is related to energy intake and nutrients. However, determining discomfort components is difficult and no information exists on the spatial location of animals during the foraging process.

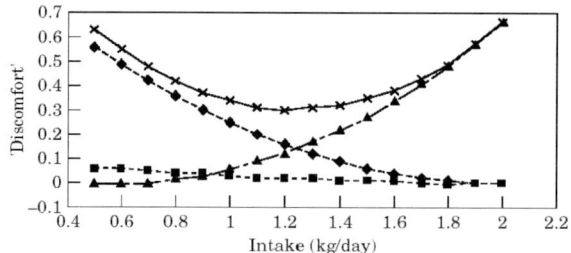

Figure 1 Relationships between intake and nutrient components in an animal diet (♦=ME; ■= CP; ▲= NDF; ×= Total) (Forbes, 2001)

Among the current foraging theories or hypothesis, a universal model than can explain the foraging process as a whole is lacking. Future studies on animal physiological state, learning and experience, and environmental constraints should be conducted to verify many of these hypotheses. A better integration between decision-making and intake, and the development of improved elementary components may help unify these hypotheses.

Foraging behaviour and plant and patch characteristics of steppes

The patch is considered as a cluster of one or more plant species with high density, and also as a basic foraging scale (i.e. the level of variability encountered within a landscape unit) (Wallis de Vries & Daleboudt, 1994; Bailey *et al.*, 1996; O'Reagain, 2001). Natural 'meadow steppe' vegetation of North-eastern China has a higher heterogeneity than that of managed grasslands, so foraging behaviour needs to match this variability.

Plant and patch location (decision-making) for herbivores

Plant and patch decision-making implies that herbivores forage directionally. It is often assumed that herbivores tend to graze patches prior to individual plants because patches contain a greater amount of nutrient and energy, but a shift in food location may occur when mixed or mosaic food distribution is encountered in heterogeneous grasslands. What factors influence herbivores food location? Controlled experiments in natural 'meadow steppes' were conducted in 2004 with yellow cattle to address the effect of various factors on herbivore's food location.

In the experiment, four cattle grazed for 10 minutes in different paddocks (paddock area = 600m^2) with patches and randomly distributing plant individuals (dominant species was *Phragmites communis*). The cattle displayed a strong preference for patches compared to plant individuals Table1). Moreover, the size of the patch was positively correlated to herbivore foraging time. Foraging time on patches increased considerably when patch size reached 50m^2. When patch size was small (10m^2), it was difficult for cattle to locate a food patch. Previous experiments have shown that herbivores have the ability to remember spatial location of food patches (Edwards *et al.*, 1996; Sibbald & Hooper, 2003) and can learn to associate visual cues with disparate food quality, and use this information to forage more efficiently (Howery *et al.*, 2000). Results from limited experiments of food location imply that a shift of foraging level from plant to patch copes with the size, number and quality of patches.

Table1 Foraging time of yellow cattle at individual plant and patch scales (patch treatments: size and number) in meadow steppes of northeastern China (mean ± sd)

Foraging time (s)	Paddock number and patch size (m^2)				
	P1 (0 m^2)	P2(10 m^2)	P3(50 m^2)	P4(100 m^2)	P5(160 m^2)
At individual plant	401 ± 56	300 ± 50	155 ± 64	27 ± 13	10 ± 7
At patch		107 ± 39	336 ± 34	469 ± 14	474 ± 26

Foraging time (s)	Paddock number and (patch number)			
	P1 (2)	P2 (3)	P3 (4)	P4 (5)
At individual plant	369 ± 56	72 ± 33	121 ± 33	38 ± 14
At patch	81 ± 26	191 ± 38	195 ± 50	197 ± 51

Influence of plant and patch on herbivore diet selection

Effect of plant diversity on intake

Vegetation in 'meadow steppes' is diverse with over 30 plant species that provide food for grazing sheep or cattle to meet their physiological requirements. In these conditions herbivores are better able to meet their dietary needs and mitigate against toxins in their intake compared to exploiting a single food source. An experiment with increasing plant species and free choice for sheep showed that sheep preference was strongly correlated to plant diversity (Figure 2).

Average daily intake of sheep was 603.7g when a single plant species (*L. chinensis*) was offered, but it increased significantly to 823.5g when four plant species were fed. Little further increase was found when the number of species increased to nine. The strong correlation between animal selection and plant species incidence was reported in a Sicilian pasture (Carpino *et al.*, 2003). Provenza *et al.* (2003) reviewed the relationships between herbivore's diet and plant biochemical diversity, and concluded that foraging diverse plant species would benefit nutrient balance and limit toxins in food. It is suggested that diverse foraging could improve satiety and modulate taste for herbivores, and stimulate the ingestion of more food. Diverse foraging in natural steppes with low vegetation production may be valuable for vegetation production and ecosystem stability during the co-evolution between plants and herbivores, because diverse foraging can help maintain high regrowth potential and species diversity of the entire community.

Vegetation (patch) height selection

Vegetation height (sward or patch surface height) is used as an indicator for plant growth and production, herbage allowance, and is a useful parameter for grazing management (Hodgson,

1981). Many grazing experiments have been conducted on artificial grasslands to determine the relationships among sward height, foraging behaviour, and defoliation pattern (Amstrong *et al.*, 1995; Barrett *et al.*, 2001; Tharmaraj *et al.*, 2003; Wang D. *et al.*, 2003).

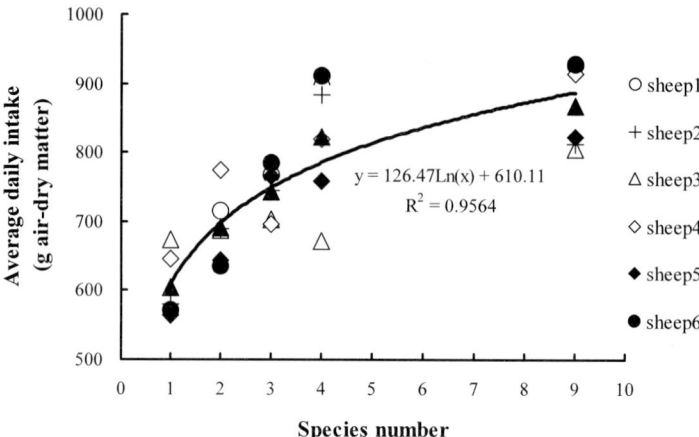

Figure 2 Change in intake for sheep with an increasing number of alternative food sources. Sheep were fed with different combinations of plants (treatment level: 1, 2, 3, 4, 9 species; six replications). Daily intake mass was enough for each sheep and the feeding duration was 15 days.

The heterogeneity of vegetation in meadow steppes has a great impact on herbivore's diet selection, expressed as average bite depth (Figure 3) (Wang X. *et al.*, 2002). Goats prefer a certain height of vegetation (20-25cm) and this preference remains unchanged during the grazing season. Vegetation height selection for herbivores indicates that there is a trade-off between energy and nutrient for herbivore foraging. When goats graze on patches with tall plants, a greater herbage allowance can be consumed or grazing time reduced.

Most patches with tall plants tend to be comprised of mature plants, thus the higher proportion of reproductive shoots and lignin content would adversely affect digestion and actual plant availability. On the other hand, goats can spend more time and energy obtaining food in lower height patches with higher carbohydrate and protein but lower lignin contents. The consistent trend of patch height selection during the full grazing season suggests that there is a greater intake mass (energy) requirement during early season grazing (less vegetation production) and a greater nutrient preference during late grazing season (adequate vegetation production).

Patch property affecting herbivore intake

Patch characteristics such as the quantity (biomass, height, density) and quality (nutrient-proportions of vegetative shoot/reproductive shoot, legume/grass, and digestive energy and toxin) influence intake, bite dimensions, and animal production (Bailey, 1995; Dumont *et al.*, 1995; Distel *et al.*, 1995; Prache & Peyraud, 2001; Griffiths *et al.*, 2003). However, in natural meadow steppes with high heterogeneous patches, patch height, but not mass and tiller density, are important factors affecting cattle behaviour, with a positive correlation between height and foraging time (Figure 4c). Bite rate of cattle did not vary with patch properties (Figure 4).

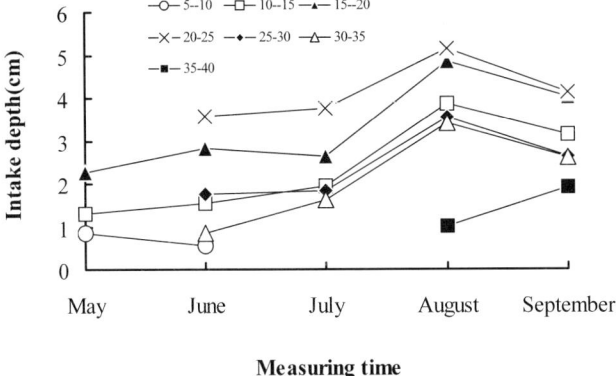

Measuring time

Figure 3 Variations in seasonal bite depth of goats with average vegetation (patch) height. Symbols indicate vegetation heights at 5 cm increments. (Experiment, involved grazing twenty sheep for six months in 'meadow' steppe dominated by *L. chinensis* (Wang D. *et al.*, 2002; Wang D., 2004).

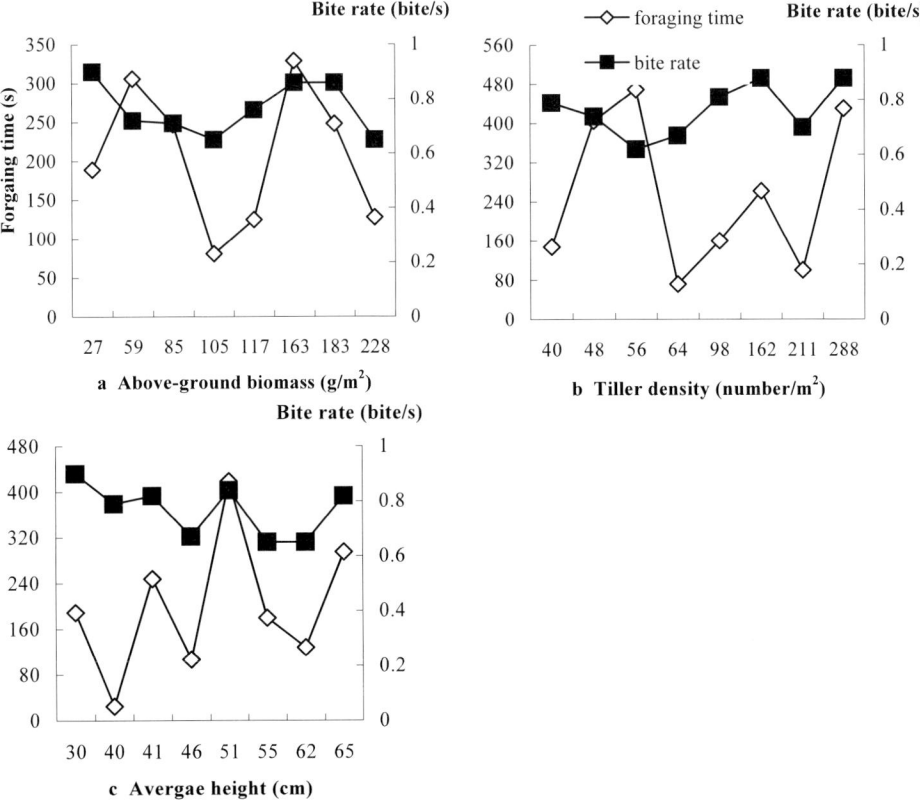

Figure 4 Relationships between patch properties (a, above ground biomass; b, tiller density; c, height) and cattle behaviour (foraging time and bite rate) on meadow steppes. Grazing time was 10 mins. with six replications for the experiment.

Studies on foraging behaviour can be at different scales of plant, patch and vegetation. Previous studies have focused more on ingestive behaviour such as instantaneous intake rate, and less on the digestive process during and after ingestion (O'Reagain, 2001). However, foraging optimisation may be better achieved within a long-term framework including digestion, which involves physiological state, experience, social organisation (grazing spatial distribution), and patch property. Herbivores often have to make trade-offs between foraging behaviour and social behaviour, quality and quantity of patch, and energy intake and time spent grazing. It may be that foraging can only be optimised with some constrains or trade-offs.

Intake of herbivores within intermediate time scale

Intermediate time is defined as the intermittent temporal scale distinguishing short term (second, minute) and long time (month, year). Although sheep tend to choose diverse plant species in steppes, actual species selection is usually concentrated on four or five plant species (Figure 2). During a 12-day experimental period, preference followed *P. tenueoflora* > (*L. chinensis* = *Ph. communis*) > *Kelimeris integrifolia*. This preference was displayed on a daily basis. The two species in the intermediate preference group inter-compensated in the sheeps diet (Figure 5).

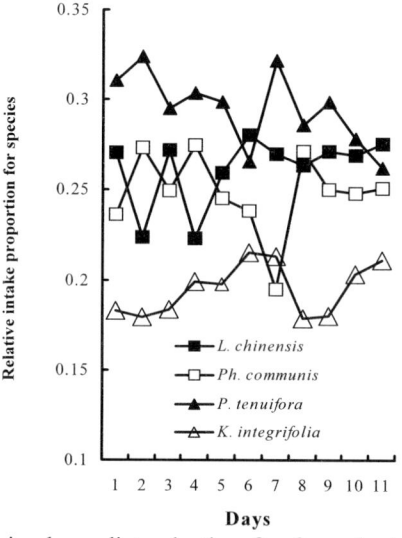

Figure 5 Daily variation in sheep-diet selection for four plant species. Sheep were fed combinations of four common plants in meadow steppes. Daily intake mass was enough for each sheep and the feeding duration was 15 days

An empirical assumption is that the inter-compensation of the two species could meet the needs of daily intake mass (half of total mass in this experiment), but it is unclear whether the displacement and transition of preference between the two species every two or three days can provide essential nutrient balance or need of food stimulus to maintain sheep's preference. The diversity of diet composition for herbivores benefits not only intake mass but also the consequence of intake or preference because of an interaction between nutrient and plant secondary metabolites (PSM) (Villalba *et al.*, 2002). However, further experiments on plant nutrient and PSM, and herbivore nutrient intake and metabolism are needed to address this.

Stocking rate of herbivores and vegetation regrowth

There exists a close relationship between vegetation regrowth and stocking rate (Li *et al.*, 1997; Kowalenko & Romo, 1998; Wang D. *et al.*, 2002). Regrowth derives from the production of new leaves of vegetative tillers, or new tillers produced either from buds on the shoot apex, or rhizome nodes situated at ground level (Davies, 1988). Regrowth is affected by grazing intensity and increasing stocking rate can enhance average leaf elongation and appearance rates and reduce senescence rate of dominant grasses in steppes (Liu Y. *et al.*, 2001, 2003). Another characteristic of plant regrowth in response to stocking rate can be expressed as tiller density and the related bud bank of plant population.

Generally, the contribution of tiller number to plant regrowth varies with time of the year, and maintains a certain pattern within the defoliation system. Vegetation tiller density can be stimulated to some extent (Liu Y. *et al.*, 2002; Wang S. *et al.*, 2003). Variations in tiller density under grazing disturbance may be attributed to triggering tiller bud expansion, which leads to compensatory growth. An observation on the bud bank of *L. chinensis* population under different stocking rates illustrates that grazing directly influenced the number of active buds on rhizomes and intermediate grazing maintained a higher level of active buds (Figure 6).

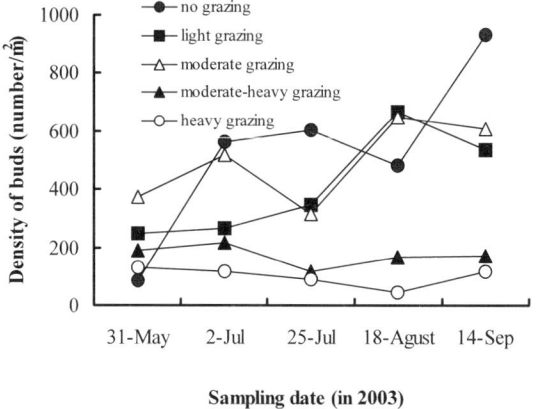

Sampling date (in 2003)

Figure 6 Seasonal bud bank of *L. chinensis* populations as affected by grazing intensity in meadow steppes

Plant regrowth post grazing is a complicated physiological process that involves carbon and nitrogen storage and remobilisation. It is known that water-soluble carbohydrate (WSC) and not carbohydrate concentration in stubble at the time of defoliating, plays the major role in plant regrowth (Volenec, 1986; Fulkerson & Slack, 1994). Work by the authors indicates that there is considerable variation in WSC of basal stems but not in WSC of leaf and root for *Ph. communis*, *P. tenuifora* and *L. chinensis* (Liu J. *et al.*, 2003). Nitrogen contents of leaf, stem and roots were not significantly different among defoliation intensities (Liu J. *et al.*, 2003). However, carbohydrate or WSC storage alone was not sufficient to explain the amount of regrowth because there was an interaction between carbon and nitrogen recycling during the regrowth of defoliated plants (Lemaire & Chapman, 1996; Thornton *et al.*, 2000). Further experiments examining the interaction between carbon and nitrogen metabolism, and quantifying remobilisation will be necessary to fully understand the physiological basis of regrowth post defoliation.

Optimal stocking rate and vegetation production

In China, where there is no standardised grazing capacity for grassland management, stocking rate is the most important management factor. The principal consideration for any rangeland grazing system is to balance livestock needs with the available forage supply through proper stocking rates.

Stocking rate affects standing crop (yield) and net primary productivity. Standing crop of *S. breviflora* desert steppes in Inner Mongolia was 78% higher under moderate as opposed to heavy grazing, and 90% higher under light than under heavy grazing (Liu *et al.*, 1996). During the growing season net aboveground productivity was 635, 580, and 535 kg/ha under light, moderate, and heavy grazing respectively (Han *et al.*, 1999). Light or moderate grazing in typical steppes is beneficial to forage production especially in dry years.

Heavy stocking rates lower animal liveweight production compared to moderate or light grazing. In a sheep grazing experiment on typical steppes of the Mongolian Plateau, average liveweight gain was 19.4, 15.6 and 11.8 kg/sheep when stocked at 0.68, 0.94 and 1.5 sheep/ha per year (Wei & Han, 1995). Losses due to sheep death were also higher under heavy grazing than under moderate and light grazing (sheep death numbers were 7, 1, and 0 sheep after 5 years heavy, moderate, and light grazing, respectively. The higher sheep mortality in heavy and moderate grazing is due to shortage of forage supply caused by stocking rate in winter and spring seasons.

Liveweight gain was 28.2 and 13.6% higher under light and moderate than under heavy grazing, respectively (Wei & Han, 1995). Liveweight gain per animal and per area unit was affected differently by stocking rate. Even though productivity per animal unit declined as stocking rate increased, productivity per area unit increased up to a point. When grass supply became limited, productivity per unit area then decreased. This is why most ranchers and local people often favour heavy grazing.

It is not possible to achieve both maximum gain per animal and per unit area concurrently. The curves of gain per animal and per unit area cross at the 'peril point' (Figure 7). At this point stocking rate is considered optimal for animal production. For *S. breviflora* desert steppe, the optimal stocking rate equates to a moderate grazing pressure. Therefore, curves of liveweight gain per animal and per unit area can be used to determine the optimal stocking rate in steppe zones of China (Han *et al.*, 2000).

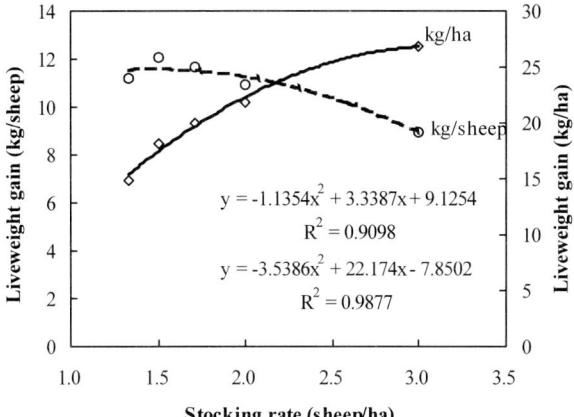

Figure 7 Changes in sheep liveweight gains with different stocking rates in *S. breviflora* desert steppe

Stocking rate or carrying capacity

Stocking rate should match the carrying capacity of grassland for grazing management to maintain its sustainability (Heady & Child, 1994). Long-term data of forage production can be used to calculate the carrying capacity of steppes. Primary production in Inner Mongolian steppes varies with season and year because of the fluctuation in temperature and precipitation (Figure 8) (Han *et al.*, 2001). Thus, carrying capacity of steppes varies with forage production. Stocking rate can be either flexible or fixed. Ideally, flexible stocking rate can meet fluctuations in forage production in different seasons and years, but is difficult in practical farming since every farm has a relatively stable livestock number. However, Martin (1975) reported that 90% of proper fixed stocking rate had good results in southern Arizona grasslands.

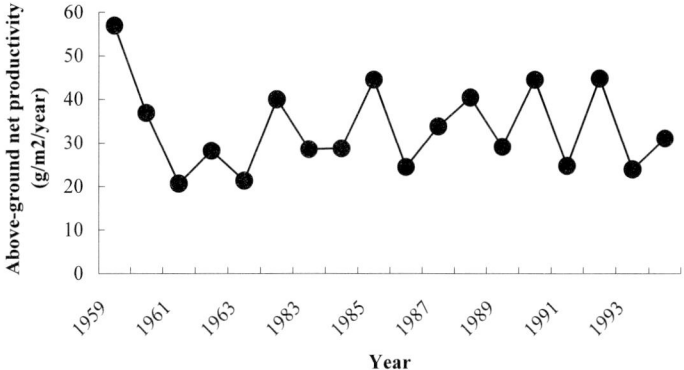

Figure 8 Yearly variation in above-ground net primary productivity of *S. kelemenzii* desert steppe

In conclusion, a series of experiments were conducted to investigate herbivore behaviour and the interaction between herbivore and plant or vegetation at various scales in the eastern Eurasian steppes. The studies on heterogeneous natural steppes produced some unexpected results and may provide new knowledge on the understanding of animal foraging behaviour which will benefit grassland management in the steppes of China, Mongolia and other countries

References

Armstrong, R.H, E. Robertson & E.A. Hunter (1995). The effect of sward height and its direction of change on the herbage intake, diet selection and performance of weaned lambs grazing ryegrass swards. *Grass and Forage Science*, 50, 389-398.

Bailey, D.W. (1995). Dailey selection of feeding areas by cattle in homogeneous and heterogeneous environments. *Applied Animal Behaviour Science*, 45,183-200.

Bailey, D.W., J.E. Gross, E.A. Laca, L.R. Rittenhouse, M.B. Coughenour, D.M. Swift & P.L. Sims (1996). Mechanisms that result in large herbivore grazing distribution patterns. *Journal of Range Management*, 49 (5), 386-400.

Bailey, D.W., L.D. Howery & D.L. Boss (2000). Effects of social facilitation for locating feeding sites by cattle in an eight-arm radial maze. *Applied Animal Behaviour Science*, 69, 93-105.

Barrett, P.D., A.S. Laidlaw, C.S. Mayne & H. Christie (2001). Pattern of herbage intake rate and bite dimensions of rotationally grazed dairy cows as sward height declines. *Grass and Forage Science*, 56 (4), 362-375.

Bazely, D.R. (1990). Rules and cues used by sheep foraging in monocultures. In: R.N. Hughes (ed.) Behavioural mechanisms of food selection. Spring-Verlag, Berlin, 343-366.

Belovsky, G.E., J. Fryxell & O.J. Schmitz (1999). Natural selection and herbivore nutrient: optimal foraging theory and what it tells us about the structure of ecological communities. In: H.J.G. Jung & G.C. Fahey Jr (eds.) Nutritional ecology of herbivores. American Society of Animal Science, Savoy, USA, 1-70.

Carpino, S., G. Licitra & P.J. Van Soest (2003). Selection of forage species by dairy cattle on complex Sicilian pasture. *Animal Feed Science and Technology*, 105, 205-214.

Charnov, E.L. (1976). Optimal foraging, the marginal value theorem. *Theoretical Population Biology*, 9, 129-136.

Crawley, M.J. (1983). Herbivory: the dynamics of animal-plant interactions. Blackwell Scientific Publications, Oxford, 403pp.

Davies, A. (1988). The regrowth of grass swards. In: M.B. Jones & A. Lazenby (eds.) The Grass Crop. Chapman and Hall Limited, London, UK, 85-117.

Distel, R.A., E.A. Laca, T.C. Griggs & M.W. Demment (1995). Patch selection by cattle: maximization of intake rate in horizontally heterogeneous pastures. *Applied Animal Behaviour Science*, 45, 11-21.

Dumont, B., M. Petii & P. D'hour (1995). Choice of sheep and cattle between vegetative and reproductive cocksfoot patches. *Applied Animal Behaviour Science*, 43, 1-15.

Edwards, G.R., J.A. Newman, A.J. Parsons & J.R. Krebs (1996). The use of spatial memory by grazing animals to locate food patches in spatially heterogeneous environments: an example with sheep. *Applied Animal Behaviour Science*, 50, 147-160.

Forbes, J.M. (1999). Minimal total discomfort as a concept of the control of food intake and selection. *Appetite*, 3, 371.

Forbes, J.M. (2001). Consequences of feeding for future feeding. *Comparative Biochemistry and Physiology*, 128, 463-470.

Fulkerson, W.J. & K. Slack (1994). Leaf number as a criterion for determining defoliation time for *Lolium perenne*. 1. Effect of water-soluble carbohydrates and senescence. *Grass and Forage Science*, 49, 373-377.

Gerber, L.R., O.J. Reichman & J. Roughgarden (2004). Food hoarding: future value in optimal foraging decisions. *Ecological-Modelling*, 175 (1), 77-85.

Griffiths, W.M., J. Hodgson & G.C. Arnold (2003). The influence of sward canopy structure on foraging decisions by grazing cattle. I. Patch selection. *Grass and Forage Science*, 58, 112-124.

Han, G.D., B. Li, Z.J. Wei, J. Yang & H. Li (2000). Liveweight change of sheep under 5 stocking rates in *Stipa breviflora* desert steppe. *Grassland of China*, 1, 4-6.

Han, G.D., B. Li, Z.J. Wei, J. Yang, X. Liu & H. Li (1999). Plant compensatory growth in the grazing system of *Stipa breviflora* desert steppe. *Acta Agrestia Sinica*, 7 (1), 1-7.

Heady, H.F. & R.D. Child (1994). Rangeland ecology and management. Westview Press, Boulder, US, 155-168.

Hodgson, J. (1981). Variations in the surface characteristics of the sward and the short-term rate of herbage intake by calves and lambs. *Grass and Forage Science*, 36, 49-57.

Hodgson, J. (1990). Grazing management: science into practice. Longman, London, p1.

Howe, H.F. & L.C. Westley (1988). Ecological relationships of plants and animals. Oxford University Press, New York and Oxford, 271pp.

Howery, L.D., D.W. Bailey, G.B. Ruyle & W.J. Renken (2000). Cattle use visual cues to track food locations. *Applied Animal Behaviour Science*, 67, 1-14.

Iwasa, Y., M. Higashi & N. Yamamura (1981). Prey distribution as a factor determining the choice of optimal foraging strategy. *The American Naturalist*, 117, 710-723.

Kowalenko, B.L. & J.T. Romo (1998). Regrowth and rest requirements of northern wheatgrass following defoliation. *Journal of Range Management*, 51, 73-78.

Lemaire, L. & D. Chapman (1996). Tissue flow in grazed plant communities. In: J. Hodgson & A.W. Illius (eds.) The ecology and management of grazing systems. CAB International Publishing, Wallingford, 3-36.

Lemon, W.C. (1991). Fitness consequences of foraging behaviour in the zebra finch. *Nature*, 352, 153-155.

Li, G.D., P.D. Kemp & J. Hodgson (1997). Regrowth, morphology and persistence of grasslands Puna chicory (*Cichorium intybus L.*) in response to grazing frequency and intensity. *Grass and Forage Science*, 52 (1), 33-41.

Liu, J., D. Wang & L. Ba (2003). Studies on the leaf regrowth dynamics in *Leymus chinensis* under different cutting condition. *Journal of Northeast Normal University*, 35 (1), 117-124.

Liu, J., G.W. Cui, Y.J. Chen & Z.T. Zhang (1996). Comparing study on various grazing intensities of *Stipa breviflora* grassland in summer and fall. *Journal of Northeastern Agriculture University*, 27 (4), 370-375.

Liu, Y., D. Wang, S. Han, Z. Cheng, J. Du & X. Wang (2003). Variations of water-soluble carbohydrate and nitrogen contents in *Leymus chinensis* and *Phragmites communis* under different stocking rates. *Chinese Journal of Applied Ecology*, 14 (12), 2167-2170.

Liu, Y., D. Wang, X. Wang, L. Ba & W. Sun (2001). Studies on the leaf regrowth dynamics of three grasses in *Leymus chinensis* grassland under different grazing intensity. *Acta Prataculturae Sinica*, 10 (4), 40-46.

Liu, Y., D. Wang, X. Wang, L. Ba & W. Sun W. (2002). Comparison of leaf turnover for *Leymus chinensis* and *Phragmites communis* after grazing. *Chinese Journal of Applied Ecology*, 13 (5), 573-576.

Loeuille, N., M. Loreau & R. Ferriere (2002). Consequences of plant-herbivore coevolution on the dynamics and functioning of ecosystems. *Journal of Theoretical Biology*, 217, 369-381.

Manning, A. & M.S. Dawkins (1998). Animal behaviour. Cambridge University Press, UK, 193-255.

Martin, S.C. (1975). Stocking strategies and net cattle sales on semi-desert range. USDA Forest Service Research Paper, RM-146.

Newman, J.A., A.J. Parsons, J.H.M. Thornley, P.D. Penning & J.R. Krebs (1995). Optimal diet selection by generalist grazing hebivore. *Functional Ecology*, 9, 255-268.

O'Reagain, P. (2001). Foraging strategies on rangeland: effects on intake and animal performance. *Proceedings of the XIX International Grassland Congress*, Sao Pedro, Brazil, 277-284.

Perry, G. & E.R. Pianka (1997). Animal foraging: past, present and future. *TREE*, 12 (9), 360-364.

Prache, S. & J.L. Peyraud (2001). Foraging behaviour and intake in temperate cultivated grasslands. *Proceedings of the XIX International Grassland Congress*, Sao Pedro, Brazil, 309-319.

Provenza, F.D. (2003). Linking herbivore experience, varied diets, and plant biochemical diversity. *Small Ruminant Research*, 49, 257-274.

Roe, E.M. (1997). Viewpoint: On rangeland carrying capacity. *Journal of Range Management*, 50, 467-472.

Roguet, C., S. Prache & M. Petit (1998). Feeding station behaviour of ewes in response to forage availability and sward phenoligical stage. *Applied Animal Behaviour Science*, 56, 187-201.

Sibbald, A.M. & R.J. Hooper (2003). Trade-offs between social behaviour and foraging by sheep in heterogeneous pastures. *Behavioural Process*, 61, 1-12.

Stephens, D.W. & J.R. Krebs (1986). Foraging theory. Princeton University Press, Princeton, 170-182.

Tharmaraj, J., W.J. Wales, D.F. Chapman & A.R. Egan (2003). Defoliation pattern, foraging behaviour and diet selection by lactating dairy cows in response to sward height and herbage allowance of a ryegrass-dominated pasture. *Grass and Forage Science*, 58 (3), 225-238.

Thornton, B., P. Millard & U. Bausenwein (2000). Reserve formation and recycling of carbon and nitrogen during regrowth of defoliated plants. In: G. Lemaire, J. Hoghson, A. de Moraes, P.C. Carvalho & C. Nabinger (eds.) Grassland ecophysiology and grazing ecology. CAB International Publishing, Wallingford, UK, 85-100.

Ungar, E.D. (1996). Ingestive behaviour. In: J. Hodgson & A.W. Illius (eds.) The ecology and management of grazing systems. CAB International Publishing, Wallingford, UK, 185-218.

Villalba, J.J., F.D. Provenza & J.P. Bryant (2002). Consequences of the interaction between nutrients and plant secondary metabolites on herbivore selectivity: benefits or detriments for plants? *Oikos*, 97, 282-292.

Volenec, J J. 1986. Nonstructural carbohydrates in stem base components of tall fescue during regrowth. *Crop Science* 26: 122-131.

Wallis de Vries, M.F. & C. Daleboudt (1994). Foraging strategy of cattle in patchy grassland. *Oecologia*, 100, 98-106.

Wang, D, X. Teng, Y. Wang & Z. Cheng (2003). Variations in the plant heights on the artificial grassland with cow grazing. *Journal of Northeast Normal University*, 35 (1), 102-109.

Wang, D, X. Wang, Y. Liu & X. Teng (2002). The study on grazing ecology for animals on the Songnen grasslands. In: L. Wenhua & W. Rusong (eds.) Ecological safety and ecological construction. Meteorological Press, Beijing, 241-247.

Wang, D. (2004). Grassland ecology and management. Chemical Industrial Press, Beijing, 407pp.

Wang, S., Y. Wang & Z. Chen (2003). Management for grazing ecological systems. Science Press, Beijing, 13-44.

Wang, X., D. Wang, Y. Liu, L. Ba, W. Sun & B. Zhang (2002). Intake mass and diet selection for grazing goats on *Leymus chinensis* grassland. *Acta Ecologica Sinica*, 22 (5), 661-667.

Ward, D. (1992). The role of satisfying in foraging theory. *Oikos*, 63, 312-317.

Wei, Z.J. & G.D. Han (1995). The influence of grazing intensity on sheep performance. *Acta Agrestia Sinica,* 1, 22-28.

Wu, Z. (1990). Vegetation of China. Science Press, Beijing, 505-582.

Zhu, T. (1993). Grasslands of China. In: R.T. Coupland (ed.) Ecosystems of the World (8B): Natural Grasslands. Elsevier, Amsterdam, 61-82.

Strategies to mitigate seasonality of production in grassland-based systems

C. Porqueddu[1], S. Maltoni[1] and J.G. McIvor[2]
[1]CNR-ISPAAM, Sezione di Sassari, Via E. De Nicola, 07100, Sassari, Italy
Email: c.porqueddu@cspm.ss.cnr.it
[2]CSIRO Sustainable Ecosystems, 306 Carmody Road, St Lucia, Queensland 4067, Australia

Key points

1. Fertilisation use and manipulation can cost-effectively alter species composition, increase seasonal herbage production and improve herbage quality.
2. Choice of suitable grassland species, varieties and mixtures offers opportunity to mitigate limitations of seasonal grassland production.
3. Special purpose fodder crops, cereals, shrubs and trees offer alternative or supplementary feed sources.
4. Manipulation of stocking rates, grazing systems, transhumance and pasture management at various times of the season are significant advantageous options.
5. Integration of different strategies is essential to mitigate seasonality in systems of animal production that must be inherently more sustainable over a longer time frame.

Keywords: sward manipulation, animal production, fertilisers, plant choice, resource integration

Introduction

Seasonal patterns of herbage production and quality, result from variations in climate (temperature, rainfall and radiation) and environmental conditions (e.g. altitude, soil depth and fertility). Seasonality is an important limit to animal production, and total annual grassland production can be less meaningful than its distribution throughout the year.

Areas with large differences in seasonal rainfall (Mediterranean areas and the semi-arid tropics) exhibit the most extreme patterns of seasonality, but is also important in humid temperate areas where in general terms, forage production is mainly limited by low winter temperatures and occasionally by summer drought. In Mediterranean environments, the most critical periods for grassland production are during summer/early autumn (moisture limited) and winter (temperature limited); moreover, forage quality may be low during summer/early autumn. In semi-arid tropical areas, the critical period is the dry season (when both pasture quality and quantity may be limiting), and the early part of the 'wet' season until sufficient new herbage is available for animals.

This paper considers the opportunities for, and limitations to reducing seasonality of forage production and quality. It highlights strategies to match forage production and animal requirements, and strategies aimed at optimising combinations of feed resources to overcome critical periods and reduce production costs e.g. for fodder conservation. These problems require a system-based, farm-specific approach, since every farm is unique in terms of environment, available resources and management objectives.

Changing how pastures grow

Fertiliser

Fertilisation can cost-effectively alter species composition, increase herbage production, and improve seasonal growth rate and herbage quality. Fertilisers are mostly used in intensive temperate grassland systems, but they are also important for some sown tropical pastures and can be used to achieve higher levels of production from natural Mediterranean pastures (Seligman, 1996).

Early spring growth for grazing can be improved by strategic application of fertiliser nitrogen (N). However, periods of active growth are also those periods of greatest response to fertiliser N. If N is applied too early there is little yield response; if it is applied too late, pasture growth is not fully exploited. Therefore application dates that produce the highest N use efficiency and give the greatest total yield response, will not necessarily deliver herbage biomass 'out of season' for extended grazing, and may produce undesired effects such as excess biomass in spring (Morrison, 1987). Moreover, following an initially large herbage mass response, negligible and finally depressed yield responses to autumn N may occur due to the effect of legume depletion of the swards (Elliott & Abbott, 2003).

Precise weather forecasts and methods for predicting the best timing of N application in late winter to provide early spring grazing (e.g. 'T-sum 200') have been suggested (Frame, 2001). In these areas, there is also potential to extend the grazing season in autumn/early winter. Time of fertiliser application is important, e.g. in Ireland, when fertiliser was applied after late September on previously grazed swards, the response was poor and the probability of N loss by leaching and denitrification increased (Laidlaw *et al.*, 2000). Herbage grown from late autumn N applications can be prone to winter damage, especially if ryegrass swards are not well grazed, or if grazing is delayed (Frame, 2000).

Phosphate is important for maintaining the legume content of pastures, thus improving sward quality, promoting N-fixation and increasing sustainability of grassland-based systems. In New Zealand superphosphate increased legume production (predominantly *Trifolium repens* – white clover) by up to 72%, with most response in summer when improved feed quality is valuable for finishing lambs (Roach *et al.*, 1996). In Mediterranean conditions, Bullitta *et al.* (1987) showed that N-P fertilisation increased autumn-winter pasture production from 20% to 50% of total annual yield, giving a 1-2 month longer period of utilisation, and reduced year-to-year yield variation.

Farmyard manure and animal slurry are important sources of N and other nutrients. In temperate pastures, nutrients from animal slurry (particularly potash and N) are most effective when applied in early spring (Humprhreys *et al.*, 1997).

Pasture species, varieties and mixtures

In areas with high production potential but undesirable sward botanical composition, productivity and seasonal distribution can be improved by species introduction through various methods of oversowing. Breeding pasture species and varieties for increased growth at the start and/or end of the growing season has been, and continues to be an important objective in selection criteria (Gonzalez-Rodriguez *et al.*, 1996).

Low temperatures usually limit early spring growth. In temperate and sub-tropical areas species or varieties able to grow faster at this time of year have been selected e.g. *Lolium perenne* (perennial ryegrass) cv. Aberalan (Wilkins, 1995) and *Cenchrus ciliaris* (buffel grass) cv. Viva (Hacker, 1994) respectively. In several countries, Recommended Variety Lists contain information on seasonal yield distribution. In temperate climates *Lolium multiflorum* (Italian ryegrass), *Lolium hybridum* (hybrid ryegrass) and *L. perenne* are the most valuable species for early bite. *Phleum pratense* (Timothy) can produce valuable early growth in cold and wet upland conditions.

Herbage production and quality can be higher in perennial species that remain growing or green longer into the late spring and early summer in Mediterranean areas, e.g. *Dactylis glomerata* (cocksfoot) or the dry season in the tropics (McIvor, 1982). Under Mediterranean conditions deep root systems are important for maintaining high growth rates at these times. Preliminary results using mixtures of different functional types (grass/legume and annual/perennial) promise to extend the growing season and reduce variability between and within years (Porqueddu & Maltoni, 2004). The full potential of perennial grasses in Mediterranean environments has yet to be realised, particularly in drier regions, and will require greater exploitation of native germplasm, with an emphasis on selecting characteristics such as summer dormancy. Warm season grasses, both in pure stands or in mixtures with cool season species, can extend forage production in Mediterranean areas, e.g. *Eragrostis curvula* (weeping lovegrass), *Panicum virgatum* (switchgrass) and *Sorgastrum nutans* (Indian grass) showed good winter survival and high summer growth rates (Bonciarelli *et al.*, 1990). However, there are problems regarding forage quality, seed availability and cost, lack of evidence on their reliability in marginal areas and possible weed potential (Humphries *et al.*, 1991). For example *E. curvula* has been listed as a weed in Australia (Batianoff & Butler, 2002).

New thinking is required about forage herbs, e.g. the inclusion of *Cichorium intybus* (chicory) in grass/legume swards increased late season production (Kunelius & McRae, 1999). *C. intybus* was more productive than *Phalaris acquatica* (bulbous canary grass) in the warmer part of the year (Kemp *et al.*, 2002), and fine-leaved forbs such as *Plantago lanceolata* (narrowleaf plantin) and *Taraxacum officinale* (dandelion) can increase and stabilise the yield of unfertilised grass-clover swards (Isselstein, 2002).

Inter-seeding perennial cool-season cultivars with winter wheat can provide grazing during the first season and still allow good establishment of the perennial species (Kindiger & Conley, 2002). In temperate climates, subtropical grasses such as *Digitaria sanguinalis* (large crabgrass) and *Paspalum dilatatum* (dallisgrass) are used in mixture with temperate species to give higher summer yields. However, in New Zealand, subtropical grasses increase summer and autumn production, but may compete with species such as *L. multiflorum* and *T. repens,* thus reducing nitrogen fixation and rates of pasture production in the cooler season (Baars *et al.*, 1991).

Retention of nutritive value at the end of the growing season is important. The rate of decline in herbage quality with maturity is slower for legumes than grasses, and legume-based pastures can support higher animal growth rates than grass pastures at this time (Gardener *et al.*, 1993). Gloag *et al.,* (2004) found that *T. vesiculosum* (arrowleaf clover) extended the growing season by 3-4 weeks, had higher digestibility than *T. diffusum* (diffuse clover) and *T. subterraneum* (subterranean clover) from mid-December onwards, and could increase lamb liveweights in late summer by more than 10%.

A possible way to extend seasonal production is to use species with secondary compounds that limit intake in early growth stages, e.g. in late spring when the herbaceous species dry up, *Bituminaria bituminosa* (pitch trefoil) remains green and is intensively grazed by cattle (Gutman *et al.*, 2000). For the same reason some annual legumes are not readily grazed at flowering but are well utilised in late spring and are preferred to the dry pasture in summer, e.g. *Biserrula pelecinus* (biserrula) and *Melilotus alba* (white sweetclover).

'Forage chains' can be devised using species and varieties with different maturities. These may include local ecotypes, although multiplication constraints sometimes hinder their wide-scale use (Roggero *et al.*, 1990). Bullitta *et al.* (1991) set up simple forage chains (e.g. 3 rings = 3 sown pastures) for sub-climatic regions of Italy. They identified local ecotypes of annual legumes *Medicago polymorpha* (burr medic) and *T. subterraneum*, and perennial legumes, *Lotus corniculatus* (birdsfoot trefoil) and *L. tenuis* (narrowleaf birdsfoot trefoil) that allowed both spring utilisation and extension of autumn utilisation. Unfortunately, such theoretical studies are sometimes difficult to transfer to farm level, and successful use of temperate perennial cultivars has only been achieved in areas with limited summer drought (e.g. France). Clearly in devising such forage chains, some consideration need to be paid to factors other than pasture species. Many species growing as weeds on cultivated land are grazed by animals. Their exploitation and introduction into forage chains could improve forage resources and feed supply e.g. use of *Chrysanthemum coronarium* (gardland chrysanthemum) (Sulas & Caredda, 1997).

Fodder crops and special purpose pastures

These are used as an alternative to permanent grasslands. A novel special purpose pasture used on a limited scale in central Queensland is the growth of flood tolerant grasses (*Brachiaria mutica, Hymenachne amplexicaulis, Echinchloa polystachya*) in shallow ponds to provide green herbage during the dry season, when dryland areas have matured and senesced, and provide only low quality herbage.

Annual forage crops can complement pasture production in Mediterranean and temperate climates, as some species show high growth rates in winter and/or in late summer/early autumn. Under Mediterranean conditions, mixtures of cereals, *L. multiflorum* and annual legumes (e.g. *Vicia spp.*, *T. incarnatum*) provide winter grazing. *Hedysarum coronarium* (sulla) has recently re-gained interest for its flexible utilisation. In dairy-sheep forage systems, using two varieties with different growth patterns allows grazing in early autumn and late spring. Sulas *et al.* (1997) suggest mixed sward management entailing winter grazing and hay or silage making during spring for the first year, and year-round grazing in the second year.

In Sardinian cereal-based farming systems, the winter growth and forage quality of local landraces of barley and oats are often exploited by grazing up to mid-February, with almost no negative effects on grain yield. Special purpose pastures with a productive life of 2-3 years may occupy a small proportion (10-20%) of a farm. Kennedy *et al.* (2004) showed that special purpose pasture mixes can increase lamb seasonal carrying capacity and improve digestibility in late spring.

Secale cereale (rye) is the most winter-hardy small grain and the first to break dormancy in spring. It shows potential to extend grazing into late autumn and winter (Samples *et al.*, 1996). Combining *S. cereale* with stockpiled perennial pastures can reduce reliance on high-cost stored feeds, but *S. cereale* is too competitive for water and minerals to be used in mixtures (A. Hopkins, *pers. comm*). Mixtures of cultivars with different maturities can be sown together to prolong the grazing season, e.g. the combination of different *S. cereale* cultivars with *X triticosecale* (triticale) in Switzerland for green forage production (Mosimann, 1998).

Forage *Brassica spp.* are highly palatable, energy rich feeds. Their high digestibility does not markedly decrease with maturity, and their ability to retain leaves makes them suitable for stockpiling, or for late summer and autumn/early-winter grazing (Koch *et al.*, 2002). Brassicas and spring cereals can be combined, e.g. *B. rapa* and spring oats for late autumn and winter grazing, and adding *S. cereale* for early spring grazing.

Fodder shrubs and trees

The contribution of fodder shrubs and trees to livestock diets varies between regions. In less favourable environments the contribution of native shrubs to animal diets is very high (Papanastasis *et al.*, 1997). In more favourable areas, shrubs are generally used on a seasonal basis. In the tropics, fodder trees and shrubs (e.g. *Leucaena*) can be particularly productive and produce high quality feed during the dry season as a result of their deep rooting characteristics.

Fodder trees have a dual role in mitigating seasonality of pasture production: they provide fodder when herbage is not available, and they complement ruminant diets with protein and energy at times when grass is rich in fibre, enabling animals to make better use of dry pastures or cereal stubble (Abel *et al.*, 1997). Moreover, their growth is more stable between years than annual or even perennial forages. In Mediterranean regions woody fodder species (including *Quercus suber* (cork oak), are key components in traditional silvopastoral systems, e.g. 'Dehesa' (Spain) and 'Montado' (Portugal). In these systems, animal requirements are met by a combination of grasslands, crops, shrubs and trees. When oaks are pruned, the leaves and fine branches can overcome summer and winter forage gaps, while autumn acorns are reserved for fattening pigs that graze rotationally on the same pastures as cattle, sheep and goats. Other woody species (e.g. *Retama sphaerocarpa*) provide some green feed, including pods and seeds, when all other feed is dry. A greater role could be played by these native shrubs and trees if they were better studied (e.g. role of anti-nutritional characteristic) and managed (Porqueddu & Sulas, 1998). Based on experimental results, several authors have suggested introducing fodder shrubs and trees (e.g. *Atriplex* spp., *Chamaecytisus palmensis, Salsola vermiculata, Opuntia ficus-indica*) into forage systems in the Mediterranean basin and other regions (McMillan *et al.*, 2002). However, in several cases they failed at field level because of establishment difficulties, high planting costs, low persistence under heavy grazing, and low palatability.

Other strategies to change how pastures grow

Irrigation can provide high quality feed to fill seasonal feed gaps, e.g. *M. sativa* for hay and summer grazing in Mediterranean areas, irrigated permanent pastures for dairy cows in southern Australia and New Zealand, and irrigated temperate pastures during the cooler dry season in sub-tropical areas. In Mediterranean regions, occasional irrigation of pastures in late summer and early-autumn can provide early autumn feed or sustain pasture production after a false start to the growing season; even low water volumes can significantly increase and stabilise autumn production. Moreover, irrigation can give opportunities for double cropping. For example *L. multiflorum* is often grown for late autumn/winter/early spring feed in rotation with a summer-annual grass e.g. *Sorghum vulgare* (forage sorghum), *Pennisetum glaucum* (pearl millet). Orloff and Drake (2001) described systems using *X triticosecale* and/or mixtures of *X triticosecale* and *L. multiflorum* for autumn and early spring grazing, to complement cool season perennial irrigated pastures for the High Plains and intermountain region of the USA. However community concerns may lead to a reduction in water available for pasture irrigation.

Fire has many uses (e.g. control of woody plants and weeds, altering grazing distribution, pest control) and can influence seasonality of animal production by changing pasture composition and the availability of high quality feed. Fire removes mature, low quality, dry material so that the young, high quality regrowth is more available to animals, and diet quality is improved. In the central Great Plains in the USA, burning in late spring reduced undesirable grasses, and increased the desired warm-season species, thus alleviating summer forage imbalances (Mitchell *et al.*, 1996).

Application of herbicides (spray-topping) to annual grasses at seed head emergence can delay loss of digestibility during late spring and summer. Gatford *et al.* (1999) in Southern Australia, showed that the loss of digestibility of stems, leaves and seed heads was slowed, but not that of leaf blades. Loss of digestibility was delayed by 5 weeks, but pasture yield decreased by 45%.

Changing how pastures are utilised

Stocking rate

The choice of stocking rate is one of the most important grazing management decisions. Low stocking rates can prolong the grazing season and also allow maximum selection by animals, so they can select the highest quality diet possible from the herbage available. This enables high production per head but production per hectare is low due to the low levels of utilisation (Troxler & Chassot, 2004). Constantly high stocking rates in areas with strong fluctuations in forage availability are dependent on forage resources from outside the farm, through transhumance or purchase of fodder or other supplements. Stocking rates can be set at moderate levels and varied seasonally by varying animal numbers, buffer grazing or harvesting surplus forage as hay or silage. Flexible management is required to closely match grass supply and stocking rate, e.g. using sward surface height guidelines (Frame, 2000). An intermediate choice is a constant moderate stocking rate, where periods of reduced pasture growth are compensated through purchase or cultivation of forage, forage storage and conservation, and use of standing crops.

Grazing systems

There is a lot of interest in grazing systems throughout the world. Generally, fertilisation, species choice and stocking rate have a much greater impact on animal productivity than changes in the grazing method (Waller *et al.*, 2001). Animal performance under rotational (intense defoliation with low frequency) and continuous grazing (frequent but light defoliation) is generally similar at low stocking rates, with rotational systems exceeding continuous grazing only at high stocking rates. Many producers claim large increases in animal production when adopting 'cell' grazing. Although this increase could be due to other factors (in addition to grazing system), the improved uniformity of utilisation with smaller paddocks is a reasonable explanation of the increases in production (Norton, 1998).

Strip grazing enables a greater control over pasture intake than longer rotations or continuous grazing. This allows more feed to be conserved to carry animals through periods of feed shortage e.g. Doyle (1999).

Saving pastures as standing hay for out-of-season grazing is a well known practice, e.g. summer/autumn-saved pasture for *in situ* winter feeding in temperate areas, and spring-saved pasture in Mediterranean areas (Sulas *et al.*, 1995). In *T. subterraneum* based pastures of the 'Dehesa' agro-silvopastoral system, summer grazing relies on dry pasture deferred from spring with pods and seeds providing an important protein source. Not all cool-season species are adapted to stockpiling. Frame (2000) indicated *D. glomerata*, *Festuca pratensis* and *P. pratense* as the most suitable species for autumn-saved pasture. In the USA, stockpiled perennial grasses were suitable for maintaining beef cattle, dry dairy cows and sheep over winter, but for animals with higher nutrient requirements or under extreme environmental conditions, protein and energy supplementation was required (Hedtcke *et al.*, 2002). While stockpiling can provide low cost feed for the autumn and early winter months, it delays spring recovery of the stockpiled plots. An alternative to stockpiling is windrow or swath grazing where animals graze directly from windrow-stored forage, usually together with packaged hay. This can lower livestock production costs, and maintain herbage yield, diet quality and meat gains (Volesky *et al.*, 2002).

There is generally a trade off between evenness of production and total herbage production. Frequency of grazing can have a major influence on the pattern of pasture growth. More intense or more frequent defoliations interrupt the reproductive development, reducing seasonal variation by lowering the spring or summer peak, but also lowering total production (Korte & Harris, 1987). To accumulate herbage to build up a reserve ('bank') for autumn/winter use, it is recommended that the rotation length increase gradually. If the rest interval is excessive, herbage mass may be reduced because the rate of senescence approaches or exceeds the rate of leaf growth. Excessive accumulation of *L. perenne* herbage (over 2 t DM/ha) into winter, adversely affected forage quality and the state of the sward in spring, resulting in a decline in tiller density (Laidlaw *et al.*, 2000). Saved pastures should be changed from year to year to avoid possible negative effects on forage quality and sward composition (Gonzales-Rodriguez *et al.*, 1996).

Autumn and winter sward management will influence earliness of spring growth. Fields designated for early grazing should be grazed first and then shut off for the rest of the winter. Although O'Kiely & O'Riordan (1994) in Ireland recorded a reduced herbage mass in spring swards either cut or grazed in early winter (December) compared with those remaining ungrazed since autumn, later studies (Laidlaw & Mayne, 2000) observed that early winter defoliation had no effect on subsequent spring herbage mass. In any case, the detrimental effect of late defoliation on spring growth may be compensated for by increasing N application (Binnie *et al.*, 2001).

Grass growth in midsummer is influenced by the intensity of defoliation earlier in the year. Close early grazing ensures the development of densely tillered swards capable of leafy growth into summer, since the ensuing tillers will be vegetative rather than reproductive.

Considerable research effort is being devoted to 'side-by-side' grazing, based on animals' free choice of adjacent grass and clover monocultures. Preliminary results by Venning *et al.* (2004) showed that lambs grazing *L. perenne* and *T. subterraneum* sown as 'side-by-side monoculture', grew faster than those on mixed grass/clover pasture (20% clover) or ryegrass monoculture, but not on clover monoculture. Side-by-side grazing allows the use of grassland species and varieties that do not grow well in mixtures and could also be used for a range of plant species.

'Zero grazing' may be used for special purpose crops or for short periods in early or late season to avoid poaching of swards (Frame, 2000). It is based on cutting and carting forage to housed or partially housed stock, and involves high capital costs of machinery but gives a higher efficiency of forage utilisation.

Transhumance

This is a well-known solution to seasonality in several regions of the world. Traditionally, the heterogeneity of the vegetation is exploited by early-season grazing on warmer slopes and sites, late-season grazing on cooler and wetter sites, and dry-season utilisation of deeper-rooted perennial species, evergreen shrubs and other woody species. There is also integrated use of high (mountain) pastures in summer and lowland forage resources at other times. Transhumance grazing systems in temperate regions of Asia have been reviewed by Suttie & Reynolds (2003). Transhumance has potential to be integrated in multifunctional grassland systems in less favoured areas e.g. by reducing biomass to prevent summer wildfires and summer grazing of skiing lines. Moreover, the integrated and strategic use of upland pasture may confer added value to traditional and high quality animal products, through the exploitation of local breeds and knowledge.

Animal management

Matching animal biology and pasture growth

Seasonal grazing with dairy cows that calve just before the spring growth peak, and conserving surplus herbage as hay or silage, can produce economic benefits by reducing costs (Winsten & Petrucci, 2003). However, the successful application of seasonal grazing requires a high level of management skills, a thorough knowledge of animal husbandry, a keen observation of pasture growth and flexible time management.

In many extensive grazing systems, producers accept that animals will gain weight during the growing season and then lose weight at other times of the year. Rapid gains following a period of weight loss may compensate for the loss. However, care must be taken that body condition does not fall below a minimum threshold, as under-nutrition in early or critical life stages may reduce later performance. There have been contrasting results according to animal species, breed and category, type of production and grazing system (Stockdale, 2001; Hennessy & Morris, 2003).

Animal breeds

Differential selection between animal types under mixed grazing can greatly modify sward state, and may benefit species with a wider growing season, e.g. *T. repens* content is greatly affected by grazing regime (cattle > mixed > sheep) influencing availability of forage during summer (Nolan *et al.*, 2001). Animal breeds in extensive farming systems need to be matched to environmental limitations. The short green season is one of the main factors for the preponderance of small ruminants in the Mediterranean basin and in most other dry Mediterranean-type regions. *Bos indicus* animals are better adapted to tropical conditions than *Bos taurus* animals – the better adapted animals forage further and have a higher forage intake (Siebert, 1982). If local breeds are to have a role in sustaining grassland-based systems a critical analysis of inputs, feeding policies and quality of the products must be carried out (Mills *et al.*, 2004).

Supplements and alternative/additional feed stuffs

Nutritional supplements (energy, protein or N, and minerals) can provide nutrients that may be limited in the herbage. In dry tropical areas, pastures growing on phosphorus (P) deficient soils based on *Stylosanthes* spp. can produce large quantities of herbage during the wet season, with reasonable levels of digestibility and protein but inadequate P levels. Thus, supplementation of animals with P during this time can greatly increase weight gains. During the dry season large quantities of herbage with low protein levels and digestibility are often available, but rumen activity of grazing animals is low. Supplementation with protein or urea (and sometimes molasses) can counteract this leading to increases in intake of dry pasture; this can reduce or eliminate dry season weight losses and reduce mortalities.

Molasses is important in Australia and elsewhere in the tropics where it is locally available and is widely used for survival feeding, and may have an increasing role in production feeding (Hunter, 2000).

The presence of tannins in many shrubs and tree leaves, limits their utilisation as animal feeds. Decandia *et al.* (2000) showed polyethylene glycol (PEG) increased the proportion of tannin-rich species in the diet of goats browsing a Mediterranean shrubland during summer, allowing better utilisation of woody species, increasing crude protein digestibility and enhancing the efficacy of concentrate supplementation.

Integration

Overcoming seasonality of grassland production has been, and will continue to be a challenge for land managers. In extensive areas (e.g. semi-arid tropics) there is often acceptance of seasonality and little attempt to overcome it. In contrast in intensive areas (e.g. moist temperate regions) there has been much more study and practice of overcoming seasonality, sometimes leading to farming systems potentially 'challenging' to the environment. However, recent trends on sustainability, product and environmental quality and animal welfare are veering towards mitigation strategies rather than overcoming seasonality.

The challenge is to integrate what is available to each farm considering local constraints, the objectives for the farm, and consumers' requirements and demands. Farmers are in the best position to do this. Since there are many management factors, climates, variable prices etc., computer models can play a role in optimising best management policy. Many models have been developed although they are little used or validated in practice.

Management practices have associated risks and costs, e.g. establishing sown pastures over large areas is expensive if cultivated seedbeds are used, compared with lower costs if species can be established by surface sowing without soil preparation. However, this decrease in costs is accompanied by an increase in the risk of establishment failure (Cook *et al.*, 1993).

Grasslands are inextricably linked with crops and their residues in many farming systems with numerous examples of mixed farms with the integration of pasture, annual forage crops and cereals (Roggero *et al.*, 1996; Talamucci & Pardini, 1999). For efficient grassland-based forage systems, integrating more than one feed resource will be necessary to mitigate seasonality of production and achieve satisfactory feed quality. This adds complexity to the farming system, but flexible though complex systems of resource use, may be inherently more stable and sustainable over a longer time frame.

Acknowledgements

The authors wish to thank Dr. John Frame and Prof. Salvatore Caredda for their critical reading of this paper and helpful suggestions.

References

Abel, N., J. Baxter, A. Campbell, H. Cleugh, J. Fargher, R. Lambeck, R. Prinsley, M. Prosser, R. Reid, G. Revell, C. Schmidt, R. Stirzaker & P. Thorburn (1997). Design principles for farm forestry: a guide to assist farmers to decide where to place trees and farm plantations on farms. Barton, ACT, Rural Industries Research and Development Corp., Canberra, 102pp.

Baars, J.A., G.J. Goold, M.F. Hawke, P.J. Kilgarriff & M.D. Rollo (1991). Seasonal patterns of pasture production in the Bay of Plenty and Waikato. *Proceedings of the New Zealand Grassland Association,* 53, 67-72.

Batianoff, G.N. & D.W. Butler (2002). Assessment of invasive naturalized plants in south-east Queensland. *Plant Protection Quarterly*, 17, 27-34.

Binnie, R.C., C.S. Mayne & A.S. Laidlaw (2001). The effects of rate and timing of application of fertiliser nitrogen in late summer on herbage mass and chemical composition of perennial ryegrass swards over the winter period in Northern Ireland. *Grass and Forage Science,* 56, 46-56.

Bonciarelli, F.Z., A. Cardinali & R. Santilocchi (1990). Prove di adattamento di specie erbacee foraggere ad accrescimento "differito" (Adaptation trials on herbaceous forage species with differentiated growth). *Agricoltura Ricerca*, 106, 81-92.

Bullitta, P., S. Caredda & G. Rivoira (1987). Influenza dell'andamento metereologico e della concimazione azotata sulla produttività totale e stagionale di un pascolo in Sardegna (Climatic trend and nitrogen fertilisation influence on total and seasonal productivity of Sardinian pastureland). *Agronomia*, 2, 146-151.

Bullitta, S., M. Falcinelli, S. Lorenzetti, V. Negri, A. Pardini, S. Piemontese, C. Porqueddu, P.P Roggero, P. Talamucci & F. Veronesi (1991). Prime osservazioni su specie perenni ed annua autoriseminanti in vista della organizzazione di catene di foraggiamento in ambienti mediterranei (First observation on perennial and annual self-reseeding species for the organisation of forage chains for the Mediterranean environment). *Rivista di Agronomia*, 2, 220-228.

Cook, S.J., R.L. Clem, N.D. MacLeod & P.A.Walsh (1993). Tropical pasture establishment. 7. Sowing methods for pasture establishment in northern Australia. *Tropical Grasslands*, 27, 335-343.

Decandia, M., M. Sitzia, A. Cabiddu, D. Kababya & G. Molle (2000). The use of polyethylene glycol to reduce the anti-nutritional effects of tannins in goats fed woody species. *Small Ruminant Research*, 38, 157-164.

Doyle, P.T. (1999). Strip grazing to control wool growth rate of sheep grazing green annual pastures. *Australian Journal of Experimental Agriculture*, 39, 247-258.

Elliott, D.E. & R.J. Abbott (2003). Nitrogen fertiliser use on rain-fed pasture in the Mt Lofty ranges, south Australia. 1. Pasture mass, composition and nutritive characteristics. *Australian Journal of Experimental Agriculture*, 43, 553-577.

Frame, J. (2000). Improved grassland management. Farming Press, Tonbridge, UK, 353p.

Frame, J. (2001). Extending the grazing season. *Forage matters*, spring 2001, 1-3.

Gardener, C.J., M.R. McCaskill & J.G. McIvor (1993). Herbage and animal production from native pastures, and pastures oversown with *Stylosanthes hamata*. 1. Fertiliser and stocking rate effects. *Australian Journal of Experimental Agriculture*, 33, 561-570.

Gatford, K.L., R.J. Simpson, C. Siever-Kelly, B.J. Leury, H. Dove & T.A. Ciaravella (1999). Spray-topping annual grass pasture with glyphosate to delay loss of feeding value during summer. I. Effects on pasture yield and nutritive value. *Australian Journal of Agricultural Research*, 50, 453-64.

Gloag, C., A. Thompson, A. Kennedy & K. Venning (2004). Long season annual legumes to increase lamb production. Proceedings of the Grassland Society of Victoria, Albury, July 2003, 93pp.

Gonzalez-Rodriguez, A., J. Pineiro & M. Cropper (1996). Extending the grazing season. In: G. Parente, J.Frame & S. Orsi (eds.) Grassland and Land use systems - *Proceedings of the 16th EGF Meeting,* Grado (Gorizia), Italy, (September 15-19), 941-948.

Gutman, M., A. Perevolotsky & M. Stenrberg (2000). Grazing effects on a perennial legume *Bituminaria bituminosa* (L.) Stirton, in a Mediterranean rangeland. In: L. Sulas (ed.) Legumes for Mediterranean forage crops, pastures and alternative uses - Proceedings of the 10th Meeting FAO-CIHEAM, Sassari, Italy, (April 4-9), 299-303.

Hacker, J.B. (1994). Buffel grass *Cenchrus ciliaris* 'Viva'. *Plant Varieties Journal*, 7, 31-32.

Hedtcke, J.L., D.J. Undersander, M.D. Casler & D.K. Combs (2002). Quality of forage stockpiled in Wisconsin. *Journal of Range Management*, 55, 33-42.

Hennessy, D.W. & S.G. Morris (2003). Effect of preweaning growth restriction on the subsequent growth and meat quality of yearling steers and heifers. *Australian Journal of Experimental Agriculture*, 43, 335-341.

Humphries, S.E., R.H. Groves & D.S. Mitchell (1991). Plant invasions: the incidence of environmental weeds in Australia. *Kowari*, 2, 1-134.

Humphreys, J., T. Jansen, N. Culleton, F. MacNaeidhe, M.B.H. Hayes & W.S. Wilson (1997). Comparative effects of organic manures on spring herbage production of a grass/white clover (cv. Grasslands Huia) sward. In: M.H.B. Hayes (ed.) Humic substances in soils, peats and waters: health and environmental aspects. *Royal Society of Chemistry; Cambridge,* UK, 446-461.

Hunter, R.A. (2000). High molasses diets for intensive finishing of steers. *Asian-Australia Journal of Animal Science*, 13 (B), p112.

Isselstein, J. (2002). Effect of perennial ryegrass variety on the performance of grass/clover and grass/clover/forb mixtures. In: J.L. Durand, J.C. Emile, C. Huyghe & G. Lemarie (eds.) Multi-function grasslands. Grassland Science in Europe, 7, 312-313.

Kemp, D.R., D.L. Michalk & M. Goodacre (2002). Productivity of pasture legumes and chicory. *Australian Journal of Experimental Agriculture*, 42, 15-25.

Kennedy A. J., C. Gloag , A.N. Thompson & L. North (2004). High performance pasture systems to increase lamb production in southwest Victoria. *Proceedings of the Australian Society of Animal Production*, Melbourne, July 2004, 273p.

Kindiger, B. & T. Conley (2002). Competition and survival of perennial cool-season grass forages seeded with winter wheat in the southern Great Plains. *Journal of Sustainable Agriculture*, 21, 27-45.

Koch, D.W., C. Kercher & R. Jones (2002). Fall and winter grazing of *brassicas*, a value-added opportunity for lamb producers. *Journal of Sheep and Goat Research*, 17:1-13.

Korte, C.J. & W. Harris (1987). Effect of grazing and cutting. In: R.W. Snaydon (ed.) Ecosystems of the World 17B. Managed grassland analytical studies. Elsevier, 71-77.

Kunelius, H.T. & B. Mc Rae (1999). Forage chicory persists in combination with cool season grasses and legumes. *Canadian Journal of Plant Science,* 79 (2), 197-200.

Laidlaw, A.S. & C.S. Mayne (2000). Setting management limits for the production and utilization of herbage for out-of-season grazing. *Grass and Forage Science*, 55, 14-25.

Laidlaw, A.S., C.J. Watson & C.S. Mayne (2000). Implications of nitrogen fertilizer applications and extended grazing for the N economy of grassland. *Grass & Forage Science*, 55, 37-46.

McIvor, J.G. (1982). Agronomic potential of introduced grasses grown in north-east Queensland. CSIRO Tropical Agronomy Technical Memorandum No. 29, p14.

McMillan, Z., C.B. Scott, C.A. Taylor & J.E., Huston (2002). Nutritional value and intake of prickly pear by goats. *Journal of Range Management*, 55, 139-143.

Mills, J., N.H. Yarrow, M. Lherm, K.U. Röver & D. Pasut (2004). Effect of livestock breed and stocking rate on sustainable grazing systems: 6. Socio-economic impacts. In: A. Lüscher, B. Jeangros, V. Kessler, O. Huguenin, M. Lobsiger, N. Millar & D. Sutter (eds.) Land use systems in grassland dominated regions. Grassland Science in Europe, 9, 629-631.

Mitchell, R.B., R.A. Masters, S.S. Waller, K.J. Moore & L.J. Young (1996). Tallgrass prairie vegetation response to spring burning dates, fertilizer, and atrazine. *Journal of Range Management*, 49, 131-136.

Morrison, J. (1987). Effects of nitrogen fertilizer. In: R.W. Snaydon (ed.) Ecosystems of the World 17B. Managed grassland analytical studies. Elsevier, 61-70.

Mosimann, E. (1998). Rye and triticale cut as green forage. *Revue Suisse d'Agriculture*, 30: 2, 57-59.

Nolan, T., J. Connolly & M. Wachendorf (2001). Mixed grazing and climatic determinants of white clover (*Trifolium repens* L.) content in a permanent pasture. *Annals of Botany*, 88 (Special Issue), 713-724.

Norton, B.E. (1998). The application of grazing management to increase sustainable livestock production. *Animal Production in Australia*, 22, 15-26.

O'Keily, P. & E.G. O'Riordan (1994). Effects of defoliating grass in winter or spring on first cut yield and grass and silage quality, and on second cut yield. *Proceedings of the 4th research conference,* British Grassland Society, Reading, UK, 55-56.

Orloff, S.B. & D.J. Drake (2001). A grazing and haying system with winter annual grasses. *Proceedings 31st California Alfalfa and Forage Symposium,* Modesto, CA, (Dec. 11-13), 143-150.

Papanastasis, V.P., P.D. Platis & O. Dini-Papanastasis (1997). Productivity of deciduous woody and fodder species in relation to air temperature and precipitation in a Mediterranean environment. *Agroforestry Systems*, 37, 187-198.

Porqueddu, C. & S. Maltoni (2004). Evaluation of a range of rainfed grass-legume mixtures in a Mediterranean environment. In: B.E. Frankow-Lindberg (ed.) *Proceedings of the Workshop on Adaptation and management of forage legumes - strategies for improved reliability in mixed swards.* Ystad, Sweden, September 20-22, 2004 (in press).

Porqueddu, C. & L. Sulas (1998). Mediterranean grassland systems. In: G. Nagy, K. Peto (eds.) Ecological aspects of grassland management. Grassland Science in Europe, Debrecen, Hungary, (May 18-213), 335-345.

Roach, C.G., E.K.K. Nemaia, S.F. Ledgard, G.J. Brier & C.D.A. McLay (1996). Effects of long-term differences in fertiliser history on hill country: seasonal pasture production, legume growth and soil phosphorus status. *Proceedings of the New Zealand Grassland Association*, 57, 105-109.

Roggero, P.P., S. Bellon & M., Rosales (1996). Sustainable feeding systems based on the use of local rersources. *Annales. Zootechnie*, Elsevier/INRA, 45 (Suppl), 105-118.

Roggero, P.P., C. Porqueddu, S. Bullitta & F. Veronesi (1990). The choice and varieties for rainfed Sardinian forage systems. *Proceeding of the 6th Meeting FAO on Mediterranean pastures and fodder crops.* Bari, Italy, 77-81.

Samples, D.H., H.M. Bartholomew & R.M. Sulc (1996). Cereal rye to extend the grazing season. *Proceedings of the New Zealand Grassland Association,* 57, 115-118.

Seligman, N.G. (1996). Management of Mediterranean grasslands. In: J. Hodgson & A.W. Illius (eds.) The ecology and management of grazing systems. CAB International, Wallingford, UK, 359-391.

Siebert, B.D. (1982). Research findings in relation to future needs. *Proceedings of the Australian Society of Animal Production*, 14, 191-196.

Stockdale, C.R. (2001). Body condition at calving and the performance of dairy cows in early lactation under Australian conditions: review. *Australian Journal of Experimental Agriculture*, 41, 823-839.

Sulas, L. & S. Caredda (1997). Introduzione in coltura di nuove specie foraggere: produttività e composizione bromatologica di *Chrysanthemum coronarium* L. (crisantemo) sottoposto a pascolamento simulato (Introduction of new species as fodder crop: forage production and quality of a *Chrysanthemum coronarium* L. population under simulated grazing). *Rivista di Agronomia*, 31 (4), 1021-1028.

Sulas, L., S. Caredda & C. Porqueddu (1995). Evolution of the standing hay chemical composition in natural and improved Mediterranean pastures. In: Sylvopastoral Systems. Cahiers Option Mediterreenennes. 12, 155-158.

Sulas L., G.A. Re, L. Ledda & S. Caredda (1997). The effect of utilization frequency on the forage production of sulla (*Hedysarum coronarium* L.). *Italian Journal of Agronomy*, 2, 89-94.

Suttie, J.M. & S.G. Reynolds (2003). Transhumant grazing systems in temperature Asia. *Plant Production and Protection Series No. 31*, Food and Agriculture Organization of the United Nations, Rome, 353pp.

Talamucci, P. & A. Pardini (1999). Pastoral systems dominated by fodder crops harvesting and grazing. In: L. Etienne (ed.) *Dynamics and sustainability of Mediterranean pastoral systems - Proceedings of the 9'th Meeting FAO-CIHEAM*, Badajoz, Spain, (Nov. 26-29), 1997, 29-44.

Troxler, J. & A. Chassot (2004). Effects of the stocking rate on steer performance and vegetation patterns on mountain pastures. In: A. Lüscher, B. Jeangros, V. Kessler, O. Huguenin, M. Lobsiger, N. Millar & D. Sutter (eds.) Land use systems in grassland dominated regions. Grassland Science in Europe, 9, 587-589.

Venning, K., A. Thompson, A. Kennedy & D. Chapman (2004). Effects of separating grass and clover on ewe and lamb production. *Proceedings of the Grassland Society of Victoria*, Albury, July 2003, 81-82.

Volesky, J.D., D.C. Adams & R.T. Clark (2002). Windrow grazing and baled-hay feeding strategies for wintering calves. *Journal of Range Management*, 55, 23-32.

Waller, R.A., P.W. Sale, G.R. Saul & G.A. Kearney (2001). Tactical versus continuous stocking in perennial ryegrass-subterranean clover pastures grazed by sheep in south-western Victoria. 1. Stocking rates and herbage production. *Australian Journal of Experimental Agriculture*, 41, 1099-1108.

Wilkins, R.J. (1995). Improved technology for production efficiency and utilisation efficiency. In: G.E. Pollot (ed.) Grassland into the 21st century. BGS Occasional Symposium no. 29, 210-224.

Winsten, J.R. & B.T. Petrucci (2003). Seasonal dairy grazing a viable alternative for the 21st century. American Farmland Trust. Washington, DC.22. <http://grassfarmer.com/papers/studies/casestudies.pdf>

Overcoming seasonality of production: opportunities offered by forage conservation technologies

P. O'Kiely[1] and A.G. Kaiser[2]

[1]Teagasc, Grange Research Centre, Dunsany, Co. Meath, Ireland
Email: pokiely@grange.teagasc.ie
[2]NSW Department of Primary Industries, Wagga Wagga Agricultural Institute, P.M.B., Wagga Wagga, N.S.W., 2650, Australia

Key points

1. Seasonality of forage supply is a key contributor to the seasonality of meat and milk production.
2. Conserving forages as silage or hay can help reduce the seasonality of feed supply.
3. Forage conservation technologies make this contribution mainly through increases in the yield or quality of suitable crops, through an improved efficiency of the conservation process or by allowing a reduction in costs.
4. Future research needs differ considerably among regions of the world.

Keywords: silage, hay, conserved forage, animal production

Scale of seasonality

The pattern of national meat and milk output from ruminants often repeats in annual cycles and the magnitude of the disparity between the months of highest and lowest output varies considerably between countries (Table 1). Furthermore, significant seasonal patterns in output can be paralleled by seasonal variation in product quality (Keane & Allen, 1998; Lynch et al., 2002; O'Brien et al., 1999). The effects of seasonality are likely to be at their greatest where grazed forage is the main feedstuff used, and where its yield and/or quality exhibits large seasonal differences (Cummins et al., 1996). This paper outlines the opportunities offered by conserved forages to reduce the magnitude of the seasonality of meat and milk production. The principles covered are pertinent to hay and silage produced under tropical, semi-arid or temperate conditions. However, since most recently published research relates to silage production under temperate conditions, such research predominates the literature cited.

Table 1 Output of beef and milk from cattle during 2000 - highest month as multiple of lowest month

	Beef	Milk
Denmark	1.5	1.2
France	1.5	-
Ireland	2.9	6.0
Netherlands	1.4	1.1
New Zealand	-	82.9
USA	1.2	-

Causes of seasonality

Seasonality of meat or milk production is influenced by (a) the pattern of feedstuff supply, quality and cost; (b) animal type factors such as species/breed, genetic merit and physiological status; (c) animal response to influences such as climate, disease challenge and management practices; (d) the quality of animal produce and the farmers response to the market price available; and (e) the impact of limitations or opportunities provided by land, labour, capital or enterprise on farms.

The seasonality of forage production and quality is one of the most important technical constraints to overcoming the seasonality of animal production. However any response by farmers to address this problem, and increase out-of-season meat or milk production, will be heavily influenced by economic and social factors. Frawley (1980) itemised the difficulties perceived by Irish dairy farmers when presented with the opportunity of producing out-of-season milk. Economic barriers presented the biggest perceived difficulties (0.40 of total), and included factors relating to the additional costs of feedstuffs, animals, facilities and labour, and an inadequate milk price. Social barriers (0.32 of total) included factors related to the loss of a predictable seasonal respite in work-load, changing from a practised to a new system, difficulties integrating the new work demands into the overall work schedule and disincentives of moving into a higher taxation bracket. Technical barriers (0.28 of total) included perceived difficulties related to scale of operation and animal facilities, to reproductive efficiency and milk quality, and to making, storing and feeding adequate silage of appropriate quality. In principle, these barriers are similar to those outlined more recently for south-east Asian farmers by Chin (2002).

Opportunities for conserved forages

Reducing the seasonality of animal production by addressing the seasonality of forage production and quality is a key challenge for livestock producers. The extent of the seasonal timing of feed gaps varies significantly from region to region, and is driven in particular by climate and environment (altitude, soil, etc.). Forage supply can be increased through the use of irrigation, improved forage species and cultivars, and fertiliser application. However, at least some of the resultant forages usually need to be conserved in order to make the most effective contribution to filling seasonal gaps in feed availability. Clearly, the conservation option only works if the land is not overstocked (a problem in some extensive grazing areas).

Forage conservation can involve the storage of ensiled or dried grass, legume, cereal (whole-crop, straw/stover, grain/cob, etc.) or other crops. It is important that it be integrated into the overall farming system to allow the profitable and sustainable production of meat or milk (Doonan et al., 2004). Depending on circumstances, the role of conserved forages can vary from providing the main forage source throughout the year, to systems where grazing and conservation of forages are integrated and conserved forages are used seasonally either as a supplement (O'Brien et al., 1996) or the primary forage. In the latter case, forages can be conserved either opportunistically or as part of a planned management strategy specifically designed to produce forage for conservation. Within these integrated systems, the process of conserving some of the forage also provides the opportunity to make an important positive contribution to effective grazing management and improved forage utilisation by grazing animals, and to effective feed budgeting by farmers. It can also contribute to weed control in pastures (and crops in mixed grazing/grain farming systems), to maintaining the content of desirable species in pastures, to livestock not succumbing to pests and diseases at sensitive times of the year, and to avoiding soil erosion and helping conserve soil water. Furthermore, the optimal recycling of nutrients collected from housed livestock can often be best achieved by spreading the manures on the land used for producing the conserved feed.

Opportunities offered by forage conservation and associated technologies to overcome the seasonality of animal production can be considered in terms of (a) increasing the supply of feed to conserve; (b) improving the quality of feed to conserve; (c) improving the efficiency of the conservation process; (d) supplementation strategies to best complement the conserved forages; (e) mixing of complementary forages to maximise efficiency of use of nutrients; and (f) reducing the cost of providing livestock with nutrients from conserved forages through the use of the above strategies, and by restricting direct and fixed inputs and using available financial incentives (premia, subsidies, grants, bonuses etc.). The remainder of this paper will focus on new technologies relating to items (a), (b) and particularly (c).

Increase supply of feed to conserve

In many cases the yield of crops for conservation can be increased by improved management of resources (soils, water, fertilisers), better control of crop pests and diseases, altering the timing of when the crop is sown or is closed for conservation, or altering harvest dates. These higher yields can help reduce the seasonality of feed supply.

Achievements in the breeding of tropical and temperate grasses and legumes have been well documented (Miles, 2001; Quenesberry & Casler, 2001; and Wilkins & Humphreys, 2003). Forage crops that are particularly suited to producing high yields within a cutting regime can be grown, and conservation characteristics under a range of conditions have been outlined for many of these (Buxton & O'Kiely, 2003; Chin, 2002, Fraser *et al.*, 2001; Kaiser & Piltz, 2002; Panciera *et al.*, 2003; Suttie, 2000; Titterton *et al.*, 2002; and Wilkins & Jones, 2000). Significant interactions between potential yield and environment are normal for all of these crops. For example, studies on forage legumes at 12 northern European sites over two consecutive seasons, found the yield of individual legumes to be favoured or restricted at particular locations or in individual years (Halling *et al.*, 2002). New technologies can overcome some limitations - technologies related to mulching *Zea mays* L. (maize) with plastic film have allowed *Z. mays* grown in marginal climatic conditions to undergo substantial increases in dry matter (DM) yield (Easson & Fearnehough, 2003; Keane *et al.*, 2003), thereby making it a potentially viable crop for conservation. Options to seek a yield advantage by inter-cropping (Ghanbari-Bonjar & Lee, 2003; Maasdorp & Titterton, 1997) or bi-cropping (Marley *et al.*, 2003; Oyen, 1989) complementary crops are also available.

Feed supply can be increased by purchasing fresh (to ensile or dry) or conserved feed from other farms, or by ensiling purchased agricultural or industrial by-products such as apple pomace, corn stover, fruit rejects/waste, potato feed (steam peel), pressed citrus pulp, pressed sugar beet pulp, rice straw, wet brewers' grains, wet corn gluten feed and wet distillers' grains (including draff) (Chin, 2002; Crawshaw, 2001).

Increase quality of feed to conserve

Optimum crop management

The judicious use of established technologies usually provides scope to improve the quality of crops at harvest time, and this in turn can contribute to reducing the seasonality of animal production. Thus, the combination and timing of plant nutrient input, water supply and conservation, control of weeds, pests and diseases, choice of harvest date, portion of crop harvested, etc., can be optimised for specific crop species and cultivars.

Conserved forage made from a highly digestible crop can support superior rates of animal production (Steen *et al.,* 2002). It would be valuable if the natural decline in nutritive value that accompanies senescence in high-yielding forage crops could be arrested. However, the evidence to support such a mechanism following the introduction of the stay-green trait to *Lolium perenne* L. (perennial ryegrass) is not compelling (Wilman *et al.,* 2004). Caution is needed when assessing forages from semi-natural grassland because the relationship between chemical composition and digestibility differs from that of *L. perenne*, sometimes leading to an underestimation of their potential nutritive value and useful role (Bruinenberg *et al.,* 2002).

Much of the genetic variation in DM digestibility (DMD) within ryegrasses is the result of variation in the concentration of water-soluble carbohydrates (WSC) (Wilkins & Humphreys, 2003). Forages of elevated WSC concentration thus offer the potential to increase animal productivity and N use efficiency, as well as being easier to preserve, and retain a higher proportion of nutrients such as protein and WSC during ensilage (Davies *et al.,* 2002b). For example, Miller *et al.,* (2001) compared zero-grazed grasses of 126 and 165 gWSC/kgDM and recorded mean daily grass DM intakes by dairy cows of 10.7 and 11.6 kgDM, *in vivo* DMD's of 0.64 and 0.71 g/g, milk yields of 12.6 and 15.3 kg/day and urinary N excretion of 0.35 and 0.25g/g feed N.

Use of plastic film as a mulch with *Z. mays* can result in a substantial increase in the starch content of crops grown under marginal conditions (Easson & Fearnehough, 2003; Keane *et al.,* 2003). This should markedly increase their feed value. Brown midrib (bmr) genotypes in *Z. mays* and *Sorghum bicolor* L. Moench (sorghum) usually contain less lignin and may have altered lignin chemical composition. This in turn can increase forage DMD. However, the subsequent effects on animal production have not been consistent (Cox & Cherney, 2001; Oliver *et al.,* 2004) and may depend on the hybrid used. Leafy hybrids of *Z. mays* contain additional leaves above the ear, which should increase stover digestibility because leaves are more digestible than stalks. However, Cox & Cherney (2001) demonstrated that leafy hybrids had a higher neutral detergent fibre (NDF) content and a higher digestibility of NDF than normal hybrids. However, the lower harvest index was considered to reflect reduced grain fill, and resulted in similar calculated milk yields. Finally stay-green hybrids of *Z. mays* have an improved resistance to disease and leaf senescence, and can have superior yields. However, Wilkinson & Hill (2003) showed no benefit from stay-green hybrids grown under marginal conditions in terms of yield or within-plant DM distribution.

Legumes frequently offer the potential to improve the quality of forage available for conservation. Halling *et al.* (2002) in a comparison of 5 temperate legumes found that *Trifolium repens* L. (white clover) had the highest content of crude protein, digestible organic matter, WSC and metabolisable energy but the lowest content of crude fibre. *Medicago sativa* L. (lucerne) was the opposite for these traits, with *Trifolium pratense* L. (red clover), *Lotus corniculatus* L. (birdsfoot trefoil) and *Galega orientalis* (galega) being intermediate. The legumes had a higher quality than grass (except for WSC), with mixtures of legume and grass being intermediate. Bertilsson *et al.* (2002) ensiled legumes and grass separately. They were then offered to dairy cows alone or in mixtures of individual legume silages with grass silage. Intakes were higher for clover and clover-grass mixtures compared to pure grass silage, and milk production was higher for clover (particularly *T. repens*) than grass silage.

Many plants seek protection from herbivory by producing substances that may be bitter-tasting, poisonous, offensively odoured or anti-nutritional. High concentrations of secondary metabolites such as tannins, protease inhibitors and lectins can have anti-nutritional effects and plants with high concentrations of such compounds, or management practices that favour them, should be avoided. In contrast, low concentrations of some compounds can confer nutritional benefits. Thus, some naturally occurring condensed tannin-containing materials found in plants such as *L. corniculatus*, *Onobrychis viciifolia* Scop. (sainfoin) and *Hedysarum coronarium* L. (sulla) can help protect plant proteins from rumen degradation, resulting in increased post-ruminal protein supply, a reduction in urinary N losses and a reduced susceptibility to bloat (Butter *et al.*, 1999). They can also confer anthelminthic properties and thus work to counter the effects of parasitism (Butter *et al.*, 1999). Finally, reducing proteolysis during ensilage is an important goal to pursue, and it has been suggested that much of the relatively low rate of proteolysis during the ensilage of *T. pratense* may be due to polyphenol oxidases present in the crop (Jones *et al.*, 1995). Thus if plant breeders can introduce the expression of appropriate amounts and forms of polyphenol oxidases into crops, the potential exists to restrict proteolysis during ensilage.

Co-ensiling

Co-ensiling compatible forages or forage and concentrates, can improve silage quality and thereby animal productivity. The addition of certain dry concentrate feeds to forage can increase the estimated nutritive value of the ensiled crop, reduce or eliminate effluent production and create conditions conducive to a lactic acid fermentation (O'Kiely, 2002). It can also facilitate meeting labour demands during feedout. Provided excellent silage-making practices are employed, most of the nutrients present in these concentrate feeds are available to livestock at feedout (O'Kiely, 1992; Ferris & Mayne, 1994).

Improve efficiency of conservation

Reduce conservation losses

Technologies that support efficient and practicable conservation of forage help reduce the seasonality of feed supply and quality. Thus, reducing quantitative conservation losses from 0.25 to 0.20 through improved management practices will provide an additional one weeks conserved feed during a 5-month feedout. Qualitative losses similarly need to be restricted. Bolsen *et al.* (2002) showed that replacing aerobically deteriorated silage by the corresponding non-deteriorated silage, increased DM intake by steers from 6.7 to 8.0 kg/day, organic matter digestibility from 0.68 to 0.76 and crude protein digestibility from 0.63 to 0.75.

Field losses are best minimised by shortening the duration between mowing a crop and removing it to where it will be stored. Prevailing weather conditions will have a major impact on this. A range of mechanical treatments are available to speed up field drying of crops (Muck & Shinners, 2001) thereby rapidly reducing the activity of plant and microbial enzymes and, if used optimally, restricting physical losses or soil contamination. In many circumstances wilting can result in successful preservation of silage, producing feedstuffs with high intake characteristics (Ingvartsen, 1992; Wright *et al.*, 2000).

Storage losses associated with silage effluent only occur where relatively wet forage is ensiled. Methods for reducing or eliminating its production are described by O'Kiely (1989).

Silage fermentation is a complex process and, compared to industrial fermentations, can be difficult to control adequately. Fermentation needs to be guided to create conditions inhibitory to plant proteolysis and to the activities of preservation saboteurs such as *Clostridia* and *Enterobacteria*. It also needs to create conditions inhibitory to the development and activity of initiators (and their successors) of aerobic deterioration such as yeast, mould or *Bacilli*, and to the viability of pathogenic micro-organisms such as *Listeria monocytogenes* and *Cryptosporidium parvum*. This can be a particular challenge with tropical forages where the inherently low DM and WSC concentrations allied to the frequently high buffering capacities can produce crops that are difficult to preserve (Kaiser & Piltz, 2002; Buxton & O'Kiely, 2003).

Acid- and sugar-based additives have a long tradition of being used to facilitate the creation of conditions inhibitory to *Clostridia* and *Enterobacteria*. Lattemae *et al*. (1996) makes a case for simultaneously applying acid and molasses to crops such as *T. pratense*. Similarly, Davies *et al*. (2002a) have suggested that co-ensiling high sugar ryegrass with a legume such as *T. pratense* provides the opportunity to improve the preservation of the legume. Both the logistics of evenly applying the volumes of additive required and, in the case of acid-based additives, consciousness of their corrosive effects, have reduced their use. However, partial neutralisation of acid additives with ammonia can considerably reduce corrosion to machine components (Forristal, 1992), while at the same time retaining the ability to achieve satisfactory preservation (Randby, 2000). Inclusion of lignosulphonates from wood pulp liquor has also been shown to reduce corrosion (Randby, 2000). In addition applying chemical additives at mowing may reduce contact between the chemical and the harvester, without any negative effects on resultant silage conservation characteristics (Slottner & Lingvall, 2002). Other additives such as molasses could also be applied at mowing and this should be reasonably successful in the absence of rainfall.

Non-corrosive chemical additives based on hexamine and sodium nitrite in combination with sodium benzoate and sodium propionate can restrict both clostridial and yeast activity (Lattemae & Lingvall, 1996; Lingvall & Lattemae, 1999) when applied evenly and at appropriate rates of application. In contrast, the application of surfactants to grass at ensiling to reduce surface tension, aid dispersion of cell contents and thus stimulate the activity of lactic acid bacteria was not found to be successful (Pauly & Lingvall, 1999). Bacterial inoculants based on *Lactobacillus plantarum* have the potential to facilitate a fast and efficient fermentation in the silo (Muck & Shinners, 2001), particularly where competition from the natural population is limited and where adequate fermentable substrate is available. Inclusion of osmotolerant lactic acid bacteria in an inoculant can result in a considerably faster rate of acidification when applied to drier forage (Pobednov *et al*., 1997), while incorporation of bacteria with fructan hydrolase activity may facilitate more rapid acidification when applied to forages with a high proportion of WSC present as fructans (Merry *et al*., 1995). The co-application of a bacterial inoculant with sufficient molasses could also be attractive with some crops low in WSC content (Kaiser & Piltz, 2002). Similarly, inclusion of lactic acid bacteria that secrete an amylase that is optimally active in the silage pH range could facilitate the preservation of crops low in WSC, but containing starch (Fitzsimons & O'Connell, 1994) - the latter being unavailable to the bacteria in most conventional silage inoculants. Pahlow *et al*. (2003) have summarised a series of other novel approaches for improving silage fermentation and aerobic stability using inoculants. Good distribution of inoculated bacteria over the forage surface is normally considered important, and in practice can require the use of relatively large volumes of water. However, Kleinmans *et al*. (2002) have reported on the success of an ultra low volume applicator for concentrated suspensions of inoculum that gave comparable results to a more conventional system.

Moist feedstuffs are invariably susceptible to aerobic deterioration and the resultant quantitative and qualitative changes can lead to serious losses of nutrients. Minimising the opportunities for such losses with silage depends primarily on rapid filling and perfect sealing of silos, followed by minimising the duration of exposure to oxygen during feedout. Tabacco & Borreani, (2002) have shown that significant aerobic deterioration of silage is common on farms, particularly where feedout takes place during hot weather. They observed that farms with smaller areas of silage feed-face per livestock unit, coupled with higher linear rates of feedout and careful management of the feed-face had less aerobic deterioration in their silage. Growth stage of a crop at harvest can also be influential - in the case of *Z. mays*, Wyss (2002a) found that relatively immature crops of high WSC and crude fibre content, and low starch concentration, were less stable during feedout than more mature crops, whereas the effects of the variety sown were much smaller. Lactic acid assimilating yeast are the primary initiators of aerobic deterioration and are more frequent where delayed sealing of a silo accompanies a low packing density of the forage (Uriarte-Archundia *et al.*, 2002).

Additives applied at ensiling can improve aerobic stability at feedout. Examples include formic acid or mixtures containing formic, propionic and acrylic acids (O'Kiely, 1993), sorbate or benzoate together with homofermentative lactic acid bacteria (Owen, 2002; Skytta *et al.*, 2002; White *et al.*, 2002) and *Lactobacillus buchneri* alone or with *Lactobacillus plantarum* (Driehuis *et al.*, 2001; Filya *et al.*, 2002). Mo *et al.* (2002) restricted mould growth by applying adequate CO_2 to wilted forage at ensiling. Lowes *et al.* (2000) demonstrated that direct application of mycocins to grass silage could delay the onset of spoilage. Thus, inoculation with mycocinogenic yeast at ensiling presents the opportunity to biologically improve silage stability at feedout.

Preservation of moist hay, and the use of additives to assist in the process have been reviewed by Benhan & Redman (1980), and Pitt (1991).

Baled silage

Baled silage permits the expansion of successful forage conservation, and so can be important in overcoming seasonality of feed supply. Its nutritive value can be similar to conventional silage (Fychan *et al.*, 2002; O'Kiely *et al.*, 1999), and it can also be used for conserving by-products (Wyss, 2002b). However, its fermentation differs from conventional silage (Slottner, 2002) - this is likely due in particular to differences in chopping/laceration of forage, with differences in the extent of anaerobiosis being more important than differences in compaction during storage (O'Kiely & Forristal, 2002). However, slicing of forage at baling has relatively minor effects on fermentation even though it can increase bale density (Borreani & Tabacco, 2002; O'Kiely *et al.*, 1999). Forages need to be rapidly and adequately wilted before baling (Heikkila *et al.*, 2002) to reduce the number of bales per hectare, and to ensure good preservation, avoid effluent accumulation and produce bales that retain their cylindrical shape during storage. Lighter colour and 6 or more layers of conventional stretch-film appear to be important when seeking to maintain anaerobic conditions in sunnier climates (Lingvall, 2002), whereas a minimum of 4 layers of black film are usually considered adequate in more overcast, cooler conditions (O'Kiely *et al.*, 2002). The integrity of the plastic wrap must be maintained through to feedout in order to prevent fungal activity (McNamara *et al.*, 2002) - otherwise the latter can be excessive (Brady, *et al.*, 2004; O'Brien, *et al.*, 2004).

Predicting ensilability

Forages vary considerably in the ease with which they will undergo a lactic acid dominant fermentation during ensilage, and successfully predicting the outcome has important economic implications for farmers. Prediction systems range from the subjective where scoring of crop, weather and harvesting characteristics is conducted (Anon., 1983) to more objective systems based on analysing representative samples of forage for DM, WSC and buffering capacity (Weissbach *et al.*, 1974). Advances on the latter allow a more reliable prediction of the risk of butyric acid production by incorporating nitrate content and clostidial spore count (Kaiser & Weiss, 2004). Finally, new technologies using NIRS for the rapid and accurate analysis of conserved feeds (Agnew & Park, 2002) provide the opportunity for practical feedback of the success of the practices employed. The rapid assessment of free amino acids in silage could also be useful (Winters *et al.*, 2002)

Further research

Future research needs differ considerably among regions of the world. Investment is required in research facilities and expertise in many countries in order to develop technologies appropriate for reducing their seasonality of production. These resources would be foci for technology dissemination, and the participation of stakeholders in the development of these technologies is essential (Chin, 2002).

Where smallholders predominate in S.E. Asia, practical, reliable and low cost technologies are needed for successfully wilting and harvesting crops, and for packing them into relatively small storage containers (Chin, 2002). Kaiser & Piltz, (2002) identified key areas for tropical forage conservation research. They suggest post-harvest sward productivity effects of early cutting, selection of forage soybeans (or multiple-cut high leaf:stem annual legumes) that combine high yield and quality, development of efficient and effective wilting systems for humid climates, more pragmatic combinations of silage additives, and reliable strategies for improving hygienic quality and aerobic stability during feedout.

In semi-arid regions of Africa, research needs include development of strategic irrigation systems and guidelines for optimising time of harvest (Titterton *et al.*, 2002). Emphasis is needed on intercropping forage tree legumes (rather than annual legumes) with perennial tropical grasses or cereal crops. The benefits of these technologies need to be demonstrated within whole-farm systems where there is sufficient conserved feed (emphasis on silage) to cope with individual years of severe drought. In order to achieve some of the benefits of scale of operation, collective or co-operative alternatives to individual farm production may need to be tested for socio-economic viability (Titterton *et al.*, 2002).

Silage and hay are already an integral part of meat and milk production systems in many regions with a temperate climate, with the emphasis being on silage under more moist conditions. Muck & Shinners, (2001) identified needs and predicted likely trends in forage conservation technologies, and these will drive a considerable research effort for a number of years. New technologies in regions where silage and hay are already well established will need to support reliably achieving high yields of good quality crops, consistently conserving crops with minimal quantitative and qualitative losses (and in some cases with upgrading of feeds during storage), having relatively low inputs of manual labour and being cost competitive. New technologies from precision-agriculture should play a role in this regard (Marcotte *et al.*, 1999). Farmers will need rapid access to quantified data (e.g. soil nutrient and water status, fertiliser uptake rates, predicted yields and digestibilities, ensilability estimates, predicted wilting conditions, etc.), and know the strengths and weaknesses (including costs) of alternative technology options. In the case of hay, Tallowin & Jefferson, (1999) noted a need for further research to examine nutrient supply from semi-natural grasslands so that their feed value for ruminant livestock could be assessed with more confidence.

References

Agnew, R.E. & R.S. Park (2002). The development of NIRS for the rapid evaluation of grass silage. In: L.G. Gechie & C. Thomas (eds.) *Proceedings of XIII International Silage Conference*, Sept. 11-13, SAC, Auchincruive, Scotland, 290-291.

Anonymous (1983). Silage. Liscombe grass bulletin No. 2. MAFF, UK, 55pp.

Benham, C.L. & P.L. Redman (1980). Preservation of moist hay - a review. ADAS quarterly review, No. 39, 212-225.

Bertilsson, J., R.J. Dewhurst & M. Tuori (2002). Effects of legume silage on feed intake, milk production and nitrogen efficiency. In: R.J. Wilkins & C. Paul (eds.). Landbauforschung Volkenrode: Sonderheft, 234, 39-45.

Bolsen, K.K., L.A. Whitlock & M.E. Uriarte-Archundia (2002). Effect of surface spoilage on the nutritive value of maize silage diets. In: L.G. Gechie & C. Thomas (eds.) *Proceedings of XIII International Silage Conference*, Sept. 11-13, SAC, Auchincruive, Scotland, 76-77.

Borreani, G. & E. Tobacco (2002). Fermentation and losses of wrapped bug bales of Lucerne as affected by bale chopper system. In: L.G. Gechie & C. Thomas (eds.) *Proceedings of XIII International Silage Conference*, Sept. 11-13, SAC, Auchincruive, Scotland, 98-99.

Brady, K.C., P. O'Kiely, P.D. Forristal & H. Fuller (2004). Schizophyllum commune on big-bale grass silage in Ireland. *Mycologist* (in press).

Bruinenberg, M.H., H. Valk, H. Korevaar & P.C. Struik (2002). Factors affecting digestibility of temperate forages from seminatural grasslands: a review. *Grass & Forage Science*, 57, 292-301.

Butter, N.L., J.M. Dawson & P.J. Buttery (1999). Effects of dietary tannins on ruminants. In: J.C. Caygill & I. Mueller-Harvey (eds.) Secondary plant products - antinutritional and beneficial actions in animal feeding. Nottingham University Press, 51-70.

Buxton, D.R. & P. O'Kiely (2003). Preharvest plant factors affecting ensiling. In: D.R. Buxton, R.E. Muck & J.H. Harrison (eds.) Silage science and technology, Agronomy monograph No. 42. *American Society of Agronomy Inc., Crop Science Society of America Inc., Soil Science Society of America Inc.*, Madison, Wisconsin, USA, 199-250.

Chin, F.Y. (2002). Ensiling tropical forages with particular reference to South East Asian systems. In: L.G. Gechie & C. Thomas (eds.) *Proceedings of XIII International Silage Conference*, Sept. 11-13, SAC, Auchincruive, Scotland, 22-36.

Cox, W.J. & D.J.R. Cherney (2001). Influence of brown midrib, leafy, and transgenic hybrids on corn forage production. *Agronomy Journal*, 93, 790-796.

Crawshaw, R. (2001). Co-product feeds - animal feeds from the food and drinks industries. Nottingham University Press, 285pp.

Cummins, L.J., A.J. Clark, B.W. Knee, L.C. Clarke, S.J. Walsh, D.J. Sparks, R.C. Phillips, R.C. Seirer, D.A. Courtney & D. Scopel (1996). Pasture based systems to finish steers in southern Australia to the Japanese ox market specifications. *Proceedings of the Australian Society of Animal Production*, 21, 85-88.

Davies, D.R., D.J. Leemans & R.J. Merry (2002a). Ensiling either high or low sugar containing perennial ryegrasses with or without red clover. In: J-L. Durand, J.C. Emile, C. Huyghe & G. Lemaire (eds.) Grassland Science in Europe, Vol. 7, *Proceedings of 19th general meeting of the European Grassland Federation*, La Rochelle, France, 27-30 May (2002), 194-195.

Davies, D.R., D.J. Leemans & R.J. Merry (2002b). Improving silage quality by ensiling perennial ryegrasses high in WSC content either with or without different additives. In: L.G. Gechie & C. Thomas (eds.) *Proceedings of XIII International Silage Conference*, Sept. 11-13, SAC, Auchincruive, Scotland, 386-387.

Doonan, B.M., A.G Kaiser, D.F. Stanley, I.F. Blackwood, J.W. Piltz & A.K. White (2004). Silage in the farming system. In: A.G. Kaiser, J.W. Piltz, H.M. Burns & N.W. Griffiths (eds.) Successful Silage, (2nd ed.), New South Wales Dept. of Primary Industry: Orange, NSW, 1-24.

Driehuis, F., S.J.W.H Oude Elferink & P.G. van Wikselaar (2001). Fermentation characteristics and aerobic stability of grass silage inoculated with *L. buchneri*, with or without homofermentative lactic acid bacteria. *Grass & Forage Science*, 56, 330-343.

Easson, D.L. & W. Fearnehough (2003). The ability of the Ontario heat unit system to model the growth and development of forage maize sown under plastic mulch. *Grass & Forage Science*, 58, 372-384.

Ferris, C.P. & C.S. Mayne (1994). Effects on milk production of feeding silage and three levels of sugar-beet pulp either as a mixed ration or as an ensiled blend. *Grass & Forage Science*, 49, 241-251.

Filya, I., A. Karabulut & E. Sucu (2002). The effect of *L. plantarum* and *L. buchneri* on the fermentation, aerobic stability and rumen degradability of maize silage in a warm climate. In: L.G. Gechie & C. Thomas (eds.) *Proceedings of XIII International Silage Conference*, Sept. 11-13, SAC, Auchincruive, Scotland, 192-193.

Fitzsimons, A. & M. O'Connell (1994). Comparative analysis of amylolytic lactobacilli and *L. plantarum* as potential silage inoculants. *FEMS Microbiology Letters*, 116, 137-146.

Forristal, D. (1992). The effect of four silage additives on corrosion and ware of silage machinery. Teagasc, Carlow, Ireland. 14pp.

Fraser, M.D., R. Fychan & R. Jones (2001). The effect of harvest date and inoculation on the yield, fermentation characteristics and feeding value of forage pea and field bean silages. *Grass & Forage Science*, 56, 218-230.

Frawley, J. (1980). Sociological aspects of seasonality. In: Dairy industry technical study group series No.14, An Foras Taluntais, Dublin, 21-52.

Fychan, R., M.D. Fraser & R. Jones (2002). Effect of ensiling method on the quality of red clover and lucerne silages. In: L.G. Gechie & C. Thomas (eds.) *Proceedings of XIII International Silage Conference*, Sept. 11-13, SAC, Auchincruive, Scotland, 104-105.

Ghanbari-Bonjar, A. & H.C. Lee (2003). Intercropping wheat and beans as a whole-crop forage: effect of harvest time on forage yield and quality. *Grass & Forage Science*, 58, 28-36.

Halling, M.A., A. Hopkins, O. Nissinen, C. Paul, M. Tuori & U. Soelter (2002). Forage legumes - productivity and composition. In: R.J. Wilkins & C. Paul (eds.) Landbauforschung Volkenrode: Sonderheft, 234, 5-25.

Heikkila, T., S. Jaakkola, E. Saarisalo, A. Suokannas & J. Helminen (2002). Effects of wilting time, silage additive and plastic layers on the quality of round big bale silage. In: L.G. Gechie & C. Thomas (eds.) *Proceedings of XIII International Silage Conference*, Sept. 11-13, SAC, Auchincruive, Scotland, 158-159.

Ingvartsen, K.L. (1992). Effect of conservation and dry matter content of forage from grassland on feed intake, daily gain and feed conversion ratio of growing cattle: a review. *Beretning fra Statens Husdyrbrugsforsog*, 71, 1-44.

Jones, B.A., R.E. Muck & R.D. Hatfield (1995). Red clover extracts inhibit legume proteolysis. *Journal of the Science of Food & Agriculture*, 67, 329-333.

Kaiser, A.G. & J.W. Piltz (2002). Silage production from tropical forages in Australia. In: L.G. Gechie & C. Thomas (eds.) *Proceedings of XIII International Silage Conference*, Sept. 11-13, SAC, Auchincruive, Scotland, 48-61.

Kaiser, E. & K. Weiss (2004). New realisations for the estimation of the ensiling potential of forages. In: A. Luscher, B. Jeangros, W. Kessler, O. Huguenin, M. Lobsiger, N. Millar & D. Suter (eds.) Grassland Science in Europe, Vol. 9, *Proceedings of 20th general meeting of the European Grassland Federation*, Luzern, Switzerland, June 21-24, 888-890.

Keane, G.P., J. Kelly, S. Lordan & K. Kelly (2003). Agronomic factors affecting yield and quality of forage maize in Ireland: effect of plastic film system and seeding rate. *Grass & Forage Science*, 58, 362-371.

Keane, M.G. & P. Allen (1998). Effects of production system intensity on performance, carcass composition and meat quality of beef cattle. *Livestock Production Science*, 56, 203-214.

Kleinmans, J., B. Ruser & G. Pahlow (2002). Distribution and activity of inoculant applied at reduced water volume with the Pioneer Appli-Pro ULV application system. In: L.G. Gechie & C. Thomas (eds.) *Proceedings of XIII International Silage Conference*, Sept. 11-13, SAC, Auchincruive, Scotland, 208-209.

Lattemae, P. & P. Lingvall (1996). Influence of hexamine and sodium nitrite in combination with sodium benzoate and sodium propionate on fermentation and storage stability of wilted and long cut grass silage. *Swedish Journal of Agricultural Research*, 26, 135-146.

Lattemae, P., C. Ohlsson & P. Lingvall (1996). The combined effect of molasses and formic acid on quality of red clover silage. *Swedish Journal of Agricultural Science*, 26, 31-41.

Lingvall, P. (2002). The bale silage technology - factors influencing fermentation, hygienic quality, storage stability and production costs. In: L.G. Gechie & C. Thomas (eds.) *Proceedings of XIII International Silage Conference*, Sept. 11-13, SAC, Auchincruive, Scotland, 162-163.

Lingvall, P. & P. Lattemae (1999). Influence of hexamine and sodium nitrite in combination with sodium benzoate and sodium propionate on fermentation and hygienic quality of wilted and long cut grass silage. *Journal of the Science of Food & Agriculture*, 79, 257-264.

Lowes, K.F., C.A. Shearman, J. Payne, D. Mackenzie, D.B. Archer, R.J. Merry & M.J. Gasson (2000). Prevention of yeast spoilage in feed and food by the yeast mycocin HMK. *Applied and Environmental Microbiology*, 66, 1066-1076.

Lynch, A., D.J. Buckley, K. Galvin, A.M. Mullen, D.J. Troy & J.P. Kerry (2002). Evaluation of rib steak colour from Friesian, Hereford and Charolais heifers pastured or overwintered prior to slaughter. *Meat Science*, 61, 227-232.

Maasdorp, B.V. & M. Titterton (1997). Nutritional improvements in maize silage for dairying: mixed crop silages from sole and intercropped legumes and a long-season variety of maize. 1. Biomass yield and nutritive value. *Journal of Animal Feed Science & Technology*, 69, 241-261.

Marcotte, D., P. Savoie, H. Martel & R. Theriault (1999). Precision-agriculture for hay and forage crops: a review of sensors and potential applications. American Society of Agricultural Engineering annual international meeting, Toronto, Canada, July 18-21, 12pp.

Marley, C.L., R. Fychan, M.D. Frazer, A. Winters & R. Jones (2003). Effect of sowing ratio and stage of maturity at harvest on yield, persistency and chemical composition of fresh and ensiled red clover/lucerne bi-crops. *Grass & Forage Science*, 58, 397-406.

McNamara, K., P. O'Kiely, J. Whelan, P.D. Forristal & J.J. Lenehan (2002). Simulated bird damage to the plastic stretch-film surrounding baled silage and its effects on conservation characteristics. *Irish Journal of Agricultural & Food Research*, 41, 29-42.

Merry, R.J., A.L. Winters, P.L. Thomas, M. Muller & T. Muller (1995). Degradation of fructans by epiphytic and inoculated lactic acid bacteria and by plant enzymes during ensilage of normal and sterile hybrid ryegrass. *Journal of Applied Bacteriology*, 79, 583-591.

Miles, J.W. (2001). Achievements and perspectives in the breeding of tropical grasses and legumes. In: J.A. Gomide, W.R.S. Mattos & S.C. da Silva (eds.) *Proceedings of XIX International Grassland Congress*, Sao Pedro, Brazil, 509-515.

Miller, L.A., J.M. Moorby, D.R. Davies, M.O. Humphreys, N.D. Scollan, J.C. MacRae & M.K. Theodorou (2001). Increased concentration of WSC in perennial ryegrass: milk production from late-lactation dairy cows. *Grass & Forage Science*, 56, 383-394.

Mo, M., A. Nes & S. Kristiansen (2002). Carbon dioxide as an additive to heavily wilted big bale silage. In: L.G. Gechie & C. Thomas (eds.) *Proceedings of XIII International Silage Conference*, Sept. 11-13, SAC, Auchincruive, Scotland, 160-161.

Muck, R.E. & K.J. Shinners (2001). Conserving forage (silage and hay): progress and priorities. In: J.A. Gomide, W.R.S. Mattos & S.C. da Silva (eds.) *Proceedings of XIX International Grassland Congress*, Sao Pedro, Brazil, 753-762.

O'Brien, B., S. Crosse & P. Dillon (1996). Effects of offering a concentrate or silage supplement to grazing dairy cows in late lactation on animal performance and on milk processability. *Irish Journal of Agricultural & Food Research*, 35, 113-125.

O'Brien, B., R. Mehra, J.F. Connolly & D. Harrington (1999). Seasonal variation in the composition of Irish manufacturing and retail milks. 1. Chemical composition and renneting properties. *Irish Journal of Agricultural & Food Research*, 38: 53-64.

O'Brien, M., P. O'Kiely, P.D. Forristal & H. Fuller (2004). Pilot survey to establish the extent of occurrence and identity of visible fungi on baled grass silage. *Proceedings of Agricultural Research Forum*, Tullamore, Ireland, March 1-2, p50.

O'Kiely, P. (1989). Control and use of silage effluent. *Journal of Irish Grassland and Animal Production Association*, 23, 3-13.

O'Kiely, P. (1992). The effect of ensiling sugarbeet pulp with grass on silage composition, effluent production and animal performance. *Irish Journal of Agricultural & Food Research*, 31, 115-128.

O'Kiely, P. (1993). Influence of a partially neutralised blend of aliphatic organic acids on fermentation, effluent production and aerobic stability on autumn grass silage. *Irish Journal of Agricultural & Food Research*, 32, 13-26.

O'Kiely, P. (2002). Soya hulls or citrus pulp as alternatives to molassed beet pulp as silage additives. In: L.G. Gechie & C. Thomas (eds.) *Proceedings of XIII International Silage Conference*, Sept. 11-13, SAC, Auchincruive, Scotland, 220-221.

O'Kiely, P. & P.D. Forristal (2002). Interaction of compaction and anaerobiosis on the conservation characteristics of ensiled grass. In: L.G. Gechie & C. Thomas (eds.) *Proceedings of XIII International Silage Conference*, Sept. 11-13, SAC, Auchincruive, Scotland, 148-149.

O'Kiely, P., P.D. Forristal, K. Brady, K. McNamara, J.J. Lenehan, H. Fuller & J. Whelan (2002). Improved technologies for baled silage. Beef Production Series No. 50, Teagasc, Grange, Co. Meath, Ireland, 128pp.

O'Kiely, P., P.D. Forristal & J.J. Lenehan (1999). Baled silage. Beef Production Series No. 11, Teagasc, Grange, Co. Meath, Ireland, 44pp.

Oliver, A.L., R.J. Grant, J.F. Pedersen & J. O'Rear (2004). Comparison of brown midrib-6 and -18 forage sorghum with conventional sorghum and corn silage in diets of lactating dairy cows. *Journal of Dairy Science*, 87, 637-644.

Owen, T.R. (2002). The effects of a combination of a silage inoculant and a chemical preservative on the fermentation and aerobic stability of whole-crop cereal and maize silage. In: L.G. Gechie & C. Thomas (eds.) *Proceedings of XIII International Silage Conference*, Sept. 11-13, SAC, Auchincruive, Scotland, 196-197.

Oyen, J. (1989). Ettarige belgvekster i blanding med bygg eller forraps (Annual legumes in mixture with barley or fodder rape). *Norsk Landbruksforsking*, 3, 61-70.

Pahlow, G., R.E. Muck, F. Driehuis, S.J.W.H. Oude Elferink & S.F. Spoelstra (2003). Microbiology of ensiling. In: D.R. Buxton, R.E. Muck & J.H. Harrison (eds.) Silage science and technology, *Agronomy monograph no. 42, American Society of Agronomy Inc., Crop Science Society of America Inc., and Soil Science Society of America Inc.*, Madison, Wisconsin, USA, 31-93.

Panciera, M.T., W.E. Kunkle & S.C. Fransen (2003). Minor silage crops. In: D.R. Buxton, R.E. Muck & J.H. Harrison (eds.) Silage science and technology, *Agronomy monograph no. 42, American Society of Agronomy Inc., Crop Science Society of America Inc., and Soil Science Society of America Inc.*, Madison, Wisconsin, USA, 781-823.

Pauly, T. & P. Lingvall (1999). Effects of mechanical forage treatment and surfactants on fermentation of grass silage. *Acta Agriculturae Scandinavica* (Section A, Animal Science), 49, 197-205.

Pitt, R.E. (1991). Hay preservation and hay additive products. In: K.K. Bolsen, J.E. Baylor & M.E. McCullough (eds.) Field guide to hay and silage management in N. America, National Feed Ingredients Association, 127-138.

Pobednov, J., F. Weissbach & G. Pahlow (1997). About the effect of LAB-preparations on the acidification rate and fermentation quality of wilted silage. *Landbauforschung Volkenrode*, 47 (3), 97-102.

Quenesberry, K.H. & M.D. Casler (2001). Achievements and perspectives in the breeding of temperate grasses and legumes. In: J.A. Gomide, W.R.S. Mattos & S.C. da Silva (eds.) *Proceedings of XIX International Grassland Congress*, Sao Pedro, Brazil, 517-524.

Randby, A.T. (2000). The effect of some acid-based additives applied to wet grass crops under various ensiling conditions. *Grass & Forage Science*, 55, 289-299.

Skytta, E., A. Haikara, E. Saarisalo & S. Jaakkola (2002). Inhibition of aerobic spoilage yeasts in silage by hurdle technology. In: L.G. Gechie & C. Thomas (eds.) *Proceedings of XIII International Silage Conference*, Sept. 11-13, SAC, Auchincruive, Scotland, 184-185.

Slottner, D. (2002). Effect of ensiling a crop in big bales or small scale silos. In: L.G. Gechie & C. Thomas (eds.) *Proceedings of XIII International Silage Conference*, Sept. 11-13, SAC, Auchincruive, Scotland, 226-227.

Slottner, D. & P. Lingvall (2002). Effect of additive application at mowing or after chopping. In: L.G. Gechie & C. Thomas (eds.) *Proceedings of XIII International Silage Conference*, Sept. 11-13, SAC, Auchincruive, Scotland, 142-143.

Steen, R.W.J., D.J. Kilpatrick & M.G. Porter (2002). Effects of the proportions of high or medium digestibility grass silage and concentrates in the diet of beef cattle on liveweight gain, carcass composition and fatty acid composition of muscle. *Grass & Forage Science*, 57, 279-291.

Suttie, J.M. (2000). Hay and straw for small-scale farming and pastoral conditions. Plant production and protection series No. 29, Food and Agriculture Organisation of the United Nations, Rome, 303pp.

Tabacco, E. & G. Borreani (2002). Extent of aerobic deterioration in farm maize silage as affected by silo management. In: L.G. Gechie & C. Thomas (eds.) *Proceedings of XIII International Silage Conference*, Sept. 11-13, SAC, Auchincruive, Scotland, 178-179.

Tallowin, J.R.B. & R.G. Jefferson (1999). Hay production from lowland semi-natural grasslands - a review of implications for ruminant livestock systems. *Grass & Forage Science*, 54, 99-115.

Titterton, M., O. Mhere, B. Maasdorp, T. Kipnis, G. Ashbel, T. Smith & Z. Weinberg (2002). Ensiling of tropical forages with particular reference to African livestock systems. In: L.G. Gechie & C. Thomas (eds.) *Proceedings of XIII International Silage Conference*, Sept. 11-13, SAC, Auchincruive, Scotland, 37-47.

Uriarte-Urchundia, M.E., K.K. Bolsen & B.E. Brent (2002). A study of the chemical and microbiological changes in whole-plant maize silage during exposure to air. In: L.G. Gechie & C. Thomas (eds.) *Proceedings of XIII International Silage Conference*, Sept. 11-13, SAC, Auchincruive, Scotland, 172-173.

Weissbach, F., L. Schmidt & E. Hein (1974). Method of anticipation of the run of fermentation in silage making, based on the chemical composition of the green fodder. *Proceedings of the XII International Grassland Congress*, Moscow, USSR, Section 2, 663-673.

White, J.S., C.J. Lin, M.K. Woolford & K.K. Bolsen (2002). A new solution to the old problem of cereal and maize silage stability. In: L.G. Gechie & C. Thomas (eds.) *Proceedings of XIII International Silage Conference*, Sept. 11-13 (2002), SAC, Auchincruive, Scotland, 78-79.

Wilkins, P.W. & M.O. Humphreys (2003). Progress in breeding perennial forage grasses for temperate agriculture. *Journal of Agricultural Science*, 140, 129-150.

Wilkins, R.J. & R. Jones (2000). Alternative home-grown protein sources for ruminants in the United Kingdom. *Animal Feed Science and Technology*, 85: 23-32.

Wilkinson, J.M. & J. Hill (2003). Effect of yield and dry matter distribution of the stay-green characteristic in cultivars of forage maize grown in England. *Grass & Forage Science*, 58, 258-264.

Wilman, D., R.A.M. Irianni & M.O. Humphreys (2004). Stay-green compared with non-stay-green *Lolium perenne* in field swards with different cutting and nitrogen treatments. *Annals of Applied Biology*, 144, 95-101.

Winters, A.L., J.D. Lloyd, R. Jones & R.J. Merry (2002). Evaluation of a rapid method for estimating free amino acids in silages. *Animal Feed Science & Technology*, 99, 177-187.

Wright, D.A., F.J. Gordon, R.W.J. Steen & D.C. Patterson (2000). Factors influencing the response in the intake of silage and animal performance after wilting of grass before ensiling: a review. *Grass & Forage Science*, 55:1-13.

Wyss, U. (2002a). Influence of different factors on aerobic stability of maize silages. In: L.G. Gechie & C. Thomas (eds.) *Proceedings of XIII International Silage Conference*, Sept. 11-13, SAC, Auchincruive, Scotland, 176-177.

Wyss, U. (2002b). Quality of pressed sugar beet pulp in big bales. In: L.G. Gechie & C. Thomas (eds.) *Proceedings of XIII International Silage Conference*, Sept. 11-13, SAC, Auchincruive, Scotland, 392-393.

Evolution of integrated crop-livestock production systems

M.H. Entz[1], W.D. Bellotti[2], J.M. Powell[3], S.V. Angadi[4], W. Chen[2], K.H. Ominski[5] and B. Boelt[6]

[1]*Department of Plant Science, University of Manitoba, Winnipeg, Canada*
Email: m_entz@umanitoba.ca
[2]*School of Agriculture and Wine, University of Adelaide, South Australia,* [3]*US Dairy Forage Research Centre, Madison, Wisconsin, USA,* [4]*Department of Soil Science, University of Manitoba, Canada,* [5]*Department of Animal Science, University of Manitoba, Canada,* [6]*Danish Institute of Agricultural Sciences, Flakkebjerg, Denmark*

Key points

1. Many factors contribute to changes in the crop-livestock systems, but no logical end-point in the evolution process exists.
2. While benefits of integrated crop-livestock systems over specialised crop and livestock systems are well documented, there has been a move to specialised crop and livestock production.
3. Sustainability issues (manure nutrient concentration, soil quality maintenance, salinity, herbicide resistance, economic instability) have created a renewed interest in integrated crop-livestock systems.
4. Farmer adaptability is as an important link in the evolution between 'states of integration'.

Keywords: forage and pasture crops, ley farming

Introduction

Integrated crop-ruminant livestock (mixed) farming systems exist in many different forms around the world. Pasture leys are common in humid and temperate zones. Conserved forages play a significant role in temperate zones where they are grown in rotation with grain crops. This paper describes some unique benefits of integrated systems compared to specialised systems, and discusses current and future trends in crop-livestock integration.

Benefits of crop-livestock integration over specialisation

Farming systems that integrate ruminant livestock and crops tend to be more sustainable because they provide opportunities for rotation diversity and perenniality, nutrient recycling and greater energy efficiency.

Crop rotation diversity

The benefits of forage legumes in rotation have been known for centuries. Sir John Lawes reported soil structural benefits of ley phases in the 1880's (Clarke & Poincelot, 1996). Benefits of forages for soil health (soil structure, nutrient status), salinity control, pest management, improved crop yield, and higher overall whole farm profitability are well recognized (Entz *et al.*, 2002).

Hay and pasture leys reduce weed and disease problems and increase productivity of the following grain crops. In Canada, for example, *Medicago sativa* L. (Lucerne) hay crops provide excellent control of several problem weeds including *Avena fatua* L. (wild oat) and

Cirsium arvense L. (Canada thistle) (Ominski *et al.*, 1999). In a survey of 250 western Canadian farmers, 83% of respondents indicated weed control benefits from forage leys 3 years after forage crop termination, while 71% of respondents indicated higher grain yields following forage than in annual crop rotations (Entz *et al.*, 1995). Using 30 years of rotation trial data from Saskatchewan, Canada, Zentner and co-workers (in Campbell *et al.*, 1990) demonstrated that integrated crop-forage rotations had a lower cost of production than annual grain production systems, and forage crops provided more income stability to grain farming than government crop insurance programs. In south Australia, pasture-grain systems provide relatively stable income even in drought years where grains result in large financial losses. When comprehensive economic analysis is carried out, grain does not look nearly as attractive as a simple (and misleading) gross margin would suggest.

Perenniality in the cropping system

A striking example of what can happen when perenniality is removed from the landscape is dryland salinity. In southern Australia, *M. sativa* is being used to restore the hydrological balance and reduce the deep drainage that eventually results in dryland salinity (Angus *et al.*, 2001; Latta *et al.*, 2001). This restoration is based on the deep rooting system of *M. sativa* and its ability to provide transpiration leaf area for the entire year. Annual crops are unable to provide the necessary 'perenniality' to better match plant water use to available rainfall.

Rising nitrate concentration in ground water has been partly attributed to the shift away from sod-based rotations. Deep-rooted perennials such as *M. sativa* (Campbell *et al.*, 1994) and grasses (Entz *et al.*, 2001a) are able to retrieve deep-leached nitrates better than annual crops, thereby reducing the nitrate contamination risk. The deep root systems of forages also enable carbon to be placed deep into the soil profile. Gentile *et al.* (2004), using a 38-year old crop rotation study in Uruguay, showed that mixed farming systems (4 year pasture, 4 year grain crops) had significantly higher levels of subsoil C than crop only systems. Annual systems, even under grazing, return less organic matter back to soil than perennial pastures (Mapfumo *et al.*, 2000).

Nutrient cycling

Integrated crop-livestock systems allow opportunities for locally controlled nutrient recycling. Nutrient cycling occurs by: 1) adding soil-N by legumes; 2) excreta from grazing animals; 3) manure from confined fed animals; and 4) nutrient transfer specifically associated with livestock movement (i.e., nighttime corralling).

It is well established that pastures, especially legume-based pastures, return a high proportion of nutrients back to the land (70 to 90%) (Entz *et al.*, (2002). Within grazing systems, the degree of nutrient recycling depends on pasture composition and grazing intensity, with more intensive grazing increasing nutrient recycling to the soil system (Mapfumo *et al.*, 2000). Removing forage from the land (i.e., hay and silage) removes nutrients and reduces rotational benefits of the pasture ley. For example, in a long-term (1920 to 1990) rotation study in NSW, Australia, Norton *et al.* (1995) reported that a change in *Trifolium pratense* (red clover) harvest management from grazing to haying contributed to a marked decline in yields of the following crops, whereas in the grazed *T. pratense* pastures, rotational crop yields were maintained. However, a long history of legume dominant pastures has resulted in many acidic soils in some of the most productive (medium-high rainfall) soils (McCown, 1996). Regular liming is the recommended practice to maintain a desirable soil pH. The problem becomes difficult to manage when the acidity progresses into the subsoil where liming is less effective.

Nutrient recycling opportunities are especially limited when conserved forage is transported and fed a long distance from its production site. Such separation of forage and livestock can reduce soil quality at forage production sites and accumulate manure nutrients at the livestock production site.

Energy efficiency

Integrated crop-livestock systems have the potential to be more energy efficient than either specialised crop or specialised livestock production (Clarke & Poincelot, 1996). Even when ruminant livestock gain a high proportion of their nutrition from human digestible sources (e.g. grains), their basal diet is typically comprised of forage (unusable to humans) giving ruminants an advantage over pigs and chickens (Loomis & Connor, 1992).

A number of studies have confirmed that integrated systems use less energy per ha (Clarke & Poincelot, 1996) and have higher energy efficiency than either specialised crop or livestock systems. In Germany, increasing off-farm feed purchases and decreasing reliance on grazing increased energy use in dairy production (from 5.9 GJ/ha to 19.1 GJ/ha) and decreased energy use efficiency (2.7 vs. 1.2 GJ/t of milk produced) (Haas *et al.*, 2001). In Canada, integrated systems were found to have 10% higher energy use efficiency than specialised crop systems (Hoeppner, 2001).

Examples of crop-livestock integration around the world

Previous workers have suggested that changes in the degree of crop-livestock integration are a function of population and economic processes (Steinfeld, 1998). In this paper, different crop-livestock systems from different regions of the world are discussed in the context of the crop-livestock integration model presented by Powell *et al.* (2004).

West Africa

In semi-arid West Africa, vast increases in human population and urbanization have increased the need for crop and animal products. Traditional cropping systems based on shifting cultivation, and livestock systems based on transhumance and communal grazing are rapidly transforming to more sedentary, intensive mixed farming enterprises. *Pennisetum glaucum* (Pearl millet), *Sorghum bicolour* (sorghum) and *Zea mays* (maize) are the principal cereals, *Digitaria exilis* (fonio) is important in some areas, and *Oryza sativa* (rice) is cultivated in delta areas and along river and stream borders. The legumes *Vigna unguiculata* (cowpea) and *Arachis hypogea* (groundnut) are both subsistence and cash crops. Cattle, sheep and goats provide food for households, cash income, and are a means of storing capital and of buffering food shortages in years of poor crop production. Most agricultural products are used for subsistence purposes. Low rural incomes and the high cost of inorganic fertilisers and feed supplements prevent the widespread use of these external nutrient sources. Problems common to these agricultural systems include insufficient high quality forages, encroachment of cropping on communal grazing lands, reduced fallow periods (and declining soil fertility), lack of access to fertiliser and feed supplements, labour shortages during the cropping season, and inadequate market opportunities (Steinfield *et al.,* 1997; Powell *et al.*, 2004).

In semi-arid West Africa, specialised and independent crop and livestock production systems are more attractive than mixed systems when population pressures are low (Figure 1). Cropland productivity is maintained through fallowing, which is preferred to land application of manure because it requires less labour. As population pressures rise, the demand for cropland increases, and because fallows occupy too high a proportion of the land, farmers look for alternatives to maintain soil fertility. Many farmers manage livestock to graze and capture nutrients from natural pastures and transfer them to cropland in the form of manure. Daytime grazing and nighttime corralling of livestock on cropland between cropping periods, returns faeces and urine to soils, and results in much higher grain yields than the application of manure alone (Powell *et al.*, 1998). For interesting examples of such systems visit (Anon., 2005). Thus, the integration of livestock into West African crop production systems is driven by the need to increase soil fertility, which in turn is driven by population pressure. Unfortunately, even this system does not currently supply sufficient nutrients, as there is insufficient manure to sustain crop production over the long term (Fernandez-Rivera *et al.*, 1995). Manure output varies with feed availability and quality, decreasing during the dry and early wet season as grazing resources diminish.

The potential for large-scale increases in forage production from natural or improved pastures appears to be limited. In many locations, most high producing pastures have been cultivated. Crop residues will likely continue to be a principal source of feed, especially during the dry season. Sustainable increases in livestock production will depend therefore on sustainable increases in crop productivity (Powell *et al.*, 2004). Improvements in crop production can be achieved by intensifying management, for example, incorporating high yielding forage and grain legumes into the cropping system, and through crop and land management techniques that provide feed, and leave sufficient crop residues for soil conservation. Using forage legume fallows, instead of utilising the natural regeneration of indigenous plant species is desirable, but constrained by tenure security and labour (McIntire *et al.*, 1992), fencing, and legume persistence when grown in association with grasses (Powell *et al.*, 2004). Substantial gains in crop and livestock production can also be made through increased judicious use of fertiliser and diet supplements, crop genetic improvement and a wider integration of animal power instead of human power, and legumes into the farming systems.

The Loess plateau of western China

On the Loess Plateau of Gansu, western China, local farming systems have evolved due to population pressure and soil sustainability issues. Soil erosion is a major constraint in the region, and mixed farms have replaced livestock grazing on sloping and terraced land. On the better flat land, intensive cropping has replaced mixed crop-livestock systems. Integration of crops and livestock in western China occur on both a local (on-farm) and area-wide basis (Figure 1).

Benefits of integrated crop-livestock systems over specialised production are well recognized in west China. Animals supply manures for soil fertility, draught power for tillage, and are important to generate cash flow, as opposed to grain production, which is largely consumed for subsistence. Recent Government policies are aimed at increasing animal production. Other policies are aimed at replacing sloping cultivated cropland with either trees or perennial forages (cut & carry) to reduce erosion.

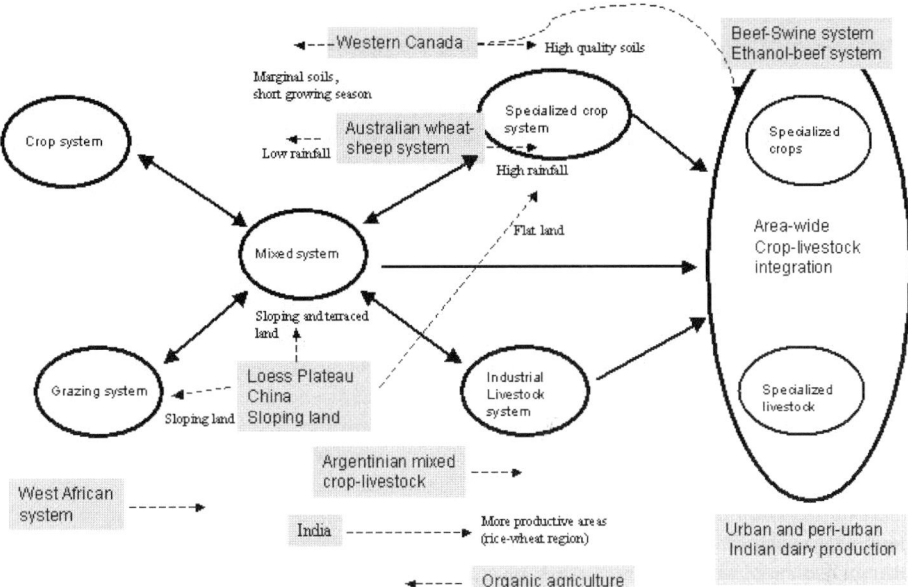

Figure 1 A 'state and transition' model describing the evolution of integrated crop-livestock systems (adapted from Powell *et al.*, 2004). Solid lines represent commonly observed changes between states of crop-livestock integration. Dashed lines represent changes between integration states using specific examples discussed in the paper

In areas where farmers have shifted production from mixed farming to specialised crop production, several risks have been identified. These include increased costs (chemicals and machinery), greater reliance on synthetic fertilisers over animal manures, the move away from largely organic systems to largely synthetic systems (risk of chemical residues, herbicide resistant weeds), and greater vulnerability to price variability in major crop commodities. Development of markets for livestock products would promote mixed systems.

Research is needed on integrating animal and crop production, both on an individual farm-scale, and on a larger industrial scale. Local farming systems are complex, intensively managed, and productive on an area basis. However, the profitability and sustainability of current systems is poor, and the integration of forage resources with cropping may provide solutions. Traditional Chinese farming systems have been highly integrated and there is opportunity for advancement through combining traditional knowledge and practice with newer technologies (GIS, computer simulation tools, etc.) for analysing resource use efficiency.

India

Livestock production in India is divided into urban, peri-urban, rural and transhumant systems (Misri, 1999). In urban and peri-urban systems cattle and buffalo are used for milk production. Dry animals are moved to rural areas where fodder is more available. Many farms around the urban centres specialise in forage production for the urban dairies, and crop-livestock interaction is minimal. Feed resources for these dairy animals include *Brassica napus* (rape), *Zea mays* (maize), *Pennisetum* spp. (millet), *Sorghum* spp., *Avena sativa* (oats), *V. unguiculata, M. sativa, Trifolium alexandrinum* (berseem clover), and crop residues (Misri, 1999). Indian urban and peri-urban dairy systems represent a form of area-wide integration (Figure 1). As milk collection from rural areas is limited by transportation and processing facilities, dairies became concentrated around urban areas.

The long-term sustainability of urban dairies is being questioned due to environmental concerns, competition for resources with people, lack of nutrient recycling (in most instances animal manure is dried into fuel cakes and sold for fuel) and the high cost of transporting forage. An alternative system for milk production has recently been implemented. Co-operatives have pooled resources for collection, processing, transportation and marketing of milk and milk products. As a result AMUL, the largest co-operative society in India (2.3 m producer members and 0.6 billion US $ annual sales; www.amul.com) is the major supplier of milk products, and has encouraged the development of smaller local co-operatives in different parts of India for fresh milk supply. The transportation of milk and milk products is less expensive than transportation of crop residue or green feed. This new system of milk production represents a shift from area-wide integration to integration at a more local level.

In rural areas, over 70% of farms are mixed enterprises (Singh *et al.*, 1997), with a higher proportion of mixed operations in less productive arid and semi-arid zones. Animals are grazed on community pastures, wastelands, forests, and crop fields post harvest. Nomadic tribes use the transhumant system for rearing goat and sheep (Misri, 1999). During the non-crop period livestock is grazed on crop stubble and weeds. In some areas, agreements are made so that transhumant sheep and goats contribute nutrients to cropland through nighttime corralling. In addition to sustainability and ecological benefits, mixed farming in India provides socio-economic benefits such as rural employment (especially for women), social status, alternate sources of income, family nutrition and maximum return from limited land and capital.

Population pressure has reduced the permanent pasture area in India to 4% of total land area (Maehl, 1997). Invasion of pasture, forest and lakes by exotic weed species like *Parthenium hysterophorus, Eupatorium odoratum, Eichhornia crassipes* (water hyacinth) has further reduced grazing area available for livestock. Reduced pasture resource increases pressure on crop residues, which already contributes 66% of feed resources in the county (*O. sativa, Triticum, Sorghum* and *Pennisetum* spp.) (Zerbini & Thomas, 2003). However feed shortages are still severe (Singh *et al.*, 1997; Kalloo, 2004). Therefore, efficiency of producing fodder, both spatially and temporally, is the top priority for sustainability of the livestock industry in India. Perennial crops or trees are also introduced in silvi-pastoral systems or silvi-horticulture systems to extend forage availability during the period when other feedstuffs are unavailable. Restoration of degraded grazing land and identification of suitable forage species for problem lands are also needed (Misri, 1999).

Crop-livestock integration is weakest in more productive agriculture regions of India (*O. sativa-Triticum* system area) (Figure 1). The use of tractor power, fertilisers and irrigation has contributed to less integration in these areas. Valuable crop residues are often burnt (Devendra & Sevilla, 2002). Small farming operations in the northern states like Punjab, Haryana, Uttar Pradesh and Rajastan are specialising in growing forages under irrigation in more than 30% of the cropped area to sell to livestock owners (Devendra & Sevilla, 2002), representing a form of area-wide integration. Sustainability problems (salinity, low fertility, weed problem, soil structure damage, shortage of organic matter) are increasing in these regions.

Southern Australia

The southern Australian crop-livestock system is driven by the need to remain profitable in the face of declining terms of trade and threats to sustainability. Local population pressure is not a factor, if anything, declining rural population and reduced labour availability is an issue.

The southern Australia wheat-sheep farming system is constantly adapting to the prevailing market and biological constraints. Indeed, this high degree of flexibility is one of the appealing features of the system. Farmers continue to practice variations of this system precisely because it has proven resilient in the face of declining terms of trade, and because it provides options for management of sustainability threats. Within the proposed model (Figure 1), the southern Australia farming system currently fits somewhere between the 'mixed system' and the 'specialised crop system'. In south Australia there is an association between average annual rainfall and the relative emphasis put on livestock and grain enterprises. As rainfall decreases and grain yield declines and becomes less reliable, livestock becomes more attractive.

In response to declining profitability of wool throughout the 1990's there has been an intensification of cropping at the expense of pastures and livestock in the medium to high rainfall zones, but not in the low-rainfall zone where reliable cropping options are not available. Since 2000, wool profitability has increased, and sheep meat has been profitable, resulting in an increase in livestock numbers and area under pasture. In addition, constraints to the sustainability of intensive cropping, such as herbicide resistant weeds and dryland salinity, have forced some growers to reintroduce pastures into their cropping systems. Pastures provide many options for non-selective herbicide management options, and can provide a high water use option if a perennial such as *M. sativa* is used.

Australian researchers are making good use of analytical tools (e.g. MIDAS, GrassGro and APSIM) to explore; 1) grazing management options where annual and seasonal climate variability provides major challenges; and 2) analysing rotational sequences where knowledge of soil water and nitrogen is critical for managing integrated crop and livestock production, under a variable climate. When used in conjunction with farmer knowledge, these tools can provide valuable insights into both the biology of the system, and farmer decision-making processes.

Uruguay and Argentina

Both Uruguay and Argentina have a long history of integrated crop-livestock production. Ley pasture systems based on *Festuca* and *Lotus* spp. are rotated with *Triticum*, *Zea mays* and other grains. A recent transition from integrated to specialised cropping (*Zea mays*, *Glycine max* (soybeans) under no-till management) has been most dramatic in the flat, productive soils in Buenos Aires province in central Argentina (Figure 1). In nearby southern Uruguay, where the land is hilly and susceptible to water erosion, mixed systems have been maintained as they have proven more sustainable than specialised cropping (Martino, *pers. comm.*).

Intensive cropping is driven largely by short-term profitability, sometimes at the expense of sustainability. An important question for Argentinian researchers is whether no-till grain cropping provides as many soil benefits as integrated grain-forage rotations. Subsoil constraints, such as compaction, may increase when deep-rooted perennial plants are absent in rotations (Martino, *pers. comm.*). Also, over-reliance on glyphosate, the herbicide that 'supports' no-till cropping systems worldwide, is of concern. However, the argument to retain mixed farming systems must be won on economic grounds. Comprehensive analysis including market and non-market values, and taking into account climate and market variability is needed.

Northern interior plains of North America

Like southern Australia, farms in western Canada and the northern USA are driven by the need to remain profitable in the face of threats to sustainability and trade. Local population pressure is not a factor. Growing season length (frost-free days) is a constraint to plant growth. The dominant forage systems are perennial grass pastures and *M. sativa* or *M. sativa*/grass mixtures grown for conserved forage.

Traditionally, mixed farming has been practiced on poorer lands while specialised crop production is practiced in longer-season areas, where soils are more productive (Figure 1). In response to strong wheat markets in the 1970's even traditional mixed farms abandoned livestock in favour of specialised grain production. Many of these farmers have switched back to integrated systems due to a prolonged period of low profitability and biological constraints in specialised cropping (e.g. herbicide resistant weeds, soil salinity etc.). Currently, approximately 50% of farms in the region are integrated crop-livestock enterprises (Small & McCaughey, 1999) and many farmers have reinvested in infrastructure to support livestock in addition to crops. Despite having both livestock and crops in the same enterprise, the degree of integration is limited by management time, labour and producer knowledge. Integrated farming receives much less attention from researchers and extension workers than specialised production systems.

Example of area-wide integration for nutrient management

Expansion of the livestock sector in western Canada has necessitated partnerships between livestock and crop producers, as the former looked for avenues to utilise the nutrients present in the manure generated from the livestock operations, and also meets government regulations on manure disposal. One example of area-wide integration in response to these drivers involves beef cattle, pastures, grain and hogs (Figure 1). Four individuals who collectively owned 6,000 sows and 900 steers established a company in 1994. Manure produced from the swine operation was utilized to fertilise not only annual crops, but also forage land owned by the company. Cereal harvested from the annual crops was utilized in the swine rations, while the forage land was rented to local producers. As the swine component of the operation continued to grow, the potential to further integrate the operation by adding in a second livestock species – cattle was recognized. This was a logical choice, given that much of the land surrounding the swine operations was marginal and suitable only for forage production. Currently, the operation consists of 40,000 sows, 100,000 finishing and isowean sites, 600 cow-calf pairs and 300 yearling heifers. The number of acres of forage land has increased considerably, such that the total land base now consists of 180 ha of annual cropland, 800 ha of hay, 4000 ha of pasture and an additional 1200 ha that remains undeveloped. A portion of this forage land is rented to local producers and supports an additional 1000 cow-calf pairs during the summer months. Grain for the hog operations is purchased and processed by two mills that are owned and operated by the company.

Example of area-wide integration for ethanol production

Integrating ruminant livestock into ethanol production facilities has presented a new opportunity for area-wide crop-livestock integration. In Saskatchewan, one such facility integrated ethanol production (15 million litre ethanol/yr) with a 20,000 head beef cattle feedlot. This facility currently buys calves, grain and forage resources from farmers in the consortium. Economic analysis has shown that a cluster of 5 ethanol/beef complexes will purchase four times more resources from area farmers (grains, forages and livestock) than a comparable sized US corn ethanol plant.

Organic agriculture in industrialised countries

Organic farming is creating a renewed interest in integrated farming in many parts of the world. The absence of synthetic chemical use means that pest management relies on crop rotation, and soil fertility maintenance relies on crop rotation and animal manures. Over 30% of the land base on Canadian organic farms is dedicated to forage crops (Entz *et al.*, 2001b), compared with less than 10% in conventional production. Therefore, organic farming systems represent a shift from specialized to integrated production at the local farm level (Figure 1).

Organic agriculture is most developed in Europe. In Scandinavia, for example, close to 50% of dairy production is organically managed. Fodder crops for these dairies (including seed crops), must be produced organically, creating new market opportunities for farmers and new challenges for researchers.

Synthesis

Previous workers have suggested that the evolution of crop-livestock integration begins with separate crop and livestock production (subsistence systems), integration (mixed farming), then specialization (current status in many industrialized countries), and finally integration on an area-wide basis (Steinfeld; 1998; Powell *et al.*, 2004). Examples in the present paper suggest that while such an evolutionary pattern clearly exists, especially in response to population pressure and economic and market development, this model does not adequately describe all types of changes in crop-livestock integration. Examples presented here also show that crop-livestock integration is a dynamic process and is practiced at different scales. It appears, therefore, that the evolution of crop-livestock systems is best described by a 'state and transition' model, where systems move to different states depending on transition events (Figure 1). The state and transition model presented here depicts how systems may respond to changes in, for example, population pressure, market opportunities and unexpected environmental changes (development of dryland salinity, herbicide resistant weeds, nutrient loading, pollution, etc.).

While there has been a strong trend to specialised crop production on better quality land in many parts of the world, interest in integrating livestock into the crop production system is gaining strength. This shift from specialized to integrated crop-livestock production is driven by biological, economic and environmental sustainability problems associated with monoculture grain systems, as well as adherence to environmental regulations. Examples presented in this paper suggest that integration occurs in both directions (left and right; Figure 1) resulting in both small-scale on-farm integration and larger scale area-wide integration (Table 1).

Table 1 A summary of the two main types of crop-livestock integration in response to biological, economic and environmental constraints of specialised crop production

Type of crop-livestock integration	Major drivers for integration and location	Requirements for successful integration
Local, on-farm integration	- Soil sustainability (e.g. soil erosion in western China); - On-farm salinity (e.g. Australia); - On-farm economic stress; - Shift to organic production; - Population pressure (e.g. western Africa); - Energy costs; - Pest resistance.	- Knowledge (education); - Labour; - Local markets; - Government support; - Access to capital.
Area-wide integration	- Excess manure nutrients at farm-scale; - Wide-spread salinization; - Necessity to share resources with urban areas; - Opportunities to recycle manure nutrients through crops.	- Co-operation between groups of specialised crop and livestock producers; - Strong environmental legislation; - Government support and facilitation: - Technology (e.g. GIS).

While both on-farm and area-wide integration provides benefits over commodity specialisation, the nature of these benefits differ. Advantages of on-farm integration over area-wide integration are related to the benefits that come with increasing on-farm diversity. On-farm diversity is often the most effective way to; 1) address site-specific problems (e.g. soil erosion on sloping land such as in western China); and 2) exploit site-specific opportunities (e.g. predator-prey relationships in ecological pest control). Local or on-farm integration will have unique social benefits in that individual farmers will retain more control of their whole production system.

Advantages of area-wide integration over on-farm integration are related mostly to labour and economic efficiency. Area-wide integration may also address regional ecological problems such as soil salinisation, better than less coordinated efforts. Area-wide integration conforms to the industrial model of centralisation and standardisation better than on-farm integration, making it a more attractive option to policy makers and investors.

Summary and conclusions

Crop and livestock producers will always be connected, either through on- or off-farm linkages between feed production and manure recycling through crops. It is argued that the spatial nature of crop-livestock integration is important, as the interaction of crops and livestock has the potential to produce large benefits for people and the environment.

Farmers, who manage crops and livestock, adapt to local land and environmental pressures and market opportunities. Examples from a number of countries (Australia, Canada, China) were provided where farms evolved from mixed systems to specialised systems, and then back. While the trend to integration from specialised production in developed countries is perhaps strongest in the organic farming sector, sustainability issues on 'conventional' farms also are driving farmers to reconsider mixed farming systems. In high population density areas such as India and China, a strong move toward specialisation and area-wide integration is apparent. These examples support the conclusions of Steinfeld (1998), that integration is often a function of population pressure, however, the examples given in this paper also demonstrate that crop-livestock systems are dynamic and farmers sometimes switch between systems.

A trend observed in many countries is that mixed crop-livestock systems have largely been abandoned on better quality soils in favour of specialised cropping. A number of sustainability issues have been highlighted in specialised crop systems, e.g. soil salinity, pests (including pest resistance), and soil health. Models such as APSIM, MIDAS and others should enable researchers to assess the combined economic and environmental costs of specialised vs. integrated systems, and enable new systems to be tested and validated.

Farmers have proved their ability to adapt their farming systems to changing social, physical and economic conditions. This adaptability is central to the evolution of crop-livestock systems, and efforts to help farmers adapt must be ongoing. There is a worrisome trend in agricultural R&D for reductionism. This is most evident in the trend for research funding towards biotechnology. Agriculture is not just about soils, plants and animals, but also about people, a consideration that is sometimes woefully lacking in policy decision-making. There is a need to strengthen knowledge and understanding (through research) of the decision-making process used by crop and livestock producers.

References

Anonymous (2005). Crop-livestock technologies. FAO, Rome.
   <http://www.fao-kyokai.or.jp/edocuments/document3/JAPMIX6CROPLIVESTOCK4.DOC.pdf>
Angus, J.F., R.R. Gault, M.B. Peoples, M. Stapper & A.F. van Herwaarden (2001). Soil water extraction by dryland crops, annual pastures, and lucerne in south-eastern Australia. *Australian Journal of Agricultural Research*, 52, 183-192.
Campbell, C.A., G.P. Lafond, R.P. Zentner & Y.W. Jame (1994). Nitrate leaching in a Udic Haploboroll as influenced by fertilization and legumes. *Journal of Environmental Quality*, 23, 195-201.
Campbell, C.A., R.P. Zentner, H.H. Janzen & K.E. Bowren (1990). Crop rotation studies on the Canadian prairies. Canadian Government Publication Centre, Ottawa, ON, Canada, Publication 1841, 133pp.
Clarke, A.E. & R.P. Poincelot (1996). The contribution of managed grasslands to sustainable agriculture in the Great Lakes Basin. The Haworth Press, Inc. New York, 189pp.
Devendra, C. & C.C. Sevilla (2002). Availability and use of feed resources in crop animal systems in Asia. *Agricultural Systems*, 71, 59-73.
Entz, M.H., V.S. Baron, P.M. Carr, D.W. Meyer, S.R. Smith Jr. & W.P. McCaughey (2002). Potential of forages to diversify cropping systems in the Northern Great Plains. *Agronomy Journal*, 94, 240-250.
Entz, M.H., W.J. Bullied, D.A. Forster, R. Gulden & J.K. Vessey (2001b). Extraction of subsoil N by alfalfa, alfalfa-wheat and perennial grass systems. *Agronomy Journal*, 93, 495-503.
Entz, M.H., W.J. Bullied, & F. Katepa-Mupondwa (1995). Rotational benefits of forage crops in Canadian prairie cropping systems. *Journal of Production Agriculture*, 8, 521-529.
Entz, M.H., R. Guilford & R. Gulden (2001a). Crop yield and soil nutrient status on 14 organic farms in the eastern portion of the Northern Great Plains. *Canadian Journal of Plant Science*, 81, 351-354.
Fernandez-Rivera, S., T.O. Williams, P. Hiernaux & J.M. Powell (1995). Livestock, feed, and manure availability for crop production in semi-arid West Africa. In: J.M. Powell, S. Fernandez-Rivera, S. Williams & C. Renard (eds.) Livestock and sustainable nutrient cycles in mixed-farming systems of sub-Sahara Africa. Volume II: Technical papers. ILCA (International Livestock Centre for Africa), Addis Ababa, Ethiopia, (November 22-26), 1993, 149-170
Gentile, R.M., D.L. Martino & M.H. Entz (2004). Influence of perennial forages on subsoil organic carbon in a long-term crop rotation in Uruguay. *Agriculture Ecosystems and the Environment*, (in press).
Haas, G., F. Wetterich & U. Kopke (2001). Comparing intensive, extensified and organic grassland farming in southern Germany by process of life cycle analysis. *Agriculture Ecosystems and the Environment*, 83, 43-53.
Hoeppner, W.G. (2001). The effects of legume green manures, perennial forages, and cover crops on non-renewable energy use in western Canadian cropping systems. MSc thesis, Department of Plant Science, University of Manitoba, Canada, (R3T 2N2), 180pp.

Kalloo, G. (2004). Forage research - new dimensions. 42nd Foundation Day Lecture, Indian Grassland and Fodder Research Institute, India, 12pp. <http://www.icar.org.in/IGFRI/ DDG-Kalloo-Lecture.pdf>

Latta R.A., L.J. Blacklow & P.S. Cocks (2001). Comparative soil water, pasture production, and crop yields in phase farming systems with lucerne and annual pasture in Western Australia. *Australian Journal of Agricultural Research*, 52, 295-303.

Loomis, R.S. & D.J. Connor (1992). Crop ecology: productivity and management in agricultural systems. Cambridge University Press, 538pp.

Maehl, J.H.H. (1997). The national perspective: a sysnthesis of country reports presented at the workshop. In: C. Renard (ed.) Crop residue in sustainable mixed crop/livestock farming systems. CABI, Wallingford, UK, 269-310.

Mapfumo, E., D.S. Chanasyk, V.S. Baron & M.A. Naeth (2000). Grazing impacts on selected soil parameters under short-term forage sequences. *Journal of Rangeland Management*, 53, 466-470.

McCown, R.L. (1996). Being realistic about no-tillage, legume ley farming for the Australian semi-arid tropics. *Australian Journal of Experimental Agriculture*, 36, 1069-1080.

McIntire J., D. Bourzat & P. Pingali (1992). Crop-livestock interactions in sub-Saharan Africa. World Bank, Washington, DC, 258pp.

Misri, B.K. (1999). Pasture/forage resource profiles for India, 11pp.
<http://www.fao.org/WAICENT/FAOINFO/AGRICULT/AGP/AGPC/doc/Counprof/India.htm>

Norton, M.R., R. Murison, I.C.R. Holford & G.G. Robinson (1995). Rotation effects on sustainability of crop production: the Glen Innes rotation experiments. *Australian Journal of Experimental Agriculture*, 35, 893-902.

Ominski, P.D., M.H. Entz & N. Kenkel (1999). Weed suppression by *Medicago sativa* in subsequent cereal crops: a comparative study. *Weed Science*, 47, 282-290.

Powell, J.M., F.N. Ikpe, Z.C. Somda & S. Fernandez-Rivera (1998). Urine effects on soil chemical properties and impact of dung and urine on pearl millet yield. *Experimental Agriculture*, 34, 259-276.

Powell, J.M., R.A. Pearson & P.H. Hiernaux (2004). Crop-livestock interactions in the West African Drylands. *Agronomy Journal*, 96, 469-483.

Singh, K., G. Habib, M.M. Siddiqui & M.N.M. Ibrahim (1997). Dynamics of feed resources in mixed farming systems of south Asia. In C. Renard (ed.) Crop residue in sustainable mixed crop/livestock farming systems. CABI, Wallingford, UK, 113-130.

Small, J.A. & W.P. McCaughey (1999). Beef cattle management in Manitoba. *Canadian Journal of Animal Science*, 79, 539-545.

Steinfeld, H. (1998). Livestock and their interaction with the environment: an overview. In: M. Gill (ed.) Foods, lands and livelihoods - setting the research agendas for animal science. Occasional Pub. No. 21. British Society of Animal Science, 53-56.

Steinfeld, H., C. de Haan & H. Blackburn (1997). Livestock - environment interactions: issues and options. European Commission Directorate-General for Development. Rome, Italy, 127pp.

Zerbini, E. & D. Thomas (2003). Opportunities for improvement of nutritive value in sorghum and pearl millet residues in south Asia through genetic improvement. *Field Crops Research*, 84, 3-15.

Adoption of tropical legume technology around the world: analysis of success

H.M. Shelton[1], S. Franzel[2] and M. Peters[3]
[1]*School of Land and Food Sciences, University of Queensland, Australia, 4072*
Email: m.shelton@uq.edu.au
[2]*World Agroforestry Centre, United Nations Ave. P.O. Box 30677-00100, Nairobi, Kenya,*
[3]*Centro Internacional de Agricultura Tropical (CIAT), Cali, Colombia*

Key points

1. Examples of successful adoption of forage legumes are reported from all continents, where they delivered profitability and often provided multipurpose benefits to farmers.
2. Factors vital to successful adoption were: meeting the needs of farmers; building relevant partnerships; understanding the socio-economic context and skills of farmers; participatory involvement with rural communities; and long-term involvement of champions.
3. Organisation of seed supply, achieving scale-up and forming partnerships to implement adoption are key features.
4. Legumes remain an important but under-exploited resource for tropical farming systems. The alternative to legumes will be greater and more costly use of N-fertilisers and purchased protein concentrates.
5. The R&D organisations will need to provide long-term support and greater investment for legume technologies to deliver benefits to farmers. Support will be needed for training and education programmes to overcome declining availability of forage legume expertise and lack of awareness of opportunity for use of tropical forage legumes.

Keywords: nitrogen fixation, animal production, technology transfer

Significance of tropical legumes in agriculture

Legumes with their associated nitrogen fixation have long been realised to have a large potential contribution to animal production in the tropics. Australian scientists in CSIRO led the early research into forage legumes for grazing (Coleman & Leslie, 1966). They understood that tropical grasses were of lower quality than their temperate counterparts, and that the introduction of adapted legumes into tropical grazing systems would simultaneously address the problem of a) low N status of leached tropical soils, and b) low dietary protein intake by grazing ruminants. The search for adapted tropical forage legumes commenced in earnest after 1950, and by 1990, >17,000 accessions of >20 genera had been introduced into Australia, largely from Central and South America but also from Asia and East Africa. This early Australian enthusiasm for tropical legumes was not shared internationally. Colman & Leslie (1966), when reviewing the IX International Grassland Congress held in Brazil in 1965, noted 'an anti-legume complex' which they said was due to the failure of legumes to provide a stable pasture under grazing either in association with grasses or in pure stands.' Nevertheless, scientists based at International Agricultural Research Centers such as ILRI, ICRAF, CIAT and ICARDA, initiated introduction and evaluation programs for herbaceous, shrub and tree legumes. A large number of germplasm accessions was collected and conserved in gene banks (Maass & Pengelly, 2001). It is timely to review the impact of forage legumes on agricultural systems over the past 50 years. Fittingly, the genesis of this paper was the XIX International Grassland Congress in Brazil in 2001, where the view was expressed that adoption of tropical legume technology may be less than anticipated.

Has the original promise of tropical forage legumes been realised?

Reviews of the uptake of tropical forage legumes around the world have revealed that the original promise of legume technology has not been fully realised (Thomas & Sumberg, 1995; Elbasha *et al.*, 1999; Peters & Lascano, 2003). Pengelly *et al.* (2003) concluded that 'despite 50 years of investment in forage research in the tropics, forage adoption has been relatively poor across all tropical farming systems'.

In Africa, Sumberg (2002) reported that fodder legumes have not achieved their potential in sub-Saharan Africa despite 70 years of R&D promoting forage legumes. He queried the long-held view that the introduction of legumes into mixed farming systems was the key to their upgrade. A similar situation existed in Latin America and the Caribbean. Between 1980 and 2000, of 14 legume cultivars that were released none was well adopted (Peters & Lascano, 2003). Miles & Lascano (1997) reported that 'the impact of *Stylosanthes* spp. (stylos) on tropical American livestock production was not proportional to the research literature generated over the past 30 years or so'. In the southern US, the impact of tropical forage legumes has also been relatively small (Williams *et al.*, 2005; Sollenberger & Kalmbacher, 2005).

However, in spite of the overall consensus that adoption has been lower than expected, there have been many examples of successful uptake of forage legumes. There are good examples of successful adoption of legumes in regions of Asia, especially the use of *Stylosanthes* spp. in India (Ramesh *et al.*, 2005), China (Guodao & Chakborty, 2005), and Thailand (Phaikaew & Hare, 2005). Multipurpose tree legumes have played an important role in southeast Asia where *Leucaena leucocephala* has been a significant forage species in the Nusa Tenggara Timor (NTT) Province of Indonesia (Piggin, 2003), and in the Batangas Province of the Philippines. *Gliricidia sepium* (Gliricidia) is widely used in Indonesia and the Philippines, and leguminous cover crops in the rubber and oil palm plantations of Malaysia have been widely used since the 1800s. In Australia, tropical legumes have also had a significant impact, although only a small number (<10) of the >70 legume cultivars that have been officially released by Government agencies since 1910 have made a noteworthy impact on the pastoral industry.

Reasons cited for poor adoption

Lack of perceived benefits of legumes

There is an emerging view in developing countries that grasses are being adopted more quickly and more strongly than legumes. Legumes were regarded as less resilient than grasses under cutting or grazing, benefits were largely long-term in nature, and grass/legume systems were more complex to manage (Peters & Lascano, 2003). Similarly, in East Africa, the rapid adoption of grasses, such as *Pennisetum purpureum* (napier grass) in cut and carry systems, contrasted with the lack of adoption of herbaceous legumes (Omore *et al.*, 1999).

Miles & Lascano, (1997) and Andrade *et al.* (2004) reported that farmers in Latin America did not appreciate the benefits of legumes. Therefore, for adoption to occur, even of the best cultivars, they argued that targeted education programmes, successful demonstrations and favourable profitability were needed. The objective of targeting low-cost improvement of grass pastures to improve dry season feeding, which worked well in Australia, was not successful in Latin America. Sumberg (2002) suggested that legumes were not just limited by adoption constraints, but that even under favourable circumstances, scientists need to accept that they may not be able to reliably deliver economic benefits to African farmers where there is no tradition of planting legumes for fodder.

Failure of technology

In many instances, lack of adoption could be related to failure of the technology for technical or socio-economic reasons, i.e. the technology did not live up to expectations and/or was not targeted at the appropriate production system.

In Latin America, a major reason for failure of the Australian *Stylosanthes* cultivars Schofield, Cook and Endeavour in commercial pastures was devastation by the anthracnose pathogen. That the cultivars did not persist under grazing was another significant disadvantage. This led to widespread disappointment among farmers, extension workers and consultants (Andrade *et al.*, 2004). Lack of persistence was also cited as a reason for lack of adoption of forage legumes in Africa (Boonman, 1993). In Florida, the slow uptake of *Aeschynomene americana* and *Desmodium heterocarpon* was due to an underestimation of the difficulty of establishing and maintaining the legumes in *Paspalum notatum* (bahia grass) pastures. Farmers found that neither legume was dependable when grown with this competitive grass (Sollenberger & Kalmbacher, 2005).

Socioeconomic factors contributed to the lack of adoption of intercropping and legumes in communal grazing. Attempts to promote intercropping of maize with legumes in East Africa failed due to the high cost of technology, variable rainfall and lack of interest in innovation by older farmers (Ndove *et al.*, 2004). Similarly, Maasdorp *et al.* (2004) found that promoting the multi-purpose use of *Mucuna pruriens* (mucuna) failed due to lack of interest in green manuring or intercropping, due partly to labour constraints of the cash-cropping farmers. Elbasha *et al.* (1999) reported that legume adoption in West Africa was constrained by lack of extension information, credit and seed, high costs of fencing, shortage of labour, insecurity of land tenure and land scarcity, livestock diseases, invasion by weeds, and fire damage. Where land tenure is uncertain, most researchers report failure of adoption. Farmers were simply not interested in investing in their land when they had no assurance of being able to reap the benefits. Pasture improvement technology applied to communal grazed lands by government supported projects, usually suffered from a lack of interest by the pastoralists involved (Pengelly *et al.*, 2003).

Failure in approach

Failure of the key stakeholders to form effective partnerships between farmers and public and private institutions was often cited as reason for lack of adoption (Miles, 2001), leading to ineffective release and follow-up procedures. Andrade *et al.* (2004) stated that, while the release of *Stylosanthes macrocephala* cv. Pioneiro overcame deficiencies of earlier stylo cultivars, the cultivar was not promoted. With no extension support there was no interest from private seed companies, as they did not see a large market.

Lack of establishment of a reliable seed production and supply system to ensure that high quality seed was available at a reasonable price was regularly cited as a key reason for adoption failure, e.g. for *Stylosanthes* in Latin America (Peters & Lascano, 2003), *Vigna unguiculata* (cowpea) in Nigeria (Kristjanson *et al.,* 2004) and *Aeschynomene americana* (aeschynomene) and *Desmodium heterocarpon* (carpon desmodium) in Florida (Sollenberger & Kalmbacher, 2005). Andrade *et al.* (2004) reported that of 3 Australian and 10 South

American *Stylosanthes* cultivars released into the South American market, seed is available for only two - Mineirão (*S. guianensis* var. *vulgaris*) and Campo Grande (a mixture of *S. capitata* & *S. macrocephala*).

Lack of a participatory approach was also cited as a reason for ineffective promotion of legume technology. Douthwaite *et al.* (2002) criticised the International Research Centres for basing their approach on scientific enquiry independent of social factors, rather than on a 'learning selection model' that builds on farmer and group experiences.

Are there any success stories?

Difficulties with promotion and use of forage legumes, and the consequent low adoption rates, are of great concern to the R&D community. Without improved levels of adoption, and explicit demonstration of the relevance and benefits of forage legumes, the good will and support of funding and donor agencies will diminish (Shelton *et al.*, 2000), preventing the realisation of much potential advantage for rural communities.

Our analysis of 19 successful case studies (Table 1) revealed that greater adoption success has been achieved in Asia and Australia than in Africa, the US or Latin America, although Brazil had some notable successes. *Stylosanthes* species and tree legume species dominated the success case studies, while species that delivered multipurpose benefits, such as *V. unguiculata* in West Africa and *Pueraria phaseoloides* (kudzu) in Brazil, were also important. *Arachis* spp. were successful in niche environments and were being adopted in three of the case studies.

Authors of the papers on successful legume adoption prioritised the adoption factors that they considered were important to success. Based on their expert opinion and knowledge of each case study, they were asked to allocate 100 points among a list of possible adoption factors to reflect the relative significance of the factors. This subjective analysis indicated that five key factors were important. The most important was that the technology met a need of farmers. The other factors (which were similar in their priority) were: the socio-economic situation of farmers was conducive to adoption; partnerships between relevant stake-holders (government, private, farmers) were in place; there was long term commitment by key players; and a farmer centred research and extension programme was implemented.

The gross economic benefits were naturally highest where large-scale adoption had occurred e.g. from adoption of *Stylosanthes* in West Africa, southern China, or northern Australia, *Leucaena leucocephala* (leucaena) in Queensland, and from adoption of *P. phaseoloides* in the Amazon of Brazil (Table 1).

Reasons for success

The technology met a need of farmers

Adoption of legumes occurred when the technology met farmers' needs, although the particular need to be met varied among farmers and regions. Examples include:
- *West Africa*; *V. unguiculata* was adopted because it provided multiple benefits, e.g. grain for human consumption, fodder for livestock, and opportunity to rotate with cereals to reduce the impact of the parasite *Striga hermonthica* which causes loss of grain yield (Tarawali *et al.*, 2005b).
- *East Africa*; farmers lacked adequate protein for their stall-fed dairy cows and goats, but did not want to spend scarce cash on expensive concentrates. They preferred instead to

plant fodder shrubs (primarily *Calliandra calothyrsus, Leucaena trichandra* and *Morus alba*). The shrubs required only small amounts of labour for planting and harvesting, and farmers found that they could establish tree legume hedges along pathways, field boundaries, and create soil conservation bunds along contours (Franzel *et al.*, 2003).

- *Northern Australia*; graziers found that dryland annual cropping on fertile clay soils was economically marginal due to uncertain rainfall and variable grain prices. In contrast, good cattle prices and the prospect of an agreeable lifestyle change for ageing farmers encouraged them to move to a lower cost but profitable cattle fattening enterprise. This led to the large-scale adoption of both *L. leucocephala* (Mullen, 2005) and *Clitoria ternatea* (butterfly pea) (Conway, 2005).

- *Gulf Coast of the USA*; there was a market for high quality hay for the horse and dairy industries. New varieties of *Arachis glabrata* cv. Florigraze and Arbrook (rhizoma peanut) were well adapted, the equipment for vegetative propagation was available, and it was profitable compared to alternative land uses (Williams *et al.*, 2005).

- *India*; establishment of *Stylosanthes* to produce leaf meal was a cheap but profitable option for infertile acid soils in arid zones (Ramesh *et al.*, 2005). Establishment was simple with no special equipment required. In southern China, there was also a need for high protein leaf meal for the large numbers of livestock in the region (ruminants and non-ruminants). *Stylosanthes* was well adapted and met this need (Guodao & Chakraborty, 2005).

- *Nusa Tenggara Timor Province of Indonesia*; there was serious land degradation (erosion and weeds) in Amarasi and Sikki Districts in the 1930s. The high population densities required a change from swidden agriculture to sedentary agriculture. Alternatives, such as hand-made terraces failed as they were too labour intensive and difficult to construct. In Amarasi District, farmers found that they could rotate *L. leucocephala* with corn to improve fertility and thus corn production, and the *L. leucocephala* could be used to feed tethered cattle and housed goats (Piggin, 2003). *Lantana camara* (Lantana) was largely eliminated as a weed problem by the system and *L. leucocephala* provided wood for a variety of uses.

The analysis indicated that success could be achieved when the technology led to profitability, on-farm environmental benefits such as fertility improvement or weed control, and other multipurpose benefits - often there was a combination of several benefits. However, most successful examples of adoption of forage legumes were unambiguously profitable for the adopter. Farmers normally choose profit, food and income security, before environmental protection (Peters *et al.*, 2001). However, many scientists and government development personnel continue to justify the extension of forage legume technology by promoting natural resource management benefits, including off-farm benefits such as carbon sequestration and watershed management. The fundamental need for the legume technology to be firstly profitable and then afford delivery of on-farm environmental services as a secondary priority, cannot be emphasised strongly enough.

The technology matched farmers' socio-economic situation and skills

It is necessary to begin with an understanding of the production system in which the legume will be promoted. From the survey, examples of legume technology matching farmers' socio-economic situation and skills include:

- *Eastern Indonesia*; farmers found that planting of *L. leucocephala* was compatible with local farming systems. It could be interplanted into maize patches without decreasing maize yield, and then rotated with maize as a soil fertility building exercise (Piggin, 2003). In China, production of leaf meal from *Stylosanthes* planted into young rubber plantation forests, or horticultural systems provided an income stream for a large and inexpensive workforce, especially women (Guodao & Chakraborty, 2005).

- *Queensland Australia*; graziers needed an intensive, highly productive pasture beef fattening system, capable of delivering similar weight gains to feedlots in order to meet different market options. Those graziers with previous dryland cropping experience had less difficulty in establishing hedge-rows of *L. leucocephala* than graziers without this experience (Mullen, 2005).
- *Asia*; the introduction of well-adapted *Stylosanthes* spp. into communal grazing lands in Northeast Thailand was an easy low cost strategy that delivered multiple benefits (improved livestock diets and improved land fertility) and was therefore well suited to the socio-economic conditions of the region. While many argue that it is not feasible to improve forage on communal lands due to lack of management (Cramb, 2000; Pengelly *et al.*, 2004), it was possible in Thailand due to Government sponsorship of the improvement. Nevertheless, due to overgrazing and changed land use, the benefits of this oversowing strategy have been less than what could be achieved on private land (Phaikaew & Hare, 2005). In India, State and Federal Governments and NGOs also had long-running programmes (25 years) of support for revegetation of village commons and watersheds (Ramesh *et al.*, 2005).
- *Nepal*; in the mid-hill farming areas, small farm size and intensive cropping practices, coupled with back-yard dairy production, created the socio-economic environment for immediate interest and adoption of *Arachis pintoi* (forage peanut). Farmers in the region were accustomed to vegetative propagation, and there were many niche environments where *A. pintoi* could be planted (Robertson, 2005). In Lombok Indonesia, rice farmers needed to improve the diets of goats and cattle fed rice straw. They found that *Sesbania grandiflora* (sesbania) was tolerant of waterlogging and grew extremely well along the rice bunds without reducing yields of the rice crop. The side branches and leaves were easily harvested for fodder and the main stem was eventually cut to provide timber and poles. Nursing mothers also found *S. grandiflora* to be a nutritious vegetable.

The experiences reviewed in Table 1 also confirm that simple innovations are more quickly adopted than complex ones, e.g. a new variety of *Stylosanthes* was more readily accepted in northeast Thailand (Phaikaew & Hare, 2005) or in northern Australia (Rains, 2005) than a new farming system such as the *L. leucocephala* system in Indonesia (Piggin, 2003), or Australia (Mullen, 2005). After 20 years of R&D into suitable *Stylosanthes* cultivars for the Brazilian savannas, Campo Grande was finally released in 2000. This cultivar overcame earlier difficulties with lack of persistence, susceptibility to anthracnose disease and poor seed production, and by 2004, more than 500 t of seed had been produced and sown on almost 150,000 ha of grass pastures (Fernandes *et al.*, 2005).

Partnerships between stakeholders (government, private, farmers) were evident

All of the successful case studies have involved the formation of critical partnerships between the significant stakeholders (Williams *et al.*, 2005; Conway, 2005). In Nusa Tenggara Timor in Indonesia, local village heads, NGOs, church groups, the Dutch Administration, and government departments all showed great commitment to *L. leucocephala* adoption (Piggin, 2003). Local administrators instituted new regulations creating a favourable policy environment for adoption to proceed. These included (a) enforcement of tethering to replace free grazing; (b) credit provided only to those who agreed to plant *L. leucocephala*; (c) promotion of erosion control programmes; (d) regulations requiring the obligatory planting of *L. leucocephala* (1932 & 1948); and (e) promotion of cattle husbandry in livestock distribution schemes.

Table 1 Summary of tropical forage legume success stories

Country or region	Species	Principal uses	Area planted (,000 ha)	No. farmers	Seed production	Estimated gross benefit (US$,000/year)	Reference
Africa							
West Africa	*Vigna unguiculata*	Multi-purpose food, fodder, soil fertility	1,400	~350,000	Not available (n.a).	n.a.	Tarawali *et al.*, (2005b)
West Africa	*Stylosanthes* spp. *Centrosema pascuorum*, *Aeschynomene hystrix*	Legume fodder banks & soil fertility restoration	19	27,000	n.a.	16,500 by 1997	Tarawali *et al.*, (2005a); Kristjanson *et al.*, (2004)
East Africa (Kenya, Uganda, Tanzania, Rwanda)	*Calliandra calothyrsus*, *Leucaena trichandra*, *Morus alba*, *Sesbania*	Cut and carry for dairy cow & goat production, fuelwood, erosion control	~4 M m of hedges	>40,000	n.a.	2.22	Franzel & Wambugu, (2005)
Asia							
Southern China	*Stylosanthes guianensis* cv. Graham & CIAT184)	Leaf meal, agroforestry, green manure, erosion control	>200	>30,000	20 t/yr not including seed sold by farmers	22315	Phaikaew *et al.*, (2004); Guodao & Chakraborty, (2005)
Thailand	*Stylosanthes humilis* (cv. Khon Kaen), *Stylosanthes hamata* (cv. Verano), *Stylosanthes guianensis* (cv. CIAT184)	Forage for cattle, dairy cow and buffaloes; cut & carry and grazing; roadside, communal and private pastures	>300	~12,000	4500 t since 1976	~750	Phaikaew & Hare, (2005)
India	*Stylosanthes scabra*, *Stylosanthes hamata*, *Stylosanthes guianensis*	Revegetation of wastelands for erosion control and fodder	>250	>5000	800 t/yr	>160/ha for stylo seed	Ramesh *et al.*, (2005)
Indonesia (Nusa Tenggara Timor)	*Leucaena leucocephala*	Erosion control, forage for cattle, wood production	70-93	n.a.	n.a.	n.a.	Piggin, (2003)
Indonesia (Lombok)	*Sesbania grandiflora*	Cut & carry forage	n.a.	65,000	Transplanted seedlings	n.a.	Dahlanuddin *et al.*, (2005)
Nepal	*Arachis pintoi*	Cut & carry forage, cover cropping, erosion control, ley farming	n.a.	~20,000 since 1999	Vegetatively propagated	n.a	Robertson, (2005)

Table 1 Continued Summary of tropical forage legume success stories

Country or region	Species	Principal uses	Area planted (.000 ha)	No. farmers	Seed production	Estimated gross benefit (US$,000/year)	Reference
Australasia							
Queensland, Australia	*Leucaena leucocephala*	Grazed pastures	100	400	~10t/yr	20	Mullen, (2005)
Queensland, Australia	*Clitoria ternatea*	Grazed pastures	100	500	300t/yr	~900	Conway, (2005)
N. Territory, Australia	*Centrosema pascuorum*	Hay and pellet production	5 sown yearly	100	35 t/yr	2,844	Cameron, (2005)
N. Australia	*Stylosanthes scabra, Stylosanthes hamata*	Grazed pastures	~1,500 naturalised up to 3,000	n.a.	n.a.	~30,000	Rains,(2005)
Latin America							
Brazil	*Stylosanthes capitata S. macrocephala* (cv. Campo Grande)	Grazed pastures	>150	n.a.	>500 t since 2001	n.a.	Fernandes *et al.*, (2005)
Brazil (Acre State in Amazonian region)	*Pueraria phaseoloides*	Grazed pastures, reclamation of degraded areas	480	5400	25 t/yr	~33,000	Valentim & Andrade, (2005a)
Brazil (Acre State in Amazonian region)	*Arachis pintoi* (cv. Belmonte)	Grazed pastures, reclamation of degraded areas	65	~1000	100 t/yr of vegetative material	~4,000	Valentim & Andrade, (2005b)
Colombia	Forage peanut (*Arachis pintoi*)	Grazed dairy brachiaria pastures	3	100	n.a.	n.a.	Lascano *et al.*, (2005)
USA							
Florida	*Arachis glabrata*	Hay production, grazed pastures, ornamental turf	8	n.a.	Vegetatively propagated	7,000	Williams *et al.*, (2005)
Florida	*Aeschynomene americana, Desmodium heterocarpon*	Grazed pastures	65 & 14	750 & 200	60 & 4 t/yr	2,400 & 540	Sollenberger & Kalmbacher, (2005)

Partnerships that integrated a mechanism for supply of good quality seed at a reasonable price were essential for success (Kristjanson *et al.,* 2004). Similarly, where successful adoption involved vegetatively propagated species such as *Arachis* spp. (Robertson, 2005; Williams *et al.*, 2005; Lascano *et al.*, 2005), an accessible supply of planting material was essential. In some developing countries, legume seed production was achieved by contracting smallholder farmers to produce the seed for government or NGO groups. This approach was first used in the 70s in northeast Thailand to produce seed of *Stylosanthes humilis* (Wickham *et al.*, 1977), and was consolidated by continuing support from the Thai Government for a further 25 years (Phaikaew & Hare, 2005). Approximately 4,500 t of seed have been produced since the scheme commenced (Phaikaew & Hare, 2005). Seed is now exported as well as being purchased for local programmes, and there are farmer to farmer seed sales. Seed producers exported 3 t of *S. guianensis* CIAT184 and 8-9 t *S. hamata* cv. Verano in 2002 and 2003 (Phaikaew *et al.*, 2004). A 'Thai club of seed producers' was formed to handle production and marketing. The Department of Land Development assists with monitoring of seed quality and testing, seed marketing, and seed packaging and storage. Successful contracting of seed production to smallholders has also occurred in India where Government has supported *Stylosanthes* seed production (Ramesh *et al.*, 2005); in Bolivia where the NGO Empresa de Semillas Forrajeras SEFO – SAM has supported production of a variety of legume species for export (G. Sauma, *pers. comm.*); and in Benin where purchase of *M. pruriens* seed from farmers by the NGO Sasakawa Global occurred (Douthwaite *et al.*, 2002). However, partnerships that link smallholder seed production with the private sector should be sought to provide long-term continuity of seed supply.

For broadacre plantings, such as are found in Latin America, the USA and Australia, seed production is normally handled by specialist private seed merchants (Conway, 2005; Williams *et al.*, 2005; Mullen, 2005). Lack of reliable seed supply has limited adoption of *Stylosanthes* spp. in Brazil, *Centrosema pascuorum* in the Northern Territory of Australia, and *Arachis pintoi* in Colombia (Lascano *et al.*, 2005). Andrade *et al.*, (2004) reported that release of new *Stylosanthes* cultivars with high seed yield potential was vital, in order to gain the support of seed producers in Brazil. The seed company that marketed the *Stylosanthes* variety Mineirão from 1996, found that low seed yields, and consequent high market prices, led to a relatively large number of buyers purchasing small amounts of seed. Consequently from 2000, the seed firm mixed Mineirão and Campo Grande (1:3) as a strategy to facilitate sales of Mineirão.

Partnerships with researchers were also an integral part of the successful case studies. Researchers needed to be available to solve problems and progress the technology. In Kenya, researchers have introduced new species (*Leucaena trichandra* and *Morus alba*) to reduce farmers' dependence on *Calliandra calothyrsus* (calliandra). Diversification is important for minimising the effect of a pest or disease attack on any one species, and also for providing a more balanced feed ration (Franzel *et al.*, 2003). In Australia, the beef industry is supporting the breeding of a psyllid resistant *Leucaena* spp., and research into the management of subclinical DHP toxicity, which was recently observed in Queensland cattle herds (Mullen, 2005). Cramb (2000), agreed that successful adoption occurred where there was a timely formation of a 'flexible' coalition of key stakeholders, whose interests converge sufficiently so that their joint resources focus on achieving the adoption outcomes.

Long-term commitment of key stakeholders

Most successful case studies have occurred over a long time period e.g. 10-50 years (Shelton *et al.*, 2000). In central Kenya, 10 years elapsed between the start of the first on-farm trial and the wide-scale uptake of fodder shrubs by farmers. Elbasha *et al.* (1999) noted that realisation of benefits from use of tropical legumes took at least 15 years in West Africa and at least 20 years in Australia. Kristjanson *et al.* (2004) indicated that 20 years were needed to extend the results of *Vigna unguiculata* research in Nigeria. Strategies that have immediate and profitable short-term benefits will be favoured. This was the case with milk production systems in Kenya where adoption of *Calliandra* occurred relatively quickly as dairy producers responded to the immediate increase in milk yield and the opportunity to reduce their use of expensive concentrates. In Brazil, *Arachis pintoi* was quickly adopted in the Amazon due to the introduction of environmental regulations preventing more clearing of forested lands.

Successful adoption was also associated with dedicated champions who were willing to commit their time to achieving a successful outcome (Williams *et al.*, 2005; Conway, 2005; Mullen, 2005; Ramesh *et al.*, 2005). Examples include northeast Thailand where interest in the promotion of *Stylosanthes* commenced in the 1970s with Thai, New Zealand and Australian input. This was followed by World Bank support, and now Japanese support. A key factor was the continuing support from the Thai Department of Livestock Development, and consistent support from key individuals. Such sustained donor support is critical to ensuring the success of the technology.

Farmer centred research and extension programs were implemented

Many workers have pointed to the need for a close interactive working relationship with farmers in order to achieve adoption. Horne *et al.* (2000) were critical of the lack of participatory involvement with farmers during 40 years of forage development programmes in southeast Asia. They proposed an intensive interactive programme of discussion, interviews and on-farm trials jointly conducted with farmers to identify the best solutions to problems identified by the farmers. Tuhulele *et al.* (2000) reporting experiences using Participatory Rural Appraisal (PRA) tools found that careful selection of participating farmers was important and that good facilitation and communication skills with farmers were essential. However, a flexible approach is necessary so that farmer innovations can be absorbed into the technology recommendations and passed on. Further improvements occurred as farmers experimented with the technology, e.g. in Kenya, researchers encouraged farmers to conduct their own experiments, called 'farmer-designed trials', in which farmers planted *Calliandra* as they wished. Several important lessons emerged from these trials, and were incorporated into extension recommendations, including planting in different niches and planting *Calliandra* between rows of *Pennisetum purpureum*, and between *Grevillea robusta* trees along field boundaries.

Braun & Hocdé (2000), referred to the need to change the orientation of existing R&D structures and to develop sustainable community based research capacity. This has happened in northern Australia where graziers have the major say in establishing priorities for research expenditure in the northern Australian beef industry (via the Northern Australia Beef Research Committee). A network of *Leucaena* growers has formed 'The *Leucaena* Network', and has played a major advocacy role promoting research, negotiating with government agencies regarding environmental issues, and conducting training courses for growers.

Within the participatory framework, it was important to ensure that accurate and practical information on the technology was readily available and transmitted to farmers using an appropriate vehicle. Wortman & Kirungu (2000), considered that smallholder farmers in sub-

Saharan Africa were influenced by government extension services, neighbours, relatives, schools and radio. Ndove *et al.* (2004) reported that adoption of legumes in maize cropping systems was assisted by training, demonstrations, tours and on-farm experiments. In Australia, the five most important information sources for graziers were rural newspapers, local Department of Agriculture, national radio, neighbours and stock & station agents (Anon., 2004).

Issues and opportunities for the future

Relevance of tropical legumes to future livestock production

The future of the tropical ruminant livestock sector seems assured with predictions of continuing strong demand for livestock products due to population increase (Kristjanson *et al.,* 2004), and to an increasingly prosperous middle class in developing countries. It will be the integrated crop-livestock smallholder systems of Africa and Asia, and to a lesser extent Latin America, that will be the major suppliers of meat and milk (Delgado *et al.*, 1999). However, production systems will need to intensify to meet demand for higher quality products, while remaining environmentally sustainable. In Africa, there is a move from pastoralism to sedentary farming, requiring greater inputs and a more sustainable production system (J. Lenné, *pers. comm*). In Asia, smallholder livestock farmers are moving from herding systems to tethering systems, or to intensive penned animal systems that require cut-and-carry forage (Fujisaka *et al.*, 2000). Most are planting high yielding grasses to supplement dry season crop residues, and many now purchase feed concentrates to supply protein, energy and minerals thereby improving productivity, especially milk production. As production systems intensify, the inability of farmers to adequately feed their livestock year round will be even more important. The outstanding value of legumes in general and of *Calliandra* in particular is needed to meet this dry season feed gap, with the additional benefit of increased intake of associated poor quality roughage (Shelton, 2004b). It is not surprising that tree legumes figure strongly among the successful case studies. They are multipurpose, and their superior rooting depth delivers excellent water use efficiency and drought tolerance (Shelton, 2004a).

Similarly the broad-scale grazed tropical grass pastures in Australia, Southern USA, and Central and South America will neither be productive or stable unless their N-nutrition is maintained. Declining N status leads to reduced productivity, reduced pasture vigour and weed invasion. Whilst use of inorganic N is feasible in the southern USA (Sollenberger & Kalmbacher, 2005), it is less economically attractive in Australia, Africa, and Latin America.

There is an emerging and significant role for legumes as a protein supplement to reduce reliance on expensive concentrates (Franzel & Wambugu, 2005), which often account for a high proportion of direct costs. Related to this is a rapidly increasing demand for legume hay and leaf meal. This is happening in India (Ramesh *et al.*, 2005), China (Guodao & Chakraborty, 2005) and in Latin America (Peters & Lascano, 2003).

Are we short of adapted legumes?

A considerable amount of adaptation research has already been completed (Pengelly *et al.*, 2004), although there remains a continuing need for germplasm evaluation and genotype X environment studies to better understand the range of environmental niches for legume accessions (Peters & Lascano, 2003). Databases are available, e.g. the CIAT Forage Database

(Barco *et al.*, 2002) and SoFT (Selection of Forages for the Tropics) (Pengelly *et al.*, 2005). The web sites of FAO (http://www.fao.org/) and PROSEA (Plant Resources of Southeast Asia) (http://www.prosea.nl/) have species information; and documentation describing the characteristics of a large number of tropical forage legumes is available (Horne & Stür, 1999). Data on forage adaptation and farmer preference have been linked to a GIS system, based on biophysical and socio-economic data for different regions (Peters *et al.*, 2000). It is hoped that it will be possible to extrapolate the forage adaptation data to new regions, by inputting information on production system, market access, and social preference into the GIS-based tool.

There are many accessions of legumes currently in world germplasm banks, although this resource is under threat due to lack of adequate funding (Maass & Pengelly, 2001). It is vital that the capability to identify new varieties to meet the continuing challenges of pests and diseases is retained, and that there is access to new accessions for niche environments. On occasion, discarded accessions have become relevant, due to the changed circumstances of farmers, e.g. the success of *Clitoria ternatea* cv. Milgarra as a ley legume to restore nitrogen fertility in cropping lands in central Queensland occurred many years after it was first evaluated (Conway, 2005).

In recent years there has been increasing interest in indigenous species as an alternative to introducing exotic species. There are many reasons for this trend: (a) farming communities have detailed knowledge of their use and value, (b) there are ecological and conservation advantages in using indigenous species, and (c) there is a risk of unwanted weed invasion from exotic species. Indigenous forage tree species have generally been used for subsistence feeding rather than commercial systems. Exotic species are usually more vigorous, and produce higher yields than indigenous species, as they have been carefully selected for use as forage and removed from the challenge of pests and diseases present in their native range (Shelton, 2004). Roothaert & Franzel (2001), noted that most fodder tree screening programmes in Africa involved exotic species, but that the local species offer great potential. The challenge is to find trees that can be propagated easily, are highly nutritious, and can be pruned intensively.

Seed production strategies

Most authors of successful case studies cite the need for readily available cheap seed or planting material of good quality. The use of smallholders for contract growing of seed has worked well in many developing countries. Small-scale production of legume seed has successfully matched the skills and resources of smallholder farmers, and has often involved rural women in seed harvesting and cleaning. Nevertheless, in Kenya despite high adoption of fodder trees, seed marketing is still problematic. Commercial firms have not shown interest in marketing seed, and individual seed growers find it difficult to link with potential buyers, who are usually smallholder farmers interested in buying minute quantities. Many NGOs give away free seed and this is a disincentive for farmers interested in selling seed. However, researchers can facilitate seed marketing in several ways: a) by helping producers to produce high quality seed, b) helping producers link with merchants in areas of high demand, and c) helping merchants to sell seed in small packets (Russell & Franzel, 2004). There is a need to improve the linkages between smallholder seed production and the private seed sector, to ensure long-term continuity of seed supply. Improved levels of adoption will help overcome the problem of low market volume for legume seed, thus encouraging private seed merchants to make investments.

In developed countries, a reliable supply of high quality affordable seed is similarly crucial to successful adoption (Conway, 2005; Mullen, 2005; Rains, 2005; Sollenberger & Kalmbacher, 2005). A number of constraints continue to hamper seed production and distribution from private seed companies including: variable environmental conditions affecting production; and variable economic conditions affecting demand (especially export demand); and declining R&D into new varieties. Miles (2001) reported that EMBRAPA (Centro da Empresa Brasileira de Pesquisa Agropecuária) and Unipasto (an association of Brazilian pasture seed firms) are collaborating to ensure pasture seed supply in the region.

Who is best qualified to implement adoption projects?

There is considerable debate concerning the respective roles of farmers, technology researchers, socio-economists, and other stakeholders in the adoption process, and the relative contributions that can be made by a traditional scientific approach and by participatory approaches. In reality, the prime movers of adoption programmes will vary. It is often not the local extension service, but may be a farmer organisation, a university, or locally or internationally funded R&D agencies (Braun & Hocdé, 2000). A major problem for all those wishing to promote the use of forage legumes is declining resources, and especially the declining number of pasture scientists in national and international agencies trained in tropical pasture science R&D. Andrade *et al*. (2004) report that CIAT and national Research agencies in Latin America have reduced their forage research budgets. The number of Australian pasture researchers has declined dramatically over the past decade. Given the increased demand for livestock products, and the clear evidence that poor animal nutrition is the major factor limiting productivity, the need to ensure sustainability of more intensive production systems, national and international agencies will need to increase their investment in education, training, research and extension of tropical pastures if the potential is to be realised.

Scale-up and sustainability

Scientists and development workers are often involved in developing and demonstrating technologies at a small-scale. Scaling-up to large numbers of farmers involves working across villages, districts and provinces. This requires alliances with a multitude of institutions working with farmers, many of which will have limited expertise on forages. The use of expert decision support systems such as SoFT - a database and selection tool for identifying forages adapted to local conditions in the tropics and subtropics, and the linked GIS-based CaNaSTA (Crop Niche Selection for Tropical Agriculture) may assist in this regard. However, these computer tools cannot replace the long-term experience of forage agronomists. Fliert *et al*. (2000) stated that participatory activities are often characterised by intensive guidance processes, which may limit capacity for up scaling. Tuhulele *et al*. (2000) recommended an intensive process of interaction with participating farmers using Participatory Research Appraisal (PRA) approaches, but these approaches can create problems in up scaling to new regions due to the heavy involvement of farmers and researchers in the process of promoting the technology (Horne *et al*., 2000). Some technologies, e.g. maize varieties, spread easily across an area, many fodder legumes require more facilitation because they are 'information-intensive' and involve the learning of new skills. The building of partnerships and coalitions of a range of stakeholders, such as government agencies, NGOs, church organisations, community groups, farmer groups and schools, is the key to successful up scaling (Franzel *et al*., 2003).

Sustaining scaling up and the adoption process is not always possible. *Mucuna* was adopted by >10,000 hillside farmers in Honduras and several thousands of farmers in Guatemala and Southern Mexico. It was used as a relay crop with maize, delivering benefits for soil fertility, soil structure and weed suppression (Peters *et al.*, 2001). However, due to farms becoming smaller and tenure less secure, much of this policy was reversed. Growth in the cattle industry reduced the area of land available for landless peasant to use the *Mucuna-Zea mays* rotation, and *Zea mays* became less attractive relative to other crops and off-farm employment opportunities (Neill & Lee, 2001). In Florida, the availability of cheap nitrogen for use on N-fertilised grass and other more profitable land use options has diminished grazier interest in use of forage legumes (Sollenberger & Kalmbacher, 2005). Market failures and problems with the legume technology can cause the technology to fail. For these reasons, the long-term sustained involvement of researchers to address technical problems is crucial for successful adoption.

Computer modelling

There is much controversy over the role of computer tools for promoting adoption. Pengelly *et al.* (2003) argued that simulation modelling combined with socio-economic research would improve adoption. A simulation model assessing year-round feeding strategies for smallholder crop-livestock systems is being developed by ILRI and their partners (Domingo, 2004). The software enables forecasting of livestock performance under varying feeding and management conditions. A similar strategy is being pursued by ACIAR in southern Africa, south Sulawesi and Indonesia, where an integrated livestock, crop, horticultural and economic model of smallholder systems is being developed.

However, there is concern that the use of computer tools as an aid to generate adoption options for forages in smallholder crop-livestock systems is a high-risk strategy. It may not be possible to achieve a credible, robust model for the smallholder farmers of Africa, Asia and Latin America, because of lack of technical information for the diversity of situations and the high skill levels and sustained commitment needed to develop and support effective models. It is concluded that computer modelling is not going to be an important contributor to improved adoption outcomes. Instead, it is vital that development workers continue to engage with rural communities in relevant and practical ways, especially since the number of professionally trained forage scientists is declining.

Conclusions

Although adoption of tropical legumes worldwide has been less than anticipated, there have been notable adoption successes, especially in Asia and Australia, and to a lesser extent in Brazil. Where data were available the economic returns from adoption have been significant. Successful legumes have included *Stylosanthes*, tree legumes and niche legumes, such as forage *Arachis* species. Their characteristics varied greatly, but with some exceptions, they demonstrated persistence, vigour and longevity under grazing or cut and carry systems, ease of establishment (with the exception of *Leucaena*), and either high seed yield or ease of vegetative propagation. They delivered profitability and multipurpose benefits to farmers, including on-farm environmental benefits.

Meeting the needs of farmers was the most significant factor leading to successful uptake of tropical forage legume technology. Other factors vital to successful adoption were (a) the building of a coalition of relevant partnerships, (b) understanding the socio-economic context and skills of farmers and their farming systems, (c) a participatory involvement with the rural communities involved, and (d) the long-term involvement of champions who ensured the process did not stall and that problems were resolved.

Nitrogen is the key-sustaining element in tropical farming systems, and as ruminant production systems are intensified, there is great potential and opportunity for exploiting tropical forage legume technology. Leaf meals in particular will become more common in the future. The alternative to legumes will be greater and more costly use of N-fertilisers and purchased protein concentrates.

If R&D organisations wish to see the technologies developed from their research programs delivering benefits to farmers, they will need to take extension work more seriously. They will need to be prepared for long-term involvement, and to build partnerships with other organisations with complementary expertise and interest but similar goals. Increased investment will be needed to support R&D programmes, including greater support for long and short-term training and education programmes to overcome declining availability of forage legume expertise and lack of awareness and opportunity for use of tropical forage legumes. Such investment will ensure adoption of tropical forage legume technology, and will increase the economic, environmental and social well being of rural communities.

Acknowledgement

We acknowledge with thanks the editorial inputs and creative suggestions of Dr Bob Clements and Dr Scott Dalzell.

References

Andrade de, R.P., C.T. Karia & A.K. Braga Ramos (2004). *Stylosanthes* as a forage legume at its centre of diversity. In: S. Chakraborty (ed.) High yielding Anthracnose resistant *Stylosanthes* for agricultural systems. ACIAR Monograph 115, 39-50.

Anonymous (2004). Producer R&D awareness and adoption research - northern beef report. Report prepared for Meat and Livestock Australia by Solutions Marketing and Research. 21pp.

Barco, F., M.A. Franco, L.H. Franco, B. Hincapié, C.E. Lascano, G. Ramirez & M. Peters (2002). Base de datos de Recursos Genéticos Multipropósito. Version 1.0. Centro Internacional de Agricultura Tropical (CIAT) Serie CD-ROM.

Boonman, J.G. (1993). East Africa's grasses and fodders: their ecology and husbandry. Kluwer Academic Publishers, Dordrecht, 343pp.

Braun, A.R. & H. Hocdé (2000). Farmer participatory research in South America: four cases. In: W.W. Stür, P.M. Horne, J.B. Hacker & P.C. Kerridge (eds.) Working with farmers: the key to adoption of forage technologies. ACIAR Proceedings No. 95, 32-53.

Cameron, A.G. (2005). *Centrosema pascuorum* in Australia's Northern Territory: a tropical forage legume success story. In: *Proceedings of the XX International Grassland Congress*, Dublin, offered paper (in press).

Coleman, R.L. & J.K. Leslie (1966). The 9th International Grassland Congress and tropical grasslands. *Journal Australian Institute of Agricultural Science*, 32, 261-270.

Conway, M.J. (2005). Butterfly pea in Queensland: a tropical forage legume success story. In: *Proceedings of the XX International Grassland Congress*, Dublin, offered paper (in press).

Cramb, R.A. (2000). Processes influencing the successful adoption of new technologies by smallholders. In: W.W. Stür, P.M. Horne, J.B. Hacker & P.C. Kerridge (eds.) Working with farmers: the key to adoption of forage technologies. ACIAR Proceedings No. 95, 11-22.

Dahlanuddin, Hasniati & M. Shelton (2005). *Sesbania grandiflora*: a successful tree legume in Lombok, Indonesia. In: *Proceedings of the XX International Grassland Congress*, Dublin, offered paper (in press).

Delgado, C., M. Rosegrant, H. Steinfeld, S. Ehui & C. Courbois (1999). Livestock to 2020: the next food revolution. Food, agriculture and the environment discussion paper No. 28. International Food Policy Research Institute, Washington, D.C., 72pp.

Domingo, S.N. (2004). Simulation modelling gaining support in Casren countries. Casren Newsletter Special Edition, April 2004, ILRI, Nirobi, Kenya, 8pp.

Douthwaite, B., V.M. Keatinge & J. Chianu (2002). The adoption of alley farming and *Mucuna*: lessons for research, development and extension. *Agroforestry Systems*, 56, 193-202.

Elbasha, E., Thornton, P.K. & G. Tarawali (1999). An *ex post* economic impact assessment of planted forages in West Africa. ILRI Impact Assessment Series 2, ILRI, Nairobi, Kenya, 68pp.

Fernandes, C.D., B. Groff, S. Chakraborty & J.R. Verzignassi (2005). Estilosantes Campo Grande in Brazil: a tropical forage legume success story. In: *Proceedings of the XX International Grassland Congress*, Dublin, offered paper (in press).

Fliert, van de E., R. Asmunati & W. Tantowijoyo (2000). Participatory approaches and scaling-up. In: W.W. Stür, P.M. Horne, J.B. Hacker & P.C. Kerridge (eds.) Working with farmers: the key to adoption of forage technologies. ACIAR Proceedings No. 95, 83-90.

Franzel, S. & C. Wambugu (2005). Fodder shrubs for improving incomes of dairy farmers in the East African highlands. In: *Proceedings of the XX International Grassland Congress, Dublin*, offered paper (in press).

Franzel, S., C. Wambugu, P. Tuwei & G. Karanja (2003). The adoption and scaling up of the use of fodder shrubs in cental Kenya. *Tropical Grasslands*, 37, 239-250.

Fujisaka, S., I.K. Rika, T. Ibrahim & L.V. An (2000). Forage tree adoption and use in Asia. In: W.W. Stür, P.M. Horne, J.B. Hacker & P.C. Kerridge (eds.) Working with farmers: the key to adoption of forage technologies. ACIAR Proceedings No. 95, 243-253.

Guodao, L. & S. Chakraborty (2005). Stylo in China: a tropical forage legume success story. In: *Proceedings of the XX International Grassland Congress*, Dublin, offered paper (in press).

Horne, P.M., E. Magboo, P.C. Kerridge, M. Tuhulele, V. Phimphachanhvongsod, F. Gabunada Jr, L.H. Binh & W.W. Stür (2000). Participatory approaches to forage technology development with smallholders in Southeast Asia. In: W.W. Stür, P.M. Horne, J.B. Hacker & P.C. Kerridge (eds.) Working with farmers: the key to adoption of forage technologies. ACIAR Proceedings No. 95, 23-31.

Horne, P.M. & W.W. Stür (1999). Developing forage technologies with smallholder farmers – how to select the best varieties to offer farmers. ACIAR Monograph No. 62, 80pp.

Kristjanson, P.M., I. Okike, S. Tarawali & B.B. Singh (2004). Adoption and impact of improved dual-purpose cowpea on food, feed and livelihoods in the West African dry savanna zone. *Agricultural Economics*, (in press).

Lascano, E.E., M. Peters & F. Holmann (2005). *Arachis pintoi* in the humid tropics of Colombia: a forage legume success story. In: *Proceedings of the XX International Grassland Congress*, Dublin, offered paper (in press).

Maasdorp, B.V., O. Jiri & E. Temba (2004). Contrasting adoption, management, productivity and utilisation of mucuna in two different smallholder farming systems in Zimbabwe. In: A.M Whibread, B.C. Pengelly (eds.) Tropical legumes for sustainable farming systems in Southern Africa. ACIAR Proceedings No 115, 15-163.

Maass, B.L. & B.C. Pengelly (2001). Tropical forage genetic resources – will any be left for future generations? In: *Proceedings of the XIX International Grassland Congress*, Sao Paulo, Brazil, 541-542.

Miles, J.W. (2001). Achievements and perspectives in the breeding of tropical grasses and legumes. In: *Proceedings of the XIX International Grassland Congress*, Sao Paulo, Brazil, 509-515.

Miles, J.W. & C.E. Lascano (1997). Status of *Stylosanthes* development in other countries. 1. *Stylosanthes* development and utilisation in South America. *Tropical Grasslands*, 31, 454-459.

Mullen, B.F. (2005). Leucaena in northern Australia: a tropical forage legume success story. In: *Proceedings of the XX International Grassland Congress*, Dublin, offered paper (in press).

Ndove, T.S., A.M. Whitbread, R.A. Clark & B.C. Pengelly (2004). Identifying the factors that contribute to the successful adoption of improved farming practices in the smallholder sector of Limpopo Province, South Africa. In: A.M Whibread, B.C. Pengelly (eds.) Tropical legumes for sustainable farming systems in southern Africa, ACIAR Proceedings No 115, 146-154.

Neill, S.P. & D.R. Lee (2001). Explaining the adoption and disadoption of sustainable agriculture: the case of cover crops in northern Honduras. *Economic Development and Cultural Change*, 49, 793-820.

Omore, A., H. Muriuki, M. Kenyanjui, M. Owango & S. Staal (1999). The Kenyan dairy sub-sector: a rapid appraisal. Ministry of Agriculture/Kenya Agricultural Research Institute/International Livestock Research Institute Smallholder Dairy Project Report, ILRI, Nirobi, Kenya, 37pp.

Pengelly, B.C., B.G. Cook, I.J. Partridge, D.A. Eagles, M. Peters, J. Hanson, S.D. Brown, J.L. Donnelly, B.F. Mullen, R. Schultze-Kraft, A. Franco & R. O'Brien (2005). Selection of forages for the tropics (SoFT) – a database and selection tool for identifying forages adapted to local conditions in the tropics and subtropics. In: *Proceedings of the XX International Grassland Congress*, Dublin, offered paper (in press).

Pengelly, B.C., A. Whitbread, P.R. Mazaiwana & N. Mukombe (2003). Tropical forage research for the future - better use of research resources to deliver adoption and benefits to farmers. *Tropical Grasslands*, 37, 207-216.

Pengelly, B.C., A. Whitbread, P.R. Mazaiwana, & N. Mukombe (2004). Tropical forage research for the future - better use of research resources to deliver adoption and benefits to farmers. In: A.M. Whibread, B.C. Pengelly (eds.) Tropical legumes for sustainable farming systems in southern Africa, ACIAR Proceedings No 115, 28-37.

Peters, M., P. Argel, C. Burgos, G.G. Hyman, H. Cruz, J. Klass, A. Braun, A. Franco & M.I. Posas (2000). Selection and targeting of forages in Central America linking participatory approaches and geographic information systems - concept and preliminary results. In: W.W. Stür, P.M. Horne, J.B. Hacker & P.C. Kerridge (eds.) Working with farmers: the key to adoption of forage technologies. ACIAR Proceedings No. 95, 63-66.

Peters, M., P. Horne, A. Schmidt, F. Holman, P.C. Kerridge, S.A. Tarawali, R. Schultz-Kraft, C.E. Lascano, P. Argel, W.W. Stür, S. Fujisaka, K. Müeller-Sämann & C. Wortman (2001). The role of forages in reducing poverty and degradation of natural resources in tropical production systems. In: Agricultural Research and Extension Network Paper No. 117, ODI, London, 12pp.
<http://www.odi.org.uk/agren/papers/agrenpaper_117.pdf>

Peters, M. & C.E. Lascano (2003). Forage technology adoption: linking on-station research with participatory methods. *Tropical Grasslands*, 37, 197-203.

Phaikaew, C. & M.D. Hare (2005). Stylo adoption in Thailand: three decades of progress. In: *Proceedings of the XX International Grassland Congress*, Dublin, offered paper (in press).

Phaikaew, C., C.R. Ramesh, Yi Kexian, W. Stür (2004). Utilisation of *Stylosanthes* as a forage crop in Asia. In: S. Chakraborty (ed.). High yielding anthracnose resistant *Stylosanthes* for agricultural systems. ACIAR Monograph no 111, 65-76.

Piggin, C. (2003). T he role of *Leucaena* in swidden cropping and livestock production in Nusa Tenggara Timur province, Indonesia. In: H. da Costa, C. Piggin, C.J. da Cruz & J.J. Fox (eds.) Agriculture: new directions for a new nation, ACIAR Proceedings No 113, 115-129.

Rains, J.P. (2005). Stylos: the broad acre legumes of northern Australian grazing systems. In: *Proceedings of the XX International Grassland Congress*, Dublin, offered paper (in press).

Ramesh, C.R., S. Chakraborty, P.S. Pathak, N. Biradar & P. Bhat (2005). Stylo in India – much more than a plant for the revegetation of wasteland. In: *Proceedings of the XX International Grassland Congress*, Dublin, offered paper (in press).

Robertson, A.D. (2005). Forage arachis in Nepal: a simple success. In: *Proceedings of the XX International Grassland Congress*, Dublin, offered paper (in press).

Roothaert, R. & S. Franzel (2001). Farmers' preferences and use of local fodder trees and shrubs in Kenya. *Agroforestry System,* 52 (3), 239-252

Russell, D. & S. Franzel (2004). Trees of prosperity: agroforestry, markets, and the African smallholder. *Agroforestry Systems,* 61-62 (1-3), 345-355.

Shelton, H.M. (2004a). Importance of tree resources for dry season feeding and the impact on productivity of livestock farms. In: L. 't Mannetje, L. Ramiriz,, M. Ibrahim, C. Sandoval, N. Ojeda & J. Ku (eds.) The importance of sylvopastoral system in rural livelihoods to provide ecosystem services. 2nd International symposium on silvopastoral systems, Merida, Mexico, 158-174.

Shelton, H.M. (2004b). Perspectives on forage tree legumes. In: S.G. Reynolds & J. Frame (eds.) Grasslands: developments opportunities perspectives. FAO and Science Pub. Inc., 87-118.

Shelton, H.M., C.M. Piggin, R. Acasio, A. Castillo, B.F. Mullen, I.K. Rika, J. Nulik & R.C. Gutteridge (2000). Case studies of locally-successful forage tree systems. In: W.W. Stür, P.M. Horne, J.B. Hacker & P.C. Kerridge (eds.) Working with farmers: the key to adoption of forage technologies, ACIAR Proceedings No. 95, 120-131.

Sollenberger, L.E. & R.S. Kalmbacher (2005) Aeschynomene and carpon desmodium: legumes for bahiagrass pasture in Florida. In: *Proceedings of the XX International Grassland Congress*, Dublin, offered paper (in press).

Sumberg, J. (2002). The logic of fodder legumes in Africa. *Food Policy*, 27, 285-300.

Tarawali, S., N. de Haan & P. Thornton (2005a). Planted forage legumes in west Africa. In: *Proceedings of the XX International Grassland Congress*, Dublin, offered paper (in press).

Tarawali, S., P. Kristjanson, B.B. Singh, A. Okike & P. Thornton (2005b). Dual purpose cowpea for west Africa. In: *Proceedings of the XX International Grassland Congress*, Dublin, offered paper (in press).

Thomas, D. & J. Sumberg (1995). A review of the evaluation and use of tropical forage legumes in sub-Saharan Africa. *Agriculture, Ecosystems and the Environment*, 54, 151-163.

Tuhulele, M., L.V. An, P. Phengsavanh, Ibrahim, W. Nacalaban, V.T. Yen Hai, T. Tan Khanh, Tugiman, P. Heriyanto, P. Asis, R. Hutasoit, H. Phimmasan, Sukan, T. Ibrahim, B.T. An, E. Magboo, & P.M. Horne (2000). Working with farmers to develop forage technologies - field experiences from the FSP. In: W.W. Stür, P.M. Horne, J.B. Hacker, & P.C. Kerridge (eds.) Working with farmers: the key to adoption of forage technologies, ACIAR Proceedings No. 95, 54-62.

Valentim, J.F. & C.M.S. Andrade (2005a). Tropical kudzu (*Pueraria phaseoloides*): a success history of its adoption in sustainable cattle production systems in the western Brazilian Amazon. In: *Proceedings of the XX International Grassland Congress*, Dublin, offered paper (in press).

Valentim, J.F. & C.M.S. Andrade (2005b). Forage peanut (*Arachis pintoi* cv. Belmonte): a high yielding and high quality tropical legume used in sustainable cattle production systems in the western Brazilian Amazon. In: *Proceedings of the XX International Grassland Congress*, Dublin, offered paper (in press).

Wickham, B., H.M. Shelton, M.D. Hare & A.J. de Boer (1977). Townsville stylo seed production in northeast Thailand. *Tropical Grasslands*, 11, 177-187.

Williams, M.J., K.H. Quesenberry, G.M. Prine & C.G. Olsen (2005). Rhizome peanut - more than a 'lucerne' for the subtropical USA. In: *Proceedings of the XX International Grassland Congress*, Dublin, offered paper (in press).

Wortman, C. & B. Kirungu (2000). Adoption of legumes for soil improvement and forage by smallholders in Africa. In: W.W. Stür, P.M. Horne, J.B. Hacker & P.C. Kerridge (eds.) Working with farmers: the key to adoption of forage technologies, ACIAR Proceedings No. 95, 140-148.

Grasses as biofactories: scoping out the opportunities

N. Roberts[1], K. Richardson[1], G. Bryan[1], C. Voisey[1], W. McNabb[2], T. Conner[3], M. Christey[3] and R. Johnson[1]

[1]AgResearch Ltd., Forage Biotechnology, Private Bag 11008, Tennent Drive, Palmerston North, 5301, New Zealand
Email: Nick.Roberts@agresearch.co.nz
[2]AgResearch Ltd., Nutrition and Metabolism, Private Bag 11008, Tennent Drive, Palmerston North, 5301, New Zealand, [3]Crop and Food Research, Private Bag 4704, Christchurch, New Zealand

Key points

1. Plant biopharming is set to dominate commercial recombinant protein expression for specific proteins.
2. The choice of plant species depends on a multitude of factors and is determined on a case-by-case basis.
3. As a leaf based expression system grasses would have to compete predominantly with tobacco and alfalfa.
4. The grass-endophyte symbiosis offers a number of unique possibilities for biopharming.

Keywords: biopharming, endophyte, forage, transformation

Introduction

Whole genome sequencing is a remarkable scientific tool. Developing tools for recombinant protein technologies is arguably one of the strongest growth areas in applied biological research. The drive for this can be attributed to both the rising demand for highly purified proteins and secondary metabolites, and the costs associated with producing them. The list of organisms being investigated as biofactories for production of pharmaceuticals, nutraceuticals, industrial enzymes and industrial polymers is expanding; although in order to be successful in this area new organisms must first compete with the traditional systems including bacteria, yeast and Chinese Hamster Ovary cells (CHO cells). To our knowledge forage grasses have been overlooked as a potential biofactory candidate. In this paper the potential of grasses as recombinant protein expression systems for the manufacture of commercially valuable products is explored.

Advantages of plant biopharming

Costs

The main reason for investigating plants as biofactories is cost. It has been estimated that to produce a recombinant protein in a mammalian cell culture can cost up to 100 times more than a plant system. Bacterial and yeast fermentation systems fare better but even they are between 10 to 50 times more expensive than plant systems (Giddings, 2001). Furthermore, current plant-based systems are amazingly flexible and are scaleable at comparatively little cost. In comparison, up-scaling cell-based production systems requires the construction of expensive equipment, and animal-based production systems are limited by the breeding cycle of the animal.

Tissue specific gene expression

Plants can generate recombinant proteins using either prokaryotic (chloroplast) or eukaryotic (cytoplasmic and secretory) machinery; they also offer a range of expression organs, e.g. seed, leaf, root, fruit and tubers. A number of these organs can also be used as the delivery system of the final product. Examples include active and passive vaccines (antibodies). Approximately 20% of biopharmaceuticals under development are antibodies, and around 200 of these are already at the clinical trial stage. It is anticipated that the use of secretory immunoglobulin-A (sIgA) will be widely used in the future as a means of passive vaccination because of the stability of sIgA in the mucosa (Schillberg *et al.*, 2003). Plants are capable of producing the most complex antibodies (sIgA) and in some cases plant produced antibodies need no further processing other than quantification.

Quality and safety considerations

Cell systems based on bacteria, yeast or animal lines have inherent risks associated with contamination of fermentation vats by a range of microorganisms. Contaminants can multiply fast, produce undesirable by-products and gases, and result in high costs for cleaning and lost production. If undetected, preparations made from infected vats impart a considerable safety risk. Plant expression systems present no such threat. Furthermore, animal cell lines may also harbour undetected viral or other infectious particles that can cause disease in humans. Again, no such concerns arise from using plant material.

Despite the advantages of cost and flexibility of using plants as mentioned above, to date there are only a relatively small number of commercial products/services manufactured through plant biopharming technologies, some of these are listed in Table 1.

Table 1 Examples of commercial biopharming in plants

Product/service	Company	Web site
Oleosin::protein fusion purification	SemBioSys Genetics Inc.	www.sembiosys.com
Human secreted alkaline phosphatase and botanical therapeutics	Phytomedics Inc.	www.phytomedics.com
Monoclonal antibodies and plasmatic proteins	MeDicaGO	www.medicago.com
CaroRx™ RhinoRx™ DoxoRx™	Planet Biotechnology Inc.	www.planetbiotechnology.com
Lipase, HSA, Lactoferrin, Collagen	Meristem ® Therapeutics Inc.	meristem-therapeutics.com
Avidin, trypsin, β-glucuronidase, aprotinin and oral vaccines	ProdiGene Inc.	www.prodigene.com

Disadvantages of plant biopharming

Expression levels of heterologous proteins

One of the most important criteria in selecting an organism as a biopharming host is adequate expression of the heterologous protein in appropriate tissues. Until recently plants (with their comparatively low expression of introduced genes), were considered to be the poor cousins of the recombinant protein expression world. The situation has changed dramatically through the use of appropriate promoters (tissue specific as well as inducible) and more astute construct design including choice of untranslated regions, insertion of introns, incorporation of Kozac's sequence for initiation of translation, sub-cellular targeting/retention sequences, and optimal use of codons and polyadenylation signal sequences. In addition, the development of chloroplast transformation systems has enabled further increases in protein expression levels. Combined, these have facilitated accumulation of heterologous proteins to 36% of total soluble protein (TSP) in seeds (De Jaeger *et al.*, 2002), 25% of TSP in chloroplast expression systems (Tregoning *et al.*, 2003) and up to 21% of TSP (but more typically 0.5-2%) in vacuole, apoplast or endoplasmic reticulum (ER) -targeted systems (Conrad & Fielder, 1998).

Glycosylation

The majority of proteins that are of high value are destined for the mammalian circulatory system. One of the biggest stumbling blocks for this market is that in their natural host, many of these proteins are decorated with specific carbohydrate residues, a process known as glycosylation. Glycosylation in plants frequently includes the residues $\beta(1,2)$ xylose and $\alpha(1,3)$ fucose that are not produced by mammals, and in some cases have been shown to be immunogenic in humans (for review see Lerouge *et al.*, 2000). A number of strategies exist today where the incorporation of these motifs has been minimised and the glycosylation pattern has been partially humanised by the over expression of $\beta(1,4)$ galactosyl transferase (Lerouge *et al.*, 2000). Prior to secretion from the endoplasmic reticulum, the final key glycosylation reaction required for the maturing polypeptide is the incorporation of Neu5Ac N-terminal sialic acid. At present this is not performed by any of the recombinant plant expression systems; indeed it has been predicted that it will require complex engineering to generate plants with this capability (Gormond & Faye, 2004). However, even CHO cells (the most commonly used commercial mammalian expression system) do not replicate the human sialic acid glycosylation step 100% faithfully. Instead they incorporate both Neu5Ac and Neu5Gc sialic acid motifs where the latter has also been shown to be immunogenic in humans (Varki, 2001). It is predicted that our ability to manipulate this vital step in recombinant protein expression in plants will improve as our understanding of glycosylation processes advance.

Environmental and food safety

Public debate regarding plant biopharming tends to focus on potential negative issues associated with containment of genetically modified organisms (GMOs) such as escapes, pollen spread, horizontal gene transfer, difficult site cleanup procedures and the possibility of contamination of the food chain (Conner *et al.*, 2003). There is general agreement in the science community that these issues need to be addressed on a case-by-case basis when considering any GM species for release. With respect to transgenic grasses, risk of dispersal of species such as *Lolium perenne* L. (perennial ryegrass), *L. multiflorum* Lam. (Italian ryegrass), *Festuca pratensis* Huds. (meadow fescue) and *F. arundinacea* Schreb. (tall fescue) have been categorised as "substantial and widespread" (Ammann *et al.*, 2001) as the species are outcrossing and the pollen dispersed by wind. Stable integration of transgenes into wild populations depends on several factors including whether the introduced gene confers any selective advantage on the progeny or if the transgenic crop is capable of being weedy in its own right. Nevertheless, there is a likelihood of gene flow from transgenic ryegrass to other grass species unless technological solutions such as chloroplast transformation (if chloroplasts are exclusively maternally inherited), terminator technology (imposes reversible sterility on reproductive plant parts), or transgenic endophytes (see later section) are developed. With regard to the issue of food contamination, the use of forage grasses as an expression system provides an attractive alterative to other food crops used directly for human consumption, as humans are unlikely to ingest the material. Equally however, grazing animals must also be considered when risk analyses are being conducted.

Crop choice

For input traits (e.g., herbicide resistance) and output traits (e.g., modified lipid profile oil crops) the trait suits the crop, in contrast, for biopharming the crop suits the product. The crop choice depends on a large variety of factors and will be determined on a case-by-case basis for each product. In many cases the products can be targeted to a variety of subcellular compartments/tissues/organs, the requirements depend on the product itself and its designated use. Currently, biopharming utilises cell culture, root culture, root expression, leaf expression, tuber expression and seed expression with the latter three dominating current industry activity. In addition, hairy root cultures produced by *Agrobacterium rhizogenes* infection have potential as an expression system (Christey & Braun, 2004).

The species chosen must be able to quickly produce a large amount of recoverable recombinant active protein or secondary metabolite and at the same time produce low levels of toxic compounds; in essence it will provide the best compromise between production and profit. Contemporary crops are likely targets for lower-value proteins due to existing agronomic practices and processing systems, e.g. avidin is now produced in *Zea mays* (maize) at 10% of the cost compared with the extraction of native avidin from egg white (Hood *et al.*, 1997).

Extraction and purification are usually the greatest cost components of biopharming. The compromise for grain products is that while it costs relatively little to store in comparison to frozen leaf or fruit material, the cost of extraction and purification from grain is higher than from leafy material. In some cases the technology developed by SymBioSys may help negate these costs. In this technology the recombinant protein is generated as a fusion with oleosin, the purification then takes advantage of the unique oleosin-oil body relationship that is formed during seed formation. A drawback is that the oleosin-fusion cannot be used directly for proteins requiring the post-translational modifications afforded in the secretory pathway.

Zea mays and other seed crops including cereals (*Oryza* spp.- rice and *Triticum* spp. -wheat) and legumes (*Pisum* spp. - pea and *Glycine* spp. - soybean) and some oil seed plants are being investigated as potential crops. While the overall protein yield/ha is lower in seeds compared with leafy systems, the proteins in seeds tend to be stable at ambient temperature and as such can make an ideal storage and delivery system for some products such as active and passive oral or topically-applied vaccines.

Current leafy plants for delivery systems include *Nicotiana tabacum* (tobacco), *Medicago sativa* (Lucerne) and *Lactuca sativa* (lettuce). The choice of *N. tabacum* is predominantly historical where the transformation systems (nuclear and chloroplast) are all well established, and there are the genetic and agronomic factors required to generate a high biomass return. Other advantages for *N. tabacum* include prolific seed yield and the fact that the risk of contamination is reduced as it is not a food crop. As a leaf crop however it must be pre-processed (frozen or dried) before transporting, or processed immediately after harvest due to the relatively unstable environment for proteins in the senescing leaf. Alternative leaf crops under investigation at this time include perennials such as *M. sativa*, which is easy to propagate; hence it is possible to generate large amounts of clonal material with relatively uniform expression. *M. sativa* also has a very consistent glycosylation pattern (Fischer *et al.*,

2004) as well as reduced fertiliser requirements due to its symbiotic relationship with nitrogen fixing bacteria. A further advantage of *M. sativa* is that it can be harvested up to nine times in a year resulting in around 12t DM/ha per year (D'Aoust *et al.*, 2004).

A modified wet fractionation of leaf material has been developed by MeDicaGo that allows the concentration (as opposed to expensive purification) of the bioactive molecule in *M. sativa* pellets to be stored for months before distribution. Combining inducible promoters with wet fractionation would make leafy crops very competitive with seed crops for certain applications such as direct feeding to animals.

Industrial enzymes, proteins and polymers: biopharming and combinatorial biopharming with output trait plants

Billion dollar pharmaceutical markets are not always needed in order to make a biopharming product commercially successful. Avidin, used in numerous laboratory-based assays and processes, until recently was purified from the egg white of chicken eggs. Prodigene Inc. has now commercialised recombinant avidin produced in *Zea mays* and is investigating the production of a number of other technical enzymes (Twyman *et al.*, 2003). *M. sativa* is being investigated as a source of recombinant phytase enzyme that is normally incorporated into animal feed. Austin-Phillips & Ziegelhoffer (2001) reported they achieved economically viable expression in the field and that it could be used directly in animal feed with minimal preparation, thus replacing the microbial enzyme. Not all biopharming will be immediately profitable. An example of this has been the failure to commercialise plants producing biodegradable plastics. Monsanto purchased the rights to this technology in 1994 but has found that even with a plant generating two separate products (plastic from the leaf and canola oil from the seed) that the overall product was not commercially viable (Gross & Kalra, 2002). However, an alternative approach that may be sufficient to tip the scales in favour of commercialisation could be the combination of an inducible promoter with a dedicated perennial leafy plant thus enabling multiple high yields in a single year.

Grasses as biofactories

If grasses are to be considered as biofactories, it is necessary to examine what it would take to engineer such a plant as well as the unique advantages it might deliver in comparison to existing crops. Work performed in New Zealand is detailed below.

Genetic transformation of ryegrass

Technologies for genetic transformation of forage grasses are now well established (Wang *et al.*, 2001). In general, the methodologies have been adapted from those developed for grain species such as *Oryza* spp., *Zea* spp. and *Triticum* spp. but by comparison, *Lolium perenne* is relatively recalcitrant to transformation. Efficient transformation is largely genotype-dependant due to the heterozygous nature of the plants. Consequently the generation of large plant numbers can be a laborious process. Until recently the emphasis has concentrated on development of protocols for stable integration of genetic constructs and regeneration of plants. However, with the advancement in gene isolation techniques and the rapid implementation of plant functional genomics programs, reports of successful gene expression studies in grass species have increased. Routine transformation systems now exist for *Lolium perenne*, *L. multiflorum* and *F. arundinacea* that focus on the manipulation of genes involved in the regulation of flowering and improvement of energy content (Figure 1).

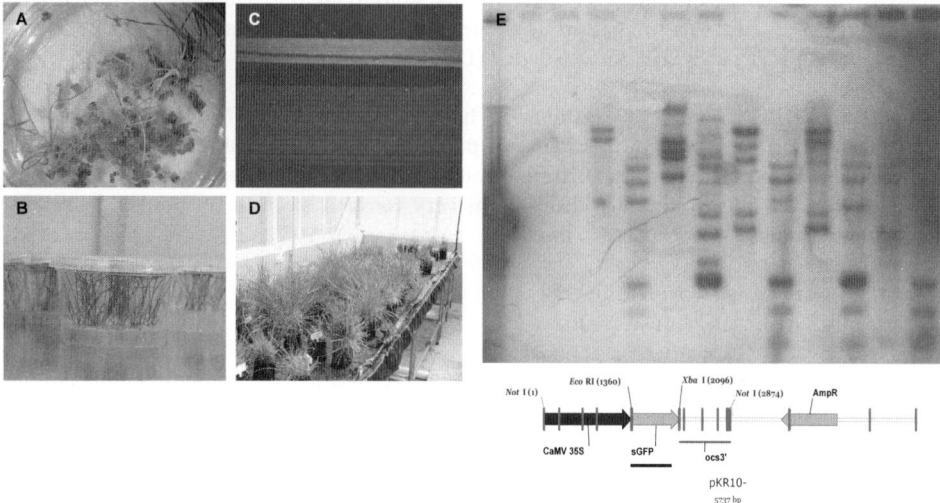

Figure 1 Production of transgenic ryegrass A) Regeneration of plants from embryogenic callus cultures. Micro-projectile bombardment was used to co-transformation calli with plasmids containing an expression cassette for the over-expression of the green fluorescent protein (GFP), and the hygromycin phosphotransferase (*hph*) gene as the selectable marker. B) Transformed *Lolium perenne* plantlets growing *in vitro* following regeneration. C) Constitutive expression of GFP in the leaf blade (top) as compared to a non-transformed plant (bottom). D) Transformed *Lolium perenne* in the greenhouse six months post transformation. E) Southern hybridisation analysis of transformants digested with *EcoRI* or *EcoRI/XbaI* and hybridised with a GFP probe.

Genotypes responsive to tissue culture and transformation remain the key component of a reproducible forage grass transformation system. The genera *Lolium* and *Festuca* are self-incompatible requiring cross-pollination; therefore a high degree of heterozygosity exists within populations that extend to responsiveness of genotypes in tissue culture. Initially this posed a limitation to transformation in *Lolium perenne* where low numbers of independently transformed lines resulted. Transformation frequencies were significantly improved by addressing the need for genotypes that perform well in tissue culture. Altpeter *et al.* (2000a) identified homozygous inbred lines of *Lolium perenne* that responded well in culture, and were subsequently used to develop an optimised transformation protocol. Manipulation of growth media components was used by Cho *et al.* (2000) to produce highly regenerative cultures for improved transformation frequency in *Fescua*. In AgResearch Ltd., the use of tissue culture responsive genotypes derived from elite *Lolium perenne* cultivars, resulted in a five-fold increase in transformation efficiency over material obtained from seedlings.

Direct DNA transfer by micro-projectile bombardment forms the primary method for transformation of many grass species. A range of transformed grasses have now been produced using this method, including *Lolium perenne* (Spangenberg *et al.*, 1995b; Dalton *et al.*, 1999; Altpeter *et al.*, 2000b), *L. multiflorum* (Ye *et al.*, 2001; Dalton *et al.*, 1999), *F. arundinacea* (Cho *et al.*, 2000; Spangenberg *et al.*, 1995a), *F. rubra* L. (red fescue) (Cho *et al.*, 2000; Spangenberg *et al.*, 1995a), *Poa pratensis* L. (Kentucky bluegrass) (Ha *et al.*, 2001), *Dactylis glomerata* L. (Orchardgrass) (Cho *et al.*, 2000) and *Elymus junceus* Fisch.

(Russian wild rye) (Wang *et al.*, 2004). Typically, embryogenic callus cultures derived from tissue culture-responsive genotypes have been used as the target for transformation.

In the future, *Agrobacterium*-mediated transformation is likely to emerge as the favoured method for transformation of grasses. It is currently the preferred method for transforming many cereal species since the successful transformation of *Oryza sativa* using this method (Hiei *et al.*, 1994). The primary advantage of *Agrobacterium* as a strategy for gene transfer is the ability to obtain plants with a simple transgene integration pattern and a higher proportion of plants containing a functional expression cassette. In general, *Agrobacterium*-mediated transformation is likely to generate plants containing 1-3 transgene loci that often contain only a single T-DNA, whereas micro-projectile bombardment frequently generates high copy number loci, complex in structure and containing up to 20 transgene copies (Kohli *et al.*, 2003). *Agrobacterium*-mediated transformation has been compared to micro-projectile bombardment in *Lolium perenne* (Altpeter *et al.*, 2003). Of the 49 plants resulting from *Agrobacterium* transformation, the majority displayed two transgene inserts at independent loci whereas micro-projectile bombardment generated higher numbers of plants with multi-copy inserts.

The potential for somaclonal variation arises following the transformation process and care is required when attributing an alteration in phenotype to the inserted transgene (Conner & Christey, 1994). Analysis of transgenic ryegrass lines in the laboratory has identified chromosomal instability, including ploidy change and aneuploidy. Cytometric analysis of ploidy in a pool of 30 independently transformed plant lines revealed 6 lines to be tetraploid or aneuploid while the remaining 24 were cytologically stable with a diploid chromosome number ($2n=2x=14$). Further to this, molecular cytogenetics has shown rearrangement of 5S and 18S rDNA in transformed lines independent of ploidy change (Ansari & Richardson, *pers. comm.*).

Transgenic approaches are now being implemented as a tool for grass improvement. Target traits include resistance to abiotic stress, improved feed quality, control of plant development, and disease resistance. Generally the altered traits either exhibit a low heritability for the character or the trait has not been identified within the existing germplasm, limiting the ability for improvement by traditional breeding techniques.

Plants encoding genes for the accumulation of fructans have been produced. Ye *et al.* (2001) used the *sacB* gene from *Bacillus* to increase fructan levels in *L. multiflorum*. Contrary to expectations, total levels of fructose were reduced and severely stunted the growth of the resulting plants. In contrast, the *Triticum* genes encoding sucrose-fructan 6-fructosyltransferase and sucrose-sucrose 1-fructosyltransferase under the expression of the CaMV35S promoter displayed a significant elevation in fructan content and increased tolerance to freezing (Hisano *et al.*, 2004).

Improvement of feed value has been approached directly via the regulation of genes associated with lignin biosynthesis to reduce lignin content of the whole plant, and also by the manipulation of flowering genes to minimise plant tissues containing a relatively higher lignin content. Down-regulation of the CAD and COMT genes from *F. arundinacea* was used to decrease lignin content (Chen *et al.*, 2003; Chen *et al.*, 2004). An increase in *in vitro* dry matter digestibility (9.8-10.8%) was associated with these plants. The vegetative phase of *F. rubra* was extended by over-expression of a *Lolium* clone of the terminal flower gene (Jensen *et al.*, 2004). The expression of this gene resulted in the full inhibition of floral development over a two-year period. However, use of this strategy to improve feed value would require the association of a switch for floral induction. In the laboratory the acceleration of flowering by up to three weeks has been demonstrated in *F. arundinacea*, by the expression of an *Arabidopsis thaliana* clone of the *FT* gene (Kardailsky, *pers. comm.*).

Introduction of a partial coat-protein gene into *Lolium perenne*, to provide resistance against ryegrass mosaic virus was achieved by Xu *et al.* (2001). Plant lines within the population of transformed lines expressing the construct displayed high, moderate, low or partial resistance.

Grass-endophyte biopharming

One of the most intriguing possibilities for biopharming temperate grasses is the presence, in many of them, of fungal endophytes, particularly those of the *Neotyphodium/Epichloë* genus. For example, *N. lolii* and *N. coenophialum* are fungal endophytes that live entirely within the intercellular spaces of *Lolium perenne* and *F. arundinacea* respectively. Infection is symptomless and the endophyte relies entirely on the host plant for dissemination via the seed or through vegetative structures (Schardl *et al.*, 2004). The association is mutually beneficial since the endophyte confers a number of biotic and abiotic advantages to the host, including enhanced plant growth, protection from certain mammalian and insect herbivores, enhanced resistance to nematodes, resistance to some fungal pathogens and in some associations, enhanced drought tolerance (Scott, 2001; Schardl, 2001). Some of these benefits are due to the production of fungal secondary metabolites such as peramine (pyrrolopyrazine) and the loline (aminopyrrolizidine) alkaloids. However, endophytes also produce additional secondary metabolites such as ergovaline (ergopeptine) and lolitrem (indole diterpene) alkaloids, which cause mammalian toxicoses.

Although the endophyte comprises a very small amount of the total grass biomass (~0.5%), certain fungal secondary metabolites have been shown to accumulate to very high levels; lolines for example can accumulate to concentrations of up to 5% dry weight (Craven *et al.*, 2001; Spiering *et al.*, 2002). In addition, the endophytes remain metabolically active throughout the growth of the host grass (Tan *et al.*, 2001); hence compounds associated with endophyte infection are continually produced which would allow for year round production. Clearly then, the potential exists to use grasses infected with endophytes to produce highly bioactive secondary metabolites.

Whilst there is considerable interest in secondary metabolites produced from endophytes, biopharming of fungal-derived secondary metabolites from infected grasses is not new. For many decades alkaloids with useful pharmacological properties were obtained from grasses infected with the ergot fungus *Claviceps purpurea* (Hoffman, 1978). Perhaps the most well-known ergot alkaloid is lysergic acid diethylamide (LSD), but other more complex ergopeptine alkaloids (similar to those produced by endophytic fungi of grasses) also accumulate to significant levels. Although nowadays these compounds can be produced directly from fermentation cultures of *C. purpurea*, the above example illustrates the potential for using infected grasses as biofactories to produce secondary metabolites not normally produced by plants themselves. Indeed, some of the enzymes associated with fungal biosynthetic pathways are novel to fungi. Crucially, although fermentation technology has been successful for producing certain secondary metabolites from some fungi, many compounds with interesting bioactive properties are only produced when the fungus is in association with its host grass. The insect feeding deterrents peramine and loline are cases in point, being barely detectable in axenic culture, but accumulating to high levels in infected grasses. This may well be true for many other as yet to be discovered compounds where biopharming of infected grasses will be the only way of producing useful quantities.

At AgResearch Ltd., a group is interested in identifying additional endophyte-derived secondary metabolites, of which it is believed there is a vast undiscovered pool, particularly compounds which may have bioactive properties. In comparison to plants, the filamentous fungal genes involved in biosynthetic secondary metabolite pathways are usually clustered, thus by identifying particular genes frequently associated with these clusters (for example non-ribosomal peptide synthetases), it is possible to quickly isolate all the genes in a particular biosynthetic pathway. A metabolomics approach is being used to help identify the compounds produced from these 'unknown' biosynthetic pathways. Ultimately, isolated gene clusters encoding novel secondary metabolites can be manipulated at the molecular level, allowing, for example, both increased expression of the secondary metabolite, in addition to modifying the metabolite for improved properties. Modified fungal gene clusters could be transferred between different endophyte strains, considerably widening the biosynthetic potential of grasses infected with these strains. With some modification, gene clusters encoding bioactive secondary metabolites from other fungi could be transferred to the endophyte, providing an alternative to more expensive fermentation systems.

Using endophytes as surrogate transformation systems

In addition to modifying existing endophyte secondary metabolites, it is also possible to express heterologous proteins in these fungi. Transformation techniques are well established for filamentous fungi, including endophytes, and as such these can be used as surrogate hosts to introduce foreign genes into *Lolium perenne* or other temperate grasses (Murray *et al.*, 1992). To demonstrate this potential, AgResearch Ltd. transformed endophyte to over-express GFP and re-introduced the endophyte into *Lolium perenne* (Figure 2). Further to this, the levels of heterologous protein produced in these associations will depend in part on the regulatory sequences used to drive the expression of the foreign genes. It is not anticipated that achieving high levels of gene expression will be a problem, since transcripts of some endophyte genes can accumulate *in planta* to levels higher than that of equivalent plant house keeping genes (Johnson *et al.*, 2003).

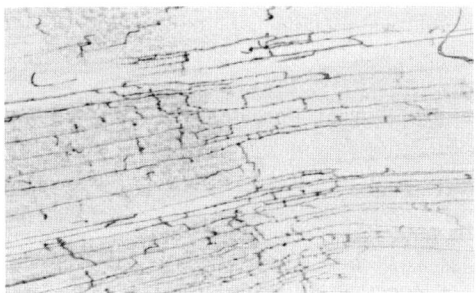

Figure 2 Endophyte (*E. fesctucae*) expressing green fluorescent protein in *Lolium perenne*

Transgenic endophytes pose reduced environmental risks

One significant factor to consider is the degree of environmental risk posed by using transgenic endophytes for biopharming. *Neotyphodium* endophytes are strictly biotrophic and live exclusively within grasses and their seed (Schardl *et al*., 2004). Since endophytes are asexual and cannot be transmitted horizontally (for example through spores) and they are not transmitted through pollen, their only mechanism for dissemination is through seed. Thus, providing seed production is controlled, the risk of large-scale spread of transgenes through other means such as wind-dispersed pollen is eliminated. The strict biotrophic relationship also assists in confining the transgenic endophyte to specific regions that are required to bulk up seed for large-scale plantings.

References

Altpeter, F. & U.K. Possel (2000a). Improved plant regeneration from cell suspensions of commercial cultivars, breeding-and inbred lines of perennial ryegrass (*Lolium perenne* L.). *Journal of Plant Physiology*, 156, 790-796.

Altpeter, F., J. Xu & S. Ahmed (2000b). Generation of large numbers of independently transformed fertile perennial ryegrass (*Lolium perenne* L.) plants of forage- and turf-type cultivars. *Molecular Breeding*, 6, 519-528.

Altpeter, F., J. Xu, Y.D. Fang, X. Ma, J. Schubert, G. Hensel, H. Baeumlein & V. Valkov (2003). Molecular improvement of perennial ryegrass by stable genetic transformation. In: I.K. Vasil (ed.) Plant biotechnology 2002 and beyond. Kluwer Academic Publishers, Dordrecht, 519-524.

Ammann K., Y. Jacot, P. Rufener & A. Mazyad (2001). Safety of genetically engineered plants: an ecological risk assessment of vertical gene flow. In: R. Custers (ed.) Safety of genetically engineered crops. VIB Publication, Flanders Interuniversity Institute for Biotechnology, 60-87.

Austin-Phillips, S. & T. Ziegelhoffer (2001). The production of value-added proteins in transgenic alfalfa. In: G. Spangenberg (ed.) Molecular breeding of forage crops. Kluwer Academic Publishers, the Netherlands, 285-301.

Chen, L., C.Y. Auh, P. Dowling, J. Bell, F. Chen, A. Hopkins, R.A. Dixon & Z.Y. Wang (2003). Improved forage digestibility of tall fescue (*Festuca arundinacea*) by transgenic down-regulation of cinnamyl alcohol dehydrogenase. *Plant Biotechnology Journal*, 1, 437-449.

Chen, L., C.Y. Auh, P. Dowling, J. Bell, D. Lehmann & Z.Y. Wang (2004). Transgenic down-regulation of caffeic acid *O*-methyltransferase (COMT) led to improved digestibility in tall fescue (*Festuca arundinacea*). *Functional Plant Biology*, 31, 235-245.

Cho, M.J., C.D. Ha & P.G. Lemaux (2000). Genetic transformation and hybridization. Production of transgenic tall fescue and red fescue plants by particle bombardment of mature seed-derived highly regenerative tissues. *Plant Cell Reports*, 19, 1084-1089.

Christey, M.C. & R.H, Braun (2004). Production of hairy root cultures and transgenic plants by *Agrobacterium rhizogenes*-mediated transformation. In: L. Pena (ed.) Methods in molecular biology. Humana Press Inc., Totowa, NJ, 286, 47-60.

Conner, A.J. & M.C. Christey (1994). Plant breeding and seed marketing options for the introduction of transgenic insect-resistant crops. *Biocontrol Science and Technology*, 4, 463-473.

Conner, A.J., T.R. Glare & J.P. Nap (2003). The release of genetically modified crops into the environment: II. Overview of ecological risk assessment. *The Plant Journal*, 33, 19-46.

Conrad, U. & U. Fiedler (1998). Compartment-specific accumulation of recombinant immunoglobulins in plant cells: an essential tool for antibody production and immunomodulation of physiological functions and pathogen activity. *Plant Molecular Biology*, 38, 101-9.

Craven, K.D., J.D. Blankenship, A. Leuchtmann, K. Hignight & C.L. Schardl (2001). Hybrid fungal endophytes symbiotic with the grass *Lolium pratense*. *Sydowia*, 53, 44–73.

Dalton, S.J., A.J.E. Bettany, E. Timms & P. Morris (1999). Co-transformed, diploid *Lolium perenne* (perennial ryegrass), *Lolium multiflorum* (Italian ryegrass) and *Lolium temulentum* (darnel) plants produced by microprojectile bombardment. *Plant Cell Reports*, 18, 721-726.

D'Aoust, M.A., U. Busse, M. Martel, P. Lerouge, D. Levesque & L.P. Vezina (2004). Perennial plants as a production system for pharmaceuticals. In: P. Christou & H. Klee. (eds.) Handbook of plant biotechnology. John Wiley and Sons, 759-767.

De Jaeger, G., S. Scheffer, A. Jacobs, M. Zambre, O. Zobell, A. Goossens , A. Depicker & G. Angenon (2002). Boosting heterologous protein production in transgenic dicotyledonous seeds using *Phaseolus vulgaris* regulatory sequences. *Nature Biotechnology*, 12, 1265-1268.

Fischer, R., E. Stoger, S. Schillberg, P. Christou & R.M. Twyman (2004). Plant-based production of biopharmaceuticals. *Current Opinion in Plant Biology*, 7, 152-158.

Giddings, G. (2001). Transgenic plants as protein biofactories. *Current Opinion in Biotechnology*, 12, 450-454.

Gomord, V. & L. Faye (2004). Posttranslational modification of therapeutic proteins in plants. *Current Opinion in Plant Biology*, 7, 171-181.

Gross R.A. & B. Kalra (2002). Biodegradable polymers for the environment. *Science*, 297, 803-807.

Ha, C.D., P.G. Lemaux & M.J. Cho (2001). Stable transformation of a recalcitrant Kentucky bluegrass (*Poa pratensis* L.) cultivar using mature seed-derived highly regenerative tissues. *In vitro Cell Development and Biology – Plant*, 37 (6), 6-11.

Hiei, Y., S. Ohta, T. Komari & T. Kumashiro (1994). Technical advance. Efficient transformation of rice (*Oryza sativa* L.) mediated by *Agrobacterium* and sequence analysis of the boundaries of the T-DNA. *The Plant Journal*, 6 (2), 271-282.

Hisano, H., A. Kanazawa, A. Kawakami, M. Yoshida, Y. Shimamoto & T. Yamada (2004). Transgenic perennial ryegrass plants expressing wheat fructosyltransferase genes accumulate increased amounts of fructan and acquire increased tolerance on a cellular level to freezing. *Plant Science*, 167, 861-868.

Hoffman, A. (1978). Historical review on ergot alkaloids. *Pharmacology*, 16 (Suppl 1), 1–11.

Hood, E.E., D. Witcher, S. Maddock, T. Meyer, C. Baszczynski, M. Bailey, P. Flynn, J. Register, L. Marshall, D. Bond, E. Kulisek, A. Kusnadi, R. Evangelista, Z. Nikolov, C. Wooge, R. Mehigh, R. Hernan, B. Kappel, D. Ritland, L. Chung-Ping & J. Howard (1997). Commercial production of Avidin from transgenic maize: characterization of transformant, production, processing, extraction and purification. *Molecular Breeding*, 3, 291-306.

Jensen, C.S., K. Salchert, C. Gao, C. Andersen, T. Didion & K.K. Nielsen (2004). Floral inhibition in red fescue (*Festuca rubra* L.) through expression of a heterologous flowering repressor from *Lolium*. *Molecular Breeding*, 13, 37-48.

Johnson, L.J., R.D. Johnson, C.L. Schardl & D.G. Panaccione (2003). Identification of differentially expressed genes in the mutualistic association of tall fescue with *Neotyphodium coenophialum*. *Physiological and Molecular Plant Pathology*, 63, 305-317.

Kohli, A., R.M. Twyman, R. Abranches, E. Wegel, E. Stoger & P. Christou (2003). Minireview. Transgene integration, organization and interaction in plants. *Plant Molecular Biology*, 52, 247-258.

Lerouge P., M. Bardor, S. Pagny, V. Gormord & L. Faye (2000). N-glycosylation of recombinant pharmaceutical glycoproteins produced in transgenic plants: Towards humanisation of plant N-glycans. *Current Pharmaceutical Biotechnology*, 1, 347-354.

Murray, F.R., G.C. Latch & D.B. Scott (1992). Surrogate transformation of perennial ryegrass, *Lolium perenne*, using genetically modified *Acremonium* endophyte. *Molecular and General Genetics*, 233, 1-9.

Schardl, C.L (2001). *Epichloë festucae* and related mutualistic symbionts of grasses. *Fungal Genetics and Biology*, 33, 69-82.

Schardl, C.L., A. Leuchtmann, & M.J. Spiering (2004). Symbioses of grasses with seedborne fungal endophytes. *Annual Revue of Plant Biology*, 55, 315–340.

Schillberg, S., R. Fischer & N. Emans (2003). Molecular farming of recombinant antibodies in plants. *Cellular and Molecular Life Sciences*, 60, 433-445.

Scott, B. (2001). *Epichloë* endophytes: fungal symbionts of grasses. *Current Opinion in Microbiology*, 4, 393-398.

Spangenberg, G., Z.Y. Wang, X.L. Wu, J. Nagel, V.A. Iglesias & I. Potrykus (1995a). Transgenic tall fescue (*Festuca arundinacea*) and red fescue (*F. rubra*) plants from microprojectile bombardment of embryogenic suspension cells. *Journal of Plant Physiology*, 145, 693-701.

Spangenberg, G., Z.Y. Wang, X.L. Wu, J. Nagel & I. Potrykus (1995b). Transgenic perennial ryegrass (*Lolium perenne*) plants from microprojectile bombardment of embryogenic suspension cells. *Plant Science*, 108, 209-217.

Spiering, M.J., H.H. Wilkinson, J.D. Blankenship & C.L. Schardl (2002). Expressed sequence tags and genes associated with loline alkaloid expression by the fungal endophyte *Neotyphodium uncinatum*. *Fungal Genetics and Biology*, 36, 242-254.

Tan, Y.Y., M.J. Spiering, V. Scott, G.A. Lane, M.J. Christensen & J. Schmid (2001). *In planta* regulation of extension of an endophytic fungus and maintenance of high metabolic rates in its mycelium in the absence of apical extension. *Applied Environmental Microbiology*, 67, 5377-5383.

Tregoning, J.S, P. Nixon, H. Kuroda, Z. Svab, S. Clare, F. Bowe, N. Fairweather, J. Ytterberg, K.J. van Wijk, G. Dougan & P. Maliga (2003). Expression of tetanus toxin Fragment C in tobacco chloroplasts. *Nucleic Acids Research*, 31,1174-1179.

Twyman, R.M., E. Stoger, S. Schillberg, P. Christou & R. Fischer (2003). Molecular farming in plants: host systems and expression technology. *TRENDS in Biotechnology*, 21 (12), 570-578.

Varki, A. (2001). Loss of N-glycolylneuraminic acid in humans: Mechanisms, consequences, and implications for hominid evolution. *Yearbook of Physical Anthropology*, 44, 54-69.

Wang, Z.Y., J. Bell & D. Lehmann (2004). Genetic Transformation and hybridization. Transgenic Russian wildrye (*Psathyrostachys juncea*) plants obtained by biolistic transformation of embryogenic suspension cells. *Plant Cell Reports*, 22, 903-909.

Wang, Z., A. Hopkins & R. Mian (2001). Forage and turf grass biotechnology. *Critical Reviews in Plant Sciences*, 20 (6), 573-619.

Xu, J., J. Schubert & F. Altpeter (2001). Dissection of RNA-mediated ryegrass mosaic virus resistance in fertile transgenic perennial ryegrass (*Lolium perenne* L.). *The Plant Journal*, 26 (3), 265-274.

Ye, X.D., X.L. Wu, H. Zhao, M. Frehner, J. Nösberger, I. Potrykus & G. Spangenberg (2001). Genetic transformation and hybridization. Altered fructan accumulation in transgenic *Lolium multiflorum* plants expressing a *Bacillus subtilis sacB* gene. *Plant Cell Reports*, 20, 205-212.

The potential of grassland and associated forages to produce fibre, biomass, energy or other feedstocks for non-food and other sectors: new uses for a global resource

M.F. Askew
Head of Agricultural and Rural Strategy, Central Science Laboratory, Sand Hutton, York, YO41 1LZ, UK
Email: m.askew@csl.gov.uk

Key points

1. In developed countries increased areas of land will become available for non-food production. Recent reforms of the Common Agricultural Policy will further intensify this trend in Europe.
2. There is potential for grassland and associated species to contribute to large tonnage markets of energy and bulk fibres, to the supply of fermentation products and to speciality markets, but processes and approaches to the market are not as yet developed.
3. There is potential for the establishment of *Graminaceous* species - specifically for non-food use. For European conditions particular attention is being given to *Miscanthus sinensis* (Miscanthus), *Arundo donax* (Giant Reed), *Phalaris arundinacea* (Reed Canary Grass) and *Spartina* spp. (Cord grass, Marsh Grass).
4. Whilst grass and forage species could be used for individual non-food uses (e.g. biomass for energy), value may be added by adopting a biorefinery approach in which a range of products are derived from the different components in the feedstock.

Keywords: grassland, non-food, climate change

Introduction

"Grassland covers some two thirds of the total agricultural land area of the earth. Its potential annual dry matter yield is at about 40 billion tonnes per annum but its yield is less than a third of that. There is, therefore, ample scope for grassland improvement and development activities.

Food shortage is one of the most serious problems of the present time in the developing countries and there is little possibility of increasing grain production through clearing new lands. Some 50 million tonnes of grain, 40% of the grain production of the world, is used as animal feed. This illustrates the enormous potential for increasing food supply through replacing feed grains by increased pasture production." (F. Riveros, undated).

In developed countries the position is quite different, grasslands are broadly categorised into grazing, conservation and amenity/sports foci. However, because of wider changes that are occurring (or will occur) in the future, these definitions are no longer appropriate for the exploitation of grassland. Rather, there will be an increasing emphasis on the integration of land use, whereby the provision of feed and feedstocks for non-food use and public good occurs simultaneously. This approach will lead to a more sustainable outcome (i.e. economic viability, and environmental/social/cultural acceptability). The rate at which this change will occur will vary politically and geographically, with areas like EU-25 being early adopters.

Whilst this paper is written primarily in the context of Northwest European agriculture, horticulture and wider land use, its content is indicative of the needs and outcomes for many other areas globally.

In developed and relatively densely populated areas as occur in Northwest Europe; surplus land (i.e. surplus relative to food production) is likely to go into integrated production of non-food crops; amenity and environment-ameliorating activities. Such a trend would offer new opportunities for grassland species.

Environment ameliorating activities

Requirements to ameliorate/prevent environmental 'decline' or improve the wider environment fall into the sector of public-good. Their impact can be enormous, but because such activities cover larger areas of land than occur on one holding or interact with other policies (e.g. water management, erosion control, biodiversity etc.), they are difficult for an individual business to influence or manage, and difficult to quantify in terms of financial cost: benefit. However, in some instances the economists' methodologies of Contingent Valuation may be valuable indicators.

Environment-ameliorating activities impinge considerably at the international level. Significant recent developments include the Kyoto Agreement (UNFCCC, 1997) and the United Nations Convention on Biological diversity (UNEP, 1992). Both treaties offer considerable opportunities for grassland and forages to be expanded as a means of ameliorating global warming and promoting environmental goods and biodiversity.

Non-food uses of forage species

Non-food uses of plants are not new, but with the exception of species like *Gossypium hirsutum* (cotton), *Hevea brasiliensis* (rubber) and *Elaeis* spp. (oil palm), have not in recent years been internationally exploited - especially in cool temperate areas. Much of this under-exploitation may be attributable to the availability of fossil oil-derived feedstocks, which have advantages of known technology, price and uniformity of feedstock. It is to be noted however that starch, from many diverse sources, is a major international food and non-food feedstock.

For grassland and other forage species to reach similar levels of knowledge and commercial exploitation will take time. Nonetheless, evidence from related species like *Triticum aestivum* (wheat) offers an indication of potential that may be exploited in preparation of non-food products.

A large number of metabolites have been identified in common UK tree species. These were catalogued by Central Science Laboratory; http://treechemicals.csl.gov.uk/review/index.cfm.

Hitherto such market opportunities were unexploited. However the position is less well documented with many forage grasses. This aspect needs action if value is to be added to forages through simple extraction or biorefining technologies. It seems likely that biorefining will be a key component of exploitation of sustainable biomass in the future since it allows fullest economic exploitation of biomass.

Current markets for forages in the non food sector

Current markets (or those close to exploitation) fall into 2 categories; 1) large tonnage commodity markets for production of energy or bulk fibre, or 2) using forage crops as a feedstock for fermentation, and speciality markets using grasses like *Nandina* spp. (bamboo) for added-value fashion fibres (in Europe) or building (in Asia), or Miscanthus for plant pots.

Evidence reported recently (Askew, 2001) indicated anticipated growth patterns, on a global scale, for all biorenewables. These estimates, which at that time excluded primary energy and biorefining markets, are shown in Tables 1 and 2.

Table 1 Production of crop derived raw materials for industrial use – million tonnes[1]

	Europe	USA	Global
Vegetable oils[2]	2.6	3.0	12.5
Starch	2.4	6.5	15.0
Non-wood fibres	0.5	3.0	23.4
Total	5.5	12.5	50.9

Source: IENICA Report UK (2000); [1]1998 Figures; [2]Practical applications (palm oil = soap; starche = paper industry; fibres = paper industry).

Table 2 Anticipated growth: production of crop derived products - million tonnes

	Global output 1998	Global output 2003	% growth
Vegetable oils[1]	12.5	19.8	58
Starch[2]	15.0	22.5	50
Non-wood fibres[3]	23.4	28.4	21
Total	50.9	70.7	38.9

Source: IENICA Report UK (2000); [1]Vegetable Oils (projection based on forecast for EU growth - Source: FEDIOL, 2000); [2]Starch (projection based on forecast EU growth - Source: National Starch, 2000); [3]Non-wood fibres (projection based on Paper Industry Research Association (Pira) global forecast for pulp and paper use combined with EU figures for non-wood fibre production).

Biomass for energy

Energy is currently a key interest area. This is due to the impact of changes in global atmosphere as reflected in the recently ratified Kyoto Agreement, and drivers for renewable energy e.g. EU White Paper on renewable energy, (White Paper for a Community Strategy and Action Plan) which identifies 135 million tonnes oil equivalent as the contribution of biomass to energy generation for heat and electricity in the former EU15 by 2010 (Tables 3 and 4), and the EU legislation on biofuels (EU Biofuels Directive, 2003) which projects an increasing contribution to substitution of fossil-oil derived gasoline and diesel oil (projections stand at 20% by 2020).

Table 3 Current and projected extent of renewable energy sources in EU-15

Type of energy	Energy units	Share in the EU in 1995	Projected share by 2010
Biomass	Mtoe[1]	44.8	135
Geothermal			
Electric	GW	0.5	1
Heat (incl. heat pumps)	GWTH[2]	1.3	5
Hydro	GW	92	105
Large	GW	82.5	91
Small	GW	9.5	14
Passive solar	mtoe	35	
Photovoltaics	GWp[3]	0.03	3
Solar thermal collectors	million m²	6.5	100
Wind	GW[4]	2.5	40
Other	GW	1	

Source: EU, (1997); [1]Mtoe = million tonne oil equivalent; [2]GWTH = giga watt thermal; [3]GWp ; giga watt photo; [4]GW = giga watt

Table 4 Current and projected contribution of renewable sources of energy to electricity generation in EU 15

Type of energy	Actual in 1995		Projected for 2010	
	TWh[1]	% of total	TWh	% of total
Total	2,366		2,870	
Wind	4	0.2	80	2.8
Total hydro	307	13	355	12.4
Large (including pumped storage)		(270)		(300)
Small		(37)		(55)
Photovoltaics	0.03	-	3	0.1
Biomass	22.5	0.95	230	8
Geothermal	3.5	0.15	7	0.2
Total renewable energies	337	14.3	675	23.5

Source: EU (1997); [1]TWh = Terrawatt hrs

Considerable emphasis has been laid on the development of *Miscanthus* spp. This is commonly, but incorrectly, called Elephant Grass. The primary market has been for production of electricity/combined with heat and power, although secondary markets, such as manufacture of plant pots, are being developed. In terms of the wider exploitation of this plant, an integration of uses needs to be achieved in order to optimise growers' financial returns - energy may not offer the most profitable market (Table 5). This probably summarises the position for most forages.

Biomass for liquid fuels

Feedstocks for gasoline replacement could be sugar (sucrose) or starch-derived, but are more likely to be derived from low value cellulosic materials, e.g. cereal straw in the immediate future (technology exists with the IOGEN company in USA at present). Whilst current technologies to produce diesel replacements from biomass are heavily focussed upon esterification of vegetable oils (e.g. rapeseed oil), it seems likely that the next generation of bio-diesel will be developed from pyrolysis of plants or animal wastes, producing a bio-oil, which could then be refined. Gasification may offer opportunities too.

Table 5 Alternative markets for *Miscanthus* spp. at a standard dry matter

Base price for *Miscanthus* spp. intended for;	£/tonne
Power generation	20-40
Equine bedding	45-70
Bagged equine bedding	160-200
Organic straw	70
Industrial fibres and composites	70

Source: D.B. Turley (*pers. comm.*)

Markets for liquid fuels are extensive and it seems unlikely that land-based industry in its totality could produce enough feedstock to satisfy them totally. Nonetheless, feedstocks from land-based industry could form an integral part of a broader feedstock stream, which would include urban/municipal and other wastes.

Examples of current potential for bioethanol production are given in Table 6. Bioethanol provides a substitute fuel for gasoline. Usage varies but will begin at or about 5% volume/volume level in Europe.

Table 6 Tonnes of feedstock crop required to produce 1 tonne of bioethanol and typical yields of bioethanol per hectare of feedstock crop

Ethanol feedstock (typical field yield)	Feedstock requirement per tonne of ethanol produced (tonne)	Estimated ethanol yield from typical UK crops (kg/ha per year)
Starch crops		
Potatoes (40 t/ha)	11[a]	3600
Wheat (8 t/ha)	2.5-3.0[a]	2600 - 3200
Sugar crops		
Sugar beet (53 t/ha)	11-12.5[a]	4240- 4818
Lignocellulosic		
Grown		
SRC* (32-35 odt/ha)	5.5-7.5[b]	1,200-1,650
Miscanthus spp. (10-12 t/ha)	5.5-7.5[c]	1,400-2,000
Waste or co-product		
Hardwood	5.5-7.5[b]	5-6
Softwood	6.25-9.75[b]	3-5
Straw	4.25-6.25[b]	750-1050

*SRC = Short Rotation Coppice is harvested every 4 years; the yield indicates the equivalent annual ethanol production potential per hectare. Source: [a]derived from Marrow *et al.* (1987); [b]derived from Marrow & Coombs (1990); [c]estimated based on material composition (Turley *et al.*, 2005)

Anaerobic digestion is a further area where a range of feedstocks from land-based industry may provide energy. This process has been used on graminaceous species in the past and currently new projects like "Greenfinch" in UK are testing potential. The essence of such processes is shown in Figure 1.

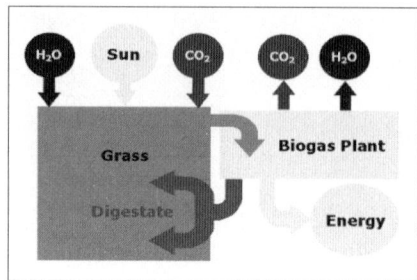

Figure 1 Anaerobic digestion of grass (Source: www.greenfinch.co.uk)

Promotional information provided by FNR in Germany (Fachagentur Nachwachsende Rohstoffe, 2004) indicated potential production of biogas from ensiled forages (Table 7). However, at present most interest in anaerobic digestion is focussed upon animal wastes.

Table 7 Potential production of biogas from ensiled forages

	Tonnes required to produce		
	55kw	330kw	500kw
Grass silage	400	1500	-
Maize silage	600	2500	1700

Source: FNR (2004)

Grass (2004) reported on a pilot unit in Switzerland intended to integrate biorefining and anaerobic digestion. The Schaffhausen installation was found not to be economically viable and the operation was stopped in summer 2003. Production-related reasons for the failure include low revenues from product sales (blow-in insulation) and the unsatisfactory performance of some plant components (e.g. fibre drying and packaging).

The plant produced fibreboard insulation, fibres and biogas from grass, for use in technical applications. Innovations regarding raw material fractionation, production of biogas in the UASB reactor, and grass washing were demonstrated successfully. Technical fibres were further processed on-site for production of a 'blow-in' insulation product marketed under the brand name *2B Gratec*. Certification and market introduction of *2B Gratec* was successful. Biogas was utilised in a combined heat and power plant. Heat was used internally for drying the fibres.

The power was certified and marketed under the label 'Nature Made Star'. The plant had a raw material throughput capacity around 0.8 t DM/h. The yield of *2B Gratec* was 500-600 kg/t DM and the biogas yield was 150-250 m^3 at around 6 kWh/m^3, depending on raw material quality. Relative costs and values of inputs and products are shown in Table 8.

Table 8 Product attractiveness

	Investment cost	Yield	Value added	Large-scale product marketing	Overall attractiveness
Fibre insulation boards	High	High	High	Possible	High
Fibre blow-in insulation	High	High	Medium	Difficult	Medium
Grass juice for animal feed	Low	High	Low	Possible*	Medium
Dried protein for animal feed	High	Medium	Medium	Easy	Low
Biogas/energy	High	Medium	Low	Easy	Low

*In small-scale production units. Source: Grass (2004)

Grass (2004), concluded that:
- Wet fractionation of grass is a powerful tool for the generation of added value from several products;
- Insulation boards combine high yield with high added value and appear as the most attractive of the products examined;
- Production of ethanol, biogas and power is cost-intensive and contributes little to the overall economics of a biorefinery;
- Linking two or more expensive new production lines results in the spreading of risks and should be avoided.

Whilst the potential market for fibres in textiles is enormous and increasing, emphasis is on polyester and cotton (IENICA, 2000).

Further, estimates for the automobile industry in EU show considerable potential to replace fibreglass with plant fibre: currently this is sourced from hemp or flax. Whether or not fibre from forage species could substitute has not been reported.

A note on some individual *Graminaceous* species under development for the non-food sector

Miscanthus sinensis (Miscanthus)

Miscanthus is a perennial graminaceous crop, growing up to 4m in height under European conditions. 'Life expectancy' can be up to 20 years. Dry matter yields vary according to location but can reach 15 t DM/annum.

Key agronomic aspects of production are:
- Not suited to drier/drought areas;
- Currently propagated by rhizomes – c. 20,000/ha;
- Weed control in the establishment phase is essential.

Arundo donax (Giant Reed)

Arundo donax (also sometimes referred to as Giant Cane, Wild Cane, Common Reed, Spanish Reed, False Bamboo or Dumb Cane) is native to south-eastern Europe, and so is already adapted to EU agro-climatic conditions. It is quite common in the Mediterranean where it occurs wild in marshy areas or by rivers. It is often planted as a windbreak at the edges of cultivated fields, on the banks of dykes etc., and can help maintain soil structure in these situations due to the abundant root system. Stems are stiff, smooth and hollow, usually around 5 cm thick, growing up to a height of 6m. The pointed leaves are greyish-green and usually 2-5 cm wide and up to 30 cm long. *Arundo donax* flowers infrequently in late summer with purple-brown flower heads borne in long, dense, plume like panicles.

The fibre produced by the crop is of high quality and has a long, thin structure making it suitable for a wide variety of uses (Table 9). In suitable conditions it has been shown to be a potentially prolific producer of biomass, capable of yielding up to 34 t DM/ha per annum for several years. However, it usually takes 3-5 years to reach its full biomass production. *Arundo donax* can tolerate severe drought conditions (yields of up to 19t DM/ha can still be achieved), and is generally found in warmer and drier regions than other reeds. Thus it appears to be more economical and environmentally favourable to grow under moderate irrigation without dramatically reduced yields.

Arundo donax can either be harvested annually or biannually depending on production expectations and growing conditions. Seed viability is currently unknown but it is clear that the crop requires replanting every 25-120 + years to maintain productivity. Unlike some novel crops, mechanical means are available for both planting and harvest of *Arundo donax*. However, two main problems arise when growing the crop; I) the interconnecting root mats form debris dams in rivers and increase the risk of flooding; ii) the crop ignites easily and can cause intense fires if not controlled with care. The requirements for fertilisers on the crop are also low due to the dry leaves returning to the soil enriching it with organic matter.

Table 9 Potential non-food uses for *Arundo donax* (Giant Reed)

Pipe organs
Basketry
Fishing rods
Pharmaceuticals
Soil erosion control
Industrial cellulose
Pulp, paper
Feedstock for electrical energy
Panels, flooring, beams

Few pests have been reported on *Arundo donax*, and so the requirement for pesticides is negligible in most cases. The crop also appeals to many growers due to the low agrochemical inputs required; this is also beneficial to the environment. Whilst there is currently an absence of demand for products from *Arundo donax*, the crop has not yet become domesticated. However it does have potential in a number of non-food market sectors (Table 9).

Phalaris arundinacea (Reed Canary Grass)

Phalaris arundinacea is a robust coarse perennial, widely distributed across temperate regions of Europe, Asia and North America. It grows to between 0.6–2.0 m high, and has hairless light green or whitish green leaves 10-35 cm long and 6-18 cm wide. Flowering occurs in June to August, and seed is produced. *Phalaris arundinacea* spreads naturally by creeping rhizomes, but plants can be raised from seed. The plant frequently occurs in wet places, along the margins of rivers, streams, lakes and pools.

Until the mid 1950s *Phalaris arundinacea* grew wild and received little scientific or commercial attention. Researchers then noticed that the plant possessed two desirable characteristics: the ability to withstand drought and conversely excessive precipitation.

More recently, the species is being evaluated in Sweden as a fibre and energy-producing crop, where there is a breeding programme evaluating *Phalaris arundinacea* grass as a potential source for fibre from pulping and for fuel. Current production and yields are shown in Table 10. (NB uses of crop vary in this data).

Table 10 Current production and yields of *Phalaris arundinacea* (Reed Canary Grass)[1]

Area	Number of cuts	Yield t DM/ha
USA	3	11
USA	1	4.4-8.6
Canada	3	9.5 – 12
Sweden	2	10
UK	1	4

[1]uses of crop vary in this data. Source: Chisholm (1994)

Spartina spp. (Cord Grass/Marsh Grass)

Spartina pectinata and *S. cynosuroides* occur naturally in western Europe, North America and Africa. *Spartina cynosuroides* is found on salt or brackish marshes from Massachusetts through Florida to Texas, and *S. pectinata* on marsh shores or wet prairies from as far north as Newfoundland through the prairie states as far south as Texas (Gleason, 1952). They are related to the native esturine species *S. anglica* and *maritima*. *Spartina* spp. spread by means of scaly creeping rhizomes to form clumps and mats.

All species are C4 pathway species and have higher carbon assimilation rates than C3 forage grasses, and are more efficient interceptors of radiation receipts. They also have a significantly higher uptake of CO_2 and are less sensitive to chilling than annual C4 species. The yield potential of these grasses may exceed 10-15 t DM/ha.

Spartina pectinata and *cynosuroides* have been shown to be adaptable to a range of growing conditions and to produce higher yields than most natural grasses with a low input of fertiliser. However yields are lower than seem possible from other biomass crops such as *Miscanthus*. The advantage of these species is their potential to be established from seed, their greater adaptability to adverse soil conditions, low fertiliser requirement and their higher dry matter content earlier in the winter. It is likely that these species will be well suited to mild wet climate areas in Europe.

Reform of Common Agricultural Policy (CAP): an example of a new scenario

The Common Agricultural Policy of the European Community (EU-25) applies to all member states (though when initial accession occurs there is usually an adoption period of several years). To a great extent therefore the following comments relate to the established 15 states in the EU.

CAP had the intention of 'providing farmers with a reasonable income'. This was achieved originally via market intervention and direct aids on many, but not all agricultural/horticultural products. However since 1992 focus has changed, initially to area or headage payments related directly to production, and more recently to decoupled single farm payments (SFP). The two most recent reviews of CAP were Agenda 2000 and the 2003 Reform (sometimes called Mid Term Review). These two reforms have begun the transition from production-orientated agriculture (Pillar 1) to a focus on provision of public goods in the environmental sector (Pillar 2).

Provided that cross-compliance measures are met then each agricultural holding will receive a SFP. Clearly the SFP allows a holding to remain viable, but does not necessarily demand agricultural/horticultural production in the traditional sense. Hence 'profitability' of individual enterprises will change radically. This will be reflected in the upland grazing situation in particular. Data from PROMAR (2004) summarised the situation; they suggest a severe downward readjustment in livestock numbers as decoupling occurs due to significant reduction in profitability.

From this it is clear that an individual must add value to his/her grazing livestock production, support the enterprise financially from other resources, or reduce or abandon the enterprise entirely. This situation demands new approaches and thinking in terms of forage grazing utilisation. At the same time, the speciality root crops - sugar beet and potatoes appear likely to decline in area, the former because of revision of the EU Sugar Regime. Areas of potato production are declining in EU, as a reflection of increased yield and static or declining demand. Hence, there is the opportunity for additional areas of land to be released for new uses, provided that sustainable alternatives can be characterised. It will be essential that this land is managed rather than abandoned; grass species probably offer the best utilisation option.

Output needs focus

Whilst development of sustainable markets for feedstocks from graminaceous and related forage species appear to have potential, they must be able to compete economically in the marketplace. Hitherto each outlet for bio-renewables has been developed in isolation; some markets may be more viable than others. Unfortunately no integration of production has been undertaken: wheat straw could be burned for heat and power, fermented to produce bioethanol to replace gasoline or used to make paper. It would appear logical that a similar situation will occur with some non-specialist products from grasses.

Ways forward

The key elements for progress are:
- Identify and prioritise opportunities – especially markets currently in production or otherwise;
- Identify and prioritise forages as feedstocks linked to top priority opportunities;
- Develop integrated sustainable production and utilisation procedures for forages.

Conclusions

Experimental evidence and scientific/technical practice have confirmed the potential of many species of plants as feedstocks for non-food products. Amongst the *Gramineae*, progress has been relatively limited with only a small number of species being exploited. Whilst there must be considerable emphasis on food production in many parts of the world (especially Asia and Africa), new policies and direction in more economically developed areas like EU-25 or North America are offering new opportunities. However in terms of the forage grasses and associated dicotyledonous species little direct progress has been made and a radical new strategy to develop forage grasses *et al.* in a sustainable environment in the absence of livestock is required.

Acknowledgments

The author is pleased to acknowledge the assistance of Mrs Denise Beardall in the typing of many drafts of this paper, also assistance from colleagues in Agricultural and Rural Strategy Group of CSL in sourcing data.

References

Askew, M.F. (2001). Biofuels: the European experience. In: Proceedings from the 5th National Symposium New Crops and New Uses Strength in Diversity, Atlanta, Georgia, USA, Published as Trends in New Crops & New Uses, ASHS Press, Alexandria, VA, USA, 598pp.

Chisholm, C.J. (1994). Reed Canary Grass. In: Towards a UK research strategy for alternative crops. Silsoe Research Institute, UK, Internal MAFF Report, 164-166.

EU (1997). Energy for the future: renewable sources of energy. White Paper for a Community Strategy & Action Plan. COM (97) 599 FINAL (26/11/1997).

EU (2003). The promotion of the use of biofuels & other renewable fuels for transport. Directive 2003/30/EC.

FNR (2004). Pamphlet FNR, Gulzow, Informal publication, 1pp.

Gleason, H.A. (1952). An illustrated flora of the Northeast of United States and adjacent Canada. Vol 1. New York Botanical Gardens, 482pp.

Grass, S. (2004). Utilisation of grass for production of fibres, protein and energy. OECD Publication Service, Paris; 169-177

Greenfinch (2005). <www.greenfinch.co.uk>

IENICA (2000). <www.ienica.net>

Marrow, J.E & J. Coombs (1990). An assessment of bio-ethanol as a transport fuel in the UK, Volume 2. HMSO, 144pp.

Marrow, J.E., J. Coombs & E.W. Lees (1987). An assessment of bioethanol as a transport fuel in the UK, HMSO, 154pp.

PROMAR (2004). Business pointers for livestock enterprises. Supplement to Farmers' Weekly, 5 Nov 2004, 1-11.

Riveros, F. (undated). FAO Grassland Group Document
    <www.fao.org/ag/AGP/AGPC/doc/PUBLICAT/GRASSLAN/3.pdf>

Turley, D.B. (undated). Liquid biofuels – prospects and potential impacts on UK agriculture, the farmed environment, landscape and rural economy. Report prepared for DEFRA, Organics, Forestry and Industrial Crops Division.

UNEP (1992). The Convention on Biological Diversity. <http://www.biodiv.org/convention/default.shtml>

UNFCCC (1997). Kyoto Protocol. <www.unfcc.int>

Section 2

Grassland & the environment

Grasslands[1] for production and the environment

D.R. Kemp[1] and D.L. Michalk[2]
[1]*Charles Sturt University & The University of Sydney, Orange, NSW 2800 Australia*
Email: David.Kemp@orange.usyd.edu.au
[2]*NSW Department of Primary Industries, Orange, NSW 2800 Australia*

Key points

1. To manage grasslands for production and enhanced environmental values requires a redefinition of the frameworks within which management decisions are made, and a tailoring of practices to suit the ways that farmers operate.
2. Improving the perenniality and permanence of grasslands usually leads to better environmental and production outcomes.
3. There is a case for a more conservative approach to utilising grasslands in order to sustain the functioning of local ecosystems and to improve water quality, nutrient and energy cycling and biodiversity.
4. A landscape rather than paddock focus is more appropriate for meeting current grassland management objectives. Grasslands can be triaged to better focus R&D and management, though this could challenge society's preferences for products from more environmentally friendly ecosystems.
5. There is a need to find payment and/or market systems that mean environmental values are enhanced and farm income does not suffer.

Keywords: ecosystems, landscape, water, nutrients, biodiversity

Introduction

Grasslands[4] occupy large areas of the world's 117m km^2 of vegetated lands and provide forage for over 1800m livestock units (World Resources Institute, 2000), and wildlife. Grasslands are a major natural resource utilised for the production of food, fibre, fuel and medicines, and are critical for maintaining a favourable global environment. Today however, human populations are exerting such pressure on grassland resources that it is not possible to continue to simply maximise production as in the past. The challenge is to replenish resources depleted by production and reduce soil losses and vegetation degradation through a better understanding of the multi-functionality of grassland, to improve the ecosystem services upon which all living organisms rely and to enhance environmental values.

The objective of this paper is to discuss ways of responding to this challenge. What levels of utilisation from grasslands enable the maintenance or enhancement of the environmental values of grasslands? What are the opportunities for advances when much of the world's grasslands are overgrazed? Overgrazing/over-utilisation arguably applies where herbivore consumption rates exceed growth and, or recovery rates of desirable plant species and the grassland ecosystem function is impaired/degraded, often necessitating reseeding, or resulting in declining productivity over time, both of which reduce environmental values. What strategies are needed to resolve how best to satisfy production and environment goals? Are these problems technical or social? These dilemmas are complex because grasslands are productive natural resources and part of the environment that interacts with other

---

[1] In this paper the term 'grasslands' includes all those forage systems used by grazing livestock.

environmental processes e.g. climate change. Ultimately a way has to be found to utilise these important resources in a sustainable manner.

This paper considers both natural and sown grasslands. Emphasis is placed on general principles that underpin the development of grassland/livestock systems that are productive and environmentally sustainable. The paper is written from a farm-centred environmental perspective, looking back at grazing livestock systems to see what may need to change in order for grasslands to remain productive and to enhance the environment. For many issues there are as yet no optimal solutions. A starting point in this discussion is to first consider the services grasslands provide and then the framework within which farmers operate.

Grassland services

From man's perspective, the primary services from grasslands are the provision of food, fibre, fuel and medicines. Secondary services are the biodiversity of animals and plants needed to maintain nutrient, water and energy flows and the functionality of ecosystems. Additional services emerging as crucial in the 21st century are the role of grasslands in maintaining air quality, as carbon sinks, supporting pollinating and symbiotic organisms and other processes that maintain or repair ecosystems and landscapes. Unfortunately much of the work through the last century concentrated upon looking at grasslands as a production system rather than as an ecosystem and the services that grasslands provide were often ignored. Where productivity must be sustained over the long-term, the key issue is now how best to use grassland resources without causing deleterious effects to the environment, or decreasing incomes. To achieve this requires more integrated thinking about the processes operative in grassland ecosystems. Decision-making typically devolves onto farmers, so what influences them?

Do farmers aim to maximise product or profit?

The management of productive grasslands and of environmental impacts is influenced by the attitudes of farmers. Management has been treated as a simple technical issue, but the reality is that farming is done in a social context and profit is only one goal. Farmers often seek maximum productivity even if it is not profitable e.g. they want to maximise the number of animals per ha (as a measure of wealth) or the size of their animals (for show) or the total animal product per ha (almost irrespective of cost). The focus on animal performance can result in high grazing pressures and the need to continually resow grasslands. Vanclay (2004), proposes 27 principles that govern farmer behaviour; a subset relevant to this paper is:

- 'It is hard to be green when you are in the red', environmental management costs money.
- 'Doing the right thing' is a strong motivational factor; what is 'right' is a social construct.
- Farmers don't distinguish environmental issues from other farm management issues, they are all part of managing the farm.
- Sustainability means staying on the farm; the concept of sustainability often has a strong social component – passing on the farm to the next generation is a central issue.
- Non-adoption of 'good' practices may reflect the fact that past recommendations were not good. Can we guarantee that new recommendations will work any better?
- Farmers construct their own knowledge; practices need to be tailored to their frameworks.
- Farmers need to feel valued, particularly if society wants them to adopt more sustainable practices that may not make more money.

Applying these principles to improve recommendations for grassland management is a difficult task. They illustrate the need to consider the social context within which farming occurs and the ways that managing for the environment and production need to align.

Environment and society

The general importance of the environment in developed countries is evident in surveys. In Australia, members of the public rank the environment third behind family and friends as an issue of importance, ahead of leisure (which includes sport), service to others, work, religion or politics (Young, 2003). The general picture is that the public is concerned; they want something done, they are willing to pay for environmental management and they are concerned about agriculture, but less so than with other sectors of the economy cf. government and manufacturing industry. This suggests that agriculture has societal support to develop and implement more environmentally friendly policies. Similar attitudes would apply in Europe, where this is reflected in the policies of the European Union and the shift in the Common Agricultural Policy towards environmental management (Hall *et al.*, 2004). However given that farmers do not necessarily distinguish between environment and farm issues (Vanclay, 2004), better environmental practices on farms need to be developed as part of production systems.

Productivity, profits and the environment

The need to maintain productivity, profitability and the environment is becoming the major interaction that farmers seek to resolve. Arguably environmental risks and costs increase with intensification, and the level of disturbance and utilisation of grasslands, leading to loss of resources and increasing costs of replacing nutrients and species and in managing the soil water balance and nutrient runoff. Exceptions to these generalisations do occur e.g. the intensively grazed, well-adapted *Lolium perenne* (perennial ryegrass) and *Trifolium repens* (white clover) pastures of New Zealand.

High utilisation rates and levels of productivity may not always be the most profitable. This is illustrated with the relative economics of sheep for wool production on grasslands in central NSW, using a gross-margin model (Vere *et al.*, 1993). A comparison was done between resowing a degraded pasture with introduced species (*Phalaris* spp. and *Trifolium subterranean* (subterranean clover), or fertilising and managing the grassland (upgraded) back to a more productive state and stocking conservatively (Figure 1), (Michalk *et al.*, 2003b).

In each case it was assumed that maximum stocking rates were delayed until year three in order to optimise the proportion of perennial grasses and improve the likely longevity of the pasture (Michalk *et al.*, 2003b). The alternate pathway shown may be more typical of patterns on farms in the region, where few farmers have sown pastures in recent years (Reeve *et al.*, 2000). The gross margin analysis of these pathways over the ten-year period for a range of conditions typical of the region (Figure 2) demonstrated that there was little difference in the net profitability of the two main pathways until higher stocking rates were attained.

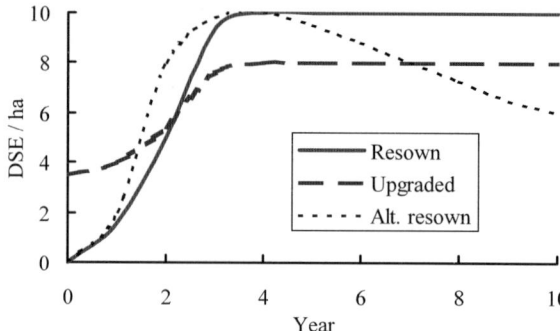

Figure 1 Modelled patterns in stocking rates (DSE: dry sheep equivalents) for grasslands managed to a more productive state (upgraded) or resown. The alternative resown pattern may be more typical of the region. Data are for an average area in central NSW (D.R. Kemp, *unpublished*)

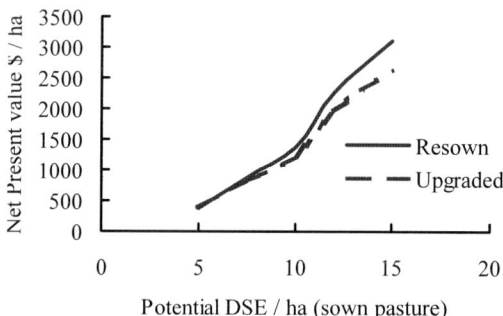

Figure 2 Net present value of alternative grassland management strategies over a ten-year period (D.R. Kemp, *unpublished data*). The 'upgraded' grassland is stocked at 80% of the sown

However, if the alternative (and more common scenario) is considered, i.e. that productivity in the region tends to decline over time (Figure 1; alternative resown pathway), the economics are worse (data not shown). This leads to the general conclusion that the most appropriate pathways for development and improved environmental outcomes arise from (conservative) grazing management - designed to retain the perennial species that are often lost, as opposed to resowing (Dowling *et al.*, 2001). The key point is that focus needs to be on net farm profits over the long-term rather than physical output or gross income. Much of the literature seems to ignore these criteria.

Revisiting concepts for productive versus sustainable grassland systems

There are often two general views of farming systems. The **first** view is that a 'factory' approach is sufficient where resources are put in one end of a pipe and products extracted at the other. In the 19th century J. von Liebig developed the concept of the 'law of the minimum' i.e. one factor is usually limiting production at any one point in space and, or time. That 'law' has helped develop the 'factory' view, but is it conceptually useful for more sustainable systems? In this mindset the focus is often only on a limited number of components, with some consideration of efficiencies, but no solutions for unplanned or adverse environmental effects. Limiting factors are applied to excess and any 'leakage' is seen only as inefficiency in primary production. Nitrate in water supplies can be seen as a legacy of this view.

The related concept of 'grasslands as crops' - developed from the early work of Stapledon, proved an excellent stimulus to better management of grasslands, but is it still useful? The crop analogy has merit for good husbandry, but as most crops aim to be monocultures, it conflicts with the reality of many types of grassland. Perennial grass and nitrogen fertiliser systems are still important in many regions, but many of those systems now aim to incorporate legumes to deliver better environmental outcomes, and to use the grassland phase to restore soil resources. If grasslands are treated as crops, does this then encourage over-use of resources and poorer environmental outcomes?

A **second** view of farming systems (coming more from managing natural grasslands) is of an ecosystem model where it is acknowledged that the system is sustained by maintaining species and the cycling of water, nutrient and energy flows. In this scenario products are harvested from ecosystems, it is acknowledged that the internal flows of energy and nutrients greatly exceed the amount harvested, and that the ecosystem needs to be maintained in order to sustain production. An ecosystem view aims to cycle nutrients internally and minimise leakage as much as possible for more judiciously applied inputs. A focus on overcoming single limiting factors is then inappropriate as this can distort the ecosystem leading to more chaotic behaviour. This approach fosters a more conservative, optimising net profit strategy.

A switch from a 'factory' to 'ecosystem' view of farming has implications. A key consideration is that within an ecosystem it would be impractical to manage each component to an optimum. Around the optimum there are a range of values that do not result in great losses in productivity, or ecosystem function; hence managing within a range is acceptable (Kemp, 1991) e.g. animal gain/ha. Farmers can manage more successfully over a range than continually chasing optimum or maximum values, thus actions only need to be initiated when near the limit of that range, rather than continuously.

Linkages between the environment and production

The close natural linkages between grazing, grassland productivity and gross environmental effects are well known e.g. normal seasonal cycles, the extremes of droughts and floods; but more subtle effects and how the resource base is sustained are less well known. Grazing can have significant negative effects on the soil water balance, depending upon the intensity of grazing, the residual herbage mass and the proportion of bare ground. Surface runoff, nutrient losses and reduced infiltration rates can increase in grazed paddocks (Elliott & Carlson, 2004). More conservative grazing strategies e.g. 10-20% below average, can increase the content of perennial grasses, reduce runoff, increase water transpired (reducing salinity and acidity risks) and reduce the need to resow (Michalk et al., 2003b). These components may have been influenced more by the standing herbage mass than directly by grassland growth rates or livestock productivity. The challenge for farmers is to determine the grazing pressure where adverse environmental impacts are low, yet productivity is satisfactory (Figure 1). These interactions will be considered from an ecosystem perspective in terms of the management of water, nutrient and energy flows, and the diversity of species required to sustain the ecosystem.

Managing water with grasslands

Sustaining clean and safe drinking water and irrigation supplies to meet the needs of the global population is a major issue for the 21st century. Water demands already exceed supplies in more than 80 countries with 40% of the world's population (Swain 2001). Grasslands are vitally important for the provision of water for agriculture, industry and domestic use.

The way that grasslands are managed can have major impacts on the water cycle, ecosystem function and on production. In Australia solutions to these problems lie in increasing the perenniality of plants in the landscape (Singh et al., 2003). Perennial-dominated grasslands fulfil two roles in water management – they efficiently utilise soil water due to deeper root systems and longer growing seasons than annual species, and they can protect the soil surface from soil and nutrient loss, particularly during high intensity short-duration storms. Unfortunately there is often a decline in the perennial component in grasslands resulting in more variable production over time, higher incidence of weed invasions, soil acidification, acute soil erosion and increasing salinity.

Harvesting water

Clean runoff and aquifer recharge from grassland catchments is crucial to maintaining water supplies for agricultural, industrial, domestic and environmental use, food security and for the progression of many societies beyond subsistence (van Wesemael et al., 1998; Raju 1998). Grassland management practices can vary the proportion of surface runoff to aquifer recharge (Bergkamp, 1998). This was evidenced in the recent Australian drought where over-grazing was controlled and less water (from the limited rain that did fall) flowed into dams and rivers. Maintaining herbage mass > 2 tDM/ha effectively controls runoff (Packer et al., 2003). Excessive clearing and high grazing pressures results in less water being retained in the landscape and reduced dry season river flows.

To improve water management, many grasslands are being planted with trees or allowed to progress to shrubland. However this may be an over-reaction that leads to other problems as is evidenced in semi-arid zones such as the Edwards Plateau of the USA (Scholes & Archer, 1997), in Spain (Bellot et al., 2001), in India (Sikka et al., 2003), and with the invasion of Australian *Acacia* spp. in South Africa (Dye & Jarmain, 2004). In each case water yield in rivers declined. In environments where grasslands have been a significant part of ecosystems, learning to manage them appropriately would be preferable to altering the ecosystem. Recharging aquifers can be better for human health as there is then a reduced risk of pathogens (David & Mazumder, 2003). As a minimum, domestic water supplies need to come from water collected over high grass cover.

Dryland salinity and soil acidification

Australia has the reputation of being the world's driest inhabited continent, yet salinity caused by excess water in the wrong parts of the landscape is a major problem in both irrigated and dryland agriculture. Up to 17m ha of Australia's more valuable crop and grassland is at risk of salinisation by 2050 (National Land and Water Resources Audit, 2001). Induced salinity has been known since land clearing, agriculture and irrigation developed, affecting environmental values (Mahmood et al., 2001).

Accelerated soil acidification is recognised as a major degradation issue that affects the sustainability of current production systems in both temperate (Ridley *et al.*, 2004) and tropical (Noble *et al.*, 1998) environments. Short-term annual pasture legume leys are now recognised as a significant cause of acidity (Ridley *et al.*, 2000; Noble *et al.*, 2002). The symptoms are less visible than those that occur with erosion and salinity and the gradual decline in production is often ascribed to other factors such as season. Declining acidity typically results in toxic levels of aluminium developing (Scott *et al.*, 2000), and some 24m ha of southern Australia are now affected (Cregan & Scott, 1999). Developing acidity under grasslands has long been recognised in Europe where lime is regularly applied.

How have these problems arisen? In general, salinity and acidity result from a substantial change in land use practice that has altered the hydrological balance from catchment to regional scales. Salinity tends to be worse in regions where annual rainfall is less than evaporation rates, allowing salts to accumulate. Excessive nitrate from legume-dominant swards and livestock urine patches, coupled with excess drainage results in nitrate leaching, all contribute to soil acidification. Management of salinity and acidity requires control over the partitioning of the water balance and the depth of water tables. Pressures to extract increasing returns from livestock caused by declining terms of trade has led to high grazing pressures and significant deterioration in the perennial species base - commonly to less than 20% composition, which exacerbates these problems (Moore, 1970; Kemp & Dowling, 1991; Kemp & Michalk 1993). Similar problems occur throughout the world.

The risks of acidification and of salinity in grasslands can be reduced by the use of perennial plants to use more water and capture nutrients throughout the growing season (Ridley *et al.*, 2000; Noble *et al.*, 2002). Management practices to increase/maintain the perenniality of grasslands are being developed (Kemp *et al.*, 2000; Michalk *et al.*, 2003b). Both perennial grasses and strategically placed trees and shrubs (if they cover 16-22% of the catchment) have a role (White *et al.*, 2002). The aim is to manage recharge or discharge to levels similar to natural ecosystems (Hatton & Nulsen, 1999). Engineering solutions (Mahmood *et al.*, 2001) are too expensive for many types of grassland. A further strategy is to plant species that can be productive and help maintain ecosystems under saline (Cocks, 2001) and acid soils (Scott *et al.*, 2000). In central New South Wales, Australia deep drainage was reduced in the order; native C3 grasses < sown C3 grasses < mixed C3/C4 grasses, when present at the same level of herbage mass (Packer *et al.*, 2003). The greater the area of perennial herbage plants the less the need for trees (Dunin, 2002). Studies of Mediterranean systems (Joffre & Rambal, 1993) have shown the proportions of perennial grasslands and trees needed to manage the water balance and to capture nitrate. The wider use of lime for acid soil management is restricted to higher rainfall, more fertile and profitable soils.

Economics of water management on grasslands

For most farmers, management of grasslands to alleviate water, salinity and acidity problems is of only secondary concern, compared to obtaining an economic return from livestock production (Pannell, 1999). The proportion of a farm that is sown to perennial plants is then more an economic than an environmental decision. Australian work has found though that the level of use of perennials promoted through traditional education and extension is less than required to halt expansion in saline areas (Bathgate & Pannell, 2002). Achieving a better environmental outcome will depend upon farmers receiving prices for their products that reflect the environmental services delivered e.g. clean water, or from direct subsidies.

Managing nutrient loss in grasslands

Intensive grassland production

Intensive grassland production has come to rely on the identification and alleviation of mineral deficiencies that limit plant growth. Fertiliser application - particularly nitrogen, has significantly increased productivity of intensively managed tropical and temperate grassland systems (ten Berge *et al.*, 2002). However, the environmental effects of eutrophication of surface and ground water due to nutrient leakage are now evident at regional and global scales (Oenema *et al.*, 2003). Nutrients in animal faeces and urine can be utilised for growing plants or are serious sources of pollution (ten Berge *et al.*, 2002). As previously argued, application of the 'law of the minimum' fails within a sustainability context. There is a need to balance these conflicting goals of profitable production and environmental protection in fertiliser management. The key issue would seem to be the better control of nutrient leakage from farms; though within a farm some nutrient movement can be accommodated.

Nitrate leakage is a major issue (McGechan & Topp, 2004, Rimski-Korsakov *et al.*, 2004). Leaching losses from grasslands are generally low compared to cropping systems, but can exceed 34 kg N/ha per year - the limit set by the EU Nitrates Directive to maintain the groundwater nitrate concentration below 50 mg/l (Anon., 1991). In Belgium, laws limit residual 'post-harvest' soil nitrate-N to <90 kg N/ha in the soil profile (0-90 cm) between 1 October to 15 November (Nevens & Rehuel, 2003). In the Netherlands, farmers have to maintain records of nutrient use and are taxed on nutrient surpluses generated from their farms, with penalties up to €400/ha for livestock producers (Ondersteijn *et al.*, 2002).

Despite this leakage, farmers are reluctant to reduce nitrogen applications because of the decline in livestock performance that results (Valk *et al.*, 2000). An analysis of fertiliser strategies in Europe suggested that focussing on efficiency provides the best means to reduce leaching losses and minimise off-site impacts (Ondersteijn *et al.*, 2002). Only 30-50% of applied N and ~45% of applied P, is typically taken up by target plants (Smil, 1999; 2000). Improved nutrient-use efficiency can be achieved. Applying fertiliser during periods of peak plant demand, placing fertiliser nearer plant roots and using smaller, more frequent applications all have the potential to reduce nutrient losses while maintaining yield and quality (Tilman *et al.*, 2002).

Integrated landscape management can complement paddock measures to effectively control nutrient fluxes and leakage within catchments. As with acidity management, perennial grasses provide a sink to effectively capture or trap nutrients, often as part of water flows (Schilling & Wolter, 2001); capturing nitrate reduces the development of acidity. Buffer strips with 5m of grassland and a line of trees can effectively trap nitrate in soil water flows (Borin & Bigon, 2002). Nutrients can be tied up in less palatable species (Myklestad, 2004); this may be useful in a landscape context, but not for productive grasslands.

Extensively managed grasslands

Extensively managed grasslands depend on nutrients released from parent material and organic matter by biological and chemical processes. These grasslands rarely receive fertiliser; hence legumes often play a key role in productivity. Nutrient release from stored organic pools depends on litter quality, environmental conditions and the level of microbial activity (Swift *et al.*, 1979); subsequently these nutrients need to be captured by desirable perennial grasses. Less palatable plant species tend to lock up nutrients and reduce grassland productivity (Moretto & Distal, 2003). Grasses with a high C:N ratio immobilise nutrients due to slow decomposition rates (Hobbie, 1992).

There are large amounts of N and P that are unavailable to plants, and developing management strategies to release more of these 'fixed' nutrient pools is an obvious area for research. Better nutrient management would require sequestration of mineral nutrients within more readily available organic pools, the retention of those pools in the system and then the managed release and capture of nutrients by desirable plant species. In practice this would involve using grazing tactics to maintain litter and retain higher quality plant species to minimise nutrient loss. Overgrazing is arguably the biggest single cause of nutrient loss in extensive grasslands. Wind erosion and nutrient loss in Chinese grasslands can be clearly linked to adverse grazing practices (Dong *et al.*, 2000; Michalk *et al.,* 2003a; Li *et al.,* 2004).

Managing carbon and energy

The energy in organic carbon compounds within grasslands is central to their productivity and environmental values. Carbon sequestration is now also a major management objective as elevated atmospheric CO_2 concentrations associated with increasing global temperatures and more variable rainfall patterns are a major concern.

Native grasslands that have never been heavily utilised usually have high organic carbon levels. In extensive grasslands 90% of organic C is below ground (Schuman *et al.*, 1999). Elevated atmospheric CO_2 is increasing, in part due to land use changes (Vitousek *et al.*, 1997) especially the conversion of large areas of grassland to crop production (Scholes & Noble, 2001). Cultivation of the grassland steppe in China decreased soil organic carbon (SOC) by 25% in 8 years, rendering the soils highly susceptible to erosion (Wu & Tiessen, 2002). The challenge for grasslands is to reverse the losses of CO_2 by storing more in plant and soil organic matter, and to be a net 'sink' for carbon to help minimise adverse environmental impacts (Vleeshouwers & Verhagen, 2002); this also aids water and nutrient management. During the restoration phase of grasslands the rate that carbon is sequestered can be high, relative to grasslands in a 'maintenance' phase (Schlesinger, 1990; Fisher *et al.*, 1994). As with earlier discussions, the more perennial plants there are in the system, the more carbon is sequestered. Grazing strategies need to retain litter and utilise forage at intervals that maximise the storage of carbon in the soil. Farming systems where pasture phases are wedged between cropping cycles pose a continuing problem to accumulate C.

Grasslands are not part of the current Kyoto protocol accounting period for 2007-2012, but their vast area makes this resource a potential sink, exceeding that of forests. Payments for carbon sequestration could make the difference between profit and loss for many livestock systems. However, this requires that a commercially realistic system for verification be developed. Grasslands not only capture and store energy, they are often significant consumers of energy and from an environmental perspective these terms need to show a positive balance. It is clear though that the energy balance becomes more negative as livestock production is intensified and manufactured inputs increase, especially N fertiliser (Kelm *et al.*, 2004), chemicals and fossil fuels. This can be offset to some extent by the use of legumes and farm manure. Ultimately accounting procedures will need to consider the net energy balance if society is to make effective judgements about the sustainability of grassland practices.

Managing biodiversity

Biodiversity and production

Biodiversity is crucial to the productivity of grasslands, though an understanding of the numbers and types of species that optimise production is still to be determined. Many agronomists assume that grass plus clover equals a good pasture; but is this so? Early work in small plots suggested that maximum productivity occurred when 6-10 plant species were present (Tilman, 1996). At small scales sometimes fewer species were sufficient to maximise grassland production (Nicholas *et al.*, 1997). Recent work in small paddocks (<2 ha) where each had > 10 species, suggested that 10-20 species may be optimal for productivity. The net primary productivity of those paddocks declined as species number increased from 20-50 (Kemp *et al.*, 2003). Arguably the optimum number of species would be expected to increase as the paddock area and number of different resource niches increase.

Not all species are equal; plant functional type is important. Perennial grasses can dominate grasslands (Tilman, 1996) consuming most of the resources. However, if the fertility is low they may be less competitive and minor species become more frequent (Kemp *et al.*, 2003). The importance of plant functional type over species *per se* supports the theory of species redundancy; e.g. *Lolium perenne* can be replaced by *Festuca arundinacea* (tall fescue) in some environments without any significant impact on grassland productivity. An implication for native grasslands is that management to conserve minor species that are part of a common plant functional type, may not be important for the productivity of the grassland.

One or two species are unlikely to dominate natural grasslands. Invariably other species invade supporting the view that stable grasslands require many species. 'Weeds' are frequently the biggest issue confronting farmers. Typical mixtures of grass and clover could be augmented with other forbs and 'gap fillers' as the later plant types frequently invade swards. The release of cultivars of *Cichorium intybus* (chicory) (Rumball, 1986) and *Plantago lanceolata* (plantain) are a revival of an old practice (Foster, 1988) that fell out of favour during the twentieth century. More diverse grasslands can reduce invasion of other species (van Ruijven *et al.*, 2003). Managing grasslands to enable existing species to utilise more resources e.g. by maintaining higher mean levels of herbage mass, can also restrict weed invasion (Badgery, 2003; Meiners *et al.*, 2004).

Nature conservation

High grazing pressures can lead to a degradation of grasslands that invariably leads to the loss of species, while conservative practices with low grazing pressures usually retain species. However, an optimal curve related to standing biomass may apply (Grime, 2002). Grassland use ranges from commercial livestock production to wildlife tourism. To attain production and biodiversity goals, the amount of forage utilised needs to satisfy the needs of all relevant herbivores, and enable all the desirable species present (including native species), to persist over the long term.

In grasslands that have been utilised for production, it is reasonable to assume that any rare and endangered species have either been lost long ago, or if still present are being maintained under current management practices. Experience in Australia has been that grassland areas that become part of the National Park system need to be grazed otherwise unwanted species invade and threaten that grassland (Lunt, 2003). For grasslands across southern Australia

(Kemp *et al.*, 2003), native plant species in mixed grasslands were maintained if the average herbage mass did not decrease below 2 tDM/ha. This provided farmers with a simple management guideline to conserve those species.

Management for wildlife is becoming an increasingly important economic activity e.g. 'Ducks Unlimited' in North America are funding habitats (there are 50 million bird watchers in the USA and Canada), South Africa now has 45,000 private game reserves. Problems can arise where a single species becomes the sole focus for management, which can have adverse consequences for the rest of the ecosystem. Limited culling of animals and resultant over-grazing so that tourists can more readily see wildlife is an increasing problem.

The biggest threats to species conservation are changing land use. Many grasslands are now dominated by non-preferred species and, or invaders from other environments. To restore rare and endangered species in these circumstances can be an expensive task e.g. transplanting individual grass plants (Hocking, 1998). Little work has been done on the restoration of rare plant species in grasslands. Some disturbance of the ecosystem is required to enable any plant to establish. Knowledge of the size of suitable micro-sites can then help design suitable management practices (Bullock *et al.*, 1995). Gap sizes can be varied with grazing tactics. Ecological theory suggests that species richness is maximised at intermediate levels of disturbance (Huston, 1979).

An increasing area of interest is the maintenance of meso- and micro-fauna within grasslands, in part because of their critical roles in cycling nutrients and energy. In a study on a range of grassland systems in central NSW, Australia, the majority of soil insect species were retained across a range of grassland systems, but the proportions changed (Reid, 2004). In Wales retaining a higher average herbage mass in summer resulted in more *Coleoptera* (beetle) species within the grassland, especially those dominated by native plant species (Dennis *et al.*, 2004). It was considered that these effects were constrained by the previous history of the grassland, e.g. drainage, fertiliser and lime inputs, and botanical composition. Studies on invertebrate communities in grazing systems indicate however that more intensively managed grazing systems can have lower invertebrate richness and abundance than ungrazed or conservatively grazed grasslands (King & Hutchinson, 1983). High fertiliser inputs can lead to species-poor swards, even years after fertiliser applications cease, which may be the cause of some of the effects noted (Walker *et al.*, 2004).

Strategies for managing biodiversity can focus at the level of the paddock, property and landscape. Given the ways farms are managed it may be preferable to focus biodiversity management at a landscape level. This approach is reflected to some extent in EU policies for grassland restoration and management (Smith *et al.*, 2003). A mosaic of different fields connected by non-cropped habitat is known to increase diversity of breeding birds, ground beetles, spiders and butterflies (Benton *et al.*, 2003). Using a mosaic approach enables species to move between sites to avoid creating islands that then limit the viability of populations (Poschlod *et al.*, 1998).

Categorisation of grasslands according to state

Given the world's need for food, fibre and fuel from grasslands, the density of humans and the difficulties of enhancing the environment in all cases, is it possible to categorise the world's grasslands to achieve realistic goals for production and the environment? There are often three broad groupings where different management practices would apply.

- *Healthy*: Those areas where grasslands are predominately in a native state; where there are minimal or readily manageable environmental threats; where current levels of production are appropriate and can be maintained with current knowledge.
- *Disturbed:* Those areas where more intensive practices often apply; where fewer native species exist and more careful management is required to manage nutrient leakage, soil water etc. Knowledge may be insufficient to achieve a good balance between the environment and production.
- *Dying:* Very disturbed areas often subject to intense use, e.g. continual reseeding and cropping, and where there may be an almost complete absence of native species and where there is a high risk of adverse effects on the neighbouring environment. Such areas may not be realistically restored to an environmentally friendly state, and the management objective may be more directed to preventing adverse impacts on neighbours. These areas though may produce much of the food for society, which raises additional issues about society wanting cheap *vs.* clean and green food.

Applying this concept to grassland management may be the way forward. In some ways this is already being done with more 'pristine' areas being incorporated into parks and farmers being subsidised to look after such areas on their farms, while the more intensively used areas are being subject to increased regulation to prevent damage to surrounding environments. If this concept is applied to grassland research, then the area of highest priority may be the 'disturbed' category i.e. those areas where the environmental values can be enhanced and where levels of production can either be sustained or even enhanced, cost-effectively. This category may currently be the less regulated and consequently is where the development of better practices and more self-regulation is possible. Unfortunately many policies focus on the 'healthy' and 'dying' lands where there can often be a sense of emergency.

To continue to provide food for the world's ever-increasing population will require acknowledgement of a cost in terms of some leakage impacts from the more intensively used grasslands. The worst cases are the cities where it would be impossible to remove all adverse environmental impacts; other categories of landscape use may need to be considered in the same light.

Conclusions

With the majority of the world's grasslands considered to be degraded to some degree, how can the need to achieve the levels of food, fibre, fuel and medicine production required be satisfied, while enhancing the environment? Well-managed grasslands based upon improving the perenniality of the ecosystem, should retain species and manage water, nutrient and energy cycles with reasonable efficiencies, and still achieve suitable levels of production. This can be assessed by monitoring species composition, the quality of water coming from the catchment and the efficiency of nutrient use over the medium to long-term. There is a requirement that these needs are translated into tools that farmers can use daily to track their progress and to know if their systems are sustainable. Evidence needs to be provided that recommended tactics and strategies are compatible with normal farm management and economically acceptable.

Many societies now expect that agriculture will look after the environment. The 'polluter pays' principle however, has not always been applied in agriculture, as many of the environmental values often reflect societal wants for enhancement e.g. biodiversity, as opposed to outright pollution. The benefits from remedying environmental problems do not necessarily return to the farmer, they often return to the community at large, even across national boundaries. These effects mean that there is a good case for direct community payments to solve environmental problems. An alternative of extracting market premiums for good environmental practices has had only limited success, and is difficult to apply universally to remedy general problems. Throughout this paper it has been argued that a case exists for direct or market-based payments for environmental/ecological services, and many grasslands could be used for these purposes e.g. storing carbon, delivering clean non-saline water, and maintaining biodiversity. Practical ways of achieving this needs to be developed, as encouraging desired environmental outcomes from grasslands often requires farmers to go beyond normal commercial boundaries, even though there are some synergies that can be explored. Such payments need to be global in approach as many of the environmental problems of grasslands occur in the developing world.

References

Anonymous, (1991). Council Directive of 12 December 1991 concerning the protection of waters against pollution caused by nitrates from agricultural sources (91/676/EEC). *Official Journal of the European Communities*, L375, 1-8.

Badgery, W. (2003). Managing competition between *Nassella trichotoma* (Serrated Tussock) and native grasses. PhD Thesis, University of Sydney, 240pp.

Bathgate, A. & D.J. Pannell (2002). Economics of deep-rooted perennials in western Australia. *Agricultural Water Management*, 53, 117-132.

Bellot, J., A. Bonet, J.R. Sanchez & E. Chirino (2001). Likely effects of land use changes on the runoff and aquifer recharge in a semiarid landscape using a hydrological model. *Landscape and Urban Planning*, 55, 41-53.

Benton, T.G., J.A. Vickery & J.D. Wilson (2003). Farmland biodiversity: is habitat heterogeneity the key? *TRENDS in Ecology and Evolution*, 18, 182-188.

ten Berge, H.F.M., H.G. van der Meer, L. Carlier, T. Baan Hofman & J.J. Neeteson (2002). Limits to nitrogen use on grassland. *Environmental Pollution*, 118, 225-238.

Bergkamp, G. (1998). A hierarchial view of interactions of runoff and infiltration with vegetation and microtopography in semiarid shrubland. *Catena*, 33, 201-220.

Borin, M, & E. Bigon (2002). Abatement of NO_3-N concentration in agricultural waters by narrow buffer strips. *Environmental Pollution*, 117, 165-168.

Bullock, J.M., B. Clear Hill, J. Silvertown & M. Sutton (1995). Gap colonization as a source of grassland community change: effects of gap size and grazing on the rate and mode of colonization by different species. *Oikos*, 72, 273-282.

Cocks, P.S. (2001). Ecology of herbaceous perennial legumes: a review of characteristics that may provide management options for the control of salinity and waterlogging in dryland cropping systems. *Australian Journal of Agricultural Research*, 52, 137-151.

Cregan, P.D. & B.J. Scott (1999). Soil acidification - an agricultural and environmental problem. In: J.E. Pratley & A. Robertson (eds.) Agriculture and the environmental imperative. CSIRO Publishing, Melbourne, Australia, 98-127.

David, J.M. & A. Mazumder (2003). Health and environmental policy issues in Canada: the role of watershed management in sustaining clean drinking water in surface sources. *Journal of Environmental Management*, 68, 273-286.

Dennis, P., J. Doering, J.A. Stockan, J.R. Jones, M.E. Rees, J.E. Vale & A.R. Sibbald (2004). Consequences for biodiversity of reducing inputs to upland temperate pastures: effects on beetles (*Coleoptera*) of cessation of nitrogen fertiliser application and reductions in stocking rates of sheep. *Grass and Forage Science*, 59, 121-135.

Dong, Z.B., X.M. Wang & L.Y. Liu (2000). Wind erosion in arid and semiarid China: an overview. *Journal of Soil and Water Conservation*, 55, 439-444.

Dowling, P.M., R.E. Jones, D.R. Kemp & D.L. Michalk (2001). Valuing the pasture resource - importance of perennials in higher rainfall regions of south-eastern Australia. In : A.M.C. Filho (ed.) Grassland ecosystems: an outlook into the 21st Century. *Proceedings XIXth International Grassland Congress*, Brazil, 963-964.

Dunin, F.X. (2002). Integrating agroforestry and perennial pastures to mitigate water logging and secondary salinity. *Agricultural Water Management*, 53, 259-270.

Dye, P., & C. Jarmain (2004). Water use by black wattle (*Acacia mearnsii*): implications for the link between removal of invading trees and catchment streamflow response. *South African Journal of Science*, 100, 40-44.

Elliot, A.H. & W.T. Carlson (2004). Effects of sheep grazing episodes on sediment and nutrient loss in overland flow. *Australian Journal of Soil Research*, 42, 213-220.

Fisher, M.J., I.M. Rao & M.A. Ayarza (1994). Carbon storage by introduced deep-rooted grasses in the South American savannas. *Science*, 371, 236-238.

Foster, L. (1988). Herbs in pastures. Development and research in Britain (1850-1984). *Biological Agriculture and Horticulture*, 5, 97-133.

Grime, J.P. (2002). Plant strategies, vegetation processes and ecosystem properties. 2nd Edition. Wiley, Chichester, 417pp.

Hall, C., A. McVittie & D. Moran (2004). What does the public want from agriculture and the countryside? A review of evidence and methods. *Journal of Rural Studies*, 20, 211-225.

Hatton, T.J. & R.A. Nulsen (1999). Towards achieving functional ecosystem mimicry with respect to water cycling in southern Australian agriculture. *Agroforestry Systems*, 45, 203-214.

Hobbie, S.E. (1992). Effects of plant species on nutrient cycling. *Trends in Ecological Evolution*, 7, 336-339.

Hocking, C. (1998). Land management of *Nassella* areas - implications for conservation areas. *Plant Protection Quarterly*, 13, 86-91.

Huston, M.A. (1979). A general hypothesis of species diversity. *The American Naturalist*, 113, 81-101.

Joffre, R. & S. Ranbal (1993). How tree cover influences the water balance of Mediterranean grasslands. *Ecology*, 74, 570-582.

Kelm, M., M. Wachendorf, H. Trott, K. Volkers & F. Taube (2004). Performance and environmental effects of forage production on sandy soils. III. Energy efficiency in forage production from grassland and maize for silage. *Grass and Forage Science*, 59, 69-79.

Kemp, D.R. (1991). Defining the boundaries and manipulating the system. *Proceedings 6th Annual Conference Grassland Society NSW*, The Grassland Society of New South Wales, Orange NSW Australia, 24-30.

Kemp, D.R. & P.M. Dowling (1991). Species distribution within improved pastures over central NSW in relation to rainfall and altitude. *Australian Journal of Agricultural Research*, 42, 647-659.

Kemp, D.R., W.McG. King, G.M. Lodge, S.R. Murphy, P. Quigley & P. Sanford (2003). SGS biodiversity theme: the impact of plant biodiversity on the productivity and stability of grazing systems. *Australian Journal of Experimental Agriculture*, 43, 962-975.

Kemp, D.R. & D.L. Michalk (1993). (eds.) Pasture management: technology for the 21st century. CSIRO, Melbourne, 177pp.

Kemp, D.R., D.L. Michalk & J. Virgona (2000). Towards more sustainable pastures: lessons learnt. *Australian Journal of Experimental Agriculture*, 40, 343-356.

King, K.L. & K.J. Hutchinson (1983). The effects of sheep grazing on invertebrate numbers and biomass in unfertilised natural pastures of the New England Tablelands (NSW). *Australian Journal of Ecology*, 8, 245-255.

Fengrui, L., L. Zhao, H. Zhang, T. Zhang & Y. Shirato (2004). Wind erosion and airborne dust deposition in farmland during spring in the Horqin Sandy Land of eastern Inner Mongolia. *Soil & Tillage*, 75, 121-130.

Lunt, I.D. (2003). A protocol for integrated management, monitoring and enhancement of degraded *Themeda triandra* grasslands based on plantings of indicator species. *Restoration Ecology*, 11, 223-230.

Mahmond, K, J. Morris, J. Collopy & P. Slavish (2001). Groundwater uptake and sustainability of farm plantations on saline sites in Punjab province, Pakistan. *Agricultural Water Management*, 48, 1-20.

McGechan, M.B. & C.F.E. Topp (2004). Modelling environmental impacts of deposition of excreted nitrogen by dairying dairy cows. *Agriculture, Ecosystems and Environment*, 103, 149-164.

Meiners, S.J., M.L. Cadenasso & S.T.A. Pickett (2004). Beyond biodiversity: individualistic controls of invasion in a self-assembled community. *Ecology Letters*, 7, 121-126.

Michalk, D.L., P.M. Dowling, D.R. Kemp, W.McG. King, I.J. Packer, P.J. Holst, S.J. Priest, G.D. Millar, S. Brisbane & D.F. Stanley (2003a). Sustainable grazing systems for the central Tablelands of New South Wales. *Australian Journal of Experimental Agriculture*, 43, 861-874.

Michalk, D.L., C.L. Liang, Q. Feng, & D.R. Kemp (2003b). Development of sustainable grazing systems for degraded grassland in Xingan League, Inner Mongolia. In: N. Allsopp, A.R. Palmer, S.J. Milton, G.I.H. Kerley, K.P. Kirkham, R. Hurt, & C. Brown (eds.) Rangelands in the new millennium. *Proceedings International Rangeland Congress*, South Africa, 906-908.

Moore, R.M. (1970). South-eastern temperate woodlands and grasslands. In: R.M. Moore (ed.) Australian grasslands (1st Edition). ANU Press: Canberra, Australia, 169-190.

Moretto, A.S. & R.A. Distal (2003). Decomposition of and nutrient dynamics in leaf litter and roots of *Poa ligularis* and *Stipa gyneriodes*. *Journal of Arid Environments*, 55, 503-514.

Myklestad, A. (2004). Soil, site and management components of variation in species composition of agricultural grasslands in western Norway. *Grass and Forage Science*, 59, 136-143.

National Land and Water Resources Audit (2001). Australian Dryland Salinity Assessment 2000: extent, impacts, processes, monitoring and management options. Land and Water Australia, Canberra, 89pp.

Nevens, F. & D. Rehuel (2003). Effects of cutting or grazing grass swards on herbage yield, nitrogen uptake and residual soil nitrate at different levels of N fertilization. *Grass and Forage Science*, 58, 431-449.

Nichols, P.K., P.D. Kemp, D.J. Barker, J.L. Brock & D.A. Grant (1997). Production, stability and biodiversity of North Island New Zealand hill pastures. In: G Sheath (ed.) Grasslands of our World, *Proceedings 18th International Grassland Congress,* Palmerston North, New Zealand, 21, 9-10.

Noble, A.D., C. Middleton, P.N. Nelson & L.G. Rogers (2002). Risk mapping of soil acidification under *Stylosanthes* in northern Australian rangelands. *Australian Journal of Soil Research,* 40, 257-267.

Noble, A.D., C.H. Thompson, R.J. Jones & R.M. Jones (1998). The long-term impact of two pasture production systems on soil acidification in southern Queensland. *Australian Journal of Experimental Agriculture*, 38, 335-343.

Oenema, O., H. Kros & W. de Vries (2003). Approaches and uncertainties in nutrient budgets: implications for nutrient management and environmental policies. *European Journal of Agronomy*, 20, 3-16.

Ondersteijn, C.J.M., A.C.G. Beldman, C.H.G. Daatselaar, G.W.J. Giesen & R.B.M. Huirne (2002). The Dutch Mineral Accounting System and the European Nitrate Directive: implications for N and P management and farm performance. *Agriculture, Ecosystems and Environment*, 92, 283-296.

Packer, I.J., D.L. Michalk, S. Brisbane, D.M. Dowling, G.D. Millar, W.McG. King, D.R. Kemp & S.M. Priest (2003). Reducing deep drainage through controlled runoff management in high recharge tablelands landscape. In: M. Unkovich & G. O'Leary (eds.) Solutions for a better environment, *Proceedings of the 11th Australian Agronomy Conference*, Geelong, Victoria, 1-4.

Pannell, D.J. (1999). Social and economic challenges in the development of complex farming systems. *Agroforestry Systems*, 45, 393-409.

Poschlod, P., S. Kiefer, U. Traenkle, S. Fischer & S. Bonn (1998). Species richness in calcareous grassland is affected by dispersability in space and time. *Applied Vegetation Science,* 1, 75-90.

Raju, K.C.M. (1998). Importance of recharging depleted aquifers: state of the art of artificial recharge in India. *Journal of the Geological Society of India*, 51, 429-454.

Reid, A.M. (2004). Effect of fertiliser and grazing on grassland invertebrates. PhD Thesis, University of Sydney, 220pp.

Reeve, I.J., G. Kaine, J.W. Lees & E. Crosby (2000). Producer perceptions of pasture decline and grazing management. *Australian Journal of Experimental Agriculture*, 40, 331-341.

Ridley, A.M., P.M. Mele & C.R. Beverly (2004). Legume-based farming in Southern Australia: developing sustainable systems to meet environmental challenges. *Soil Biology & Biochemistry,* 36, 1213-1221.

Ridley, A.M., R.E. White, K.R. Helyar, G.R. Morrison, L.K. Heng & R. Fisher (2000). Nitrate leaching loss under annual and perennial pastures with and without applied lime on an acid Sodosol in humid South eastern Australia. *European Journal of Soil Science*, 52, 237-252.

Rimski-Korsakov, H., G. Rubio & R.S. Lavado (2004). Potential nitrate losses under different agricultural practices in the pampas region, Argentina. *Agricultural Water Management*, 65, 84-94.

van Ruijven, J., G.B. de Deyn & F. Berendse (2003). Diversity reduces invasibility in experimental plant communities: the role of plant species. *Ecology Letters*, 6, 910-918.

Rumball, W. (1986). 'Grasslands Puna' chicory (*Cichorium intybus* L.). *New Zealand Journal of Experimental Agriculture*. 14, 105-107.

Schilling, K.E. & C.F. Wolter (2001). Contribution of base flow to non-point source pollution loads in an agricultural watershed. *Ground Water*, 39, 49-58.

Schlesinger, W.H. (1990). Evidence from chronosequence studies for low carbon-storage potential of soil. *Nature*, 348, 267-276.

Scholes, R.J. & S.R. Archer (1997). Tree-grass interactions in savannas. *Annual Review of Ecological Systems,* 28, 517-544.

Scholes, R.J. & I.R. Noble (2001). Storing carbon on land. *Science*, 294, 1012-1013.

Schuman, G.E., J.D. Reeder, J.T. Manley, R.H. Hart & W.A. Manley (1999). Impact of grazing management on the carbon and nitrogen balance of mixed-grass rangelands. *Environmental Applications*, 9, 65-71.

Scott, B.J., A.M. Ridley & M.K. Conyers (2000). Management of soil acidity in long-term pastures of south-eastern Australia: a review. *Australian Journal of Experimental Agriculture*, 40, 1173-1198.

Sikka, A.K., J.S. Samra, V.N. Sharda, P. Samraj & V. Lakshmanan (2003). Low flow and high flow responses to converting natural grassland into bluegum (*Eucalyptus globulus*) in Nilgiris watersheds of South India. *Journal of Hydrology*, 270, 12-26.

Singh, D.K., B.R. Bird & G.R. Saul (2003). Maximising the use of soil water by herbaceous species in the high rainfall zone of southern Australia; a review. *Australian Journal of Agricultural Research*, 54, 677-691.

Smil, V. (1999). Nitrogen in crop production: an account of global flows. *Global Biogeochemical Cycles*, 13, 647-662.

Smil, V. (2000). Phosphorus in the environment: natural flows and human interference. *Annual Review of Energy and Environment*, 25, 53-88.

Smith, R.S., R.S. Shiel, R.D. Bardgett, D. Millward, P. Corkhill, G. Rolph, P.J. Hobbs & S. Peacock (2003). Soil microbial community, fertility, vegetation and diversity as targets in the restoration management of a meadow grassland. *Journal of Applied Ecology*, 40, 51-64.

Swain, Ashok (2001). Water wars: fact or fiction? *Futures* **33**, 769-781.

Swift, M.J., O.W. Heal & J.M. Anderson (1979). (eds.) Decomposition in terrestrial ecosystems. Blackwell Scientific Publications, Oxford, 372 pp.

Tilman, D. (1996). Biodiversity: population versus ecosystem stability. *Ecology*, **77**,350-63.

Tilman, D., K.G. Cassman, P.A. Matson, R. Naylor & S. Polasky (2002). Agricultural sustainability and intensive production practices. *Nature*, 418, 671-677.

Valk, H., I.E. Leusink-Kappers & A.M. van Vuuren (2000). Effect of reducing nitrogen fertilizer on grassland on grass intake, digestibility and milk production of dairy cows. *Livestock Production Science*, 63, 27-38.

Vanclay, F. (2004). Social principles for agricultural extension to assist in the promotion of natural resource management. *Australian Journal of Experimental Agriculture*, 44, 213-222.

Vere, D.T., M.H. Campbell & D.R. Kemp (1993). Pasture improvement budgets for the central and southern tablelands of New South Wales. *NSW Agriculture Bulletin,* 32pp.

Vleeshouwers, L.M. & A. Verhagen (2002). Carbon emission and sequestration by agricultural land use: a model study for Europe. *Global Change Biology*, 8, 519-530.

Vitousek, P.M., H.A. Mooney, J. Lubchenco & J.M. Melillo (1997). Human domination of the earth's ecosystem. *Science*, 277, 494-499.

Walker, K.J., P.A. Stevens, D.P. Stevens, J.O. Mountford, S.J. Manchester & R.F. Pywell (2004). The restoration and re-creation of species-rich lowland grassland on land formerly managed for intensive agriculture in the UK. *Biological Conservation*, 119, 1-18.

van Wesemael, B., J. Poesen, A.S. Benet, L.C. Barrionuevo & J. Puigdefabregas (1998). Collection and storage of runoff from hillslopes in a semi-arid environment: geomorphic and hydrologic aspects of the aljibe system in Almeria Province, Spain. *Journal of Arid Environments*, 40, 1-14.

White, D.A., F.X. Dunin, N.C. Turner, B.H. Ward & J.H. Galbraith (2002). Water use by contour-planted belts of trees comprised of four *Eucalyptus* species. *Agricultural Water Management*, 53, 133-152.

World Resources Institute (2000). World Resources 2000-2001. World Resources Institute, Washington DC, 389p.

Wu, R. & H. Tiessen (2002). Effect of land use on soil degradation in Alpine grassland soils, China. *Soil Science Society of America Journal,* 66, 1648-1655.

Young, G. (2003). <http://www.environment.nsw.gov.au/whocares>

Soil microbial community: understanding the belowground network for sustainable grassland management

Y.G. Zhu[1], W.D. Kong[1], B.D. Chen[1], Z.B. Nan[2] and P. Christie[3]
[1]*Research Center for Eco-environmental Sciences, Chinese Academy of Sciences, Beijing 100085, China*
Email: ygzhu@mail.rcees.ac.cn
[2]*College of Grassland Ecology, Lanzhou University, Lanzhou, China,* [3]*Agricultural and Environmental Science Department, Queen's University Belfast, Newforge Lane, Belfast BT9 5PX, UK*

Key points

1. In addition to the use of conventional methodologies in soil microbial research, molecular techniques are now being applied to gain insights into the soil microbial community;
2. Plant diversity can exert impacts on soil microbial diversity (through root activities and plant litter etc.), but may in itself be significantly altered by soil properties;
3. Soil microbial diversity largely determines the stability of soil ecosystems under biotic and abiotic perturbations.
4. Management of soil microbial diversity can only be achieved through better understanding their structures and functions.

Keywords: microorganisms, decomposition, soil-plant systems

Introduction

Soil microorganisms play important roles in many ecosystem processes such as biogeochemical cycling of nutrients (nutrient transformations) and soil structural and hydrological properties (Lesser *et al.*, 2004; Driver *et al.*, 2005). A decline in soil productivity can often be attributed to a number of factors such as nutrient depletion, pests and diseases, drought, salinity and changes in climatic conditions. However, with the development of DNA and RNA methodologies there is an increasing understanding of the importance of the microbial community for soil health (Stocking, 2003). Thus, soil microbial properties are often used as indicators of soil quality. Traditionally, the belowground and aboveground components of soil-plant systems have been studied in isolation from one another (Wardle *et al.*, 2004), due in the main to a lack of proper techniques. Consequently our understanding of the belowground ecosystem has been far poorer than of the aboveground ecosystem. Over the last ten years or so there have been substantial advances in soil microbial ecology, including the development of new methodologies, and the recognition of the linkage between belowground and aboveground components.

Methodologies

Progress in examining microbial diversity in soils has been limited due to difficulties in the accurate measurement of species richness and evenness in soil microbial communities (Wardle & Giller, 1997). Microbial diversity can be categorized into two groups: species diversity and functional diversity. Functional diversity is a component of overall soil diversity, which may provide a more practical and ecologically relevant measure of microbial diversity (Zak *et al.*, 1994).

Species diversity consists of species richness, the total number of species present, species evenness, and the distribution of species (Øvreås, 2000). Methods measuring species diversity in soil include fatty acid methyl ester (FAME) and molecular-based techniques. The FAME method provides information on the microbial composition based on groupings of fatty acids (Frostegård *et al.*, 1993). Treonis *et al.* (2004) combined microbial community phospholipid fatty acid (PLFA) analyses with an *in situ* stable isotope $^{13}CO_2$ labelling approach, to identify microbial groups actively involved in assimilation of root-derived C in limed grassland soils. Ritz *et al.* (2004) studied the relationship between vegetation composition and soil microbial structure in upland grassland, and found that PLFA and community level physiological profiles (CLPP) showed some association with vegetation composition, but denaturing gradient gel electrophoresis (DGGE) profiling did not.

Alternative methods for studying microbial species diversity are based on molecular techniques. Polymerase chain reaction (PCR) - based approaches, i.e. DGGE and terminal restriction fragment length polymorphism (T-RFLP), have been used extensively. Increasingly PCR targeting of the 16S rDNA are been used to study prokaryote diversity, whilst 18S rDNA and internal transcribed spacer regions (ITS) are used to study fungal communities in soils. In these methods, genomic DNA is extracted from soil samples and purified. Target DNA (16S, 18S or ITS) is amplified using universal or specific primers, and then differences among microbial communities in soil are further analyzed. Nicol *et al.* (2003) used a PCR-DGGE method to study spatial differences in community structure in grassland soils. Griffiths *et al.* (2003) investigated the effects of water stress upon the diversity of bacterial communities in the rhizosphere of established upland grassland, subjected to different watering regimes over a two-month period. Profiling using DGGE analyses of community 16S rRNA genes and rRNA transcripts did not reveal any changes relating to the moisture regimes, suggesting that bacterial communities in this soil were resistant to water stress. The technique is thought to be accurate and sensitive for assessment of species diversity of arbuscular mycorrhizal fungi.

The use of T-RFLP for the study of microbial diversity in soil has become widespread. This technique allows the detection of only the labeled terminal restriction fragment (Liu *et al.*, 1997). It provides information on diversity, as each visible band represents a single operational taxonomic unit (Tiedje *et al.*, 1999). The method can be automated to allow analysis of larger numbers of soil samples, but it can also lead to overestimation due to incomplete digestion by restriction digestion (Osborn *et al.*, 2000). Klamer *et al.* (2002) used the T-RFLP technique to assess the effects of elevated atmospheric CO_2 levels. Although fungal biomass was enhanced in the high CO_2 concentration treatment, there was no significant influence on fungal composition and richness in the soil. The same technique (T-RFLP analysis) was also useful for identifying broad-scale, consistent differences in the bacterial communities in different soil locations over the natural micro scale heterogeneity of the soil in arid grassland (Kuske *et al.*, 2002).

Microbial functional diversity is different from that obtained by measuring species diversity, in that it concerns the range and evenness of functions expressed *in situ* by the microbial community, rather than the species present in soils. Microbial functional diversity includes a vast range of activities comprising nutrient transformations, decomposition, plant growth promotion/suppression, modification of soil physical processes (Wardle *et al.*, 1999), resistance/resilience to exotic stress or disturbance (Griffiths *et al.*, 2001a), potential carbon utilization (Garland & Mills, 1991), and catabolic response profile (Degens & Harris, 1997).

Thus, different methods measuring the functions of the microbial community in soil have been developed for different functions.

Among the various approaches, Biolog substrate decomposition and catabolic profiling have become popular. Mäder *et al.* (2002), used the Biolog method to demonstrate that enhanced soil biodiversity and fertility in organic plots, might render these systems less dependent on external inputs such as fertilisers and pesticides. In temperate upland grassland ecosystems, potential substrate utilisation using the Biolog method was assessed across a gradient of three upland grassland types (unimproved, semi-improved and improved) (Grayston *et al.*, 2001). Greater total carbon utilization by microbial communities from the improved grassland, suggested that there was more readily available carbon present in improved grasslands which stimulated bacterial growth. Therefore, plant species composition is an important factor governing microbial community structure in upland grasslands. With the Biolog method, Papatheodorou & Argyropoulou (2004), indicated that there were significant seasonal differences in functional diversity and evenness, and that the substrates responsible for the monthly differences were mainly carbohydrates and carboxylic acids.

Another popular approach is catabolic response profiles (CRPs) developed by Degens & Harris (1997). The method provides an easily interpretable and practical indicator of the diversity component of the soil microbiota (Sparling *et al.*, 2000). Stevenson *et al.* (2004) used the method to show that in each pasture or forest category, catabolic responses showed a similar pattern, suggesting similarities in functional catabolic capability. They found that pasture soil communities had significantly higher relative responses to carbohydrate and amino acid substrates, and significantly lower relative response to carboxylic acid substrates than microbial communities from forest soils. Changes caused by exotic stresses in the catabolic response of the soil microbial community were determined in crop and pasture soils (Degens *et al.*, 2001). The results showed that the crop soil with lower catabolic diversity reduced the resilience to applied stress compared with the pasture soil with higher catabolic diversity.

Substrate decomposition is also extensively used to assess functions of the microbial community in soil. The decomposition of a range of substrates added to soil is multiphasic. Hu & van Bruggenm (1997), demonstrated that multiphasic decomposition of cellulose is controlled by C and N availability, and the structure of the microbial community. Thus, given similar nutrient status, the pattern of substrate decomposition should reflect microbial community function. Griffiths *et al.* (2001a) found an interesting result that decomposition of grass shoot residues added to a petrol-contaminated soil, were significantly greater than in the uncontaminated soil, suggesting that microbial communities in soils contaminated with a particular pollutant could decompose that compound more rapidly than those in uncontaminated soils.

Soil microbial diversity and sustainability

The linkage between biodiversity and sustainability of terrestrial ecosystems has been studied mainly in aboveground systems. Diversity-sustainability relationships take various forms, e.g. hump-shaped, U-shaped, positive, negative and flat (non-significant) patterns, with no pattern predominating. Until recently, the hump-shaped relationship was the most widely observed pattern, in which plant diversity peaks at intermediate productivity levels under given conditions (Loreau *et al.*, 2001). Sustainability is the maintenance of system productivity and its continued function, whilst using its functions to establish sustainable production in the face of continued increases in human population (Pankhurst, 1997; Kennedy, 1995). Soil sustainability focuses more on soil functional stability, comprising resilience and resistance. Soil resistance is the inherent capacity to withstand exotic disturbance, whereas soil resilience is the capacity to recover after disturbance (Seybold *et al.*, 1999).

There is some evidence that lower diversity is less resistant to stress and disturbance. Degens *et al.*, (2001) compared soil capacity to withstand stress and disturbance in arable soil with lower catabolic evenness, and pasture soil with higher catabolic evenness. They found that increasing Cu concentration, salt stress, acidification, and wet-dry and freeze-thaw cycles caused greater decline in catabolic evenness in the arable soil than the pasture soil. Soil microbial diversity followed the classical 'hump-back' responses to the three stresses and two disturbances. Their results suggest that land use resulting in reduced catabolic evenness in soils, such as long-term cropping, might similarly exhibit a microbial community with reduced resistance to stress or disturbance. Griffiths *et al.* (2000) obtained similar results using progressive fumigation to reduce grassland soil microbial biodiversity, to measure the effects of different diversity upon the stability of key soil processes. Results suggest that more specific parameters decreased as biodiversity decreased, e.g. nitrification, denitrification and methane oxidation. The soils with the highest biodiversity were more resistant to the stress (brief heating to 40 °C and the addition of $CuSO_4$) than soils with impaired biodiversity. Griffiths *et al.* (2001a) determined the stability of three pairs of soils with different microbial diversities. The first pair of soils were planted for 5 years with either a single annual grass species or six annual grass species; a second pair were from 3 cm depth of a petrol-polluted soil, one undergoing remediation and the unpolluted control; the last two were agricultural and organic soils from intensively and extensively managed horticultural farms. The six soils were subjected to perturbation by adding 500µg Cu/g dry soil and heating to 40°C for 18 hours. The results demonstrated that both grassland soils and the organically managed agricultural soil with high diversity had the greatest resistance to both Cu and heat; the intensively managed agricultural soil had intermediate resistance to the two applied perturbations; the two petrol-polluted soils had the least resistance. Zhou *et al.* (2002) also found no reduction in soils with extremely high chromium (III) contents (e.g. >20%). They suggested that the carbon-rich soils with high microbial diversity were able to counteract stress to maintain soil stability.

There are examples of experimental reduction in microbial diversity having no direct effects on soil functions. Griffiths *et al.* (2001b) studied soil diversity relationships for a range of soil processes by inoculating sterile agricultural soil with serially diluted soil suspensions prepared from the parent soil. No consistent effect of biodiversity on the soil processes (e.g. incorporation of thymidine, potential nitrification, nitrate accumulation, respiratory growth response, community level physiological profile and decomposition) was observed. One possible reason is that there was selection of microbial communities due to the initial dilution. Degens (1998), also investigated whether reducing the microbial diversity of a pasture soil using a fumigation treatment compromises the decomposition function of the microbial community. The results indicated that the decomposition function of the soil with reduced functional diversity can be diminished under optimum moisture conditions, but is not invariably reduced when assessed under sub-optimal moisture conditions. This suggested that decreases in the functional diversity of soil microbial communities may not consistently result in declines in soil functioning.

Top-down influence

All terrestrial ecosystems consist of aboveground and belowground components, but they have traditionally been investigated separately from each other (Wardle *et al.*, 2004), except for the symbiotic relationship between plant roots and mycorrhizal fungi/nitrogen fixing bacteria (see next section). Top-down feedback is regarded as the impact of the plant community on soil microorganisms. Generally speaking, soil microorganisms are responsive to the nature of organic matter input (i.e. substrate quality and quantity). Plant species differ in their patterns of organic matter return to soils, thus individual plant species may have dominant impacts on soil microbial community and related biogeochemical processes (Wardle *et al.*, 2004). For example, size, activity and catabolic diversity of the soil microbial biomass could be substantially affected by agricultural land use (i.e. vegetation composition), and microbial diversity is greater in native grassland than in arable and forest soils (Nsabimana *et al.*, 2004). Grassland plant species may differ in the microbial communities around their roots, and this may help to explain why soils with different grassland species support different abundance of soil microbes and microbe-feeding fauna. Different traits of plant cultivars (genetic diversity) will also likely affect microbial community in the rhizosphere. However, there is little information available on grassland plant cultivars and soil microbial activities. The topic is important for two reasons: 1) the qualities of forage plants (chemical composition) may influence the soil microbial community, but this will also affect their usefulness as animal feeds and thus efficiency in nutrient cycling in grassland ecosystems; 2) the combination of different cultivars of forage plants may increase their tolerance to some diseases, particularly root diseases through the changes in microbial community. The later has been demonstrated in rice cultivation (Zhu *et al.*, 2000b), but has not as yet been extensively investigated in grassland. It is recommended that the possible functions of genetic diversity in managed grassland ecosystem, such as soil microbial diversity, nutrient cycling and feed quality should be investigated.

Plant community structure affects the soil microbial community mainly through rhizosphere deposition (e.g. root exudates and mucilage). An *in vitro* study using artificial root exudates demonstrated that the addition of root exudates enhanced bacterial density and changed the community-level profiles (Baudoin *et al.*, 2003). In temperate upland grassland, Bardgett *et al.*, (1999) demonstrated that in the short term, the abundance and activity of soil microorganisms were regulated more by plant species traits than by the direct effect of nitrogen input. It was suggested that these effects were possibly linked to variations in root exudates and nutrient acquisition of different plant species. However, the effects of individual plant species on the soil microbial community may also be dependent on environmental factors such as soil fertility. By comparing improved (optimum fertiliser application) and unimproved (with minimal nutrient input) soil types, Innes *et al.* (2004) found that all tested plant species grown in the improved soil, enhanced fatty acids synthesised by bacteria, but negatively impacted the same group of fatty acids in the unimproved soil. Soil fertility-dependent effects of plant species on soil microbes, make general predictions on how plant communities influence soil microbial properties more difficult. Other biotic and abiotic factors may also contribute to the spatial variability of the association between plant and microbial communities (Ritz *et al.*, 2004), and this needs further investigation.

Bottom-up influence

To understand the feedback mechanisms between aboveground and belowground components of the ecosystem, it is important to elucidate how soil organisms can influence the aboveground community structure and functioning (de Deyn, 2003; Wardle *et al.*, 2004). The bottom-up influence can be best illustrated by the symbiotic relationship between arbuscular mycorrhizal fungi (AMF) and plant roots. Over 90% of land plants can form symbiotic associations with AMF (Smith & Read, 1997), and these associations can facilitate plant growth by increasing resource acquisition from the soil. It has been generally found that AMF can form symbiosis with a broad range of host plants (i.e. with low specificity) (Smith & Read, 1997). However, in recent years there has been increasing evidence of some degree of ecological specificity or functional compatibility existing between host plants and AMF (Zhu *et al.*, 2000a; Smith *et al.*, 2003). The responsiveness of host plants to AMF depends on the specific plant and fungal combinations (van der Heijden *et al.*, 1998).

The ecological specificity (or functional compatibility) of AMF can have profound impacts on the aboveground plant community. Despite the specific mycorrhizal effects, it is clear that mycorrhizas exert a significant influence on plant community structure and dynamics in grasslands and other terrestrial ecosystems. The effects of mycorrhizal fungi on patterns of plant diversity and their underlying mechanisms are varied. In some plant communities, the presence and abundance of mycorrhizal fungi may regulate structure and the patterns of variation in mycorrhizal dependency among co-occurring plant species (Hartnett & Wilson, 2002). It is a frequent phenomenon that tolerant grasses prevail in poor soil conditions, but when mycorrhizal fungi are added, a broader plant community could overrun the grasses. By using two independent, but complementary ecological experiments, van der Heijden *et al.* (1998) demonstrated that the diversity of AMF is a major factor contributing to the maintenance of plant biodiversity and to ecosystem functioning. The plant species composition of microcosms that simulate European calcareous grassland fluctuate greatly with changing AMF taxa. Plant biodiversity, nutrient capture and productivity in macrocosms that simulate North American old-fields increase significantly with increasing AMF species richness. Seedling establishment within grassland communities can also be promoted by AMF. This was recently demonstrated by van der Heijden (2004), in a microcosm experiment in which the seedlings grew larger and obtained more P in the presence of AMF. These results emphasize the need to protect AMF and to consider these fungi in future management practices in order to maintain diverse ecosystems, and illustrates the potential role of belowground symbiotic systems in the promotion of species recruitment in grassland.

Biological nitrogen (N_2) fixation is another important component of grassland ecosystems, and the symbiotic association between legumes and rhizobia can provide substantial amounts of N to grassland. It has been estimated that the amounts of N fixation by various legumes can be over 500 kg N/ha per year (Carlsson & Huss-Danell, 2003). The actual amount of N fixed depends largely on the host plant species, environmental conditions, and management regimes as well as methods used to estimate N fixation. In grassland ecosystem, legumes often co-exist with non-legume plant species such as *Lolium perenne* (perennial ryegrass), and there may be an interspecific transfer of N from legumes to non-legumes. Due to its importance in the N economy of grassland, the belowground N transfer has received considerable attention in the last two decades (Laidlaw *et al.*, 1996; Hogh-Jensen & Schjoerring, 2000). In a field experiment, Hogh-Jensen & Schjoerring (2000) demonstrated that as much as 50% of total aboveground N of *Trifolium repens* (white clover) could be transferred to associated *L. perenne*, with a corresponding value of 10% for *Trifolium pratense* (red clover). Interspecific N flow may occur through several pathways. Principally, N can be transferred from legumes to associated non-legumes through decomposition of donor plant debris, root exudation and sloughing-off of cortex cells from donor plants (Paynel *et al.*, 2001), and directly through AMF interconnecting the root systems of the co-existing plants (Haystead *et al.*, 1988). By using a micro-lysimeter system, Paynel *et al.*, (2001) showed that, irrespective of direct contact between *T. repens* and *L. perenne*, about 0.076 mg of N were transferred per plant from *T. repens* and *L. perenne* during a two-month experimental period. This N transfer is believed to be mainly through root exudates. However, in the long term, the decomposition of root debris from donor plants can also be substantial. The other major pathway for interspecific N transfer is through AMF hyphal networks connecting plant roots. Nevertheless, the evidence for hyphae-mediated N transfer between plants is inconclusive (Rogers *et al.*, 2001), and may be partly due to a certain degree of ecological specificity between AMF and host plant species (Zhu *et al.*, 2000a). Novel experimental designs and detection methods have yet to be developed to provide convincing evidence of N transfer through hyphal networks.

Challenges ahead

Although there is increasing evidence that belowground and aboveground community components are closely associated, and that belowground communities play an important role in grassland productivity and sustainability, the following issues need further investigation:
* Novel molecular techniques to quantify the key functional groups of soil microbiota response to key biogeochemical processes;
* Extrapolating laboratory-based results to field conditions and scaling-up factors;
* Novel agents to manipulate specific microbial groups to understand the functions of these groups;
* Linking long-term observational data to grassland sustainability under changing environments (including management practices and global changes).

Acknowledgements

Our study is financially supported by the Natural Science Foundation of China (40321101).

References

Bardgett, R.D., J.L. Mawdsley, S.E. Dwards, P.J. Hobbs, J.S. Rodwell & W.J. Davies (1999). Plant species and nitrogen effects on soil biological properties of temperate upland grasslands. *Functional Ecology*, 13, 650-660.
Baudoin E., E. Benizri & A. Guckert (2003). Impact of artificial root exudates on the bacterial community structure in bulk soil and maize rhizosphere. *Soil Biology Biochemistry*, 35, (9), 1183-1192.
Carlsson, G. & K. Huss-Danell (2003). Nitrigen fixation in perennial forage legumes in the field. *Plant Soil*, 253, 353-372.
de Deyn, G.B., C.E. Raaijmakers, H.R. Zoome, M.P. Berg, P.C. de Ruiter, H.A. Verhoef, T.M. Bezemer & W.H. van der Puttenet (2003). Soil invertebrate fauna enhances grassland succession and diversity. *Nature*, 422, 711-713.
Degens, B.P. (1998). Decreases in microbial functional diversity do not result in corresponding changes in decomposition under different moisture conditions. *Soil Biology Biochemistry*, 30, 1989-1920.
Degens, B.P. & J.A. Harris (1997). Development of a physiological approach to measuring the metabolic diversity of soil microbial communities. *Soil Biology Biochemistry*, 29, 1309-1920.
Degens, B.P., L.A. Schipper, G.P. Sparling & L.C. Duncan (2001). Is the microbial community in a soil with reduced catabolic diversity less resistant to stress or disturbance. *Soil Biology Biochemistry*, 33, 1143-1153
Driver, J.D., W.E. Holeben & M.C. Rillig (2005). Characterization of glomalin as a hyphal wall component of arbuscular mycorrhizal fungi. *Soil Biology Biochemistry*, 37 (1), 101-106.
Frostegård, Å., A. Tulid & E. Bååth (1993). Phospholipid fatty acid composition, biomass, and activity of microbial communities from two soil types experimentally exposed to different heavy metals. *Applied Environmental Microbiology*, 59, 3605-3617.
Garland, J.L. & A.L. Mills (1991). Classification and characterization of heterotrophic microbial communities on the basis patterns of community-level sole-carbon-source utilization. *Applied Environmental Microbiology*, 57, 2351-2359.

Grayston, S.J., G.S. Griffith & J.L. Mawdsley (2001). Accounting for variability in soil microbial communities of temperate upland grassland ecosystems. *Soil Biology Biochemistry*, 33, 533-551.

Griffiths, B.S., M. Bonkowski, J. Roy & K. Ritz (2001a). Functional stability, substrate utilisation and biological indicators of soils following environmental impacts. *Applied Soil Ecology*, 16, 49–61

Griffiths, B.S., K. Ritz, R.D. Bardgett, R. Cook, S. Christensen, F. Ekelund, S.J. Sørensen, E. Bååth, J. Bloem, P.C. de Ruiter, J. Dolfing & B. Nicolardot (2000). Ecosystem response of pasture soil communities to fumigation-induced microbial diversity reductions, an examination of the biodiversity-ecosystem function relationship. *Oikos*, 90, 279-294.

Griffiths, B.S., K. Ritz, R. Wheatley, H.L. Kuan, B. Boag, S. Christensen, F. Ekelund, S.J. Sørensen, S. Muller & J. Boem (2001b). An examination of the biodiversity–ecosystem function relationship in arable soil microbial communities. *Soil Biology Biochemistry*, 33, 1713-1722.

Griffiths, R.I., A.S. Whiteley, A.G. O'Donnell & M.J. Bailey (2003). Physiological and community responses of established grassland bacterial populations to water stress. *Applied Environmental Microbiology*, 69, 6961–6968.

Hartnett, D.C. & G.W.T. Wilson (2002). The role of mycorrhizas in plant community structure and dynamics: lessons from grasslands. *Plant Soil*, 244, 319-331.

Haystead, A., N. Malajczuk & T.S. Grove (1988). Underground transfer of nitrogen between pasture plants infected with vesicular arbuscular mycorrhizal fungi. *New Phytologist*, 108, 417-423.

Hogh-Jensen, H. & J.K. Schjoerring (2000). Below-ground nitrogen transfer between different grassland species. Direct quantification by ^{15}N leaf labeling compared with indirect dilution of soil ^{15}N. *Plant Soil*, 227, 171-183.

Hu, S., A.H.C. van Bruggenm (1997). Microbial dynamics associated with multiphasic decomposition of 14C-labelled cellulose in soil. *Microbial Ecology*, 33, 134-143

Innes, L., P.J. Hobbs & R.D. Bardgett (2004). The impacts of individual plant species on rhizospher microbial communities in soils of different fertility. *Biology Fertility Soil*, 40, 7-13.

Kennedy, A.C. & K.L. Smith (1995). Microbial diversity and sustainability of agricultural soils. *Plant Soil*, 23 (2), 69-79.

Klamer, M., M.S. Robert & L.H. Levine (2002). Influence of elevated CO_2 on the fungal community in a coastal scrub oak forest soil investigated with terminal-restriction fragment length polymorphism analysis. *Applied Environmental Microbiology*, 68, 4370-4376.

Kuske, C.R., L.O. Ticknor, M.E. Miller, J.M. Dunbar, J.A. Davis, S.M. Barns & J. Belnap (2002). Comparison of soil bacterial communities in rhizospheres of three plant species and the interspaces in an arid grassland. *Applied Environmental Microbiology*, 68, 1854-1863.

Laidlaw, A.S., P. Christie & H.W. Lee (1996). Effect of white clover cultivar on apparent transfer of nitrogen from clover to grass and estimation of relative turnover rates of nitrogen in roots. *Plant Soil*, 179, 243-253.

Lesser, M.P., C.Z. Mazel, M.Y. Gorbunov & P.G. Falkowski (2004). Discovery of symbiotic nitrogen-fixing cyanobacteria in corals. *Science*, 305, 997-1000.

Liu, W.T., T.L. Marsh, H. Cheng & L.J. Forney (1997). Characterization of microbial diversity by determining terminal restriction fragment length polymorphisms of genes encoding 16S rRNA. *Applied Environmental Microbiology*, 63, 4516-4522.

Loreau, M., S. Naeem, P. Inchausti, J. Bengtsson, J.P. Grime, A. Hector, D.U. Hooper, M.A. Huston, D. Raffaelli, B. Schmid, D. Tilman & D.A. Wardle (2001). Biodiversity and ecosystem functioning: current knowledge and future challenges. *Science*, 294, 804-808.

Mäder, P., A. Flieβbach, D. Dubois, L. Gunst, P. Fried & U. Niggli. (2002). Soil fertility and biodiversity in organic farming. *Science*, 296, 1694-1697.

Nicol, G.W., L.A. Glover & J.I. Prosser (2003). Spatial analysis of archaeal community structure in grassland soil. *Applied Environmental Microbiology*, 69, 7420–7429

Nsbimana, D., R.J. Haynes & F.M. Wallis (2004). Size, activity and catabolic diversity of the soil microbial biomass as affected by land use. *Applied Soil Ecology*, 26, 81-92.

Osborn, A.M., E.R. Moore & K.N. Timmis (2004). An evaluation of terminal-restriction fragment length polymorphism (T-RFLP) analysis for the study of microbial community structure and dynamics. *Environmental Microbiology*, 21 (1), 39-50.

Øvreås, L. (2000). Population and community level approaches for analyzing microbial diversity in natural environments. *Ecology Letters*, 3, 236-251.

Pankhurst, C.E. (1997). Biodiversity of soil organisms as an indicator of soil health. In: C.E. Pankhurst, B.M. Doube & V.V.S.R. Gupta (eds.) Biological Indicators of Soil Health, CAB International, Wallingford, 297–324.

Papatheodorou, E.M., M.D. Argyropoulou & G.P. Stamou (2004). The effects of large- and small-scale differences in soil temperature and moisture on bacterial functional diversity and the community of bacterivorous nematodes. *Applied Soil Ecology*, 25, 37–49.

Paynel, F., Murray P.J. & J.B. Cliquet (2001). Root exudates: a pathway for short-term N transfer from clover to ryegrass. *Plant Soil*, 229, 235-243.

Ritz, K., J.W. McNicol, N. Nunan, S. Grayston, P. Millard, D. Atkinson, A. Gollotte, D. Habeshaw, B. Boag, C.D. Clegg, B.S. Griffiths, R.E. Wheatley, L.A. Glover, A.E. McCaig & J.I. Prosser (2004). Spatial stricture in soil chemical and microbiological properties in an upland grassland. *FEMS Microbiology Ecology*, 49, 191-205.

Rogers, J.B., A.S. Laidlaw & P. Christie (2001). The role of arbuscular mycorrhizal fungi in the transfer of nutrients between white clover and perennial ryegrass. *Chemosphere*, 42, 153-159.

Seybold, C.A., J.E. Herrick & J.J. Brejda (1999). Soil resilience, a fundamental component of soil quality. *Soil Science*, 164, 224–234.

Smith S.E. & D.J. Read (1997). Mycorrhizal symbiosis. London: Academic press, 605pp.

Smith S.E., F.A. Smith & I. Jakobsen (2003). Mycorrhizal fingi can dominate phosphate supply to plants irrespective of growth responses. *Plant Physiology*, 133, 16-20.

Sparling, G.P., L.A. Schipper, A.E. Hewitt & B.P. Degens (2000). Resistance to cropping pressure of two New Zealand soils with contrasting mineralogy. *Australian Journal of Soil Research*, 38, 85-100.

Stevenson, B.A., G.P. Sparling, L.A. Schipper, B.P. Degens & L.C. Duncan (2004). Pasture and forest soil microbial communities show distinct patterns in their catabolic respiration responses at a landscape scale. *Soil Biology & Biochemistry*, 36, 49–55

Stocking M.A. (2003). Tropical soils and food security: the next 50 years. *Science*, 302, 1356-1359.

Tiedje, J.M., S. Asuming-Brempong, K. Nusslein, T.L. Marsh & S.J. Flynn (1999). Open the black box of soil microbial diversity. *Applied Soil Ecology*, 13, 109-122.

Treonis, A.M., N.J. Ostle & A.W. Stott (2004). Identification of groups of metabolically-active rhizosphere microorganisms by stable isotope probing of PLFAs. *Oecologia*, 138, 275–284.

van der Heijden, M.G.A. (2004). Arbuscular mycorrhizal fungi as support systems for seedling establishment in grassland. *Ecology Letters*, 7, 293-303.

van der Heijden, M.G.A., J. N. Klironomos, M. Ursic, P. Moutoglis, R. Streitwolf-Engel, T. Boller, A. Wiemken & I.R. Sanders (1998). Mycorrhizal fungal diversity determines plant biodiversity, ecosystem variability and productivity. *Nature*, 396, 69-72.

Wardle, D.A, R.D. Bardgett, J.N. Klironomos, H. Setälä, W.H. van der Putten & D.H. Wall (2004). Ecological linkages between aboveground and belowground biota. *Science*, 304, 1629-1633.

Wardle, D.A. & K.E. Giller (1997). The quest for a contemporary ecological dimension to soil biology. *Soil Biology Biochemistry*, 28, 1549-1554.

Wardle, D.A., K.E. Giller & G.M. Barker (1999). The regulation and functional significance of soil biodiversity in ago-ecosystems. In: D. Wood & J.M. Lenné (eds.) Agrobiodiversity: Characteriation, Utilization and Management, CAB International, London, 87-121.

Zak, J.C., M.R. Willing, D.L. Moorhead, H.G. Wildmand (1994). Functional diversity of microbial communities: a quantitative approach. *Soil Biology Biochemistry*, 26, 1101-1108.

Zhou, J.Z., B.C. Xia., D. Treves, L.Y. Wu, T.L. Marsh, R. O'Neill, A.V. Palumbo & J.M. Tiedje (2002). Spatial and resource factors influencing high microbial diversity in soil. *Applied Environmental Microbiology*, 68, 326-334.

Zhu, Y.G., A.S. Laidlaw, P. Christie & M.E.R. Hammond (2000a). The specificity of arbuscular mycorrhizal fungi in perennial ryegrass–white clover pasture. *Agriculture, Ecosystems and Environment*, 77, 211–218.

Zhu, Y.Y., H.R. Chen, J.H. Fan, Y.Y. Wang, Y. Li, J.B. Chen, J.X. Fan, S.S. Yang, L.P. Hu, H. Leung, T.W. Mew, P.S. Teng, Z.H. Wang & C.C. Mundt (2000b). Genetic diversity and disease control in rice. *Nature*, 406 (6797): 718-722.

Soil quality assessment and management

M.G. Kibblewhite
National Soil Resources Institute, Cranfield University, Silsoe, Bedfordshire MK45 4DT, UK
Email: m.kibblewhite@cranfield.ac.uk

Key points

1. Soil quality is related to the capacity of soil to deliver ecosystem services on a sustainable basis.
2. Effective management of soil within grasslands can deliver many benefits to mankind but poor management may cause loss of soil quality from erosion, loss of organic matter, physical deterioration etc.
3. Services are delivered from soil by biological processes. Soil quality depends on the form and condition of the soil habitat. Fixed factors (e.g. texture) are useful for assigning soil to types. Variable factors (e.g. organic carbon) can then be used to assess quality within soil types, by reference to percentiles of the distribution of values for a given type.
4. Systematic monitoring of soil quality is useful for identifying the possible need for field level actions at regional to landscape scales. Assessment and management of soil quality at a local scale is supported more efficiently and effectively by in-field observation of soil profiles.

Keywords: soil protection, habitat, ecosystem services

Introduction

There is a new focus on soil protection (e.g. European Commission, 2002) driven by a realisation that soil provides much of the ecological capacity within land to support the growing footprint of human activities. This puts soil management at the heart of sustainable land management. As a key compartment in the terrestrial environment, soil needs to be given the same levels of protection as air and water, especially since soil can take orders of magnitude longer to recover from damage.

In 2002, 27% of global soil resources were in grassland (FAO, 2004). A key concern is the extent of more intensively managed grassland where the risk to soil quality is likely to be greater. For example in the United Kingdom, where grassland occupies almost half the land area, there is both extensively managed semi-natural grassland and intensively-managed permanent grass, as well grass-arable rotations. However, in all circumstances, it is clear that the effective management of soil within grassland systems has a leading part to play in the overall management of soil resources.

When considering soil, it is important to distinguish it from land. Land represents space within which different activities, human and otherwise, can take place. It is a finite resource. Soil is a habitat within land, containing ecological communities that deliver a range of services that are critical to sustainability. One key ecosystem service delivered by soil is support for food and fibre production, but there is increasing recognition that the range of services is much wider. Six types of ecosystem service are identifiable; 1) food and fibre production, 2) biodiversity conservation, 3) environmental services, 4) cultural heritage (e.g. archaeology and landscape conservation), 5) provision of platforms for built infrastructure, and 6) supply of primary materials (e.g. peat). Moreover, the economic value of these ecosystem services is large. For example, the environmental services appear to have the same order of value as agricultural outputs for some soils (Environment Agency, 2002), although this value is not included within current financial transactions. Thus semi-natural grassland soils that are of lower agricultural value may have significant overall economic value, when account is taken of their delivery of environmental and other services relating to water, atmospheric and biodiversity management.

When the effects of population increases and anticipated economic growth are combined, a tightening global food market is anticipated, especially for animal-derived protein. Soil management within grasslands is critical to meeting additional demand for protein in a sustainable framework. The productivity of grassland soils must be increased, while simultaneously protecting, and where possible raising, their capacity to deliver other ecosystem services. This requires measures of soil quality that can be used to target and monitor the efficacy of soil management measures. The key question that arises is *"What is good quality soil - taking full account of all the ecosystem services that it delivers?"*

Definition and assessment of soil quality is challenging for several reasons, including difficulties with defining both soil-based services and the variety of soil types, as well as limited scientific understanding of soil systems. The result is a gap between the current needs of policy makers and the scientific development of necessary, well-founded options for soil quality definition and monitoring. In the absence of new scientific approaches, measures of soil quality may be adopted that are convenient but lack a robust rationale in terms of indicator choice and target setting. This paper explores some of the impacts of grassland management on soil quality; it then describes a possible conceptual framework for arriving at a set of quality indicators and targets for soil, and explains how this framework can be applied to the assessment of soil quality in grasslands in support of their sustainable management.

Impacts of grassland management on soil quality

Trends in land-use are a key influence on overall soil quality. Conversion of natural and semi-natural forest to grassland causes losses of soil organic matter and alters soil physical properties. Improvement of semi-natural grassland or rough grazing land by drainage or irrigation, tillage, nutrient additions, burning, or re-seeding alters soil conditions. Where the conversion or subsequent management or both are not appropriate or effective, irreversible losses of soil quality may occur. By contrast, conversion of arable land to permanent grassland or the inclusion of grass in crop rotations increases soil organic matter levels, and may assist recovery of soil quality lost during previous cultivation.

Erosion is a natural process, which can be accelerated by poor management of grasslands. High stocking rates may lead to bare soil exposure and surface compaction. This increases the intensity of surface water runoff during storms and the consequent risk of soil erosion, especially on sloping land with soil that is susceptible to erosion. The loss of soil quality from gully erosion is obvious, but more insidious and widespread are sheet erosion losses, commonly rising to 10 to 40 t/ha per year. These are important because of the slow rate of soil formation.

Soil organic matter provides a store of soil nutrients and contributes to good soil physical conditions. The intensity of management of grasslands and their longevity between periods of arable cultivation affect both soil organic matter composition and levels, and thus soil quality. Intensification may lead to losses or gains in soil organic matter. Nutrient additions, improvements in species composition (including legume introduction), and optimisation of the soil-water regime for plant growth all increase net primary production and so tend to increase inputs of organic carbon to soil. Conversely, increased frequencies of grazing, mowing and burning all reduce these inputs. A further complication arises from the different temporal cycles of different pools of organic carbon in soil. Levels of less degraded plant litter in the surface horizon of soil profiles vary seasonally and with changing sward management. Changes in the soil organic carbon in more-recalcitrant microbial products are less rapid. Yet these products of microbial degradation and conversion of plant materials are perhaps more strongly associated with soil quality improvement, in terms of better soil structure and aggregate stability. This suggests that short-term conversion of arable land to grassland, including ley-arable rotations and set-aside (where nutrient additions and grazing are discontinued), may bring only relatively weak and ephemeral benefits to soil quality.

Animal, machinery and human traffic can cause compaction at the surface and within the soil profile, particularly when soil moisture is above the plastic limit. This impedes root growth and drainage, and these effects may extend over years, unless corrected by remedial intervention. The potential for this loss in soil quality depends on soil texture, natural or artificial drainage and climate.

Intensification of grassland management sometimes introduces materials that may cause contamination and a consequent loss of soil quality. A particular hazard arises from excessive spreading of slurries and manures. These may contain contaminants, such as copper and zinc, or veterinary products, which can accumulate in soil and may pose a threat to soil quality if spreading is frequent and over long periods.

Soil system performance and soil quality

Figure 1 shows an idealised performance curve for systems in general. As inputs are increased, outputs increase proportionally within the 'working-range' of the system. The efficiency of the system falls progressively above the working-range until the peak rate is reached, after which it becomes overloaded and further increases of input rates cause a reduction in output rates. Above the working range and up to peak capacity, the system remains resilient, so that when output demand temporarily exceeds the working range limit, the system 'recovers' without any permanent damage once excess demand is removed. The sustainable capacity of the system is defined by the conversion efficiency within the working range and the extent of that range.

A common definition of quality is that it is a measure of the extent to which a system is 'fit for purpose'. In the context of soil systems, the corresponding definition of a soil's quality is the extent that it is able to provide delivery of a portfolio of ecosystem services in a sustainable manner, with conversion efficiencies that meet wider ecosystem requirements, without exceeding working ranges.

Soil is a multi-functional system that delivers different outputs simultaneously. A set of performance curves can be imagined corresponding to these different outputs. These curves are not independent of each other. Maximising capacity to deliver one type of output will reduce capacity for other outputs, because common supporting processes have finite capacities. The management challenge is to devise a means of optimising capacity to support the sustainable delivery of a varying mix of services.

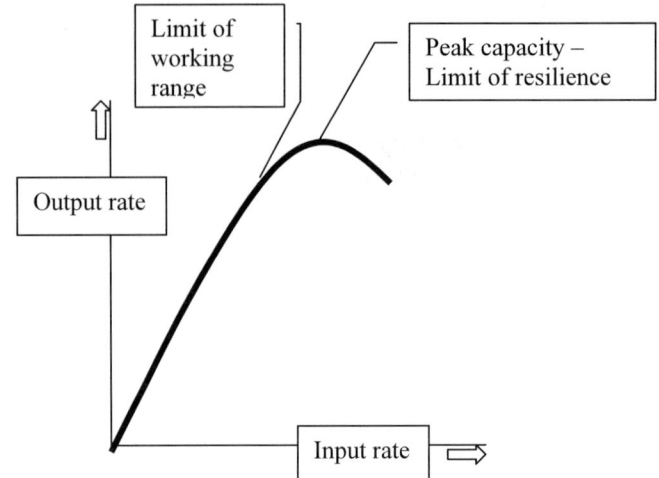

Figure 1 Idealised performance curve for systems

Assessment of soil quality should start with definition of the possible kinds and levels of soil-based ecosystem services, but the understanding necessary to complete this is not yet available. For example, the nature of biodiversity in soil is still being explored; a full description of soil services relating to air and water quality is not yet complete; and, cultural services (e.g. those related to landscape aesthetics) are not easily quantified. Without adequate definition of the ecosystem services provided by soil, it is clearly not possible to define their optimal mix. Nonetheless, a workable means of defining soil quality holistically and for setting quality targets is essential to support better soil management.

Soil is first and foremost a living system made up of biological communities existing within a habitat characterised by a highly complex architecture, that is moderated by soil-water physical processes (Young & Crawford, 2004; Ritz & Young, 2004). This perspective demands that assessment of soil quality focuses on the nature and condition of the soil habitat, as this determines fitness for purpose to support systems that deliver ecosystem services.

A basis for defining and assessing soil quality even in the absence of a complete description of services and a definition of their optimal mix, is to consider soil processes and the capacity for delivery of services as inherently biological. If the soil system is viewed as a physical habitat within which sets of biologically-mediated processes convert or control fluxes of air, water, carbon, nutrients, contaminants, etc. into products and services, the overall quality of the soil system - in terms of sustainable capacity for delivering different ecosystem services, should depend on the type and condition of the physical habitat. Although this emphasis on habitat condition in relation to soil quality is relatively new, it can also be seen as a re-interpreted rationale for using traditional soil quality measures, specifically physical and chemical parameters, such as texture, bulk density, organic carbon, nutrients or contaminant concentrations.

A wide range of soil habitat conditions can exist depending on fixed and variable factors (see Table 1). The fixed factors define the envelope of possible habitat conditions and contribute to the definition of soil type within a continuum of possibilities. The variable factors define the habitat condition within an envelope of possibilities set by the fixed factors and characteristics of the soil type. Sub-envelopes corresponding to different land use can also be set, e.g. permanent grass, grass-arable rotations, etc.

Table 1 Fixed and variable factors defining soil habitat condition

Fixed factors	Variable factors
Texture, stone content, slope angle, topographic position, annual precipitation, mean temperature, etc.	Water holding capacity, bulk density, soil organic carbon level, pH and depth (affected by tillage and erosion), plant cover, etc.

By comparing observed values for these variable factors with reference values for the soil type, the condition of the soil habitat can be assessed as an indicator of soil quality, and used as a basis for decisions about soil management. In principle, reference values for variable factors for a given soil type can be defined either by making measurements on 'benchmark' soil profiles that are judged to be of good quality, or by choosing percentile values within the distribution of values observed for the whole population of profiles within a given soil type.

Assessment of soil quality

Assignment to soil type

For soil quality assessment based on soil habitat condition, populations of soil profiles need to be identified corresponding to different soil types. These may be selected by reference to fixed factors (Table 1). The soil types identified will correspond closely to existing soil profile taxonomic classes because the same factors are used in classification systems. However, developments in remote sensing and in information sciences are allowing better capture and description of spatial and temporal variations in natural systems, and this is changing ideas about soil variety and variability. In the last century, deterministic taxonomic systems were developed to describe characteristic types of soils that evolve in response to particular combinations of soil forming factors (pedogenesis). Application of digital information systems is confirming the correctness of the concept of a three-dimensional continuum of soil properties (Fitzpatrick, 1971), with the properties associated with particular soils even within narrow spatial limits being better described in a stochastic fashion than an absolute one. The result is a realisation in the policy as well as the scientific community that a 'one-size-fits-all' approach to soil quality criteria is inadequate.

Digital information techniques have an important advantage over traditional soil taxonomy, because assignment to type is both objective and can also be qualified by an estimate of the confidence of assignment. The application of these new techniques is just emerging from the development phase and will lead to new digital soil information systems over the coming years, which can be used to assign soil profiles (including those in grassland), to populations appropriate for soil quality assessment. In the meantime, soil maps can be used to make assignments, albeit less effectively.

Assessment of soil quality within type

Measurement methods for the variable factors proposed in Table 1 and others are well established, and data are available on their spatial (and in some cases temporal) variability, at regional, landscape and sometimes field levels. The key problem is to define targets for the factors that indicate a soil habitat is 'fit for purpose' for delivery of ecosystem services.

In principle, habitat condition in 'working soils' could be assessed by comparison of measured values for variable factors, with those observed for 'benchmark' soils that are in a 'good' semi-natural condition. The case for this rests on the assumption that the latter should have evolved to optimise natural overall ecological capacity. Certainly, this approach would be appropriate when the purpose of soil management is to restore soils to their natural condition, for example as part of natural grassland restoration. However, within sustainable development, modification of soils to optimise their delivery of services to support human quality of life is as important as biodiversity conservation. Thus highly modified soils can be of good quality because they have a greater capacity to deliver a desirable mix of services than did the natural soils from which they were created, and to which they might be restored. For example, artificial drainage and sustained additions of organic matter over many years are capable of producing a modified soil that has an increased capacity to support both highly productive grassland and additional capacity to deliver other ecosystem services.

An alternative to assessing soil quality by benchmarking against semi-natural soils, is to compare values for variable factors found for a particular soil profile with measured ones for a representative population of the relevant soil type. For each factor and soil type, percentile values can be chosen to indicate good, fair or poor soil quality. It should be stressed that unless there is clear evidence for any threshold value below which soil performance is impaired, the choice of percentile values that correspond to different quality assessments is arbitrary. Such evidence is generally absent, so the choice of percentile value will be informed mainly by policy considerations, such as the proportion of soils that are to be targeted for improvement, perhaps taking account of available support funding.

One variable factor that has already been proposed (MAFF, 2000) as an indicator of soil quality is soil organic matter. The definition of soil organic matter may need elaboration, but for illustration, organic carbon may be used as a proxy. Figure 2 shows the distribution of organic carbon in shallow lime-rich soils over chalk and limestone in England and Wales. This soil type was mainly in permanent grass in the first half of the last century, but is now largely in arable production. For purposes of assessment and management, it could be proposed that all values of organic carbon that fall within the highest quartile are 'good', those in the two middle quartiles are 'satisfactory', and those in the lowest quartile 'need improvement'. Options for improving the quality of soil in the lowest quartile will include returning arable land to permanent grassland.

As another example, bulk densities in the lowest quartile for a soil type could be assessed as 'good', those in the second and third lowest quartiles as 'satisfactory' and those in the highest quartile as 'needing improvement'. The overall quality of the soil could then be assessed by completing similar assessments for a set of variable factors, including depth of surface horizon and pH, as well as organic carbon and bulk density.

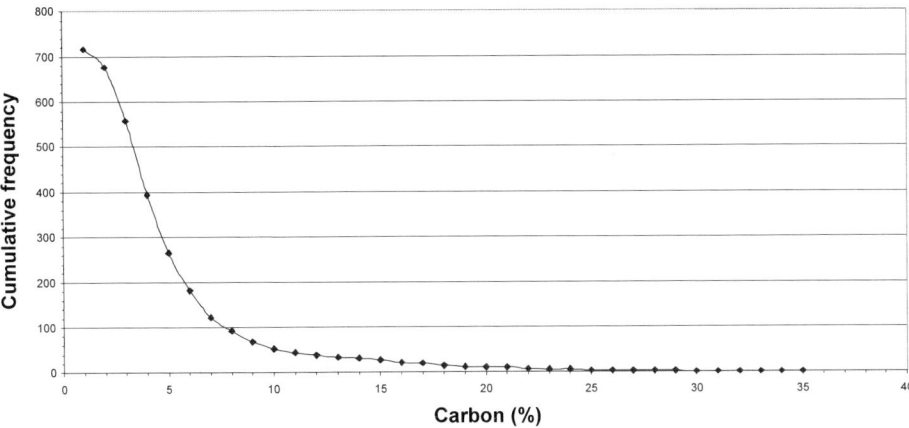

Figure 2 Cumulative frequency of measured values for organic carbon in shallow lime-rich soils over chalk and limestone in England and Wales (data extracted from the 5x5 km grid sample data held in the National Soil Inventory, NSRI, Cranfield University)

Management of soil quality

Sustainable management of grassland soil requires that the demand for ecosystem services does not exceed sustainable capacity within the soil (see Figure 1). In particular, stocking rates must be set at levels that do not place excessive demand on the soil system. Loss of plant cover through overgrazing (especially when combined with burning) puts soil quality at risk by reducing organic carbon (and so substrate) inputs to the soil system, and by increasing the risk of soil erosion. When the sustainable capacity of a soil is exceeded continuously in this way, and especially if the peak capacity is exceeded, permanent damage to the soil habitat and loss of soil quality are probable.

Monitoring at regional and landscape scales of variable factors can indicate whether soil quality is being compromised generally. For example, a high proportion of soils with organic carbon contents that are below target may indicate excessive grazing pressures. Where there is an indication that soil quality is at risk in a landscape (because values for variable factors fall within percentiles identified as unsatisfactory), an appropriate initial response is farm-level investigations to explore the true nature, extent and level of possible soil damage. High costs mean that it is normally not feasible to extend systematic monitoring of variable factors down to field scale. The best tool for assessing grassland soil quality at field scale is the spade. Opening of soil profiles and basic field observation of soil conditions allows the land manager to identify soil quality problems directly (see NSRI, 2001). However, effective responses to adverse trends in soil quality identified in soil monitoring require action to be taken throughout the wider landscape. The key to achieving this is field scale demonstration of solutions in support of effective uptake of advice and training, ideally with area-based financial incentives to changed practices.

Acknowledgements

This paper has benefited greatly from discussions with colleagues at Cranfield University, Silsoe. I thank them all and especially Ian Bradley, Professor Guy Kirk, Professor Jim Harris, John Hollis and Professor Karl Ritz.

References

Commission of the European Communities (2002). COM (2002), 179 Final. Towards a thematic strategy for soil protection. Commission of the European Communities, Brussels, 35pp.

Environment Agency (2002). Agriculture and natural resources: benefits, costs and potential solutions. Environment Agency, Bristol, 23pp.

FAO (2004). FAOSTAT data. <http://apps.fao.org/>

Fitzpatrick, (1971). Pedology. In: A systematic approach to soil sciences. Oliver & Boyd, Edinburgh, 306pp.

MAFF (2000). Towards sustainable agriculture: a pilot set of indicators. MAFF, HMSO, London, 74pp.

NSRI (2001). A guide to better soil structure. Cranfield University, Silsoe, 19pp.

Ritz, K. & I.M Young (2004). Interactions between soil structure and fungi. *Mycologist*, 18, 52-59.

Young, I.M. & J.W. Crawford (2004). Interactions and self-organisation in the soil-microbe complex. *Science*, 304, 1634-1637.

Water resources, agriculture and pasture: implications of growing demand and increasing scarcity

M.W. Rosegrant, R.A. Valmonte-Santos, S.A. Cline, C. Ringler and W. Li
International Food Policy Research Institute, 2033 K St. NW, Washington, DC 20006 USA
Email: m.rosegrant@cgiar.org

Key points

1. Water availability for irrigation is threatened in many regions by rapidly increasing demand for nonagricultural water uses in industry, households, and the environment. The scarcity of irrigation water will not only impact crop production, but also meat production, as much of the pasture used to feed livestock is irrigated.
2. Grassland is caught between two countervailing forces: a requirement for increasing meat demand that boosts the need for additional pasture to support livestock production, and rapidly increasing water scarcity that makes pasture irrigation uneconomical.
3. The most effective means of dealing with water scarcity is likely to be conserving water in existing water uses. Improvements in the irrigation sector to increase water use efficiency must be made at the technical, managerial, and institutional levels.
4. Innovative water pricing policies that increase the prices for domestic and industrial water while preserving incomes for farmers and the rural poor will encourage water-saving innovation.

Keywords: irrigation, rainfed, policy

Introduction

The world's farmers will likely need to produce enough food to feed 8 billion people by 2025, and to do so they must have enough water to raise their crops, including pasture to feed animals for human consumption. Yet farmers are already competing with industry, domestic water users, and the environment for access to the world's finite supply of water. Irrigation, which consumes far more water than any other use, has generated enormous benefits. By helping raise farmers' yields and stabilize food production and prices, irrigation has been a key to achieving food security in many parts of the world. About 250 million hectares are irrigated worldwide today, nearly five times more than at the beginning of the 20th century. Yet inappropriate water and agricultural policies and poor irrigation management have also lowered groundwater tables, damaged soils, and reduced water quality. Moreover, growing populations with rising incomes will further increase the demand for irrigation water to meet food needs.

Other users also have important claims on water. Although the domestic and industrial sectors use far less water than agriculture, water consumption in these sectors is growing rapidly. Access to safe drinking water and sanitation is critical for health - particularly children, and the importance of reserving water for environmental purposes has only recently been recognized: during the 20th century, more than half of the world's wetlands were lost. Even as demand for water increases, groundwater is being depleted and other water ecosystems are becoming polluted and degraded, developing new sources of water is getting more costly. Will available freshwater meet the rapidly growing demands for household, industrial, and environmental needs, and still provide enough water to produce food and feed crops and sustain pasture development to produce meat for a burgeoning population? What will be the impact of these trends on water for grasslands and pasture? Given the multiple uses of water, what policy recommendations can be implemented in order to attain sustainable management of water resources in relation to its multi-faceted use globally and regionally?

This paper takes three complementary approaches to look at the relationships between water scarcity and pasture. First, an integrated global water and food modelling framework, IMPACT-WATER is applied to simulate the complex relationships among water availability and demand, food supply and demand, international food prices, and trade under three different future scenarios. Next, a set of 'soft-linked' global models used in the Millennium Ecosystem Assessment is employed to examine future changes in land use and ecosystem services to 2050. Finally, a synthesis of empirical evidence on the impact of water scarcity and increasing value of water on irrigation and water allocation is presented, to examine the impact of increasing water scarcity on irrigated pasture, and assess the implications of land use pressure and water scarcity on pasture production systems.

Analytical approach

No single model incorporates the range of interactions across crops, livestock, pasture, and water that are needed to fully address these questions. Insights are gained through three approaches: (1) a global model of water and food supply - IMPACT-WATER, that combines an extension of the International Model for Policy Analysis of Agricultural Commodities and Trade (IMPACT) with a Water Simulation Model (WSM) to simultaneously assess food supply and demand, and water supply and demand to 2025 (Rosegrant *et al.*, 2002); (2) a set of 'soft-linked' global models used in the Millennium Ecosystem Assessment to examine future changes in land use and ecosystem services to 2050. These models include IMAGE (Integrated Model to Assess the Global Environment), which has been developed to study climate change and global change issues; AIM (Asia-Pacific Integrated Model), which has a similar focus; IMPACT, which has been developed to analyse the world food situation; and WaterGAP, which examines global water supply and demand; and (3) synthesis of empirical evidence on the impact of water scarcity and increasing value of water on irrigation and water allocation.

Current uses of water

Globally, agriculture accounts for about 69% of water withdrawals - mostly for irrigation, followed by industry and energy at 23%, and domestic consumption for household, drinking water and sanitation at 8% (UN, 2003; Rosegrant *et al.*, 2002; Gardner-Outlaw & Engelman, 1997). Regionally, water utilisation varies enormously. Agricultural regions such as Africa use up 88% of all water withdrawn for agriculture, domestic use accounts for 7% and industry accounts for only 5%. In contrast, in Europe, water for industry accounts for 54% of withdrawals, while agriculture's share is only 33% and domestic use is 13% (UN, 2003). In agriculture, irrigation is applied not only to crops but also to pastureland, particularly in major livestock-producing countries like US, Brazil, France, China, Australia, UK and Germany. Pastureland and crop areas take up around 37% of the earth's land area (UN, 2003) and hence compete for water withdrawal with other non-agricultural sectors.

Alternative futures for water and food

Alternative global scenarios are developed to examine how future water policies and investments will affect water use for all sectors, and water availability and food production, including business-as-usual, water crisis, and sustainable water use scenarios. IMPACT-WATER is utilised to assess these alternative futures. This model combines an extension of the IMPACT with WSM. IMPACT is a partial equilibrium model of the agricultural sector. Demand is a function of prices, income, and population growth. Growth in crop production in each country is determined by crop and input prices and the rate of productivity growth. World agricultural commodity prices are determined annually at levels that clear international markets. IMPACT generates projections for crop area; yield; production; demand for food, feed, and other uses; prices; and trade. For livestock, IMPACT projects numbers, yield, production, demand, prices, and trade. The WSM is a basin-scale model of water resource use.

The IMPACT-WATER linkage was made possible by (1) incorporating water in the crop area and yield functions of IMPACT; and (2) simultaneously determining water availability at the river basin scale, and water demand from irrigation and other uses in the WSM. IMPACT-WATER divides the world into 69 spatial units, including macro river basins in China, India, and the United States and aggregated basins for other countries and regions. Domestic and industrial water demands are estimated as a function of population, income, and water prices. Water demand in agriculture is projected - based on irrigation and livestock production growth, water prices, climate, and water use efficiency for irrigation at the basin level. Water demand is then incorporated as a variable in the crop yield and area functions for each of eight major food crops: *Triticum* spp. (wheat), *Oryza sativa* (rice), *Zea mays* (maize), other coarse grains, *Glycine* spp. (soybeans), *Solanum tuberosum* (potatoes), *Dioscorea* spp. (yams) and *Ipomea batatus* (sweet potatoes), and *Manihot esculenta* (cassava) and other roots and tubers. Water requirements for all other crops are estimated in an aggregated form.

Water availability is a stochastic variable with observable probability distributions. WSM simulates water availability for crops at the river basin scale, taking into account precipitation and runoff, water use efficiency, flow regulation through reservoir and groundwater storage, non-agricultural water demand, water supply infrastructure and withdrawal capacity, and environmental requirements at the river basin, country, and regional levels. Environmental impacts can be explored through scenario analysis of committed instream (such as recreation, hydropower generation and navigation) and environmental flows, salt leaching requirements for soil salinity control, and alternative rates of groundwater pumping. Rosegrant *et al.* (2002) provides detailed methodology for IMPACT-WATER.

The primary drivers used in the model as the building blocks of the three scenarios are:
- *Economic and demographic drivers* - population growth, rate of urbanisation, and rate of growth in GDP (gross domestic product) and GDP per capita; projected outcomes on economic and demographic drivers are held constant across scenarios;
- *Climate and hydrological parameters* - precipitation, evapotranspiration, runoff, and groundwater discharge; held constant across three scenarios;
- *Technological, management and infrastructural drivers* - river basin efficiency, reservoir storage, water withdrawal capacity, potential physical irrigated area, and crop and animal yield growth;
- *Policy drivers* - water prices, water allocation priorities among sectors, committed water flows for environmental purposes, interbasin water shares, and commodity price policy as defined by taxes and subsidies on commodities.

Based on the analysis, three scenarios were illustrated: 1) BAU (business-as-usual); 2) water crisis (pessimistic); and 3) sustainable water use (optimistic). These three scenarios are further described below.

In the first scenario (business as usual) - current trends in water and food policy, management, and investment persist. International donors and national governments continue to reduce their investments in agriculture and irrigation. Governments and water users reform institutions and management in a limited and piecemeal fashion. The demand for water for non-irrigation purposes - household, industry, and livestock - will double in developing counties and increase by two-thirds in the world as a whole (Figure 1).

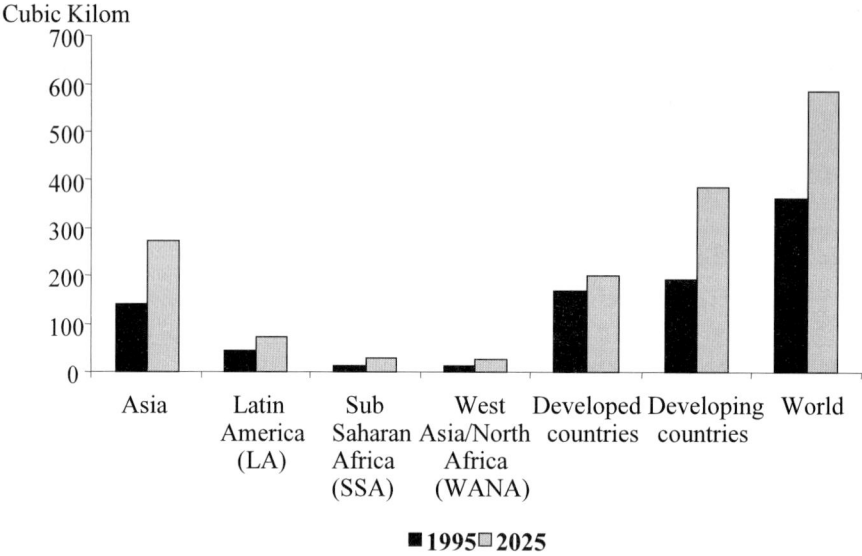

Cubic Kilom

Figure 1 Total non-irrigation water consumption by region, 1995 and 2025

Industrial water use will grow much faster in developing countries than in developed countries. Domestic water demand will also grow rapidly, especially in developing countries, as a result of urbanisation and income and population growth. Farmers will consume only about 4% more irrigation water in 2025 than in 1995, unable to increase demand as rapidly as desired due to competition for water from other sectors (Figure 1). The result will be slower growth of food production and significant shifts in where the world's food is grown. In the face of water scarcity, farmers will find themselves unable to raise crop yields as quickly as in the past, and by 2025 annual irrigated cereal production will be 300 million metric tons less than it would have been with adequate water - a difference nearly as large as the US cereal crop in 2000 (Rosegrant *et al.*, 2002). Faced with rising food demand and slowing production growth, developing countries will dramatically increase their reliance on food imports from 107 million tons in 1995 to 245 million tons in 2025. Some countries may finance these imports from economic growth in sectors other than agriculture, but when high food imports are the result of slow economic development, many countries may find it impossible to maintain the required imports, further worsening food security. Much of sub-Saharan Africa (SSA) and the non-oil-producing Middle Eastern and North African countries could be hit particularly hard. Competition from other users means that the share of water devoted to environmental uses will not increase.

Water crisis

If current trends in water usage, existing food policy and present investment levels were to worsen (even moderately), then the result could be a genuine water crisis. In such a scenario governments will tend to further cut their spending allocation on irrigation systems and rapidly turn over these irrigation systems to resource-poor farmers and farmer groups without the necessary reforms in water rights. Governments and international donors reduce their investments in crop breeding for rainfed agriculture in developing countries, especially for staple crops.

Total worldwide water consumption in 2025 is estimated to be 261 km^3 higher than under the business as usual scenario - a 13% increase - but much of this water will be wasted. Virtually all of the increase in demand will go to irrigation, mainly because farmers will use water less efficiently and withdraw more water to compensate for water losses. In search of adequate water supplies, farmers will extract increasing amounts of groundwater, driving down water tables and leading ultimately to the failure of key aquifers. Farmers will also tap environmental water flows, further reducing wetlands and compromising the integrity and health of aquatic ecosystems. Owing to inadequate water pricing and regulation reform, and slow adoption of improved technology, industrial water demand will be 33% higher in 2025 than under the business as usual scenario, without generating additional industrial production. The rapid increase in urban populations will quickly raise demand for domestic water, but without fundamental water pricing reforms, governments will lack the funds to extend piped water and sewage disposal to newcomers.

Naturally, such a scenario will have severe consequences for food harvests. Overall, farmers will produce 10% less cereal in 2025 than under 'business-as-usual' because of declines in both the amount of land cultivated and yields. This reduction is the equivalent of annually losing the entire cereal crop of India. The decline in food production will help push up food prices sharply under the water crisis scenario. The price of *Oryza sativa* will rise by 40%, *Triticum* by 80%, *Zea mays* by 120%, and other coarse grains by 85%. The ultimate result of this scenario is growing food insecurity, especially in developing countries. Per capita cereal consumption in 2025 in the developing world will actually decline compared with 1995 levels.

Sustainable water scenario

Fortunately, it is possible to envision a sustainable water scenario that would dramatically increase the amount of water allocated to environmental uses, connect all urban households to piped water, and achieve higher per capita domestic water consumption, while maintaining food production at the levels described in the 'business-as-usual' scenario.

Governments and international donors will increase their investments in crop research, technological change, and reform of water management to boost water productivity and the growth of crop yields in rainfed agriculture. Improved policies and increased investment in rural infrastructure will help link remote farmers to markets and reduce the risks of rainfed farming. To stimulate water conservation, the effective price of water to the agricultural sector will be gradually increased. Governments in many regions will shift water rights and management responsibilities to water users, and offer users training and support. As a result, farmers will increase their own investments in water-saving technologies. The over drafting of groundwater will be phased out as governments assign users rights to groundwater, while also toughening and better enforcing regulations. Domestic and industrial water use will also be subject to higher prices and stricter regulation. With strong societal pressure for improved environmental quality, allocations for environmental uses of water will increase, reducing pressure on wetlands.

In the 'sustainable water scenario' the world consumes 20% less water than under 'business as usual' but reaps greater benefits, especially in developing countries. These water savings will increase environmental flows by 1,030 km^3 globally, well over triple the annual flow of the Mississippi River. A key finding in this scenario is that, with higher public investment in crop breeding for rainfed areas, together with improved farm management (including increasing water harvesting, conservation tillage, and precision farming), rainfed production increases significantly. Faster growth in rainfed yields will make up for slower growth in harvested area and irrigated yields, and as a result total cereal production in 2025 is 1% greater than under 'business-as-usual'.

Climate change, pasture and land use

Hopkins (2004) showed that grassland production is strongly influenced by climatic variability, particularly temperature and rainfall. Variation in pasture yield and production can be over 100% between different localities depending on length of growing season, rainfall distribution and soil type. Under the climate change scenarios that were developed by Hopkins (2004), grassland production is likely to be influenced by increasing temperatures and changing seasonal patterns of precipitation.

The impact of climate change on pasture and other land use is also examined by the Millennium Ecosystem Assessment (MA), an international effort to provide scientific information to policymakers and the public on the effect of ecosystem change on human well being, and to offer options for dealing with those changes. As noted above, the MA scenarios were analysed using several global models; greater consistency between the calculations of the different models was achieved by 'soft-linking' the models, in the sense that output files from one model were used as inputs to other models. The scenarios were implemented by:
- *specifying a consistent set of model inputs* based on the scenario storylines;
- '*soft-linking' the models* by using the output from one model as input to another;
- *compiling and analysing model outputs* about changes in future ecosystem services and implications for human well-being.

As a part of the MA, a Global Scenarios group assessed plausible scenarios for land use on a global scale, using the 'soft-linked' models mentioned above, and incorporating the impact of climate change on land use and production.

These scenarios account for the impact of climate change, and represent plausible alternative futures of the world. They explore the outcomes of increased globalisation versus increased regionalisation on the one hand, and increased economic growth versus increased emphasis on local adaptive management of ecosystems and their services on the other hand. Both the Global Orchestration (GO) and Techno Garden (TG) scenarios focus on increased globalisation, with GO emphasizing economic growth and public goods provision, while TG strives for greener technologies. The Order from Strength (OS) scenario has a regionalised approach focusing on national security and self-sustenance, whereas the Adapting Mosaic (AM) scenario focuses on local adaptation and flexible governance. The GO scenario assumes low population growth, high income growth, high investments in human and physical capital, medium to high levels of development in technology, rapid irrigation efficiency and yield improvements, high meat demand, full trade liberalization, and medium to low controls on environmental pollution.

Under the TG scenario, assumptions include medium to low population growth, income growth slightly lower than GO, high investments in human, physical and natural capital, medium to high levels of development in technology and irrigation efficiency, medium meat demand, full trade liberalization, and substantial controls on environmental pollution.

The OS scenario assumes high population growth, low to medium income growth, medium to low investments in human, physical, or natural capital, low levels of development in technology, irrigation efficiency and yield improvements, high meat demand in developed countries and low demand in developing countries, increased protectionism, and little control on environmental pollution.

Finally, the AM scenario assumes relatively high population growth, low but improving income growth, medium but increasing investments, medium increasing levels of development in technology, irrigation efficiency and yield improvements, low meat demand, no irrigated area expansion, current levels of protection, and medium but improving environmental pollution controls. The scenarios are distinct from earlier global scenario exercises through their focus on alternative pathways for sustaining ecosystem services. While the GO and OS scenarios are cast as taking a reactive approach to environmental issues, the TG and AM scenarios are formulated as being proactive, embracing environmental issues (Millennium Ecosystem Assessment, 2005).

Land use change, pastures and deforestation

In the first decades of the scenario period, all scenarios show an ongoing expansion of agricultural land (including pasture and cropland) in developing countries replacing current forests, while agricultural land actually declines in the OECD (Organization for Economic Cooperation and Development) and FSU (Former Soviet Union) regions. Differences among the scenarios remain somewhat limited due to counteracting forces embedded in the drivers (for example, low population growth coupled with high economic growth).

Deforestation is fastest under the OS scenario. The rate of loss of undisturbed forests actually increases from the historic rate of 0.4% to 0.6% per year, fuelled among other factors by rapid population growth and the largest expansion in agricultural area among the four scenarios (Figure 2). Crop and pasture area continues to grow rapidly in the developing regions. Under the GO scenario, deforestation continues at historic rates, while it slows somewhat under the AM and TG scenarios.

Under the GO scenario, agricultural area expands as a result of rapid income growth and stronger preferences for meat. In the developed countries, there is no net global increase of pastureland as low-input extensive grazing systems, are replaced by more intensive forms of grazing, but pastureland grows significantly in developing countries. Undisturbed forests disappear at near current global rates. About 50% of forests in SSA disappear between 2000 and 2050.

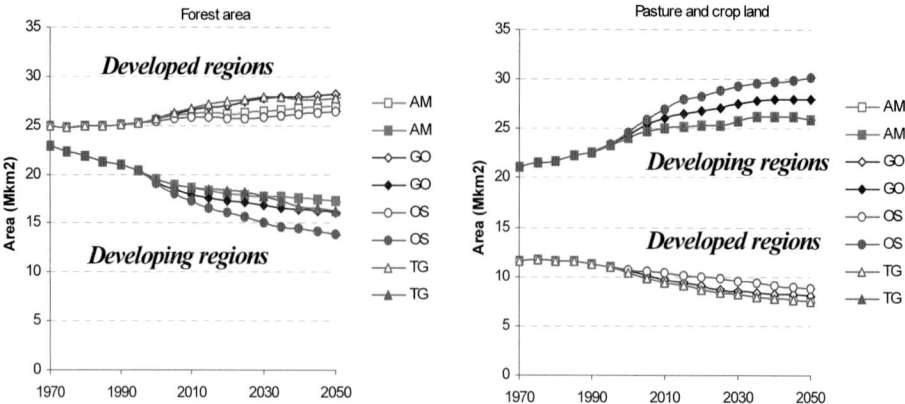

Figure 2 Change in land use, agriculture versus forests (AM = Adapting Mosaic; GO = Global Orchestration; OS = Order from Strength; TG = Techno Garden). Source: IMAGE 2.2 Projections

The smallest change in land use occurs under the TG scenario, where net forest cover is expected to increase. Production of biofuels becomes an important land use category, especially in FSU, OECD and Latin America. Due to the much lower growth for meat products, increases in irrigation area and crop yields, there is a small decrease in pastureland, and only a small increase in arable land or food production in developing regions. Deforestation in SSA and Southeast Asia is still substantial.

Under the AM scenario, changes in land use are similar to the TG scenario. Due to the application of agroecological approaches in SSA, deforestation in the region is lowest among the four scenarios. However, deforestation continues apace elsewhere, particularly in South Asia, and pastureland increases in the developing countries.

Irrigated pasture under increasing water scarcity

What are the likely implications of these alternative water and land use changes for irrigated and rainfed pasture? The models utilised do not separate irrigated and rainfed pastureland, but an understanding of the economics of irrigated pasture juxtaposed with the trends described above provide insights into the likely consequences for water and pasture development.

The literature suggests that farmers find alternative ways to respond to increased scarcity of water. Adjustments can be made through decreased water usage on a given crop, adoption of water-conserving irrigation techniques, shifting water applications to more water-efficient crops, and changing the crop mix to favour high valued crops (Gardner, 1983; Rosegrant *et al.*, 1995). The available evidence shows that the short run elasticity (responsiveness of consumers) of water demand in terms of water prices is relatively low, particularly in the agricultural sector. The longer-term response of beneficial irrigation water demand to water prices is also determined by the response of water use efficiency to water prices. Farmers respond to higher water prices not only by a direct reduction in water withdrawals and consumption, but by improving water use efficiency so that a greater portion of it is used beneficially for crop production (Caswell & Zilberman, 1985: 1986; Shah *et al.*, 1995; Varela-Ortega *et al.*, 1998;

Grassland: a global resource

Zilberman *et al.*, 1997). Water use efficiency can be increased by investment in water-conserving irrigation technology, such as drip and sprinkler irrigation, or by improving the on-farm management of the water to reduce losses to non-beneficial consumption.

Both types of responses in cropping patterns and increasing water use efficiency, were induced by reform of the Chilean water policy in 1975, to a system of tradable water rights that increased the value of water. With the increasing value of water due to tradable water rights, the area planted to fruits and vegetables, which require more water per hectare, but far less water per value of output, than most field crops, increased during the period 1975-1982 by 206,000 hectares, replacing traditional crops and irrigated pastures that needed less water (Schleyer & Rosegrant, 1996). Studies (Frías, 1992; Munita, 1994) attempted to measure the increase in aggregate water use efficiency in agriculture from 1975 to 1992, reported 22 and 26 % increase in efficiency respectively.

Irrigated pasture area

FAO (2004) land-use statistics indicate a global total land area of 13.43 billion hectares, with 5.64 billion hectares (42%) in the developed countries and 7.79 billion hectares (58%) in developing countries in 2002. Out of the total land area, 3.48 billion hectares (26%) belongs to pastureland. The developed world contributes around 34% (1.20 billion hectares) pastureland compared with 66% (2.29 billion hectares) from the developing. However, no global estimate for irrigated land area could be found. While many countries have a small amount of irrigated pasture, the majority is found in developed countries like the USA and Australia. Statistics from US Department of Agriculture show that out of 121 million hectares of harvested cropland, 21.2% (26 million hectares) belongs to forage area with 16% (4 million hectares) irrigated (USDA, 2002). Irrigated pasture together with orchards, cotton and other hay consume over 0.615 million hectare-meter (MHM) of total water applications. Irrigated water application is the water application rate per hectare times hectares irrigated. Among crops in the western US, *Medicago sativa* (Lucerne) hay has the most water applied at 1.78 MHM, followed by corn for grain at 1.29 MHM.

The total area of crops and pasture under irrigation in Australia, expanded from 1.5 million hectares in 1984, to over 2.5 million hectares in 1994 (Hamblin, 2001). About half the total volume of water used in agriculture in Australia is for irrigated pasture, but these irrigated pastures return only one-tenth the value of irrigated fruit, vegetable and vine crops (Hall *et al.*, 1993). Quiggin (2001) provides a more comprehensive comparison of the value of water in alternative agricultural use in Australia (Table 1)

Table 1 Water required for A$1,000 gross profit, Australia

Commodity (Ml)	Water use
Fruit	2.0
Vegetables	4.6
Dairy products	5.0
Gossypium hirsutum (cotton)	7.6
Oryza sativa (rice)	18.5
Pasture	27.8

Source: Quiggin (2001), adapted from Hall *et al.*, (1993)

As can be seen in Table 1, irrigation of pasture is extremely water-inefficient. Implications of changes in water prices for profitability can be drawn from the table. For example, if the price of water increased by $40 per Ml, the use of irrigation for pasture would become unprofitable, and the gross margin from irrigated *Oryza* production would fall by nearly 75%. By contrast, the profitability of fruit and vegetable production would barely be affected (Quiggin, 2001). With water scarcity projected to increase dramatically in the future, there will be significant shifts in the pattern of water use, away from low value agricultural uses to higher valued agricultural (and non-agricultural) uses. In such a situation, the first big agricultural adjustment is likely to be a shift away from the use of irrigation for pastures.

Can growth in rainfed pasture production compensate for the likely global decline in irrigated pasture area? There appears to be considerable potential for improvement in rainfed pasture production through intensification. Improved or intensive grassland includes pasture that is treated regularly with artificial fertilizer and/or herbicides, often following reseeding. Intensively grazed pasture systems comprise a number of paddocks that are grazed for one to four days with some period of rest between grazing. This is one of the most cost effective management strategies for pastures. In the UK, areas of improved grassland have increased by approximately 90% due to increased intensification in farming over the last 50 years (Marshall, 2001).

Water consumption by livestock

The projected rapid growth in livestock production is a significant factor in increasing water demand, particularly due to the demand for water to grow crops that are used as livestock feed, such as *Z. mays*, other coarse grains, and *Glycine* spp. However, extreme estimates of livestock water consumption e.g. the 100,000 l of water/kg of beef production estimated by Pimentel *et al.* (1997) for the US are not realistic. A careful and well-documented analysis by Beckett & Oltjen (1993), of the highly water-intensive US feedlot beef production system shows that 3,682 l of water is required to produce 1 kg of boneless beef. This figure is much lower compared to other studies (10,060 l/kg Chapagain & Hoekstra, 2003; 20,559 l/kg Robbins, 1987; Kreith, 1991). Based on Beckett & Oltjen (1993), direct consumption of water accounts for only 145 l/kg of boneless beef; with the vast majority of water for beef production consumed by irrigated pasture and feed crops.

Peden *et al.* (2003) examine the opposite extreme of livestock intensity - extensive beef production in Northern Africa. Intake of water by livestock depends on biological make-up, nutrition and environmental conditions where the animals are being reared, including feed and salt ingested, lactation, temperature and the animal's genetic adaptation to its environment. In a typical Northern African system (over a two-year period), one head of cattle consumes 25 l of water per day to produce 125 kg of dress weight, and consumes crop residues for which no additional water input is required. This equates to a direct water consumption of 146 l/kg. Under the most extreme hot/dry conditions, direct consumption could double to nearly 300 lkg. Even these values overstate the actual consumptive use of livestock, since much of the water consumed by livestock is released into the soil as urine providing soil nutrients and soil moisture.

Livestock water use is accounted for in the IMPACT-WATER model projections discussed previously, with the irrigated feed and pasture water consumption included in the figures for crop water demand. Direct water usage by livestock for 1995 baseline estimates was; 15.3 km^3 for developed countries, 21.8 km^3 for developing countries and 37 km^3 worldwide (Rosegrant *et al.*, 2002). Thus, compared to other uses of water, direct consumption by livestock is relatively small, accounting for only about 2% of total water consumption. Driven particularly by the rapid increase of livestock production in developing countries, global livestock water demand is projected to increase by 71% (from 37 km^3 in 1995) to 63.4 km^3 in 2025.

Implications for the future

Grassland is caught between two countervailing forces: a requirement for increasing meat demand that boosts the need for additional pasture to support livestock production, and rapidly increasing water scarcity that makes pasture irrigation uneconomical. Pasture production must therefore increase through extensive expansion into previously unused grassland areas, and intensification of rainfed pasture systems. The balance between these sources of growth will determine the extent to which expansion of pasture leads to negative environmental impacts. Excessive area expansion can lead to declines in biodiversity, a reduction in carbon storage and a reduction in ecosystem services.

The scenarios described here point to appropriate strategies for national governments, international donors, and water users that could minimise the negative environmental consequences of expanded pasture production. It is crucial to invest in expanding household and industrial water supplies, but rising financial and environmental costs will limit the expansion of irrigation water supply. Overall, the most effective means of dealing with water scarcity is likely to be conserving water in existing water uses. Improvements in the irrigation sector to increase water use efficiency must be made at the technical, managerial, and institutional levels. Technical improvements bring advanced irrigation systems, including drip irrigation, sprinklers, conjunctive or collective use of surface and groundwater, and precision agriculture, such as computer monitoring of crop water demand. Irrigation management can be improved by the adoption of demand-based irrigation scheduling systems and improved equipment maintenance. The establishment of effective water user associations and water rights, the introduction of water pricing, and improvements in the legal environment for water allocation are all examples of institutional improvements in the irrigation sector. Industrial water recycling can also be a major source of water savings in many countries to reduce water scarcity. Domestic water use can be made more efficient by steps ranging from repairing leaks in municipal systems to installing low-flow showerheads. Innovative water pricing policies that increase the prices for domestic and industrial water while preserving incomes for farmers and the rural poor will encourage water-saving innovation.

Rainfed agriculture - including rainfed pasture also emerges as a potential key to the sustainable development of water and food. Improved water management and crop productivity in rainfed areas would help relieve pressure on irrigated agriculture and on water resources. Exploiting the full potential of rainfed agriculture will require investing in; water harvesting and conservation tillage technologies, expanded investment in crop breeding targeted to rainfed environments, agricultural extension services, and improved access to markets, credit, and input supplies in rainfed areas.

Key improvements such as those mentioned above are necessary in order to address the pressures facing pasture production. The appropriate mix of water policy and management reform and investments, and feasible institutional arrangements and policy instruments employed, must be tailored to specific countries and basins. Specific plans to address these issues will vary based on underlying conditions in the regions, including levels of development, agroclimatic conditions, relative water scarcity, level of agricultural intensification, and degree of competition for water.

References

Beckett, J.L. & J.W. Oltjen (1993). Estimation of the water requirement for beef production in the United States. *Journal of Animal Science,* 71 (4), 818-826.

Caswell, M. & D. Zilberman (1985). The choices of irrigation technologies in California. *American Journal of Agricultural Economics*, 67, 224-234.

Caswell, M. & D. Zilberman (1986). The effects of well depth and land quality on the choice of irrigation technology. *American Journal of Agricultural Economics*, 68, 798-811.

Chapagain, A. & A. Hoekstra (2003). Virtual water flows between nations in relation to trade in livestock and livestock products. Water Research Report, Series No. 13, Delf, the Netherlands. 198pp.

FAO (Food and Agriculture Organization of the UN), (2004). FAO Stat-Agriculture. FAO statistical database. <http://www.fao.org/waicent/portal/statistics_en.asp>

Frías, J.L. (1992). Evolución reciente de la industria del agua en Inglaterra, Francia y Chile (Recent evolution of water industry in England, France and Chile). McKinsey & Company Inc., Mexico.

Gardner, B. (1983). Efficient redistribution through commodity markets. *American Journal of Agricultural Economics*, 65, 225-234.

Gardner-Outlaw, T. & R. Engelman (1997). Sustaining water: a second update. Population Action International, Washington DC, USA, 20pp.

Hall, N., D. Poulter & R. Curtotti (1993). ABARE model of irrigation farming in the southern Murray-Darling Basin, ABARE Research Report 94.4, Canberra, Australia.

Hamblin, A. (2001). Australia state of the environment report 2001 (theme report). CSIRO Publishing, Department of Environment and Heritage, Commonwealth of Australia, 155pp.

Hopkins, A. (2004). Climate change impacts and grassland. IGER Innovations, 47-51. <http://www.iger.bbsrc.ac.uk/Publications/Innovations/incon04.htm>

Kreith, M (1991). Water inputs in California food production. Water Education Foundation, Sacramento, California.

Marshall, B. (2001). Grassland. Hull Biodiversity Action Plan. Hull Biodiversity Partnership. <http://www.hull.ac.uk/HBP/Action Plan/Grassland.htm>

Millennium Ecosystem Assessment (2005). Chapter 9: Changes in Provisioning and Regulating Ecosystem Goods and Services and their Drivers Across the Scenarios. Millennium Ecosystem Assessment Volume 2: Scenarios Assessment. Ecosystems and human well-being. Island Press, Washington, DC, (in press)

Munita, J. (1994). Aumento de Eficiencia en el Uso Del Agua por Incorporación de Nuevas Técnicas y Arrendamientos Temporales (Increase in efficiency in the use of water by the incorporation of new technology and temporary rentals). Universidad de Chile, Santiago.

Peden, D., G. Tadesse & M. Mammo (2003). Improving the water productivity of livestock: an opportunity for poverty reduction. Paper presented during the Integrated Water and Land Management Research and Capacity Building Priorities for Ethiopia Conference, December 2-4, 2004, Addis Ababa, Ethiopia, 57-65. <http://www.ilri.cgiar.org/InfoServ/Webpub/Fulldocs/IntegratedWater/iwmi/Documents/Papers/Don.htm>

Pimentel, D., J. Houser, E. Preiss, O. White, H. Fang, L. Mesnick, T. Barsky, S. Tariche, J. Schreck & S. Alpert (1997). Water resources: agriculture, the environment and society. *BioScience*, 47 (2), 97-106.

Quiggin, J. (2001). Environmental economics and the Murray-Darling river system. *Australian Journal of Agricultural and Resource Economics*, 45 (1), 67-94.

Robbins, J. (1987). Diet for a new America. Stillpoint Press, Walpole, NH, 23pp.

Rosegrant, M.W., X. Cai & S. Cline. (2002). World water and food to 2025: dealing with scarcity. International Food Policy Research Institute, Washington DC, USA, 322pp.

Rosegrant, M.W., R.G. Schleyer & S. Yadav (1995). Water policy for efficient agricultural diversification: market-based approaches. *Food Policy*, 20 (3), 203-233.

Schleyer, R.G. & M.W. Rosegrant (1996). Chilean water policy: the role of water rights, institutions, and markets. *International Journal of Water Resources Development* 12 (1), 33-48.

Shah, F, D. Zilberman & U. Chakravorty (1995). Technology adoption in the presence of an exhaustible resource: the case of groundwater extraction. *American Journal of Agricultural Economics*, 77, 291-299.

UN World Water Development Report (2003). Water for people, water for life. <http://www.unesco.org/water/wwap/wwdr/>

USDA (United States Department of Agriculture) (2002). National agricultural statistics service database. Census of Agriculture - state data, USA. <http://www.nass.usda.gov/QuickStats/>

Varela-Ortega, C., J.M. Sumpsi, A. Garrido, M. Blanco & E. Iglesias (1998). Water pricing, public decision making and farmers' response: implications for water policy. *Agricultural Economics*, 19, 193-202.

Zilberman, D., U. Chakroavorty & F. Shah (1997). Efficient management of water in agriculture. Chapter 22. In: D. Parker & Y. Tsur (eds.) Decentralization and coordination of water resources. Boston: Kluwer Academic Publishers, 221-246.

Grassland: a global resource

Grassland productivity and water quality: a 21st Century issue

D.M. Nash[1] and P.M. Haygarth[2]

[1]Primary Industries Research Victoria (PIRVic), Department of Primary Industries-Ellinbank Centre, RMB2460 Hazeldean Road, Ellinbank Victoria, Australia 3821
Email: david.nash@dpi.vic.gov.au
[2]Institute of Grassland and Environmental Research, Soil Environmental and Ecological Sciences, North Wyke Research Station, Okehampton, Devon, EX20 2SB, UK

Key points

1. Irrigation and other changes to the hydrological cycle can increase soil and water salinity.
2. Primary salinisation is a natural process that affects much of Europe, Asia, Africa, the Americas and Australia. Secondary salinisation is caused by human activities such as irrigation and land clearing that mobilise salt stored in the soil.
3. The critical water contaminants exported from grasslands are nitrogen, phosphorus, potential pathogens and sediment.
4. The mechanisms responsible for diffuse pollution from grasslands and mitigation strategies are most effectively investigated using a 'source-mobilisation-transport' framework.
5. There is a lack of coherent interaction across discipline boundaries that links pollutant sources to impact. Grassland scientists need to work hand-in-hand with hydrologists and limnologists, to understand the water flows and the intricacies of ecological response, in stream or lake, in order to achieve a more coordinated and inclusive, holistic platform of research.

Keywords: salinisation, nitrogen, phosphorus, sediment, pathogens

Introduction

Grass and forage production systems are major users and exporters of water. For example, of ca. 3000 km^3 of freshwater used by agriculture, over 80% is applied to more than 250 million hectares of irrigated crops and pastures. Grasslands modify global hydrological cycles and influence the quality of the water that passes through them. This paper will consider 1) the effects of water quality on grassland and forage production, with particular emphasis on salinisation and 2) the wider effects of grassland farming on water quality for aquatic ecosystems and for other uses.

The effect of water quality on grassland and forage production, with particular emphasis on salinisation

To meet plant water needs for food and fibre production, in many parts of the world, irrigation is required to supplement natural rainfall. While primary salinisation is a natural process affecting ca. 955 Mha of Europe, Asia, Africa, the Americas and Australia, irrigation and man-made changes to the hydrological cycle have the potential to increase soil and water salinity in a process commonly referred to as secondary salinisation (Ghassemi et al., 1995). Often, the consequential process of sodification (alkalinsation) in which clays become increasingly saturated with sodium ions exacerbates secondary salinisation, adversely affecting soil physical properties such as structure and infiltration rates. It follows that secondary salinisation is often associated with water logging (Ghassemi et al., 1995).

Saline soils adversely affect plant growth by establishing an osmotic gradient that favours desiccation of plant roots (i.e. osmotic effect) and through specific ion toxicities. Some plant species have adapted to growing in saline conditions by excluding salt at the root surface, removing salt from their cytoplasm and placing it in vacuoles (where few metabolic processes occur) and excreting salt from their leaves. All three mechanisms are prevalent in highly salt tolerant plants (halophytes) (Barrett-Lennard, 2003), which can help ameliorate the effects of salinity (Glenn *et al.*, 1999). In Australia, for example, revegetation of saline land with salt tolerant forage species has increased the carrying capacity of grassland, lowered water tables and lowered soil salinity (Barrett-Lennard, 2003). There is reasonable literature on grassland and forage production on saline soils (Abrol *et al.*, 1988; Anon., 2004).

The quality of water supplies

While the proportion of freshwater used by agriculture worldwide is <3% of that available for ecosystem and human services, gaining access to good quality water for forage production is becoming increasingly difficult. The result has been a shift towards the use of lower quality waters, particularly saline drainage and groundwater, and industrial and municipal wastewaters (Al-Attar, 2002).

Grassland and forage production using saline irrigation water
Irrigation using saline water has been extensively studied (Richards, 1954). Limitations on the use of saline water for irrigation include the accumulation of salts in the root zone, excessive concentrations of phyto-toxic elements such as sodium, chloride, selenium, molybdenum and boron in the soil, and increased percentages of sodium adsorbed by clays.

Evapotranspiration of saline irrigation water from soils increases salt concentrations in the root zone of crops, adversely affecting their productivity. It follows that the 'salinity hazard' or risk to productivity increases with the salt concentration of irrigation water and irrigation using saline water requires that additional water beyond the needs of the crop (a leaching fraction) be applied in order to remove accumulated salts. Considerable effort has been devoted to calculating the leaching fraction (i.e. proportion of irrigation water draining below the root zone) required to maintain satisfactory growing conditions (Richards, 1954). Where drainage is restricted (i.e. clay soils) the hazard of using water of particular salinities generally increases. Consequently, irrigation water classification systems often consider both the salt concentration of the water and the characteristics of the soil to which it is applied (Table 1).

Table 1 General salinity guidelines for irrigation water[1]

Water	EC[2] (dS/m)	TDS[3] (mg/l)	Application
Fresh	<0.7	<450	No restrictions on use, potable, all crops
Brackish	0.7-3.0	450-2000	Slight to moderate restrictions on use, livestock water, most crops
Saline	>3.0	>2000	Severe restrictions on use, salt-tolerant crops, not used on soils with restricted drainage.

[1]adapted from Ayers & Westcot, 1994; [2]Electrical conductivity; [3]Total dissolved salts

Plants vary in their tolerance of salt, and as a consequence, of the salinity in irrigation water. Grasses and fodder crops tend to be amongst the more tolerant plants, although there are species and cultivar differences. For example, *Meticago sativa* (lucerne), *Agrostis stolonifera palustris* (bent grass), most *Trifolium* spp. (clovers) and *Zea Mays* (corn) are moderately sensitive to salt (i.e. 20% yield decline at >2.5 EC_e dS/m), whereas *Phalaris arundinacea* (canary grass), *Festuca* spp. (fescues), *Brassica napus* (rape) and most *Triticum* spp. (wheats) are moderately tolerant to tolerant (i.e. 20% yield decline at >10 or 14 EC_e dS/m respectively) (Maas & Hoffman, 1977; Maas, 1996).

Irrigation waters often contain nutrients such as chloride, sodium and boron that are essential for plant growth, but at higher concentrations are potentially toxic. As irrigation water constituents vary depending on the water source, the risks associated with specific ion toxicities need to be assessed on a case-by-case basis (Ayers & Westcot, 1994; Maas, 1996).

Salt in the root zone also affects soil structure. Compared to calcium and magnesium, sodium adsorption increases the tendency of clay aggregates to swell and disperse. The Exchangeable Sodium Percentage (ESP) is related to the Sodium Adsorption Ratio (SAR, Equation 1) of the soil solution (or applied water). At the ESP range most common in agricultural soils (ESP<30), the numerical values of ESP and SAR are almost equal. The 'sodium hazard' of irrigation water is therefore related to the concentration of sodium relative to calcium and magnesium concentrations, expressed by the SAR. It is of note that where salts are concentrated by evapotranspiration and as a result, the proportions of sodium, calcium and magnesium in solution are preserved, the SAR and sodium hazard of that water increases. Consequently, some authors have proposed leaching requirements to assist with sodium management (Rhoades, 1968).

$$SAR= [Na^+]/\sqrt{(0.5([Ca^{2+}]+[Mg^{2+}]))} \tag{1}$$

Where $[Na^+]$ = Sodium concentration (activity) in the solution in meq/l
$[Ca^{2+}]$ = Calcium concentration (activity) in the solution in meq/l
$[Mg^{2+}]$ = Magnesium concentration (activity) in the solution in meq/l

Even small changes in the ESP (i.e. <5%) can adversely affect soil physical properties. As the SAR of irrigation water increases, sodium is initially adsorbed to the surface of clay domains (i.e. quasi-crystals, groups of clay platelets) within aggregates. This increases the tendency for water to penetrate the spaces between the domains, increasing the distance between the domains and adversely affecting the ability of short-range attractive forces to hold the domains together, resulting in potentially dispersive behaviour. With further increases in ESP, sodium can penetrate the domains themselves increasing electrostatic repulsion between the platelets. As a result the clays swell and disperse. This phenomenon of 'demixing' is thought to account for dispersion being the dominant process at low ESP values (<15-25) and swelling at high ESP values.

The effects of a particular SAR/ESP depend on soil properties including textural class (i.e. clay percentage), clay type, quantity and type of entities that stabilise clay aggregates (i.e. iron and aluminium or organic matter). By far the most important factor affecting the expression of a particular ESP is the electrolyte concentration in the soil water. Electrostatic repulsion between moieties is greatly increased by lower electrolyte concentrations. Consequently, it is only when high quality irrigation water (i.e. low salt) is applied or rainfall occurs that the full effects of adsorbed sodium are expressed in the form of swelling and dispersion.

The effects of adsorbed sodium are particularly important in grassland and forage production systems where physical stresses placed on the soil by animal and vehicular traffic, cultivation and the like can exacerbate adverse changes in soil structure, especially where soils are waterlogged. Increased organic matter can often lessen such effects (Nelson *et al.*, 1999).

Wastewater irrigation
Industrial and municipal effluents and agricultural drainage (wastewaters) are important sources of water and nutrients. With increasing urbanisation and the associated demand for potable water, municipal (i.e. primarily domestic) wastewaters, in particular, are a useful substitute for freshwater that would otherwise be used for irrigation. Due to the highly variable quality arising from source differences and the treatment prior to land application, wastewater use in agriculture has not been without peril for either agricultural workers or consumers of the agricultural produce. Guidelines have been developed to protect against such occurrences (Blumenthal *et al.*, 2000). In addition, wastewater irrigation carries the risk of introducing potentially toxic substances, including heavy metals, into the human food chain and microbiological health risks (Mara & Caincross, 1989).

Many studies have reported decreases in soil infiltration rates and hydraulic conductivities, following wastewater irrigation. Some of the mechanisms that could be responsible for these changes include solvent-solute effects on clays (Anandarajah, 2003), accumulation of suspended solids at the soil surface or blockage of the inter-soil spaces by suspended material such as colloidal clay and cells from microorganisms (Bouwer & Chaney, 1974; Metzger *et al.*, 1983), entrapped air (Rice, 1974), the formation of a biological mat or crust including the production of microbial extracellular polymeric materials such as polysaccharides (Balks *et al.*, 1997; Taylor & Jaffe, 1990), collapse of soil structure due to organic matter dissolution (Lieffering & McLay, 1996) and physico-chemical changes to pore geometry and micro-fabric that are related to sodicity, cation exchange reactions and exchange hysteresis (Shainberg & Letey, 1984).

The swelling and dispersion of clay aggregates is likely to be particularly important in decreasing soil infiltration rates as many wastewater constituents increase the effective SAR of irrigation water. This can arise in various ways. During wastewater irrigation inorganic anions, especially SO_4^{2-}, CO_3^{2-}, HCO_3^- precipitate Ca^{2+} and Mg^{2+} from solution (Suarez, 1981), organic matter can lower the activity of Ca^{2+} and Mg^{2+} and cations that would otherwise form bridges between minerals, due to changes in soil pH, for example, the capacity of organic matter to bind particles can decrease and anaerobic or other conditions can be created that result in the chemical reduction of species that would otherwise form bridges between minerals (i.e. Fe^{3+}) (Reid *et al.*, 1982; Shainberg & Letey, 1984; Visser & Caillier, 1988; Piccolo & Mbagwu, 1989). The ability of wastewater constituents to act synergistically and impair the longer-term productive capacity of grassland production systems warrants further investigation.

Contaminants are an additional risk associated with the use of wastewater for irrigation. Some contaminants, such as zinc and copper, can accumulate to phytotoxic concentrations (Abdelrahman & Al-Ajmi, 1994). However, of greater concern are wastewater contaminants that while not phytotoxic, may be introduced into the human food chain via grassland or fodder through the grazing animals (Cooper, 1991). These include heavy metals such as cadmium and various organic chemicals that can be classed as endocrine disruptors (i.e. DDT, dieldrin, endosulfin) and biological toxins (i.e. algal toxins). These contaminants can access the human food chain in forage or as a contaminant of forage, or by the ingestion of contaminated soil by grazing animals. There is a pressing need to understand the risks associated with wastewater contaminants and develop preventative measures that allay consumer concerns.

The effects of grassland farming on water quality

Water exported from grassland production systems often contains pollutants such as sediment nitrogen, phosphorus, labile organic materials and microorganisms. For a pollutant to pose a threat to water quality there needs to be (1) a source of the pollutant that is (2) mobilised into water and (3) transported to a location where it has (4) an adverse impact. The first three of these components are often associated with grassland production systems and a conceptual framework for characterising pollutant exports is presented in Figure 1. Most pollutant exports are episodic and underlying pollutant sources and mobilisation and transport processes are 'incidental' factors associated with climate and management. This framework is useful in that it breaks down diffuse pollution into a simple *source*, *mobilisation* and *transport* framework, highlighting the importance of all stages in the diffuse pollution 'continuum' (Haygarth *et al.*, 2005). This approach can help us focus on the real cause of the problems, where intervention is most likely to be effective, and help prioritise potential mitigation strategies.

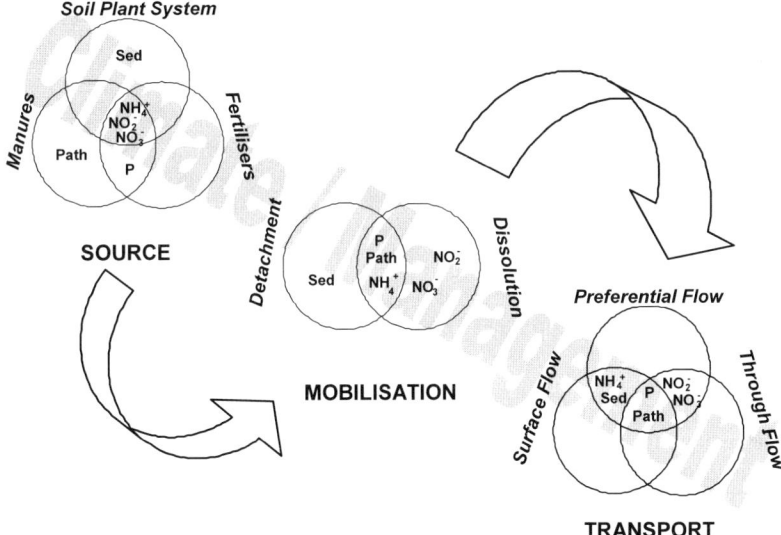

Figure 1 A conceptual framework for characterising diffuse pollutant exports from grasslands (Sed = sediment, Path = potential pathogens, NH_4^+ = ammonium, NO_3^- = nitrate, NO_2^- = nitrite, P = phosphorus) (adapted from Haygarth *et al.*, 2004)

In grassland systems, pollutant sources include agrochemicals such as fertilisers, herbicides and pesticides, the soil fabric and organic matter, live and decomposing vegetation, animals and animal defecation products deposited direct to pasture or in various forms of manure and slurry that are collected and subsequently redistributed on the land. Critical incidents such as fertiliser or manure application can increase pollutant availability and, as a consequence, exports. These 'incidental' exports are often 'preventable' and source modification has traditionally been the focus of much research effort. However, unless all pollutant sources are eliminated, even with optimal management pollutants will be exported from grassland production systems. This 'systematic' component of pollutant exports has received little study compared with the 'incidental', and often 'preventable' component. In part this is because factors such as manure application rates and timing are readily examined with standard experimental techniques, as the treatment effects are often large compared with the variability between replicates.

Pollutants can be mobilised either as particulate (>0.45μm) or dissolved (<0.45μm) materials. Particulate materials and the pollutants they may contain are mobilised by predominantly physical processes that detach sediment from the soil fabric. Dissolution (sometimes otherwise referred to as 'solubilisation' (Haygarth *et al.*, 2005)) of pollutants, on the other hand, is largely a biochemical, and often time dependant process. Again critical incidents ('incidentals') can enhance pollutant mobilisation. These effects are most pronounced for detachment processes where physical perturbation of the land (i.e. poaching and pugging of the land surface and overgrazing) can enhance the mobilisation of particulate pollutants. River siltation is a direct consequence of some grassland production systems. Dissolution processes being subtle, are generally considered less effected by 'incidentals'. However, management that changes soil properties, such as pH, may well affect dissolution processes.

Pollutants can be transported in both surface and subsurface pathways. The effectiveness of the transport process will depend on the pollutant in question and the particular pathway (Nash *et al.*, 2002). Transport processes are particularly susceptible to incidental factors, especially climatic variables such as rainfall intensity and management decisions such as the installation of drainage that increases the hydrological connectivity between the source and receiving waters. Many strategies have been used to interrupt the transport of pollutants originating from grassland production systems. These include agricultural drainage recycling systems and grassed buffer strips and other installations that lower water velocity (and kinetic energy), facilitating the sedimentation of particulate materials.

In terms of preventing unacceptable water pollution, understanding the interactions between pollutant sources, mobilisation and transport processes and how critical incidents ('incidentals') affect these relationships offers our best opportunity for developing effective remedial strategies. A good example is detached sediment. Detached sediments can 'adsorb' dissolved pollutants in transit (Sharpley *et al.*, 1981) and through their deposition or entrapment, remove dissolved pollutants from the water column. A number of studies has shown that methods used to decrease detachment and transport of sediment actually increased the concentrations of dissolved, and in some cases more potent forms of the same pollutant in the water. There is therefore a clear need for a multi-dimensional approach to this research.

Phosphorus

Elevated concentrations of phosphorus in streams, rivers and lakes may contribute to eutrophication (Pierzynski *et al.*, 2000). Eutrophication limits water use for drinking, fishing, industry and recreation (Carpenter *et al.*, 1998). Dissolved phosphorus concentrations have increased markedly in recent decades in Australia, Europe, New Zealand, and the USA. Such changes have been linked to large-scale fish kills in estuaries caused by increased populations of the dinoflagellate *Pfiesteria piscicida*, which in turn have the potential to adversely affect human health (Burkholder *et al.*, 1992). The European Union Water Framework Directive (2000/60/EC) aims to restore all waters to 'good ecological status' by 2015.

Many of the current remedial strategies rely on decreasing source availability, especially through improved management of phosphorus in manures, slurries and fertilisers.

Mobilisation control measures have generally focussed on stopping detachment processes by increasing groundcover, improving soil physical properties and minimising incidentals by ensuring management decisions take account of the prevailing climatic conditions. Transport (sometimes called 'delivery') control measures focus on preventing the movement of mobilised phosphorus across the landscape. They include the use of buffers, wetlands and drainage recycling systems. Although delivery control measures can produce a reasonable decrease in phosphorus loads, these measures only have a limited lifetime and generally require 'landscape wide' application that is often difficult to achieve in practice. Overviews of phosphorus and water quality issues relevant to grasslands can be found in Leinweber *et al.* (2002) and Haygarth *et al.* (2005).

Nitrogen

The various forms of nitrogen affect water quality differently. Nitrate is the most prevalent soluble form affecting freshwater and has generally been considered the most 'troublesome' contributing to acidification of waters and eutrophication, both of which result in major changes to aquatic bio-diversity and community structure. In potable water, nitrate may be associated with methaemoglobinaemia in very young infants and with stomach cancer, although more recently, there is some controversy over the extent of these associations (Addiscott, 1999).

Though monitored less often than nitrate, nitrite also adversely affects receiving waters. Nitrite is highly toxic to fish and invertebrates because it impairs their ability to take up oxygen and bonds with haemoglobin, thereby decreasing the oxygen carrying capacity of the blood. The European Union Freshwater Fish Directive guidelines for salmonid and coarse fish are 3.0 and 9.0 µg nitrite-N/l, respectively (European Community, 1978), which are below the concentrations in many grassland farming catchments (Haygarth *et al.*, 2004).

Ammonium transfers to watercourses can also affect water quality. While dissolved ammonia is directly toxic to fresh water fish, the transformation (nitrification) of ammonium to nitrate in water can contribute to oxygen depletion (Haygarth *et al.*, 2004).

Mitigation of nitrogen exports from grasslands has focused on source and transport control, especially in winter. Source control options are centred on maintaining the efficiency of plant uptake and trying to avoid excess application. The latter is best achieved through application of fertiliser in regular but small amounts, strategic use of fertiliser in relation to available mineral nitrogen (this includes soil testing), seeking to breed for plants that are more efficient and have deeper rooting depth, and improving efficiency and thus nitrogen requirements of the grazing animal (Scholefield *et al.*, 1993). Transport based mitigation options are to avoid drainage of land because this increases both aeration of the soil and mineralization of nitrogen, and enhances hydrological transport. Sandy textured soils need to be particularly carefully managed. Further information on nitrogen cycling and potential mitigation is provided in Scholefield *et al.* (1993) and Hatch *et al.* (2002).

Sediment

Sediment has for a long time been 'passed over' as a minor water quality problem associated with agricultural soils and grasslands in particular. However, the transfer of sediment from grasslands is one of the most significant water quality and stream health issues. For example, siltation of gravel-bed rivers prevents spawning of salmonids to the extent that reproduction is severely impaired and salmon populations in key areas of the world (e.g. South West England and Mississippi USA) have declined, with consequential impacts on ecological balance and recreational fishing.

Sediment transfers from agricultural land arise from poaching (grazing stock treading on the surface of moist or wet soil) (McDowell *et al.*, 2003) and traffic by both animals and machinery (Harrod & Theurer, 2002). Such perturbations reduce infiltration, increasing the potential for overland flow of water. The outcome is that water flow concentrates on the surface and at times of intense runoff, may become channelised with the resulting increase in kinetic energy detaching soil particles. Additional problems arise because sediment carries contaminants, particularly phosphorus and pathogens. It follows that there are mechanistic overlaps between the mitigation measures used to prevent the mobilisation of sediment and particulate (detached) phosphorus. Such mitigation measures try to avoid scenarios that give rise to compaction of the grassland surface, especially over wintering and a general presence of animals or traffic at times when the surface is saturated and vulnerable to reduced infiltration. More information on sediment transfers and impacts from grasslands is presented in Harrod & Theurer (2002).

Potential pathogens

The rumen and digestive tract of agricultural livestock is host to a diversity of microflora and can act as a reservoir for micro-organisms (Rasmussen *et al.*, 1993) that may be pathogenic to humans. As a result organisms such as *E. coli* 0157, *Salmonella* spp., *Campylobacter jejuni*, *Listeria monocytogenes*, *Cryptosporidium parvum* and *Giardia intestinalis* may contaminate livestock wastes (Oliver *et al.*, 2005). Contamination of grassland surfaces can result from either direct defecation from livestock or as a result of spreading recycled manures. Since many of these organisms persist in manure and in soil, they are vulnerable to transfer to water where they can present a serious threat to humans.

Opportunities for decreasing pathogen transfer to water include controlling pathogen sources and lessening the persistence (increasing the 'die off') of the organisms in soil and manure, in combination with minimising transport factors. Pathogen transfer is one of the least studied and emerging issues for grasslands and water quality, but presents many problems and challenges for the 21st century because of the potential to *directly* affect the health of human beings. Relevant reviews on this subject can be found in Jones (2002), and Oliver *et al.* (2005).

Conclusions and recommendations for future research

As the demand for livestock and livestock production continues to rise, grassland productivity and water quality will become truly 21st Century issues. There are many water quality problems that emanate from grassland production and to a large extent their seriousness and consequences have yet to be fully acknowledged. Nitrogen, because of its sheer volume of usage, is the most researched and understood contaminant. However, its impact in some regions of the world is now thought to be of less importance than phosphorus, which controls the productivity of many inland freshwater lakes and waterways. Sediment and pathogen transfers represent the 'newer' challenge for the 21st century. Many of the processes and mitigating strategies, particularly for micro-organisms, lack mechanistic information and further research into pathogen survival and mitigation is required.

In future, grassland research will also need to embrace a wider and 'scaled-up' view of water quality problems, with greater emphasis on landscape 'delivery' and transport controls (Beven *et al.*, 2004), to balance the relatively large research effort on finer scale processes that has been the priority for many years. There is a pressing need to integrate results from these finer scale process studies into mathematical and conceptual frameworks on which policy decisions can be based. Software that combines probability distributions through various simulation procedures (i.e. Monte Carlo simulations) and Bayesian networks are tools that can be used, in conjunction with, for example, source decay (i.e. the availability of pollutants from fertiliser with time after application) to investigate the probability that a particular mitigation strategy will achieve its objectives.

There are also other pressing deficiencies in research that must be addressed in order for progress to be made. The first is the lack of coherent interaction across discipline boundaries that links source to impact. Grassland scientists need to work hand-in-hand with hydrologists and limnologists, to understand the water flows and the intricacies of ecological response, in stream or lake, in order to achieve a more coordinated and inclusive, holistic platform of research. Secondly, although not to be treated separate to the previous issue, is the urgency for the testing of mitigation options and best management practices (BMPs) in more coordinated and localised platforms. It is not known quantitatively, how effective such management practices will be, though their performance is sure to differ in different catchments around the world. It will also be necessary to consider the economic viability of such options and offset mitigation effectiveness with potential cost effectiveness. There remains much to do.

While grasslands contribute to water quality problems they can also be part of the solution. There is considerable scope for developing irrigation systems with a reduced environmental footprint capable of mitigating past mistakes. For example, 'conjunctive water use', the mixing of saline ground water with higher quality (i.e. lower salt) waters has been used to lower water tables in irrigation regions (Prendergast *et al.*, 1994; Jury *et al.*, 2003) and, to a lesser extent, alleviate the effects of adsorbed sodium. 'Serial biological concentration' (Heuperman, 1999; Jury *et al.*, 2003) has been used to concentrate pollutants, including nutrient and other salts, which would otherwise have been released into the aquatic environment. Adapting and optimising systems for different climatic and physical environments, and social and cultural systems will be important if the quantity of water available for grassland and fodder production is to increase, the environmental impact of altered hydrology and the associated secondary salinisation is minimised, and off-site water quality is improved.

Acknowledgements

The Institute of Grassland and Environmental Research (IGER) is in part supported by the UK Biotechnology and Biological Sciences Research Council (BBSRC). Defra projects ES0121, NT2511 and PE0118 and the associated teams. Others to whom thanks are due include Philomena Gangaiya, Fiona Robertson, Kirsten Barlow and Will Gates (DPI) and Louise Heathwaite (University of Lancaster). The contributions of the Department of Primary Industries to this paper were supported by the Victorian Government, Dairy Australia and GippsDairy.

References

Abdelrahman, H.A., & H. Al-Ajmi (1994). Heavy metals in some water- and wastewater- irrigated soils of Oman. *Communications in Soil Science and Plant Analysis*, 25, 605-613.

Abrol, I.P., J.S.P. Yadav & F.I. Massoud (1988). Salt-affected soils and their management. Food and Agriculture Organisation of the United Nations, Rome, Italy, 131pp.

Addiscott, T.M. (1999). Nitrate and health: introductory comments. In: R.H. Hinton (ed.) Managing risks of nitrates to humans and the environment. The Royal Society of Chemistry, Cambridge, 247-299.

Al-Attar, M. (2002). Role of biosaline agriculture in managing freshwater shortages and improving water security. Paper presented at the World Food Prize Symposium 'From Middle East to Middle West: Managing Freshwater Shortages and Regional Water Security',Des Moines, USA, October 24-25.

Anandarajah, A. (2003). Mechanism controlling permeability change in clays due to changes in pore fluid. *Journal of Geotechnical and Geoenvironmental Engineering*, 129, 163-172.

Anonymous (2004). Biosalinity awareness project. Understanding the impact of salinisation and implications for future agriculture. Biosalinity Awareness Project <www.biosalinity.org>

Ayers, R.S. & D.W. Westcot (1994). Water quality for agriculture. Food and Agriculture Organization of the United Nations, Rome, Italy, 173pp.

Balks, M.R., C.D.A. McLay C.G. & Harfoot (1997). Determination of the progression in soil microbial response, and changes in soil permeability, following application of meat processing effluent to soil. *Applied Soil Ecology*, 6, 109-116.

Barrett-Lennard, E.G. (2003). Saltland pastures in Australia - a practical guide. Land, water and wool sustainable grazing on saline land project. Land and Water Australia, Canberra, Australia, 176pp.

Beven, K.B., L. Heathwaite, P.M. Haygarth, D. Walling, R. Brazier & P. Withers (2004). On the concept of delivery of sediment and nutrients to stream channels. *Hydrological Processes*, (in press).

Blumenthal, U.J., A. Peasey, G. Ruiz-Palacios & D.D. Mara (2000). Guidelines for wastewater reuse in agriculture and aquaculture: recommended revisions based on new research evidence. Loughborough University, Loughborough, 87pp.

Bouwer, H. & R.L. Chaney (1974). Land treatment of wastewater. *Advances in Agronomy*, 26, 133-176.

Burkholder, J.M., E.J. Noga, C.W. Hobbs, H.B. Glasgow Jr. & S.A. Smith (1992). New 'phantom' dinoflagellate is the causative agent of major estuarine fish kills. *Nature*, 358, 407-410.

Carpenter, S.R., N.F. Caraco, D.L. Correll, R.W. Howarth, A.N. Sharpley & V.H. Smith (1998). Nonpoint pollution of surface waters with phosphorus and nitrogen. *Ecological Applications*, 8, 559-568.

Cooper, R.C. (1991). Public health concerns in wastewater reuse. *Water Science and Technology*, 24, 55-65.

European Community (1978). Council directive of 18 July 1978 on the quality of fresh waters needing protection or improvement in order to support fish life. *78/659/EEC Official Journal of the European Communities*, 21, 1-10.

Ghassemi, F., A.J. Jakeman & H.A. Nix (1995). Salinisation of land and water resources: human causes, extent, management and case studies. University of New South Wales Press, Sydney, Australia, 526pp.

Glenn, E.P., J.J. Brown & E. Blumwald (1999). Salt tolerance and crop potential of halophytes. *Critical Reviews in Plant Sciences*, 18, 227-255.

Harrod, T.R. & F.D. Theurer (2002). Sediment. In: S.C. Jarvis (ed.) Agriculture, hydrology and water quality. CAB International, Wallingford, UK, 155-170.

Hatch, D., K. Goulding & D. Murphy (2002). Nitrogen. In: P.M. Haygarth & S.C. Jarvis (eds.) Agriculture, hydrology and water quality. CAB International, Wallingford, UK, 7-18.

Haygarth, P.M., D.C. Chadwick, S. Granger, B. Chambers, S. Anthony, K. Smith & D. Harris (2004). Characteristic behaviour and potential mitigation of some diffuse pollutants in England and Wales: a review of biological oxygen demand, ammonium, nitrite and some potential pathogens. Institute of Grassland and Environmental Research, Okehampton, Devon, 124pp.

Haygarth, P.M., L.M. Condron, A.L. Heathwaite, B.L. Turner & G.P. Harris (2005). The phosphorus transfer continuum: linking source to impact with an interdisciplinary and multi-scaled approach. *Science of the Total Environment*, (in press).

Heuperman, A. (1999). Serial biological concentration: a new management system for saline drainage water. *Irrigation Australia*, 14, 21-22, 27.

Jones, D. (2002). Human enteric pathogens. In: P.M. Haygarth & S.C. Jarvis (eds.) Agriculture, hydrology and water quality. CAB International, Wallingford, UK, 133-153.

Jury, W.A., A. Tuli &J. Letey (2003). Effect of travel time on management of a sequential reuse drainage operation. *Soil Science Society of America Journal*, 67, 1122-1126.

Leinweber, P., B.L. Turner & R. Meissner (2002). Phosphorus. In: P.M. Haygarth & S.C. Jarvis (eds.) Agriculture, hydrology and water quality. CAB International, Wallingford, UK, 29-56.

Lieffering, R.E. & C.D.A. McLay (1996). The effects of strong hydroxide solutions on the stability of aggregates and hydraulic conductivity of soil. *European Journal of Soil Science*, 47, 43-50.

Maas, E.V. (1996). Crop salt tolerance. In: K.K. Tanji (ed.) Agricultural salinity assessment and management. American Society of Civil Engineers. New York, 262-304.

Maas, E.V. & G.J. Hoffman (1977). Crop salt tolerance - current assessment. Journal of the Irrigation and Drainage Division, *Proceedings of the American Society of Civil Engineers*, 103, 115-134.

Mara, D. & S. Caincross (1989). Guidelines for the safe use of wastewater and excreta in agriculture and aquaculture: measures for public health protection. World Health Organization and UNEP, Geneva, 74pp.

McDowell, R.W., J.J. Drewery, R.W. Muirhead & R.J. Paton (2003). Cattle treading and phosphorus and sediment loss in overland flow from grazed cropland. *Australian Journal of Soil Research*, 41, 1521-1532.

Metzger, L., B. Yaron & U. Mingelgrin (1983). Soil hydraulic conductivity as affected by physical and chemical properties of effluents. *Agronomie*, 8, 771-778.

Nash, D.M., D. Halliwell & J. Cox (2002). Hydrological mobilisation of pollutants at the slope/field scale. In: P.M. Haygarth & S.C. Jarvis (eds.) Agriculture, hydrology and water quality. CAB International, Wallingford, UK, 225-242.

Nelson, P.N., J.A. Baldock, P. Clarke, J.M. Oades & G.J. Churchman (1999). Dispersed clay and organic matter in soil: their nature and associations. *Australian Journal of Soil Research*, 37, 289-315.

Oliver, D.M., C.D. Clegg, P.M. Haygarth &A.L. Heathwaite (2005). Assessing the potential for pathogen transfer from grassland soils to surface waters. *Advances in Agronomy*, 85, 125-180.

Piccolo, A., &J.S.C. Mbagwu (1989). Effect of humic substances and surfactants on the stability of soil aggregates. *Soil Science*, 147, 47-54.

Pierzynski, G.M., J.T. Sims & G.F. Vance (2000). Soils and environmental quality. CRC press, Boca Raton, London, New York, Washington DC, 459p.

Prendergast, J.B., C.W. Rose & W.L. Hogarth (1994). A model for conjunctive use of ground water and surface waters for control of irrigation salinity. *Irrigation Science*, 14, 167-175.

Rasmussen, M.A., W.C. Cray, T.A. Casey & S.C. Whipp (1993). Rumen contents as a reservoir of enterohemorrhagic *Escherichia coli*. *FEMS Microbiology Letters*, 114, 79-84.

Reid, J.B., M.J. Goss & P.D. Robertson (1982). Relationship between the decreases in soil stability effected by the growth of maize roots and changes in organically bound iron and aluminium. *Journal of Soil Science*, 33, 397-410.

Rhoades, J.D. (1968). Leaching requirement for exchangeable sodium control. *Soil Science Society of America Journal*, 32, 652-656.

Rice, R.C. (1974). Soil clogging during infiltration of secondary effluent. *Journal of Water Pollution Control Federation*, 46, 708-716.

Richards, L.A. (1954). Diagnosis and improvements of saline and alkaline soils. Agriculture Handbook 60. USDA, Washington, DC, 160p.

Scholefield, D., K.T. Tyson, E.A. Garwood, A.C. Armstrong, J. Hawkins & A.C. Stone (1993). Nitrate leaching from grazed grassland lysimeters: effects of fertilizer input, field drainage and patterns of weather. *Journal of Soil Science*, 44, 601-614.

Shainberg, I., & J. Letey (1984). Response of soils to sodic and saline conditions. *Hilgardia*, 52, 1-57.

Sharpley, A.N., R. Menzel, S. Smith, E. Rhoades & A. Olness (1981). The sorption of soluble phosphorus by soil material during transport in runoff from cropped and grassed watersheds. *Journal of Environmental Quality*, 10, 211-215.

Suarez, D.L. (1981). Relation between pH_c and sodium adsorption ratio (SAR) and an alternative method of estimating SAR of soil or drainage water. *Soil Science Society of America Journal*, 45, 469-475.

Taylor, S.W. & P.R. Jaffe (1990). Biofilm growth and the related changes in the physical properties of a porous medium. 1. Experimental investigation. *Water Resources Research*, 26, 2153-2169.

Visser, S.A. & M. Caillier (1988). Observations on the dispersion and aggregation of clays by humic substances. I. Dispersive effects of humic acids. *Geoderma*, 42, 331-337.

Global atmospheric change and its effect on managed grassland systems

A. Lüscher[1], J. Fuhrer[1] and P.C.D. Newton[2]
[1]*Agroscope FAL Reckenholz, Swiss Federal Research Station for Agroecology and Agriculture, Reckenholzstrasse 191, CH-8046 Zurich, Switzerland*
Email: andreas.luescher@fal.admin.ch
[2]*AgResearch, Grassland Research Centre, Private Bag 11008, Palmerston North, New Zealand*

Key points

1. Increasing atmospheric CO_2 concentration and a trend to warmer mean temperatures are the most reliable aspects of global atmospheric change. Projections of the extent of climate change and the frequency of extreme weather conditions remain uncertain.
2. Research has considerably reduced the uncertainty about effects of global atmospheric change on physiology of plants, productivity and species composition of plant communities.
3. Other factors (e.g. nutrient availability, soil type) and long-term adaptation of the ecosystem (e.g. nutrient cycling and sequestration) influence the response of plant communities to global atmospheric change. Generalisation is not possible with respect to the response of different pasture and rangeland systems.
4. In temperate grasslands with regular fertilisation and defoliation, the effects of elevated CO_2 may be smaller than those of climate and/or management. Adaptations in management can help to mitigate effects of global atmospheric change.

Keywords: elevated atmospheric CO_2, climate change, grassland systems, species, yield

Introduction

Global atmospheric change consists of (i) an increase of the main greenhouse gases - carbon dioxide (CO_2,), methane (CH_4,), and nitrous oxide (N_2O); and (ii) transient changes in temperature, precipitation and other climatic elements over the coming decades and centuries. Using general circulation models, global projections of the earth's climate have been developed for a set of emission scenarios related to a range of assumptions regarding future socio-economic development (IPCC, 2001). The increase in atmospheric CO_2 concentration is the most reliable aspect of global atmospheric change. The projections of climate change carry with them a range of uncertainty, but they consistently show a warming trend in most regions. At smaller scales, and for changes other than temperature e.g. rainfall, and the frequency of extreme weather events, uncertainties in the projections increase. However, a discussion of uncertainties of climate projections is outside the scope of this paper.

The objective of this paper is to review current knowledge on the nature and direction of global atmospheric change, and the potential impact of this change on managed grassland ecosystems in terms of herbage production, herbage composition and animal production. Management approaches to counteract effects of atmospheric change will also be considered.

The response to atmospheric change of individual plants grown in controlled conditions with ample nutrient supply was elucidated in the 1990s, and in the last decade a great deal has been learned about the response of plant communities under natural field conditions. It has become clear, that long-term adaptation processes in the soil are important for the response of the plant communities to atmospheric change. The most important gaps in knowledge still existing concern: (i) the response of low-input plant communities in non-temperate climate zones, (ii) the effects of soil type and management on the response of plant communities to atmospheric change and (iii) the consequences for animal performance. These gaps restrict our ability to make generalisations about different climatic zones and grassland types, and the whole soil-plant-animal system including socio-economic aspects.

Projection of global change

Atmospheric CO₂ concentration

The average increase of atmospheric CO_2 concentration has been about 1.5 ppm (0.4%) per year over the past two decades. The IPCC in its 3rd assessment report (IPCC, 2001), projects CO_2 concentration may rise up to between 540 and 970 ppm by 2100 (depending on future global socio-economic development and associated emission patterns).

Temperature

Annual mean surface air temperature is expected to increase by 1.4 to 5.8 C by 2100 - with distinct regional differences (depending on emission scenario and climate model assumptions) (IPCC, 2001). Nearly all land areas warm more rapidly than the global average. Maximum warming is expected to occur in the high latitudes of the northern hemisphere and minimum warming in the southern ocean region (Figure 1).

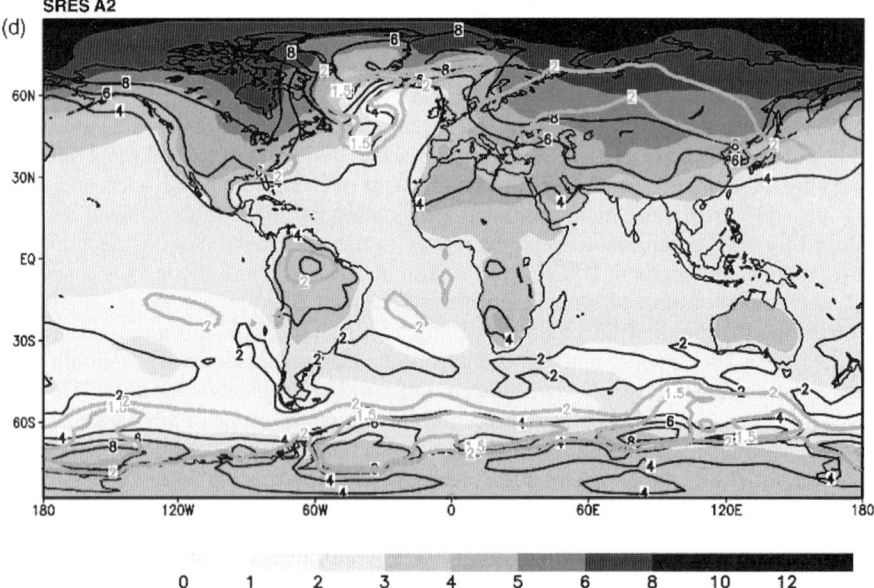

Figure 1 The annual mean change of temperature (grey shading and digits), its multi-model ensemble range (black isolines and digits) and the ensemble mean divided by its standard deviation (grey isolines) (Unit: °C) for the SRES scenario A2 (Period 2071 to 2100 relative to period 1961 to 1990) (IPCC, 2001)

Grasslands will have no major role in regions with the strongest predicted warming; in contrast, those areas that experience warming of 2-5 °C will continue to sustain grassland ecosystems; e.g. for Ireland, warming of 2-3 °C is projected (Figure 1, grey shading) with a range of 4 °C between models (Figure 1, black isoline). Increased inter-annual variability may also be a significant aspect of climate change, which will have high ecological significance. Using a regional climate model, Schär et al. (2004) predicted that year-to-year temperature variability for central Europe would increase by up to 100% by 2071-2100. This would lead to more frequent heat waves and droughts during the growing season.

Precipitation

All climate models indicate that average precipitation will increase, although in some regions reductions are likely (IPCC, 2001). A reduction in precipitation may be due to changes in synoptic scale features (e.g. changes in storm track characteristics) and/or local feedback processes (e.g. between soil moisture and precipitation). These changes in annual precipitation are associated with a shift in the seasonal distribution, e.g. in many parts of Europe it is anticipated that winter precipitation will increase and summer precipitation decline (Haylock & Goodess, 2004; Schär et al., 2004).

The magnitude of regional precipitation change varies considerably amongst models, with typical ranges between 0 to 50% where the direction of change is indicated, and around -30 to +30% where it is not. Larger ranges are projected in some regions (e.g. -30 to +60% in southern Africa for the summer) (IPCC, 2001), but this occurs mainly in regions of low seasonal precipitation where the implied range in absolute terms would not be large.

Effects of elevated CO_2 on grassland vegetation

Physiological effects

The physiological effects of elevated CO_2 are well known. Drake et al. (1997) reviewed 60 studies where plants were grown separately under controlled conditions with ample nutrient supply, and reported an average increase in leaf photosynthesis of 58% when grown at approximately double the pre-industrial concentration of CO_2. These increases in photosynthesis translate into increased dry matter production of about 30% (Newton, 1991). However, when plants are grown in communities in the field, competition and limiting growth resources can restrict the plant's response to elevated CO_2. These differences have prompted researchers to design experimental systems that more effectively simulate natural conditions. The most advanced systems for this work use a technology called Free Air Carbon Dioxide Enrichment (FACE), which does not require any enclosures and, so does not alter the microclimate of the experimental area. Examples from two of these systems (one in Switzerland (Swiss FACE) and one in New Zealand (NZ FACE)) will be used in this review.

Total yield of swards

Under field conditions Ainsworth et al. (2003), confirmed the strong increase in leaf photosynthesis in response to elevated CO_2. However yield response is much weaker; allocation of assimilates to non-harvested plant parts being increased relative to controlled conditions (Hebeisen et al., 1997; Suter et al., 2002). Results from the grassland ecosystems of the 'GCTE (Global Change and Terrestrial Ecosystems) pastures and rangelands core research project network' showed that the stimulatory effect of elevated CO_2 on sward productivity was only 15% (Figure 2A). This response was independent of the above-ground productivity of the ecosystem (Campbell et al., 2000; Figure 2A) and was similar to the effects reported by Mooney et al. (1999). In the GCTE network most of the sites are improved pastures under humid and cool/temperate climates; information on grasslands in other climatic zones is much more sparse.

Inter- and intraspecific differences in yield response

Much of the world's grasslands are characterised by swards that are botanically diverse. Thus, the response of individual functional groups and species within the sward are of interest, as changes in these components can alter nutrient cycling, the quality of the diet presented to animals and the maintenance of biodiversity. Significant differences in the yield response to elevated CO_2 are consistently found between legumes and C3 grasses. In the first six years of the Swiss FACE experiment, the yield response to elevated CO_2 of *Lolium perenne* (perennial ryegrass) ranged from -11 to +25%, compared to +10 to + 49% for *Trifolium repens* (white clover) when grown in pure swards (Hebeisen et al., 1997; Daepp et al., 2000). These differences were confirmed in separate studies for other legume and grass species (Lüscher et al., 2004). Both, the New Zealand and the Swiss FACE experiments revealed that forbs showed a stronger response to elevated CO_2 than grasses (Lüscher et al., 1998; Allard et al., 2003).

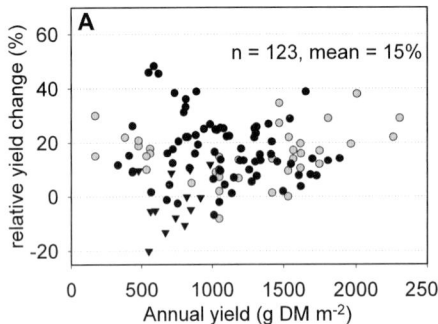 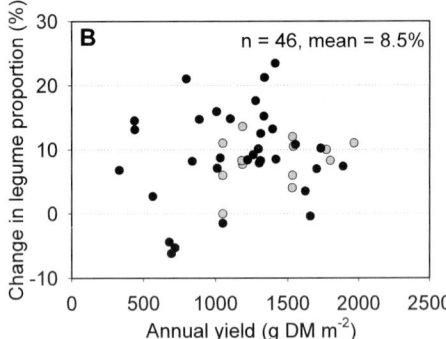

Figure 2 Effect of elevated CO_2 on (A) total production of harvestable biomass in grassland ecosystems, and (B) proportion of legumes in relation to the annual yield. Black symbols: NZ and Swiss FACE experiments; grey symbols: other experiments of the 'GCTE pasture and rangelands core research project network'; ▼ = pure grass swards at low N fertilisation; • = all other swards (mixed swards or pure grass swards with high N fertilisation). Newton et al. (1994; 1995); Ross et al. (1995; 1996); Casella et al. (1996); Soussana et al. (1996); Clark et al. (1997); Schenk et al. (1997); Owensby et al. (1999); Campbell et al. (2000).

Changes in species composition

Interspecific differences in the yield response to elevated CO_2 were accentuated in bi-species mixtures of the Swiss FACE experiment compared to pure swards (Hebeisen et al., 1997), indicating changes in competition between species. In the first 6 years the mean yield response to elevated CO_2 of *T. repens* averaged over all management treatments increased from 25% in pure stands to 874% in the mixture. Significant increases in both annual and perennial legume content at elevated CO_2 have also been measured in the NZ FACE experiment (Edwards *et al.*, 2001; Ross *et al.*, 2004) (Figure 2b).

CO$_2$ response depends on nutrient availability

Limitations to nutrient availability can limit the sward's response to elevated CO_2. This is evident from comparing the response of *L. perenne* to elevated CO_2, under limiting N fertilisation, where there was a 3% reduction in yield (Figure 2a); this compares to a 25% increase in yield when luxury levels of N-fertiliser were applied (Daepp *et al.*, 2001). Since mineral N in the soil was a major limiting factor in this system, the crucial advantage of legumes is their access to non-limiting atmospheric N through symbiotic N_2 fixation (Zanetti *et al.*, 1996; 1997). Non N-fixing mutants of *Medicago sativa* (Lucerne) consequently loose this advantage in response to elevated CO_2 (Lüscher *et al.*, 2000). Similarly, phosphorus (P)-limitation of *T. repens* has been shown to significantly reduce its response to elevated CO_2 in a growth room experiment (Almeida *et al.*, 1999) and with micro swards from calcareous grassland (Stöcklin *et al.*, 1998).

Nutrient cycling and long-term adaptation

Since grasslands cover about 20% of the land surface of the earth, potential changes in the cycling and storage of C and N in grassland soils at elevated CO_2, is of importance to future trends in soil fertility and to C sequestration.

Changes in the pools of C and N in the soil are determined by the amount and quality of material returned to the soil through rhizodeposition, via litter or animals. Only small increases in aboveground litter inputs have been found in response to elevated CO_2, but there is evidence for increased rates of input from roots (Loiseau & Soussana, 1999; Allard *et al.*, 2004a,b).

Changes in soil microbial populations (structure and/or size) may also play a part in the long-term adaptation of decomposition to elevated CO_2 (Sowerby *et al.*, 2000; Montealegre *et al.*, 2000; 2002). These responses may in the long run increase plant available N at elevated CO2 (Schneider *et al.*, 2004), but this outcome is not inevitable as it depends upon rates of N fertilisation. In lower N status systems, there is some evidence for a decrease in plant available N at high levels of CO_2 (Gill *et al.*, 2002). In low N systems, adaptation of C and N pools and thus N-availability to elevated CO_2 is expected to require much longer time periods (100 years) (Cannell & Thornley, 1998).

Several temperate grassland experiments have shown an increase in coarse particulate organic matter under monocultures (Loiseau & Soussana, 1999) and mixed swards (Allard *et al.*, 2004b; Gill *et al.*, 2002) at elevated CO_2. This can be interpreted as the first stage of incorporation of organic matter into older, finer and more recalcitrant pools, and may imply a greater sink capacity for atmospheric CO_2, but could also represent a more rapid approach to a similar C storage capacity (Hassink, 1997; Van Kessel *et al.*, 2000a,b). Potential interactions with increased temperature and changes in rainfall require further study (Thornley & Cannell, 1997), as do the dynamics of non-temperate grasslands.

Effects of climate change on grassland vegetation

Physiological effects

Direct effects of increasing temperature (T) can occur though stimulation of light-saturated photosynthetic rate (A_{max}) up to an optimum (T_{opt}), which is relevant for species growing at suboptimal temperatures. At supra-optimal temperatures, A_{max} declines mainly as a result of increased photorespiration. Thus, without an effect of increased CO_2 on T_{opt}, an increase by 2-3 °C by 2050 would reduce A_{max} in many C3 species, compared to C4 species. At the canopy level, increased photosynthesis at higher T may be counteracted by a reduction in green leaf area index (LAI) (Casella & Soussana, 1997; Long *et al.*, 2004).

At higher T, the start of the growing season is shifted, and plants develop faster. The growing season is usually limited to the period with daily mean T above a threshold of 5 °C. A survey of experimental data worldwide suggested that a mild warming generally increases grassland productivity, with the strongest positive response in currently colder regions (Rustad *et al.*, 2001). In cut and grazed grasslands, there may be a requirement to harvest additional cuts during the season, or extension of the grazing season. Earlier simulations using two climate scenarios suggested reduced production in the cold desert Steppes (-20-30% by 2100), but confirmed increased production ($\sim$10%) in the humid temperate ecoregion, and small positive effects in Mediterranean grasslands, and the dry and humid savannas (Ojima *et al.*, 1996).

Effects of temperature on plant communities

Since increasing T affects individual species or functional types differently, species shifts in frequency and cover can be expected. When transplanting turf and soil from cooler to warmer sites, Bruelheide (2003) found after seven years, that the community had changed into a different plant association. Using a multiple regression model established from observations in regions of north to Southwest Europe, Duckworth *et al.* (2000) concluded that T+2 °C caused only minor shifts. Alpine vegetation was found to have a natural inertia and could tolerate a 1-2 °C warming, but was prone to profound changes at T+3 °C or higher (Theurillat & Guisan, 2001).

Warming may affect the seasonal growth patterns of C3 and C4 species differently. Winslow *et al.* (2003) found that the location-specific temperatures used to delineate the start and end of the growing season of C3 and C4 plants was a more important determinant of final biomass than the photosynthetic response to T. All other factors being equal, C4 plants tend to be favoured over C3 plants in warm humid climates, the reverse being the case in cool climates. Epstein *et al.* (2002) in response to climate change scenarios, suggest that the relative abundance of C4 grasses in temperate grasslands will increase throughout most regions of North and South America. In New Zealand, C4 species are favoured over C3 species where there is likely to be exposure to events with extreme heating events (White *et al.*, 2000; 2001). This is an example of how increased inter-annual climatic variability, and the frequency of extreme events, are expected to suppress C3 competitive dominance and promote the invasion of C4 species.

The indirect effects of increased T can operate through altered nutrient cycling. If mineralisation under non-limiting soil moisture is stimulated by increased soil CO_2 efflux and net N mineralisation rates (Rustad *et al.*, 2001), this will lead to higher levels of available soil nutrients and N (Loiseau & Soussana, 1999). However, Kandeler *et al.* (1998) found that the effects of warming on microbial communities and processes are minimal in soils of low nutrient availability.

Effects of altered precipitation

Altered rainfall patterns, storm intensity and more frequent droughts may cause severe perturbations in grasslands. Productivity and plant composition in rangelands are highly correlated with precipitation (Knapp & Smith, 2001). Lack of precipitation operates through reductions in available soil moisture, which is further accentuated by increased evapotranspiration due to higher net radiation and T (Jasper et al., 2004). Due to the influence of texture on soil hydrology, effects of increased or decreased precipitation on plant productivity and decomposition vary with the proportion of sand and clay content (Epstein et al., 1997) and may result in distinct patterns of change across landscapes (Riedo et al., 2001).

Simulations for Switzerland using an ensemble of downscaled IPCC climate scenarios, projected significant decreases in average soil moisture towards the end of this century (Jasper et al., 2004). This would reduce aboveground net primary productivity in intensively managed grasslands. Increased variability in rainfall may cause even more severe soil moisture limitation and reduced root growth, as observed in subdominant warm-season C4 grasses in native Great Plain grasslands (Fay et al., 2003). In grasslands of northern Ontario comprised of *Trifolium hybridum*, *T. pratense* and *Phleum pratense*, a reduction in the number of precipitation events, while not altering the total rainfall, reduced productivity because of lower average soil moisture, as compared to a treatment with more evenly distributed precipitation (Laporte et al., 2002).

Reduced soil moisture may also cause adjustments in species composition in permanent, less intensive grasslands. Observations along a climatic gradient in the Hungarian Plain showed how species with Continental and sub-Mediterranean distribution increase with increasing aridity (Kovacs-Lang et al., 2000). Shifts in grass-forb relationships in response to weather variability are well documented. In C4 dominated tallgrass prairies, C3 forbs generally have an advantage over grasses during drought (Briggs & Knapp, 1995), because they respond less to variation in seasonal and inter-annual rainfall (Briggs & Knapp, 2001).

Effect of shifts in climate variability on succession

Climate variability and change is likely to affect succession. As evidenced for limestone grasslands in the UK, the more fertile, early successional grasslands are most responsive (Grime et al., 2000). In calcareous grassland, warmer winters with increased summer droughts increased the amount of plant litter and delayed succession and gap formation, which favour colonisation by annuals (Sternberg et al., 1999). Under wetter summertime conditions, sward closure by perennial grasses inhibited establishment of later successional species. Gap re-colonisation by annuals with a persistent seed bank was observed in sown grasslands after the severe drought in 1995 in Oxfordshire, UK. Hence, effects on gap formation and species invasion during extreme summer droughts could be a major aspect of future climatic conditions, and would make the maintenance of a desired species mixture more problematic.

Interactive effects of climate change and elevated CO_2

Photosynthesis

Rising T increases the stimulation of leaf photosynthesis by elevated CO_2 mainly because photorespiration strongly increases with T at ambient CO_2, but is strongly reduced at elevated CO_2. Whereas at 10 °C, doubling the CO_2 increased the light-saturated photosynthesis (A_{max}) by only 4%, it increased by 35% at 30 °C (Drake *et al.*, 1997). As a result, the optimum temperature of A_{max} increased with increasing CO_2 by several degrees. The maximum T at which photosynthesis is still positive also increased.

Water use efficiency and soil moisture

Plants growing at elevated CO_2 partially close their stomata, thus reducing stomatal conductivity (g_s) and leaf transpiration (Field *et al.*, 1995). The responses of species can differ, but on average, g_s is reduced by 20% at elevated CO_2 (Drake *et al.*, 1997). As this reduction does not reduce the rate of photosynthesis below that at ambient CO_2, and as the rate of transpiration decreases, then plants fix carbon using less water i.e. they have higher water use efficiency at elevated CO_2. This increase in water use efficiency also occurs in C4 species, and is sometimes greater than in C3 species (Drake *et al.*, 1997) providing a mechanism for positive responses of C4 species to elevated CO_2, particularly in conditions of low soil moisture (Owensby *et al.*, 1997).

Through its effect on plant water relations, elevated CO_2 can reduce soil water depletion in different native and semi-native temperate and Mediterranean grasslands, with the extent, timing and duration of this effect depending on ecosystem, year, and species (Morgan *et al.*, 2004). Based on a comprehensive assessment, the authors consider that this indirect effect of elevated CO_2 might even dominate the biomass and plant community responses of drier ecosystems, provided that their plant canopy has a high degree of coupling to the atmosphere. This could counteract expected increases in evapotranspiration due to increased temperature. In Mediterranean annual grassland warming alone increased forb production and abundance, but did not strongly affect diversity or grass response. The largest change (+50% forbs) after 3 years was found in response to the combination of warming, elevated CO_2 and increased precipitation (Zavaleta *et al.*, 2003). In Swiss calcareous grassland, Niklaus & Körner (2004) found over 6 years a significant negative correlation between the amount of precipitation during 6 weeks prior to harvest, and the relative CO_2 stimulation of biomass. Such a direct link demonstrates that future reductions in summer precipitation and higher potential evapotranspiration (PET) could increase the relative CO_2-stimulation of productivity in precipitation-sensitive grasslands. In other words, increased water use efficiency under elevated CO_2 may compensate for reduced summer precipitation and increased PET.

Effects of atmospheric change on nutritive value of herbage

Elevated CO_2 may alter the nutritive composition of animal diets directly through altering the chemical composition of individual plants, or indirectly through changes in botanical composition. A reduced concentration of N in plants and therefore a lower crude protein content is a frequently observed response to elevated CO_2 (Campbell *et al.*, 2000). In both, the Swiss and NZ FACE experiments there were significant reductions in the N content of *L. perenne* (-18% and -19%, respectively). However in *L. perenne*, the concentration of N under elevated CO_2 observed in pure stands, was considerably lower than that found in a bi-species mixture with *T. repens* (Zanetti *et al.*, 1997; Hartwig *et al.*, 2000). Furthermore, under elevated CO_2 the proportion of N-rich *T. repens* (40 mg N/g DM) increased in the mixture at the expense of the N-poor *L. perenne* (24 mg N/g DM when grown in monoculture). The same form of compensation was measured under grazing in the NZ FACE (Allard *et al.*, 2003).

Animal requirements for crude protein (CP) from pasture range from 7 to 8% for animals at maintenance, up to 24% for the highest producing dairy cows. Average values in the NZ FACE over 7 years were 15% CP with little difference between treatments. Clearly in conditions of very low N status the reduction in CP may put a system into a sub-maintenance level for animal performance. Such systems exist in more extensive rangeland situations where elevated CO_2 may reduce protein levels below 6% (Owensby *et al.*, 1993; 1996).

Under elevated CO_2, the concentration of carbohydrates in the leaves of *L. perenne* increased (Fischer *et al.*, 1997). However, the proportion of digestible organic matter and the concentration of net energy in *T. repens* (Frehner *et al.*, 1997) and *L. perenne* (Hebeisen, unpublished) monocultures were not affected by elevated CO_2. This is consistent with results from the NZ FACE where the digestibility of seven species covering C3 and C4 grasses, forbs and legumes was unaltered by CO_2 (Allard *et al.*, 2003). However, changes in botanical composition towards a higher content of the more digestible legumes resulted in a significantly higher digestibility of the herbage on offer. It must be remembered that comprehensive animal performance trials have yet to be carried out.

Less information is available on the effects of climate change on forage quality. Various traits such as digestibility, protein and energy content may all be influenced differently by climate change (Seligman & Sinclair, 1995). Extremely dry summers could increase lignification in some species, but also lead to changes in plant composition with more drought resistant species becoming more dominant. In grass-clover mixtures, the clover fraction increased in a drier and warmer year by up to 80%, presumably because of lower drought sensitivity of N_2 fixation relative to the uptake of mineral N in the grasses (Hebeisen *et al.*, 1997; Figure 3, year 2). Also, species composition change is likely to be important in altering grassland production and its value for grazing livestock, especially in drier rangelands with woody shrub invasion (Campbell *et al.*, 2000). If increasing temperatures result in increased proportions of C4 species in pastures, then a further reduction in forage quality is likely (Barbehenn *et al.*, 2004).

Potential for adaptation of management to climate change and elevated CO_2

It would seem from the above that atmospheric change could alter the quantity and quality of forage available to animals. The most predictable change in the environment - an increase in atmospheric concentration of CO_2, is likely to result in greater production of herbage with a lower protein content; where systems include legumes then an increase in the legume content is likely (Figure 2b), with beneficial effects on the availability of dietary components such as protein, water soluble carbohydrates and N-inputs to the system. Moderate increases in temperature will in general increase productivity. The effects of an increase in temperature together with an increase in CO_2 on sward composition have not been tested in a factorial field experiment. However, from the parallel seasonal fluctuations of temperature and clover proportion (Figure 3) it becomes evident, that warming and elevated CO_2 both increase clover proportion.

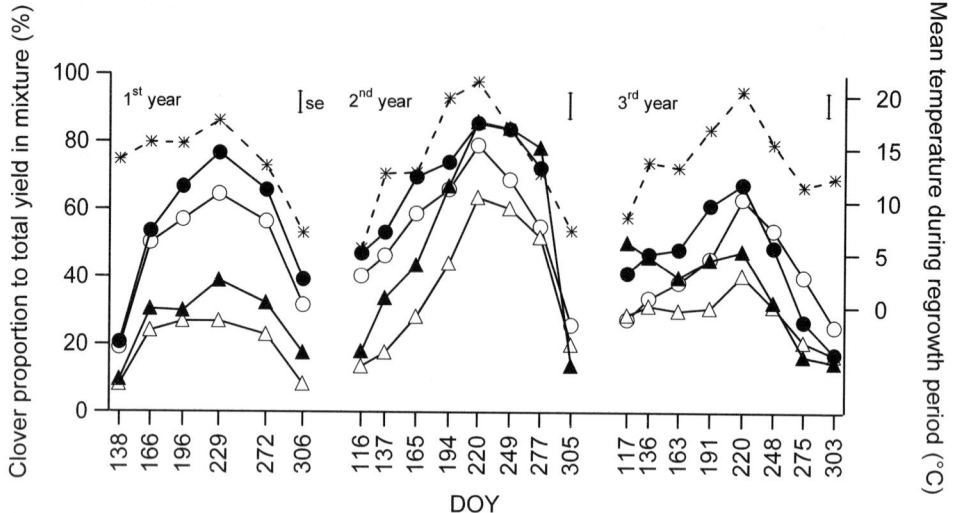

Figure 3 Yield proportion (%) of *Trifolium repens* in bi-species mixtures with *Lolium perenne* during the first three years of the Swiss FACE experiment. Swards were exposed to ambient (350 ppm; white) and elevated (600 ppm; black) CO_2 concentration under low (•=140 kg/ha per annum) and high (▲=560 kg/ha per annum) N fertilisation. Bars = standard error of mean; * = mean daily temperature; DOY = day of the year.

In intensively managed systems, the capacity of management to change herbage quality and quantity is substantial. In addition, there is strong evidence for interactions between management options and climate change drivers such as elevated CO_2 and temperature. For example, the response of grass species to CO_2 can be controlled by the level of N fertiliser input (Daepp *et al.*, 2001; Figure 2a) and the response of legume species by the level of P input (Stöcklin *et al.*, 1998; Almeida *et al.*, 1999).

CO_2-induced changes in the botanical composition of a sward can be modified by management decisions such as the choice of cutting or grazing (Newton *et al.*, 2005), or the frequency of cutting (Hebeisen *et al.*, 1997; Teyssonneyre *et al.*, 2002) and the level of N fertilisation (Figure 3). Large differences in the proportion of *T. repens* (14 vs. 57%) between extreme management regimes (infrequent defoliation combined with high N fertilisation; frequent defoliation combined with low N) were observed during the first three years of the Swiss FACE experiment. These management effects were much stronger than the effect of elevated CO_2 (increase of *T. repens* proportion from 21% to 33%). This and the strong seasonal variation in clover proportion (Figure 3) demonstrate that effects of elevated CO_2 may be smaller than those of climate and/or management. Adaptations in management can thus help to mitigate effects of global atmospheric change.

In sown grassland, there is also the potential to select plant cultivars that will take advantage of the changed environment (Lüscher *et al.*, 1998; Baker 2004; Ziska *et al.*, 2004; Wright *et al.*, 2005) although this aspect of mitigation has received little attention. Consequently, where management flexibility is possible, land managers are in a strong position to buffer the negative effects of climate change and to capitalise on the positive effects. In more extensive situations where the systems are operating close to thresholds of sustainability, management options are fewer and consequently these systems remain vulnerable to global change. A better definition of likely changes is a research priority.

Acknowledgements

We are grateful to J. Nösberger, H. Blum and U. Aeschlimann for helpful comments in the compilation of this manuscript. The work was partially supported (JF) by the Swiss National Science Foundation (NCCR Climate).

References

Ainsworth, E.A., P.A. Davey, G.J. Hymus, C.P. Osborne, A. Rogers, H. Blum, J. Nösberger & S.P. Long (2003). Is stimulation of leaf photosynthesis by elevated carbon dioxide concentration maintained in the long term? A test with *Lolium perenne* grown for 10 years at two nitrogen fertilization levels under Free Air CO_2 Enrichment (FACE). *Plant, Cell and Environment*, 26, 705-714.

Allard, V., P.C.D. Newton, M. Lieffering, H. Clark, C. Matthew & Y. Gray (2003). Nutrient cycling in grazed pastures at elevated CO_2: N returns by animals. *Global Change Biology*, 9, 1731-1742.

Allard, V., P.C.D. Newton, M. Lieffering, J.F. Soussana, P. Grieu & C. Matthew (2004a). Elevated CO_2 effects on decomposition processes in a grazed grassland. *Global Change Biology*, 10, 1553-1564.

Allard, V., P.C.D. Newton, J.F. Soussana, R.A. Carran & C. Matthew (2004b). Increased quantity and quality of coarse soil organic matter fractions at elevated CO_2 in a grazed grassland are a consequence of enhanced root growth and turnover. *Plant and Soil*, (in press).

Almeida, J.P.F., A. Lüscher, M. Frehner, A. Oberson & J. Nösberger (1999). Partitioning of P and the activity of root acid phosphatase in white clover (*Trifolium repens* L.) are modified by increased atmospheric CO_2 and P fertilisation. *Plant and Soil*, 210, 159-166.

Baker, J.T. (2004). Yield responses of southern US rice cultivars to CO_2 and temperature. *Agricultural and Forest Meteorology*, 122, 129-137.

Barbehenn, R.V., Z. Chen, D.N. Karowe & A. Spickard (2004). C_3 grasses have higher nutritional quality than C_4 grasses under ambient and elevated atmospheric CO_2. *Global Change Biology*, 10, 1565-1575.

Briggs, J.M. & A.K. Knapp (1995). Interannual variability in primary production in tallgrass prairie: climate, soil moisture, topographic position and fire as determinants of aboveground biomass. *American Journal of Botany*, 82, 1024-1030.

Briggs, J.M. & A.K. Knapp (2001). Determinants of C3 forb growth and production in a C4 dominated grassland. *Plant Ecology*, 152, 93-100.

Bruelheide, H. (2003). Translocation of a montane meadow to simulate the potential impact of climate change. *Applied Vegetation Science*, 6, 23-34.

Campbell, B.D., D.M. Stafford Smith, J. Ash, J. Fuhrer, R.M. Gifford, P. Hiernaux, S.M. Howden, M.B. Jones, J.A. Ludwig, R. Manderscheid, J.A. Morgan, P.C.D. Newton, J. Nösberger, C.E. Owensby, J.F. Soussana, Z. Tuba & C. ZuoZhong (2000). A synthesis of recent global change research on pasture and rangeland production: reduced uncertainties and their management implications. *Agriculture, Ecosystems and Environment*, 82, 39-55.

Cannell, M.G.R. & J.H.M. Thornley (1998). N-poor ecosystems may respond more to elevated [CO_2] than N-rich ones in the long-term. A model analysis of grassland. *Global Change Biology*, 4, 431-442.

Casella, E. & J.F. Soussana (1997). Long-term effects of CO_2 enrichment and temperature increase on the carbon balance of a temperate grass sward. *Journal of Experimental Botany*, 48, 1309-1321.

Casella, E., J.F. Soussana & P. Loiseau (1996). Long-term effects of CO_2 enrichment and temperature increase on a temperate grass sward. I. Productivity and water use. *Plant and Soil*, 182, 83-99.

Clark, H., P.C.D. Newton, C.C. Bell & E.M. Glasgow (1997). Dry matter yield, leaf growth and population dynamics in *Lolium perenne/Trifolium repens* dominated pasture turves exposed to two levels of elevated CO_2. *Journal of Applied Ecology*, 34, 304-316.

Daepp, M., J. Nösberger & A. Lüscher (2001). Nitrogen fertilisation and developmental stage affect the response of yield, biomass partitioning and morphology of *Lolium perenne* L. swards to elevated pCO$_2$. *New Phytologist*, 150, 347-358.

Daepp, M., D. Suter, J.P.F. Almeida, H. Isopp, U.A. Hartwig, M. Frehner, H. Blum, J. Nösberger & A. Lüscher (2000). Yield response of *Lolium perenne* swards to free air CO_2 enrichment increased over six years in a high-N-input system on fertile soil. *Global Change Biology*, 6, 805-816.

Drake, B.G., M.A. Gonzàlez-Meler & S.P. Long (1997). More efficient plants: a consequence of rising atmospheric CO_2? *Annual Review of Plant Physiology and Plant Molecular Biology*, 48, 609-639.

Duckworth, J.C., R.G.H. Bunce & A.J.C. Malloch (2000). Modelling the potential effects of climate change on calcareous grasslands in atlantic Europe. *Journal of Biogeography*, 27, 347-358.

Edwards, G.R., H. Clark & P.C.D Newton (2001). The effects of elevated CO_2 on seed production and seedling recruitment in a sheep-grazed pasture. *Oecologia*, 127, 383-394.

Epstein, H.E., R.A. Gill, J.M. Paruelo, W.K. Lauenroth, G.J. Jia & I.C. Burke (2002). The relative abundance of three plant functional types in temperate grasslands and shrublands of North and South America: effects of projected climate change. *Journal of Biogeography*, 29, 875-888.

Epstein, H.E., W.K. Lauenroth & I.C. Burke (1997). Effects of temperature and soil texture on ANPP in the US Great Plains. *Ecology*, 78, 2628-2631.

Fay, P.A., J.D. Carlisle, A.K. Knapp, J.M. Blair & S.L. Collins (2003). Productivity responses to altered rainfall patterns in a C4-dominated grassland. *Oecologia*, 137, 245-251.

Field, C.B., R.B. Jackson & H.A. Mooney (1995). Stomatal responses to increased CO_2: implications from the plant to the global scale. *Plant Cell and Environment*, 18, 1214-1225.

Fischer, B.U., M. Frehner, T. Hebeisen, S. Zanetti, F. Stadelmann, A. Lüscher, U.A. Hartwig, G.R. Hendrey, H. Blum & J. Nösberger (1997). Source-sink relations in *Lolium perenne* L. as reflected by carbohydrate concentrations in leaves and pseudo-stems during regrowth in a free air carbon dioxide enrichment (FACE) experiment. *Plant, Cell and Environment*, 20, 945-952.

Frehner, M., A. Lüscher, T. Hebeisen, S. Zanetti, F. Schubiger & M. Scalet (1997). Effects of elevated partial pressure of carbon dioxide and season of the year on forage quality and cyanide concentration of *Trifolium repens* L. from a FACE experiment. *Acta Oecologica*, 18, 297-304.

Gill, R.A., H.W. Polley, H.B. Johnson, L.J. Anderson, H. Maherall & R.B. Jackson (2002). Nonlinear grassland responses to past and future atmospheric CO_2. *Nature*, 417, 279-282.

Grime, J.P., V.K. Brown, K. Thompson, G.J. Masters, S.H. Hillier, I.P. Clarke, A.P. Askew, D. Corker & J.P. Kielty (2000). The response of two contrasting limestone grasslands to simulated climate change. *Science*, 289, 762-765.

Hartwig, U.A., A. Lüscher, M. Daepp, H. Blum, J.F. Soussana & J. Nösberger (2000). Due to symbiotic N_2 fixation, five years of elevated atmospheric pCO_2 had no effect on litter N concentration in a fertile grassland ecosystem. *Plant and Soil*, 224, 43-50.

Hassink, J. (1997). The capacity of soils to preserve organic C and N by their association with clay and silt particles. *Plant and Soil*, 191, 77-87.

Haylock, M.R. & C.M. Goodess (2004). Interannual variability of extreme European winter rainfall and links with mean large-scale circulation. *International Journal of Climatology*, 24, 759-776.

Hebeisen, T., A. Lüscher, S. Zanetti, B.U. Fischer, U.A. Hartwig, M. Frehner, G.R. Hendrey, H. Blum & J. Nösberger (1997). The different responses of *Trifolium repens* L. and *Lolium perenne* L. grassland to free air CO_2 enrichment and management. *Global Change Biology*, 3, 149-160.

IPCC, (2001). Climate Change 2001: The Scientific Basis. J.T. Houghton, Y. Ding, D.J. Griggs, M. Noguer, P.J. van der Linden & D. Xiaosu (eds.) Cambridge University Press, UK, 881pp. <http://www.ipcc.ch>

Jasper, K., P.L. Calanca, D. Gyalistras & J. Fuhrer (2004). Differential impacts of climate change on the hydrology of two alpine river basis. *Climate Research*, 26, 113-129.

Kandeler, E., D. Tscherko, R.D. Bardgett, P.J. Hobbs, C. Kampichler & T.H. Jones (1998). The response of soil microorganisms and roots to elevated CO_2 and temperature in a terrestrial model ecosystem. *Plant and Soil*, 202, 251-262.

Knapp, A.K. & M. Smith (2001). Variation among biomes in temporal dynamics of aboveground primary production. *Science*, 291, 481-484.

Kovacs-Lang, E., G. Kroel-Dulay, M. Kertesz, G. Fekete, S. Bartha, J. Mika, I. Dobi-Wantuch, T. Redei, K. Rajkai & I. Hahn (2000). Changes in the composition of sand grasslands along a climatic gradient in Hungary and implications for climate change. *Phytocoenologia*, 30, 385-407.

Laporte, M., L.C. Duchesne & S. Wetzel (2002). Effect of rainfall patterns on soil surface CO_2 efflux, soil moisture, soil temperature and plant growth in a grassland ecosystem of northern Ontario, Canada: implications for climate change. *BioMed Central Ecology*, 2: 10(doi:10.1186/1472-6785-2-10).

Loiseau, P. & J.F. Soussana (1999). Elevated [CO_2], temperature increase and N supply effects on the accumulation of below-ground carbon in a temperate grassland ecosystem. *Plant and Soil*, 212, 123-134.

Long, S.P., E.A. Ainsworth, A. Rogers & D.R. Ort (2004). Rising atmospheric carbon dioxide: plants face the future. *Annual Review of Plant Biology*, 55, 591-628.

Lüscher, A., M. Daepp, H. Blum, U.A. Hartwig & J. Nösberger (2004). Fertile temperate grassland under elevated atmospheric CO_2 - role of feed-back mechanisms and availability of growth resources. *European Journal of Agronomy*, (in press).

Lüscher, A., U.A. Hartwig, D. Suter & J. Nösberger (2000). Direct evidence that symbiotic N_2 fixation in fertile grassland is an important trait for a strong response of plants to elevated atmospheric CO_2. *Global Change Biology*, 6, 655-662.

Lüscher, A., G.R. Hendrey & J. Nösberger (1998). Long-term responsiveness to free air CO$_2$ enrichment of functional types, species and genotypes of plants from fertile permanent grassland. *Oecologia*, 113, 37-45.

Montealegre, C.M., C. Van Kessel, J.M. Blumenthal, H.G. Hur, U.A. Hartwig & M.J. Sadowsky (2000). Elevated atmospheric CO$_2$ alters microbial population structure in a pasture ecosystem. *Global Change Biology*, 6, 475-482.

Montealegre, C.M., C. Van Kessel, M.P. Russelle & M.J. Sadowsky (2002). Changes in microbial activity and composition in a pasture ecosystem exposed to elevated atmospheric carbon dioxide. *Plant and Soil*, 243, 197-207.

Mooney, H.A., J. Canadell, F.S. Chapin, J.R. Ehleringer, C. Körner, R.E. McMurtrie, W.J. Parton, L.F. Pitelka & E.D. Schulze (1999). Ecosystem physiology responses to global change. In: B. Walker, W. Steffen, J. Canadell & J. Ingram (eds.) International geosphere - biosphere programme book series. Cambridge University Press, Cambridge, 141-189.

Morgan, J.A., D.E. Pataki, C. Körner, H. Clark, S.J. Del Grosso, J.M. Grünzweig, A.K. Knapp, A.R. Mosier, P.C.D. Newton, P.A. Niklaus, J.B. Nippert, R.S. Nowak, W.J. Parton, H.W. Polley & M.R. Shaw (2004). Water relations in grassland and desert ecosystems exposed to elevated CO$_2$. *Oecologia*, 140, 11-25.

Newton, P.C.D. (1991). Review: direct effects of increasing carbon dioxide on pasture plants and communities. *New Zealand Journal of Agricultural Research*, 34, 1-24.

Newton, P.C.D., V. Allard, R.A. Carran & M. Lieffering (2005). Grazed grasslands. In J. Nösberger, S.P. Long, G.R. Hendrey, M. Stitt, R.J. Norby & H. Blum (eds.) Managed ecosystems and CO$_2$: case studies, processes and perspectives. Ecological Studies Series, Springer Verlag, (in press).

Newton, P.C.D., H. Clark, C.C. Bell, E.M. Glasgow, B.D. Campbell (1994). Effects of elevated CO$_2$ and simulated seasonal-changes in temperature on species composition and growth-rates of pasture turves. *Annals of Botany*, 73, 53-59.

Newton, P.C.D., H. Clark, C.C. Bell, E.M. Glasgow, K.R. Tate, D.J. Ross, G.W. Yeates, S. Saggar (1995). Plant growth and soil processes in temperate grassland communities at elevated CO$_2$. *Journal of Biogeography*, 22, 235-240.

Niklaus, P. & C. Körner (2004). Synthesis of a six year study of calcareous grassland under elevated CO$_2$. *Ecological Monographs*, 74, (in press).

Ojima, D.S., W.J. Parton, M.B. Coughenour, J.M.O. Scurlock, T.B. Kirchner, T.G.F. Kittel, D.O. Hall, D.S. Schimel, E. Garcia Moya, T.G. Gilmanov, T.R. Seastedt, A. Kamnalrut, J.I. Kinyamario, S.P. Long, J.C. Menaut, O.E. Sala, R.J. Scholes & J.A. van Veen (1996). Impact of climate and atmospheric carbon dioxide changes on grasslands of the world. In: A.I. Breymeyer, D.O. Hall, J.M. Mellilo & G.I. Agren (eds.) Global change: effects on coniferous forests and grasslands. John Wiley, Chichester, UK, 271-311.

Owensby, C.E., R.C. Cochran & L.M. Auen (1996). Effects of elevated carbon dioxide on forage quality for ruminants. In: C. Körner & F. Bazzaz (eds.) Carbon dioxide, populations, and communities. Academic Press, San Diego, CA, 363-371.

Owensby, C.E., P.I. Coyne & L.M. Auen (1993). Nitrogen and phosphorus dynamics of a tallgrass prairie ecosystem exposed to elevated carbon dioxide. *Plant, Cell and Environment*, 16, 843-850.

Owensby, C.E., J.M. Ham, A.K. Knapp & L.M. Auen (1999). Biomass production and species composition in a tallgrass prairie ecosystem after long-term exposure to elevated atmospheric CO$_2$. *Global Change Biology*, 5, 497-506.

Owensby, C.E., J.M. Ham, A.K. Knapp, D. Bremer & L.M. Auen (1997). Water vapour fluxes and their impact under elevated CO$_2$ in a C4-tallgrass prairie. *Global Change Biology*, 3, 189-195.

Riedo, M., D. Gyalistras & J. Fuhrer (2001). Pasture responses to elevated temperature and doubled CO$_2$ concentration: assessing the spatial pattern across an alpine landscape. *Climate Research*, 17, 19-31.

Ross, D.J., P.C.D. Newton & K.R. Tate (2004). Elevated [CO$_2$] effects on herbage production and soil carbon and nitrogen pools and mineralization in a species-rich, grazed pasture on a seasonally dry sand. *Plant and Soil*, 260, 183-196.

Ross, D.J., S. Saggar, K.R. Tate, C.W. Feltham & P.C.D. Newton (1996). Elevated CO$_2$ effects on carbon and nitrogen cycling in grass/clover turves of a Psammaquent soil. *Plant and Soil*, 182, 185-198.

Ross, D.J., K.R. Tate & P.C.D. Newton (1995). Elevated CO$_2$ and temperature effects on soil carbon and nitrogen cycling in ryegrass/white clover turves of an Endoaquept soil. *Plant and Soil*, 176, 37-49.

Rustad, L.E., J.L. Campbell, G.M. Marion, R.J. Norby, M.J. Mitchell, A.E. Hartley, J.H.C. Cornelissen & J. Gurevitch (2001). A meta-analysis of the response of soil respiration, net nitrogen mineralization, and aboveground plant growth to experimental ecosystem warming. *Oecologia*, 126, 543-562.

Schär, C., P.L. Vidale, D. Lüthi, C. Frei, C. Häberli, M. Liniger & C. Appenzeller (2004). The role of increasing temperature variability for European summer heat waves. *Nature*, 427, 332-336.

Schenk, U., H.J. Jäger & H.J. Weigel (1997). The response of perennial ryegrass white clover mini-swards to elevated atmospheric CO$_2$ concentrations: effects on yield and fodder quality. *Grass and Forage Science*, 52, 232-241.

Schneider, M.K., A. Lüscher, M. Richter, U. Aeschlimann, U.A. Hartwig, H. Blum, E. Frossard & J. Nösberger (2004). Ten years of free-air CO_2 enrichment altered the mobilization of N from soil in *Lolium perenne* L. swards. *Global Change Biology*, 10, 1377-1388.

Seligman, N.G. & T.R. Sinclair (1995). Climate change, interannual weather differences and conflicting responses among crop characteristics: the case of forage quality. *Global Change Biology*, 1, 157-160.

Soussana, J.F., E. Casella & P. Loiseau (1996). Long-term effects of CO_2 enrichment and temperature increase on a temperate grass sward. II. Plant nitrogen budgets and root fraction. *Plant and Soil*, 182, 101-114.

Sowerby, A., H. Blum, T.R.G. Gray & A.S. Ball (2000). The decomposition of *Lolium perenne* in soils exposed to elevated CO_2: comparisons of mass loss of litter with soil respiration and soil microbial biomass. *Soil Biology and Biochemistry*, 32, 1359-1366.

Sternberg, M., V.K. Brown, G.J. Masters & I.P. Clarke (1999). Plant community dynamics in a calcareous grassland under climate change manipulations. *Plant Ecology*, 143, 29-37.

Stöcklin, J., K. Schweizer & C. Körner (1998). Effects of elevated CO_2 and phosphorus addition on productivity and community composition of intact monoliths from calcareous grassland. *Oecologia*, 116, 50-56.

Suter, D., M. Frehner, B.U. Fischer, J. Nösberger & A. Lüscher (2002). Elevated CO_2 increases carbon allocation to the roots of *Lolium perenne* under free-air CO_2 enrichment but not in a controlled environment. *New Phytologist*, 154, 65-75.

Teyssonneyre, F., C. Picon-Couchard, R. Falcimagne & J.F. Soussana (2002). Effects of elevated CO_2 and cutting frequency on plant community structure in a temperate grassland. *Global Change Biology*, 8, 1034-1046.

Theurillat, J.P. & A. Guisan (2001). Potential impact of climate change on vegetation in the European Alps: a review. *Climatic Change*, 50, 77-109.

Thornley, J.H.M. & M.G.R. Cannell (1997). Temperate grassland responses to climate change: an analysis using the Hurley pasture model. *Annals of Botany*, 80, 205-221.

Van Kessel, C., W.R. Horwath, U.A. Hartwig, D. Harris & A. Lüscher (2000a). Net soil carbon input under ambient and elevated CO_2 concentrations: isotopic evidence after four years. *Global Change Biology*, 6, 435-444.

Van Kessel, C., J. Nitschelm, W.R. Horwath, D. Harris, F. Walley, A. Lüscher & U.A. Hartwig (2000b). Carbon-13 input and turnover in a pasture soil exposed to long-term elevated atmospheric CO_2. *Global Change Biology*, 6, 123-135.

White, T.A., B.D. Campbell, P.D. Kemp & C.L. Hunt (2000). Sensitivity of three grassland communities to simulated extreme temperature and rainfall events. *Global Change Biology*, 6, 671-684.

White, T.A., B.D. Campbell, P.D. Kemp & C.L. Hunt (2001). Impacts of extreme climatic events on competition during grassland invasions. *Global Change Biology*, 7, 1-13.

Winslow, J.C., E.R. Hunt & S.C. Piper (2003). The influence of seasonal water availability on global C3 versus C4 grassland biomass and its implications for climate change research. *Ecological Modelling*, 163, 153-173.

Wright, E., J. Connolly & A. Lüscher (2005). Catch-up in response to elevated CO_2 - a study of genotypes of 12 grassland species. In: *Proceedings of the XX International Grassland Congress*, Dublin, offered paper (in press).

Zanetti, S., U.A. Hartwig, A. Lüscher, T. Hebeisen, M. Frehner, B.U. Fischer, G.R. Hendrey, H. Blum & J. Nösberger (1996). Stimulation of symbiotic N_2 fixation in *Trifolium repens* L. under elevated atmospheric pCO_2 in a grassland ecosystem. *Plant Physiology*, 112, 575-583.

Zanetti, S., U.A. Hartwig, C. Van Kessel, A. Lüscher, T. Hebeisen, M. Frehner, B.U. Fischer, G.R. Hendrey, H. Blum & J. Nösberger (1997). Does nitrogen nutrition restrict the CO_2 response of fertile grassland lacking legumes? *Oecologia*, 112, 17-25.

Zavaleta, E.S., B.D. Thomas, N.R. Chiariello, G.P. Asner, M.R. Shaw & C.B. Field (2003). Plants reverse warming effect on ecosystem water balance. In: *Proceedings of the National Academy of Sciences USA*, 100, 9892-9893.

Ziska, L.H., C.F. Morris & E.W. Goins (2004). Quantitative and qualitative evaluation of selected wheat varieties released since 1903 to increasing atmospheric carbon dioxide: can yield sensitivity to carbon dioxide be a factor in wheat performance? *Global Change Biology*, 10, 1810-1819.

Grazing land contributions to carbon sequestration

R.F. Follett[1] and G.E. Schuman[2]
[1]*Soil Plant Nutrient Research, USDA-ARS, 2150 Centre Av., Bldg D, Ste. 100, Fort Collins, CO 80526, USA*
Email: Ronald.Follett@ars.usda.gov
[2]*Rangeland Resources Research, USDA-ARS, 8408 Hildreth Rd, Cheyenne, WY 82009, USA*

Key points

1. Grazing management can be used to increase soil organic carbon sequestration.
2. Grazing land soils contain large amounts of carbon with depth, and can store it for centuries.
3. Policies to encourage terrestrial carbon sequestration through conservation and good management of grazing lands are critical for many countries and the world.

Keywords: soil, rangeland, pasture, climate, policies

Introduction

The objectives of this paper are to provide a review of: (a) the influence of climatic factors and management practices on rates of carbon (C) accumulation and on long-term sequestration; (b) the potential contribution that grazing lands can make to C sequestration and soil C storage; and (c) policy issues and a current perspective of grazing lands in carbon credit trading. This paper will provide information about soil C accumulation and its long-term retention and as well as the influence of climate and management.

The term 'grazing lands' refers more to a set of highly diverse land resources than it does to a land use. Grazing lands include humid pastures that generally have more favourable climatic conditions and a greater potential to respond to management inputs. Grazing land also includes 'rangelands' which often include arid to semi-arid grazing lands, savannas, and shrub lands that are less responsive to management inputs. The term 'carbon sequestration' refers to the long-term storage of C in the terrestrial biosphere, underground, or the oceans, so that the build up of carbon dioxide (CO_2) in the earth's atmosphere will reduce or slow (http://cdiac2.esd.ornl.gov/, 2004).

Management and climate effects on carbon sequestration

Knowledge of soil organic carbon (SOC) sequestration information from grazing lands is limited, and predicting their sequestration potential is complicated by several factors including: wide regional and yearly climate variation; complexity of plant communities that vary from monocultures to an array of species; the presence and proportions of N fixing plants; type, species, and numbers of grazing animals; and management intensity and inputs (i.e. grazing and or mowing frequency, fertilising, and use of soil amendments).

Follett *et al.* (2001a) estimated that improving pastures in the US through fertiliser and/or manure application, liming, or planting improved forage species and proper grazing management, could result in the sequestration of 10.5 to 34.3 million metric tons (MMt) C/yr. These authors also predicted that improved rangeland management in the US could result in the additional sequestration of 5.4 to 16.0 MMt C/yr. A meta-analysis of 115 studies in pastures and other grazing lands worldwide (Conant *et al.*, 2001), indicated that soil carbon levels increased with improved management (primarily fertilisation, grazing management, and conversion from cultivation or native vegetation) in 74% of the studies considered, and that the greatest C sequestration occurred during the first 40 years following implementation of the management practice. With exception of a single irrigation study, the meta-analysis indicated that conversion of cultivated land to grazing land resulted in an average increase in SOC of 3-5%/yr (Conant et al., 2001). Soil C sequestration is generally greater under grazed pastures than under hayed pastures (Franzluebbers et al., 2000a). To increase SOC sequestration on rangelands generally requires improved grazing management, introduction of legumes, and control of undesirable species.

Grazing facilitates litter decomposition through the effects of grazing and animal traffic (Schuman et al., 1999). Removal of excess standing dead material by grazing increases the onset of spring growth and photosynthesis by enhancing sunlight penetration, and soil warming (LeCain et al., 2000). Good grazing management enhances rangeland productivity and maintains healthier rangelands (Schuman et al., 1999). Despite inherently low SOC sequestration rates, improved management of the world's extensive grazing lands provide a large C sequestration potential. Accurately measuring quantities of SOC in grazed pastures however is complicated by high spatial variability of the soil and vegetation, extensive land area, and the disproportionate redistribution of the nutrients in the dung and urine by grazing animals (Follett & Wilkinson, 1995; Franzluebbers et al., 2000b; West et al., 1989).

Soil management factors that improve plant productivity also contribute to increased SOC sequestration. Liming and the elimination of P deficiency have been shown to increase SOC (Ridley et al., 1990; Haynes & Williams, 1992). However, despite most rangelands being N-deficient and responding to N additions with increased production and water use efficiency, SOC often does not increase with increased N applications. The CCGRASS model (Van den Pol-van Dasselaar & Lantinga, 1995) predicts that low to moderate N applications lead to larger increases in SOC. Large N applications can result in less root production (Schnabel et al., 2001). Warm (C4) and cool (C3) season grasses often respond differently to N. In more humid climates, warm-season grasses produce more biomass and deposit more organic C than cool-season grasses with no or moderate N levels. Little difference is observed between the warm and cool season grasses with high fertility levels (Stout 1992; Stout & Jung, 1992, 1995; Wedin & Tilman, 1996). However, in more arid/semiarid climates, cool-season species respond to low N additions better than do the warm-season species. Adding legumes to grazing lands can increase N availability and plant biomass production, but SOC increases may be smaller than expected due to production of low C:N organic matter which is more readily decomposed (Schnabel et al., 2001). This response is influenced by the climatic regime of the region. Mortenson et al. (2004) found that Medicago sativa ssp. falcata interseeded into rangelands in the US Northern Great Plains significantly increased SOC sequestration and soil N levels.

A review of the literature and evaluation of SOC status and dynamics in grazing land soils of the eastern US led Schnabel et al. (2001) to an evaluation of the general effects of various conditions and pasture management practices on SOC sequestration. The magnitude and duration of management effects shown in Table 1 depend on several factors such as climate, soils, previous management or potential net primary productivity (NPP). Converting marginal cropland to pastureland will increase SOC. Changes in how animals, plants, and soils are managed can also affect the balance between C inputs to the soil via plant fixation and losses of SOC to the atmosphere via decomposition. Where pasturelands are highly productive and SOC is already high, small or no increases in C storage may be expected. Larger increases can be made on marginally productive pasturelands by improving soil fertility or animal management to enhance plant productivity (Table 1).

Table 1 The effects of pasture management methods on C storage in soil of eastern US pasturelands (Schnabel *et al.*, 2001)

Factor	Measured effect on pastureland	Measured or inferred effect on C storage in soil
Animal management		
Grazing lands	More C returned to soil for rapid incorporation.	Increase SOC[1].
Intensive grazing	With adequate moisture, intensive management increases NPP[2], increased foot traffic breaks down residue.	Increase SOC.
	With limited moisture, increased stocking can damage stands.	Decrease SOC.
Forage management		
Replacing C3 grasses with C4 grasses	At low to moderate fertility, increase NPP and reduce forage quality.	Increase SOC.
	At high fertility, little change in NPP.	Little change in SOC. May not be sustainable.
Replace endophyte infected fescue with uninfected fescue	Increase forage quality.	Decrease SOC.
Increase harvest frequency	Reduce NPP, increase forage quality.	Decrease SOC.
Delay harvest or grazing	Reduce forage quality.	Increase SOC.
Soil Management		
Liming	Increases P availability and NPP.	Increases SOC.
P fertilisation	If P deficient, increase NPP.	Increase SOC.
	If P is adequate or in excess, no change.	No change.
N fertilisation	Low inherent fertility, increase NPP and forage quality.	Increase SOC.
	High inherent fertility; NPP, and decomposition of SOC, no change or increase.	No change, decrease, or increase in SOC, depending on relative change in NPP and decomposition.
Manuring	Increases NPP if fertility limits growth.	Increases SOC.
Drainage	Increases NPP, increases SOC decomposition.	Decreases SOC.

[1]SOC = Soil organic carbon; [2]NPP = Net primary productivity

Information from US Great Plains rangelands indicates that stocking rates can impact on soil characteristics and plant species composition and, in turn, SOC storage (Bauer *et al.*, 1987; Frank *et al.*, 1995; Schuman *et al.*, 1999; Potter *et al.*, 2001). Schuman & Derner (2004) recently summarized the current state of knowledge about the effects of management practices on SOC sequestration in rangelands (Table 2).

Grazing of semi-arid rangelands generally benefits or has a neutral effect upon SOC sequestration in both shortgrass steppe and mixed-grass prairie regions. However, some grasslands in the arid southwestern US, such as *Bouteloua eriopoda* (Torr.) Torr. (black grama) dominated communities, do not tolerate grazing disturbance and when lost will not return to their previous condition even with human intervention (Kephart *et al.*, 1995).

Northern mixed-grass rangelands in Wyoming exhibited plant community composition shifts toward warm-season (C4) species when season-long heavy stocking rates were imposed

(Manley *et al.*, 1995; Schuman *et al.*, 1999). Aboveground biomass also decreased after 10 years in those pastures grazed at heavy stocking rates, mainly as a result of the loss of the more productive cool-season (C3) grasses. Increases in SOC can result from this plant community shift (Reeder & Schuman 2002; Smoliak *et al.*, 1972; Frank *et al.*, 1995) because of the more dense and shallow root system of the dominant C4 species *Bouteloua gracilis*. Interseeding of *Medicago sativa* spp. *falcata* (yellow flowered alfalfa) into northern mixed-grass rangeland, significantly increased SOC sequestration (Mortensen *et al.*, 2004). Fire, a naturally occurring ecological process in tallgrass prairie, controls the spread of woody species and benefits SOC sequestration (Rice, 2000; Sampson & Scholes, 2000).

Table 2 Rangeland management effects on soil carbon sequestration rates

Management	Ecosystem	Soil C sequestration	Location	Citation
Grazing	Shortgrass steppe.	0.07-0.12 Mt C/ha per year.	Colorado	(1)
	Northern mixed-grass prairie.	0.30 Mt C/ha per year.	Wyoming N. Dakota	(2, 3, 4)
	Southern mixed-grass prairie.	No change in C.	Oklahoma	(5)
	Canadian prairie.	Higher SOC[1] when grazed.	Canada	(6)
Nitrogen inputs	N-fertilization of tall grass prairie.	1.6 Mt C/ha per year.	Kansas	(7)
	N and S fertilization of northern prairie.	Increases of 0.45-0.72 Mt C/ha per year.	Saskatchewan	(8)
	Legume interseeded mixed-grass prairie.	0.33-1.56 Mt C/ha per year.	South Dakota	(9)
Fire	Tall grass prairie.	0.22 Mt C/ha per year.	Kansas	(7)
Restoration of degraded lands	Southern mixed-grass prairie.	Moderate grazing, no change; Heavy grazing, 65% decrease.	Oklahoma	(5)
	Marginal cropland to grazing land.	Restored soil C to 80% of native rangeland in100 years.	Sudan	(10)
	Restored semiarid savanna.	Increases of 1.9-2.75 Mt C/ha per year.	Argentina	(11)
	Southern tall grass prairie.	Avg. = 0.45 Mt C/ha per year.	Texas	(12)
	US Great Plains (CRP[2]). Introduced grasses.	0.8-1.1 Mt C/ha per year.	Texas, Kansas, Nebraska, N. Dakota	(13, 14)
	Mined land reclamation.	1.95 Mt C/ha per year.	Wyoming	(15)
Woody plant encroachment	Southern mixed-grass prairie.	Remove *Prosopis glandulosa* but no affect on soil C.	Texas	(16)
	Subtropical savanna.	0.23 Mt C/ha per year under woody plants (model predictions).	Texas	(17)
	Mesquite-acacia savanna.	0.14 Mt C/ha per year.	Texas	(18)

[1]SOC = Soil organic carbon; [2]CRP = Conservation Reserve Program
(1) Derner *et al.,* 1997; (2) Reeder & Schuman, 2002; (3) Schuman *et al.*, 1999; (4) Frank, 2004; (5) Fuhlendorf *et al.*, 2002; (6) Henderson *et al.*, 2004; (7) Rice, 2000; (8) Nyborg *et al.*, 1994; (9) Mortensen *et al.*, 2004; (10) Olsson & Ardo, 2002; (11) Abril & Bucher, 2001; (12) Potter *et al.*, 1999; (13) Gebhart *et al.*, 1994; (14) Follett *et al.*, 2001b; (15) Stahl *et al.*, 2003; (16) Teague *et al.*, 1999; (17) Hibbard *et al.*, 2001; (18) Liao, 2004.

Degraded rangelands can lose significant amounts of SOC, but in some ecosystems have substantial potential (with proper management) to sequester SOC to replace that lost. The C4-dominated grazing lands/savannas of southern Texas have been transformed to 'thorn woodlands' as tree and shrub abundance (primarily *Prosopis glandulosa* 'honey mesquite') has markedly increased in the past century (Liao, 2004). Model predictions estimate a 2-fold increase in SOC in the top 10 cm of the soil profile beneath clusters of woody plants (23.5 Mt C/ha) compared to open herbaceous vegetation (11.7 Mt C/ha) (Hibbard, *et al.*, 2001). They estimated a soil C sequestration rate of 0.23 Mt C/ha/yr for the woody plants. Field research addressing SOC storage following woody plant invasion in the Rio Grande Plains of southern Texas, generally corroborates the above model predictions (soil organic C to a depth of 30 cm was >50 Mt C/ha for woody plants and about 20 Mt C/ha in grasslands) (Liao, 2004).

Large stocks of SOC are present in the historic grazing land soils in the US Great Plains. Reseeding with grasses as a permanent cover on previously cropped soil, is carried out on an extensive basis in the US, in response to the Conservation Reserve Program (CRP), run by the US Department of Agriculture (Food Security Act of 1985 (P.L. 98-198)). The CRP is a voluntary program where landowners enrol in contracts to take vulnerable croplands or environmentally sensitive lands out of agricultural production. The current cap is 15.9 M ha, with approximately 13.6 M ha of this land currently planted to grasses.

Follett *et al.* (2001b) compared cropped and CRP land across the mid-continental US region (extending from near the Canadian border into Texas), representing approximately 5.6 M ha. The sites chosen were fields that had been reseeded as part of the CRP programme for a minimum of 5 years (average 7.9 years), and were paired with adjacent fields that had been in long-term cropping. A rapid rebuilding of the near-surface soil C stocks was observed (Table 3), averaging 570, 740, and 910 kg C/ha per year for the 0-to 5, 0-to 10, and 0-to 20 cm depth increments respectively. However, they did not reach the level of C stocks observed in paired native grass fields. In addition, research under turf grass by Qian & Follett (2002), indicates that total C sequestration increased at a nearly constant rate for 30 years.

Table 3 Average weight at different depths of soil organic carbon (Mt/ha) across 14 sites in 9 Great Plains states in the US. Reseeded conservation reserve programme sites were in permanent grass cover for at least 5 (average of 7.9) years (Follett *et al.*, 2001b)

Depth (cm)	Mt/ha		
	Native	Reseeded	Cropped
0-5	16.7	13.0	9.0
0-10	29.0	23.2	18.3
0-20	45.0	39.7	32.2
0-30	60.2	52.8	45.0
0-60	84.9	75.8	67.8
0-100	91.4	89.8	81.0

Sequestered SOC can if undisturbed, remain in the soil for centuries. Data in Figure 1 is for three native prairie sites along a 1300 km north to south transect of the US Great Plains where SOC was ^{14}C dated (Follett *et al.*, 2004). These data show that during the Holocene (beginning ~ 12,000 calendar years before present (YBP)), except where eolian disturbance had resulted in over- and under- layering of soil horizons, mean residence time (MRT) of SOC in the soil increased, but its concentration decreased with depth. Even though it decreased with depth, substantial amounts of SOC still remained, even after thousands of years.

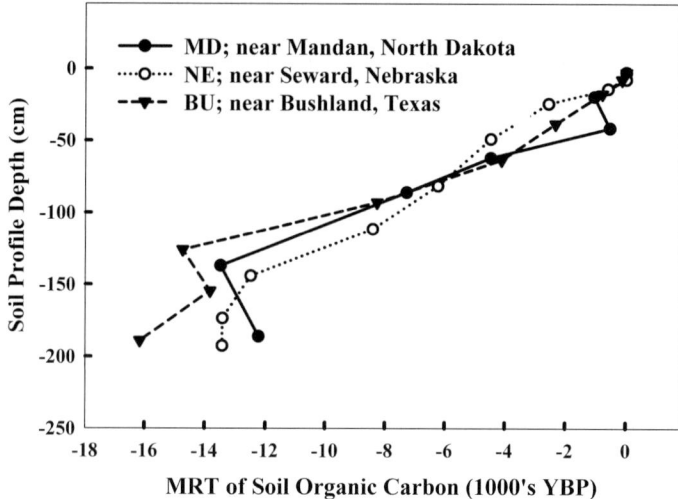

Figure 1 Mean residence time (MRT) of the soil organic carbon (SOC) with soil profile depth across a north to south transect of the US Great Plain from the State of North Dakota to the State of Texas

Potential contribution of world grazing lands to soil C sequestration and storage

This paper aims to present a worldwide view of the topic of SOC sequestration in grazing lands. However, the availability of such information around the world is quite limited, particularly in less-developed countries. Nonetheless, much information and insight can be obtained from FAO (2004) for the various regions of the world. In contrast to the previous discussion of pastures and rangelands, FAO describes 'permanent pasture' (Table 4) as land used permanently (≥ 5 years) for herbaceous forage crops, either cultivated or growing wild (wild prairie or grazing land). Permanent pasture and the category 'forests and woodland'; is vague - especially for shrub land and savanna which may be reported under either category. The National Institute of Public Health and the Environment (RIVM) in the Netherlands grouped 238 countries into 19 regions (Kreileman et al., 1998) that are used in the following discussion. The reader is referred to this report for countries included in each region, but RIVM region 19 (Antarctica) is excluded (Table 4 and 5).

Conant et al. (2001) reviewed results from 115 studies in 17 countries on the effects on SOC, of grazing management and conversion into grazing land. Evidence that SOC sequestration under grasslands can be enhanced by fertilisation, improved grazing management and grass species, conversion of cultivated land to permanent grass cover, presence of legumes, introduction of earthworm, and irrigation was identified (Conant et al., 2001; Follett et al., 2001a; Schuman & Derner, 2004). Although many important grassland areas were not represented, reported rates of sequestration ranged from –0.2 to +3.0 Mt C/ha per year.

Table 4 Agricultural land and pasture area, percent pasture to agricultural land area, N-fertiliser consumption, and N-consumption per unit of agricultural land area for World RIVM regions (Kreileman *et al.*, 1998)

RIVM No.	RIVM Region	Agricultural land area (1000 ha)	Permanent pasture (1000 ha)	Pastr/agric land area (%)	N-fertiliser consumption (t)	N-consumption/ Ag. land area (kg/ha)	Region/ World pasture area (%)
	World	5,011,700	3,488,120	69.6	80,948,904	16.2	-
1	Canada	74,700	29,000	38.8	1,564,348	20.9	0.8
2	United States	412,880	234,000	56.7	10,464,065	25.3	6.7
3	Central America	144,481	98,915	68.5	1,920,422	13.3	2.8
4	South America	639,388	513,950	80.4	3,246,436	5.1	14.7
5	North Africa	118,639	75,210	63.4	1,419,116	12.0	2.2
6	West Africa	143,964	103,663	72.0	124,840	0.9	3.0
7	East Africa	287,736	240,373	83.5	209,314	0.7	6.9
8	Southern Africa	373,462	332,778	89.1	580,743	1.6	9.5
9	OECD Europe	145,915	59,614	40.9	9,328,614	63.9	1.7
10	Eastern Europe	65,775	19,540	29.7	2,300,168	35.0	0.6
11	Former USSR	568,801	360,309	63.4	2,553,327	4.5	10.3
12	Middle East	318,898	257,400	80.7	3,089,686	9.7	7.4
13	South Asia	262,540	49,123	18.7	14,394,896	54.8	1.4
14	East Asia	683,951	529,399	77.4	22,719,085	33.2	15.2
15	Southeast Asia	110,409	17,032	15.4	5,222,997	47.3	0.5
16	Oceania	475,117	419,470	88.3	1,192,868	2.5	12.0
17	Japan	5,235	405	7.7	487,400	93.1	0.0
18	Greenland	235	235	100.0	-	0.0	0.0

Data obtained from FAO (2004).

Table 5 Total meat, meat from all grazing livestock, and meat from only sheep + goat + beef + buffalo (grazing animals) in metric tons (Mt) and as percentages for the World and within 18 RIVM regions (Kreileman *et al.*, 1998)

RIVM No.	RIVM Region	Total meat produced (Mt)	Meat from grazing livestock (Mt)	Meat from beef, sheep, goat and buffalo (Mt)	Meat from beef, sheep, goat and buffalo/meat from ALL grazing livestock (%)	Meat from beef, sheep, goat and buffalo/TOTAL meat (%)
	World	233,962,789	73,557,786	71,220,329	96.8	30.4
1	Canada	3,999,816	1,293,298	1,275,298	98.6	31.9
2	United States	37,640,452	12,627,850	12,404,100	98.2	33.0
3	Central America	6,359,620	2,161,055	2,070,839	95.8	32.6
4	South America	25,064,433	12,323,472	12,156,423	98.6	48.5
5	North Africa	2,906,326	1,499,308	1,412,589	94.2	48.6
6	West Africa	1,237,310	819,796	620,711	75.7	50.2
7	East Africa	2,517,863	1,944,790	1,745,619	89.8	69.3
8	Southern Africa	2,809,418	1,582,838	1,466,038	92.6	52.2
9	OECD Europe	36,688,554	9,066,084	8,840,280	97.5	24.1
10	Eastern Europe	7,997,711	1,320,213	1,281,061	97.0	16.0
11	Former USSR	8,827,797	4,601,328	4,510,872	98.0	51.1
12	Middle East	5,375,214	2,352,198	2,271,414	96.6	42.3
13	South Asia	7,867,328	5,510,959	5,510,959	100.0	70.0
14	East Asia	65,303,182	9,115,466	8,668,966	95.1	13.3
15	Southeast Asia	9,250,641	1,438,898	1,433,864	99.6	15.5
16	Oceania	5,111,619	3,876,097	3,807,027	98.2	74.5
17	Japan	3,005,930	537,709	530,705	98.7	17.7
18	Greenland	608	402	402	100.0	66.1

Data obtained from FAO (2004).

Total meat production from domestic livestock; meat from grazing livestock (sheep, goat, beef, buffalo, game animal, horse, ass, mule, camel, and other camelid); and meat only from sheep, goat, beef, and buffalo for the year 2000 are shown in Table 5. Sheep, goat, beef, and buffalo are the dominant grazing animals from which the world's meat is produced and account for 76 - 100% of grazing livestock meat, and between 13 and 69% of all meat production. Production of sheep, goat, beef, and buffalo ('grazing animal') meat per unit area of permanent pasture should provide a proxy of pasture productivity as influenced by climatic potential, forage productivity, and the husbandry skills of the indigenous grazing animal operations by RIVM region. A general measure of production improvement in RIVM regions may be indicative of N-fertiliser use per unit area of agricultural land in those regions.

The regression of grazing animal meat production on N-fertiliser use produces an r^2 of 0.88 (Figure 2). Japan (RIVM region 17) is not included in the analysis because of its high rate of N consumption per unit of agricultural land, and high rate of grazing animal meat production per unit of permanent pasture (Tables 4 and 5). Regions are identified (Figure 2) wherein grazing livestock meat production was >20 kg/ha, representing conditions (climate, soils, grazing management, and other management inputs) potentially favourable for SOC sequestration. The area of permanent pasture in these ten RIVM regions (1, 2, 3, 4, 5, 9, 10, 13, 14, and 15) includes that from 119 countries (FAO, 2004) and represents approximately 1,625 million ha (46.6%) of the world's permanent pasture land.

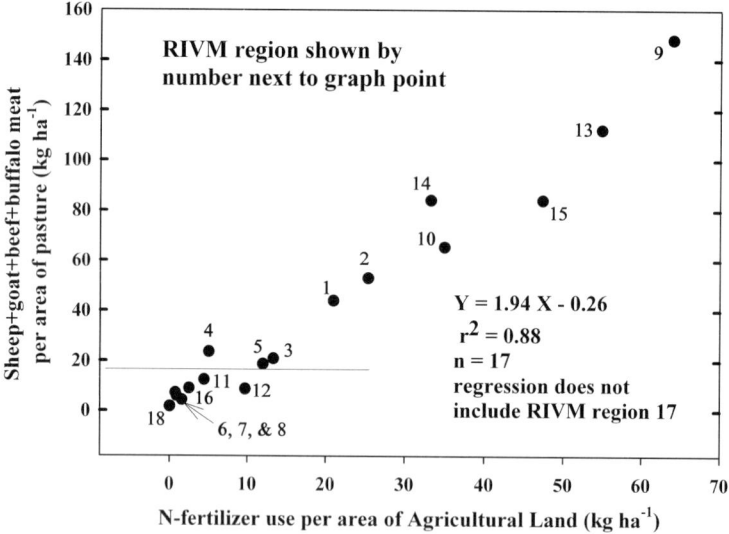

Figure 2 Meat production from grazing animals (sheep, goat, beef, and buffalo) per unit area of permanent pasture (FAO, 2004) as a function of N-fertilizer use per unit area of agricultural land

Within an individual RIVM region, there can be a wide diversity of climate, soils, and other factors. Based upon grazing animal meat production, it might be assumed that forage production per unit of pasture area would be higher with improved inputs and grazing management, which would increase the potential SOC sequestration rate. For example, in RIVM 3 (Central America), N-fertiliser consumption/ha of agricultural land and, grazing animal meat production/ha are lower than in many other RIVM regions (Figure 2, Tables 4 and 5). However, there are a number of forages that, when introduced and managed, produce high forage yields with high digestibility. It has been known for a long time (Vicente-Chandler et al., 1964) that the potential for forage, and hence cattle production in vast areas of the humid tropics with year-round warm weather, high rainfall, and deep porous soils is not being achieved. Vicente-Chandler et al. (1964, 1983) in Puerto Rico increased yields of dry forage from approximately 5.6 Mt/ha of poor quality forage to 45 Mt/ha of excellent-quality forage, by using improved grasses, heavy fertilization, liming, weed control, intensive utilisation systems, and careful management. The year-round carrying capacity reported by (Vicente-Chandler et al., 1983) was 10 cows/ha and 5 cattle/ha (average weight 270kg). Development of such systems with improved fertility and plant species has the potential to provide high amounts of meat from grazing animals and yet to conserve soil and water resources. Vicente-Chandler et al. (1983) considered that their techniques, and the potential of the system for forage and grazing were "applicable to similar vast humid tropical areas of South America, Africa, southeast Asia, Borneo, Australia, and the Antilles." Even though forage and beef production were high in these studies, soil organic matter content decreased from about 6.7% from the virgin forest soil, to 3.7% after 15 years under unfertilised volunteer pasture (Vicente-Chandler et al., 1964; 1983). Unfortunately, no measures of rates of SOC change under the high-fertility, well managed and grazed pastures were reported.

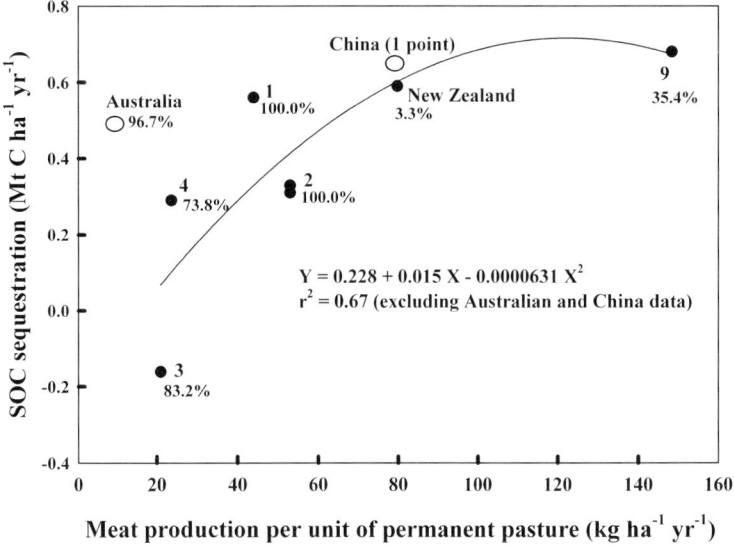

Figure 3 Average rates of SOC sequestration for countries in RIVM regions reported by Conant et al. (2001) as a function of meat production per unit of permanent pasture calculated from FAO (2004) data

Conant et al. (2001) reviewed research literature, and reported rates of SOC sequestration from individual countries that were located in RIVM regions 1, 2, 3, 4, 9, 14, and 16. In addition, Follett *et al.* (2001a) obtained estimates of C sequestration rates in US grazing lands (RIVM 2). Figure 3 shows the average rate of SOC sequestration from Conant *et al.* (2001) and the corresponding RIVM region number, plotted against meat production from grazing animals per unit of permanent pasture (FAO 2004). The % shown near the region number is the amount of pasture area within the RIVM region that is represented by the countries for which Conant *et al.* (2001) had data. For the USA (RIVM 2), Conant's data is shown as the lower point and the data for Follett *et al.* (2001a) the slightly higher upper point. Only four data points (from China) were available for RIVM 14, of which 3 were deep cores that could be expected to mask near surface C sequestration rates, and one point that lay near the regression line, Figure 3. As so few points exist for RIVM 14, they were not included in the regression. Australia and New Zealand comprise RIVM 16. The Australian data showed reasonably high rates of SOC sequestration, but meat production per unit of permanent pasture was low (indicating a breakdown in the relationship), possibly because reported experimental C sequestration data was from productive pastures that were not representative of the large dry-climate areas of Australian permanent pasture reported by FAO (2004). In contrast, New Zealand data with wetter-climate pasture areas was consistent with the relationship (Figure 3) and was included.

The quadratic regression equation ($r^2 = 0.67$) that is shown in Figure 3 does not include the Australian or Chinese data. Also, this equation does not necessarily represent sequestration rates in *all* countries and regions of the world. However, the world's rate of grazing animal production is 0.02 Mt meat/ha per year, which if used in the quadratic equation in Figure 3 results in an estimate of the world's SOC sequestration rate on permanent pasture areas of 0.06 Mt C/ha per year. If multiplied by the world's area of permanent pasture (from Table 4), amounts to 0.2 Gt SOC sequestration/yr on 3.5 billion ha of permanent pasture. For comparison, the world's SOC sequestration potential is estimated to be 0.01 to 0.3 Gt SOC/yr on 3.7 billion ha of permanent pasture (Lal, 2004). Thus SOC sequestration by the world's permanent pastures can potentially offset up to 4% of global greenhouse emissions.

Policies

Broadly speaking, possible strategies to lower or control atmospheric carbon dioxide (CO_2) concentration might include reduction in; fossil fuel use, 'separation and capture', deep ocean sequestration, injection into oil and gas geologic formations where the resource has been removed, or advanced biological processes for atmospheric CO_2 removal. Associated policies for CO_2 reduction programmes may include regulation. Some of the above strategies are currently cost prohibitive or may have declining cost:benefit relations. Markets trading in environmental commodities have existed since 2002, especially in the European Union countries. Such markets are now developing in the US, but the low value of C (< $1 Mt CO_2) results in it being mostly an experimental market that has the potential to increase if future caps or regulation were to occur (Lal, 2004). In the mean time, international negotiations and agreements, such as the 'Framework Convention on Climate Change (1994)', the Kyoto Protocol and the World Trade Organization will continue to influence the direction of international policies and agreements.

Policies to encourage SOC sequestration by agriculture (including grassland soils) have several advantages for atmospheric CO_2 reduction. Chief amongst these are that they can be implemented quickly with existing technology, and with minimal impact on the economic system. It also has high co-benefits for the environment such as improved soil quality, soil erosion reduction, and improved water quality. An additional attraction is as a potential source of income to agricultural producers, especially if the economic value of the sequestered SOC had a modest trading cost of $20/Mt (Lal, *et al.*, 2003). This higher value would rapidly stimulate the market and the rate at which co-benefits to the environment would be realized. However, these co-benefits and the amounts of SOC that can be sequestered with better management are likely to be greater for cultivated soils than for grazing lands (Lal, 2004). Sperow *et al.* (2003) estimate that US cropland soils could sequester 60-70 million Mt C/yr, and Follett *et al.* (2001a) estimate an average sequestration rate of 54 million Mt C/yr for USA grasslands. Policies and programs that encourage conservation of current grazing lands and conversion of marginally productive croplands to perennial vegetation are critical factors that can contribute to the sequestration of the potentially large amounts of SOC stored by these ecosystems.

Finally, it is highly probable that greenhouse gas emission policies in the US will differ from those of countries who agree to the Kyoto protocol. However, the US will encourage reduction in atmospheric CO_2 levels. Policy initiatives such as 'Clear Skies' are designed to reduce emissions of sulphur dioxide, mercury and nitrogen oxides; allow 'Cap and trade - Market-based approaches'; and reduce greenhouse gas intensity. Rates of greenhouse gas (GHG) emission and atmospheric CO_2 reduction will probably be market driven. Programmes within the US Department of Agriculture (USDA) will likely be implemented under USDA's conservation programmes, and will result in the development of methods to estimate sources and sinks of agricultural and forestry GHGs. These programmes will support voluntary agreements with the private sector and support development of technologies and practices that enhance SOC sequestration. An example is the Environmental Quality Incentive Programme (EQIP) of USDA which is a cost-share programme to promote resource conservation on working farms, and allows USDA to provide national guidance to make GHGs a priority resource concern, and which has the goal of reducing GHG emissions by 9.1 million Mt carbon equivalent by 2012 (Brown & Shafer, 2004).

Conclusions

On a global basis, grazing lands sequester substantial amounts of SOC annually. With improved management, the potential to sequester C on many of these lands can be greatly enhanced. Improved soil C sequestration on grasslands is an important strategy to assist in mitigating the greenhouse gas effect. Policies and programmes that encourage the conservation of current grazing lands and conversion of marginally productive croplands to perennial vegetation are critical to the implementation of strategies to sequester soil C. Additionally, efforts to increase SOC sequestration in grasslands worldwide will have important co-benefits to the environment.

References

Abril, A., & E.H. Bucher (2001). Overgrazing and soil carbon dynamics in the western Chaco of Argentina. *Applied Soil Ecology*, 16, 243-249.
Bauer, A., C.V. Cole & A.L. Black (1987). Soil property comparisons in virgin grazing lands between grazed and nongrazed management systems. *Soil Science Society of America Journal*, 51, 176-182.
Brown, J.R. & S. Shafer (2004). Mitigating global climate change through terrestrial carbon sequestration. In: Proceedings Carbon Sequestration: science, policy and marketing in Wyoming, June 22-23, Casper, WY, Wyoming Carbon Sequestration Advisory Committee, Cheyenne, WY, (CD Power Point Proceedings).
Conant, R.T., K. Paustian & E.T. Elliott (2001). Grazing land management and conversion into grazing land: effects on soil carbon. *Ecological Application*, 11, 343-355.
Derner, J.D., D.D. Briske & T.W. Boutton (1997). Does grazing mediate soil carbon and nitrogen accumulation beneath C4 perennial grasses along an environmental gradient? *Plant and Soil*, 191, 147-156.

FAO (2004). FAO Statistical Databases; Agriculture. <http://apps.fao.org/default.jsp>

Follett, R.F., J.M. Kimble & R. Lal (2001a). The potential of U.S. grazing lands to sequester soil carbon. In: R.F. Follett, J.M. Kimble & R. Lal (eds.) The potential of U.S. Grazing lands to sequester carbon and mitigate the greenhouse effect. Lewis Publishers, Boca Raton, FL, 401-430.

Follett, R.F., J.M. Kimble, S. Leavitt & E. Pruessner (2004). The potential use of soil C isotope analyses to evaluate paleoclimate. *Soil Science*, 169, 471-488.

Follett, R.F., S.E. Samson-Liebig, J.M. Kimble, E.G. Pruessner & S.W. Waltman (2001b). Carbon sequestration under the conservation reserve program in the historic grazing land soils of the United States of America. In: R. Lal (ed.) Soil carbon sequestration and the greenhouse effect. *Soil Science Society of America*, Madison, WI, 57, 27–40.

Follett, R.F.& S.R. Wilkinson (1995). Nutrient management of forages. In R.F. Barnes, D.A. Miller & C.J. Nelson (eds.) Forages: the science of grazing land agriculture. Vol II, IA State University Press, Ames, IA, 55-82.

Frank, A.B. (2004). Six years of CO2 flux measurements for moderately grazed mixed-grass prairie. *Environmental Management,* 33 (Supplement), (In press).

Frank, A.B., D.L. Tanaka, L. Hofmann & R.F. Follett (1995). Soil carbon and nitrogen of Northern Great Plains grazing lands as influenced by long-term grazing. *Journal of Range Management*, 48, 470-474.

Franzluebbers, A.J., J.A. Stuedeman & H.H. Schomberg (2000a). Soil organic C and N pools under long-term pasture management in the Southern Piedmont, USA. *Soil Biology Biochemistry*, 32, 469-478.

Franzluebbers, A.J., J.A. Stuedeman & H.H. Schomberg (2000b). Spatial distribution of soil carbon and nitrogen pools under grazed tall fescue. *Soil Science Society of America Journal*, 64, 635-639.

Fuhlendorf, S.D., H. Zhang, T.R. Tunnell, D.M. Engle & A.F. Cross (2002). Effects of grazing on restoration of southern mixed prairie soils. *Restoration Ecology*, 10, 401-407.

Gebhart, D.L., H.B. Johnson, H.S. Mayeux & H.W. Polley (1994). The CRP increases soil organic carbon. *Journal of Soil and Water Conservation,* 49, 374-377.

Haynes, R.J. & P.H. Williams (1992). Accumulations of soil organic matter and the forms, mineralisation potential and plant-availability of accumulated organic sulphur: effects of pasture improvement and intensive cultivation. *Soil Biology and Biochemistry*, 24, 209-217.

Henderson, D.C., B.H. Ellert & M.A. Naeth (2004). Grazing and soil carbon along a gradient of Alberta rangelands. *Journal of Range Management*, 57, 402-410.

Hibbard, K.A., S. Archer, D.S. Schimel & D.W. Valentine (2001). Biogeochemical changes accompanying woody plant encroachment in a subtropical savannah. *Ecology*, 82, 1999-2011.

Kephart, K.D., C.P. West & D.A. Wedin (1995). Grazing land ecosystems and their improvement. In: R.F. Barnes, D.A. Miller & C.J. Nelson (eds.) Forages (V 1), 5th Edition, Iowa State University Press, Ames, IA, 141-153.

Kreileman, E. van, J. Woerden & J. Bakkes (1998). RIVM environmental research 1998 World Regions and Subregions. National Institute of Public Health and the Environment. Bilthoven, The Netherlands. CIM Report no MO25/98, 20p.

Lal, R. (2004). Soil carbon sequestration impacts on global climate change and food security. *Science*, 304, 1623-1627.

Lal, R., R.F. Follett & J.M. Kimble (2003). Achieving soil carbon sequestration in the United States: a challenge to policy makers. *Soil Science*, 168, 827-845.

LeCain, D.R., J.A. Morgan, G.E. Schuman, J.D. Reeder & R.H. Hart (2000). Carbon exchange of grazed and ungrazed pastures of a mixed grass prairie. *Journal of Range Management,* 53, 199-206.

Liao, J.D. (004). Woodland development and soil carbon and nitrogen dynamics and storage in a subtropical savanna ecosystem. Ph.D. Dissertation, Texas A&M University, College Station, TX.

Manley, J.T., G.E. Schuman, J.D, Reeder & R.H. Hart (1995). Rangeland soil carbon and nitrogen responses to grazing. *Journal of Soil and Water Conservation*, 50, 294-298.

Mortenson, M.C., G.E. Schuman & L.J. Ingram (2004). Carbon sequestration in rangelands interseeded with yellow-flowering alfalfa (Medicago sativa ssp. falcata). *Environmental Management*, 33 (Supplement), S475-S481.

Nyborg, M., E.D. Solberg & S.S. Malhi (1994). Soil C content under bromegrass increased by N and S fertilizer applications. In: Proceedings 31st Annual Alberta Soil Science Workshop, Edmonton, Alberta, 325-328.

Olsson, L. & J. Ardö (2002). Soil carbon sequestration in degraded semiarid agro-ecosystems – perils and potentials. *Ambio*, 31, 471-477.

Potter, K.N., J.A. Daniel, W. Altom & H.A. Torbert (2001). Stocking rate effect on soil carbon and nitrogen in degraded soils. *Journal of Soil Water Conservation*, 56, 233-236.

Potter, K.N., H.A. Torbert, H.B. Johnson & C.R. Tischler (1999). Carbon storage after long-term grass establishment on degraded soils. *Soil Science*, 164, 718-725.

Qian, Y.L. & R.F. Follett (2002). Assessing carbon sequestration in turf grass soil using long-term soil testing data. *Agronomy Journal*, 94, 930-935.

Reeder, J.D. & G.E. Schuman (2002). Influence of livestock grazing on C sequestration in semi-arid mixed-grass and short-grass rangelands. *Environmental Pollution*, 116, 457-463.

Rice, C.W. (2000). Soil organic C and N in rangeland soils under elevated CO2 and land management. In: Proceedings, advances in terrestrial ecosystem carbon inventory, measurements, and monitoring, October 3-5, 2000. USDA-ARS, USDA-FS, USDA-NRCS, U.S. Dept of Energy, NASA, and National Council for Air and Stream Improvement. Raleigh, NC, p 83pp.

Ridley, A.M., W.J. Slattery, K.R. Helyar & A. Cowling (1990). The importance of the carbon cycle to acidification of a grazed annual pasture. *Australian Journal of Experimental Agricultural Research*, 30, 529-537.

Sampson, R.N. & R.J. Scholes. (2000). Additional human-induced activities-article 3.4. In: R.T. Watson, I.R. Noble, B. Bolin, N.H. Ravindranath, D.J. Verardo & D.J. Dokken (eds.) Land use, land-use change and forestry. A Special Report of the IPCC, Intergovernmental Panel on Climate Change. Cambridge University Press, Cambridge, UK, 183-281.

Schnabel, R.R., A.J. Franzluebbers, W.L. Stout, M.A. Sanderson & J.A. Stuedeman (2001). Effects of pasture management practices. In: R.F. Follett, J.M. Kimble & R. Lal (eds.) The potential of U.S. grazing lands to sequester carbon and mitigate the greenhouse effect. Lewis Publishing, Boca Raton, FL, 291-322.

Schuman, G.E. & J.D. Derner (2004). Carbon sequestration by rangelands: management effects and potential. In: Proceedings, Western Regional Cooperative Soil Survey Conference, June 13-17, 2004, Jackson, WY. Natural Resources Conservation Service, USDA, Casper, WY, (in press).

Schuman, G.E., J.D. Reeder, J.T. Manley, R.H. Hart & W.A. Manley (1999). Impact of grazing management on the carbon and nitrogen balance of a mixed-grass rangeland. *Ecological Applications*, 9, 65-71.

Smoliak, S., J.F. Dormaar & A. Johnston (1972). Long-term grazing effects on *Stipa-Bouteloua* prairie soils. *Journal of Range Management*, 25, 246-250.

Sperow, M., M. Eve & K. Paustian (2003). Potential soil C sequestration on US agricultural soils. *Climatic Change*, 57, 319-339.

Stahl, P.D., J.D. Anderson, L.J. Ingram, G.E. Schuman & D.L. Mummey (2003). Accumulation of organic carbon in reclaimed coal mine soils of Wyoming. In: R.I. Barnhisel (ed.) Working together for innovative reclamation, 20th Annual Meeting, American Society of Mining and Reclamation, June 3-6, Billings, MT. American Society of Mining and Reclamation, Lexington, KY, 1206-1215.

Stout, W.L. (1992). Water-use efficiency of grasses as affected by soil, nitrogen, and temperature. *Soil Science Society of America Journal*, 56, 897-902.

Stout, W.L. &. G.A. Jung (1995). Effects of soil and environment on biomass accumulation of switchgrass. *Agronomy Journal*, 87, 663-669.

Stout, W.L. & G.A. Jung (1992). Influences of soil environment on biomass and nitrogen accumulation rates of orchard grass. *Agronomy Journal*, 84, 1011-1019.

Teague, W.R., J.K. Foy, B.T. Cross & S.L. Dowhower (1999). Soil carbon and nitrogen changes following root-plowing of rangeland. *Journal of Range Management*, 52, 666-670.

Van den Pol-van Dasselaar, A., & E.A. Lantinga (1995). Modelling the carbon cycle of grazing lands in the Netherlands under various management strategies and environmental conditions. *Netherlands Journal of Agricultural Science*, 43, 183-194.

Vicente-Chandler, J., R. Caro-Costas, F. Abruna & S. Silva (1983). Produccion y utilizacion intensiva de las forrajeras en Puerto Rico (Production and utilization of intensively managed forages in Puerto Rico). Universidad de Puerto Rico. Estacion Experimental Agricola Boletin, 271, 226pp.

Vicente-Chandler, J., R. Caro-Costas, R.W. Pearson, F. Abruna, J. Figarella & S. Silva (1964). The intensive management of tropical forages in Puerto Rico. University of Puerto Rico. Agriculture Experiment Station Bulletin 187, 152pp.

Weden, D.A. & D. Tilman (1996). Influence of nitrogen loading and species composition on the carbon balance of grazing lands. *Science*, 274, 1720-1723.

West, C.P., A.P. Mallarino, W.F. Wedin & D.B. Marx (1989). Spatial variability of soil chemical properties in grazed pasture. *Soil Science Society of America Journal*, 53, 784-789.

Methane and nitrous oxide emissions from grazed grasslands

H. Clark, C. Pinares-Patiño and C. deKlein
AgResearch Ltd, New Zealand
Email: Harry.Clark@AgResearch.co.nz

Key points

1. Emissions of methane (CH_4) and nitrous oxide (N_2O) from grasslands make a substantial contribution to total agricultural emissions of these two gases.
2. At present practical mitigation options that relate to grazing ruminants and grazed pastures are limited.
3. Research into agricultural greenhouse gas emissions is of low priority in most developed countries.
4. Direct manipulation of the rumen ecosystem provides the best opportunity for large reductions in CH_4 in the long term.
5. Reducing the amount of nitrogen (N) excreted by grazing animals is a priority in N_2O research, as this source of N_2O constitutes almost 90% of the total global N_2O emissions from grasslands.

Keywords: greenhouse gas, climate change, ruminant, grassland mitigation

Introduction

In its third assessment report, the Inter Governmental Panel on Climate Change (IPCC) stated "The earth's climate system has demonstrably changed on both global and regional scales since the pre-industrial era, with some of these changes attributable to human activities" (IPCC, 2001a). Human activities have increased the atmospheric concentrations of greenhouse gases (GHG) and the key anthropogenic gases (carbon dioxide (CO_2), methane (CH_4), nitrous oxide (N_2O) and tropospheric ozone (O_3)), reaching their highest ever-recorded levels in the 1990's (IPCC 2001a). At the same time there is increasing evidence that the world's climate is getting warmer and that, judged from the 1861-2000 instrumental record, the 1990's were the warmest decade in recent history (IPCC 2001a). Faced with this situation there is now a major international effort to reduce anthropogenic GHG emissions to the atmosphere through such mechanisms as the United Nations Framework Convention on Climate Change (UNFCCC) and, most notably, the Kyoto Protocol. The latter treaty, which at present covers only the developed nations, is a landmark treaty in that those countries ratifying have agreed to legally binding reductions in GHG emissions compared to a 1990 baseline.

The principal agricultural GHGs are CH_4 and N_2O, and it is estimated that agriculturally derived emissions account for >55% and >75% of the world's anthropogenic CH_4 and N_2O emissions respectively (IPCC, 2001b). On a mass basis, global anthropogenic emissions of CH_4 and N_2O are small compared to CO_2 emissions, but because their global warming potentials are greater than CO_2 (CO_2 =1, CH_4 = 23 and N_2O = 296) they play an important role in the radiactive balance of the atmosphere (IPCC, 2001b).

Since agriculture is an important source of GHG there has been considerable focus in the last decade on methods to mitigate CH_4 and N_2O emissions associated with agricultural activity. Ruminant livestock production systems have received particular attention since ruminant animals directly emit CH_4 *via* the breath, and provide the substrate for CH_4 and N_2O

emissions arising from stored and pasture deposited animal excreta. In addition, nitrogenous fertiliser applications, a further source of N_2O emissions, have been a focus of mitigation studies as they are a feature of ruminant livestock production systems in many countries.

In this paper, we will concentrate on the particular problem of mitigating GHG emissions from grazing animals and from extensively grazed pastures. In these situations mitigation options have to be appropriate to systems where, in many cases, animals are handled infrequently, where there may be limited opportunities to manipulate or supplement the diet, where manipulations of the soil are constrained by terrain and accessibility and where synthetic nitrogenous fertiliser inputs are low or non-existent.

Sources of methane and nitrous oxide from grazed livestock

Methane

The principle source of CH_4 from ruminants is enteric methane arising as a by-product of the fermentation of feed in the rumen and, to a lesser extent, the large intestine. The rumen contains a large and diverse population of microorganisms and these break down feed to produce volatile fatty acids (VFA's), CO_2 and CH_4. The VFA's produced in the rumen are absorbed and used as an energy source, but most of the CO_2 and CH_4 are removed from the rumen by eructation. Typically >80% of the CH_4 is produced in the rumen and the rest in the lower digestive tract (Immig, 1996; Murray *et al.*, 1976). In sheep 98% of the CH_4 produced is released via the mouth and 2% via the flatus (Murray *et al.*, 1976). The microorganisms responsible for the production of CH_4 synthesise it from hydrogen, although they do have the ability to use other substrates (Miller, 1995). The removal of hydrogen by methanogens helps maintain a low partial pressure of hydrogen in the rumen without which microbial growth and forage digestion are inhibited (Wolin *et al.*, 1997). As a percentage of the gross energy consumed, 2 - 15% can be lost as CH_4 (Johnson & Ward, 1996), although in temperate forages the range is typically 3.5 – 7.5% (O'Hara *et al.*, 2003).

A secondary source of CH_4 is that arising from voided faecal material. In grazing animals where faecal material is deposited directly onto pastures, only small amounts of CH_4 arise from this source. For example, in New Zealand pastoral agriculture, 99% of CH_4 emissions arise from enteric sources and only 1% from faecal material (New Zealand Climate Change Office, 2004). In this paper only enteric sources of CH_4 will be considered.

Nitrous oxide

Nitrous oxide emissions from agricultural soils arise from nitrification and denitrification processes (Figure 1). Denitrification is the stepwise reduction of soil nitrate (NO_3) (to gaseous nitrogen compounds, with N_2O being one of the intermediate products (Haynes & Sherlock, 1986). It is an anaerobic process that requires a NO_3 substrate, a restricted oxygen supply and suitable pH and temperature conditions (Firestone, 1982; Mosier *et al.*, 1996). Nitrification is an aerobic process, and in most soils is controlled by the availability of ammonium (NH_4) (Schmidt, 1982).

Nitrous oxide
↑
Ammonia →Nitrite→Nitrate→Nitrite→Nitric oxide→Nitrous oxide→Nitrogen
(gas) (gas) (gas)

Figure 1 The production of nitrous oxide by nitrification and denitrification (adapted from O'Hara *et al.*, 2003)

There are two principle sources of nitrogen (N) substrate in grazed pastoral systems; recycled dietary N and applied synthetic fertilisers. Ruminants are relatively poor converters of ingested dietary N into products, and the retention of N in meat, wool or milk ranges from 3 - 25% of the N ingested (Whitehead, 1995). As a result large quantities of N are re-cycled via excreta deposited directly onto pastures by grazing livestock. The relative importance of these two sources of N substrate to nitrous oxide production is likely to vary markedly from country to country. In New Zealand pastoral agriculture, where there is a strong reliance on the biologically fixation of N by forage legumes rather than synthetic fertiliser N, approximately 90% of N_2O emissions arise from excreta N deposited by grazing animals (New Zealand Climate Change Office, 2004). This may well be typical of many developing countries, although not necessarily northern Europe where N fertiliser use is much higher.

How much agricultural methane and nitrous oxide are produced by the world's grasslands?

The IPCC publish estimates of global agricultural emissions of N_2O and CH_4, and data on a country-by-country basis are available from the UNFCCC (IPCC, 2001b; UNFCCC, 2004). In this section we present a 2003 inventory of CH_4 and N_2O emissions that relates solely to the grassland component of ruminant livestock diets.

Methane

Estimated CH_4 emissions from grasslands for the year 2003 are shown in Table 1.

Table 1 Estimates of methane production by ruminant livestock from grassland forage intake (Tg CH_4/yr)

Regions[1]	OECD	O Dev	EE+CIS	CSA	WANA	SSA	ASIA	Total
Dairy cows	2.0	<0.1	2.1	1.8	0.3	0.4	2.3	8.9
Other cattle	6.4	0.4	1.2	11.6	0.2	1.6	5.0	26.3
Buffalo	<0.1	<0.1	<0.1	<0.1	0.1	<0.1	3.2	3.3
Sheep and goats	1.3	0.2	0.4	0.5	0.3	0.5	1.5	4.6
Camelids	<0.1	<0.1	<0.1	<0.1	0.1	0.6	0.1	0.9
Total	9.7	0.6	3.6	14.0	0.9	3.0	12.2	44.0

[1]Regions: OECD, Organisation for Economic Cooperation and Development; O Dev, other developed countries (e.g. South Africa); EE+CIS, Eastern Europe and former URSS countries; CSA, Central and South America; WANA, West Asia and North Africa; SSA, Sub-Saharan Africa.

The methodology adopted to arrive at these estimates is consistent with IPCC good practice guidelines (IPCC, 2000). Feed intake for different classes of livestock was estimated from performance and population data (FAOSTAT, 2003; GLIPHA, 2003) and converted into a CH_4 output using a CH_4 yield factor (% of gross energy (GE) lost as CH_4). The methods adopted were a combination of those of Wheeler et al. (1981); Hendy et al. (1995) and USEPA, (1995). Briefly, livestock production systems were separated into nine different types, and the world into seven different geographical regions as described by Seré & Steinfeld, (1996). Daily feed intake for each animal class was assumed to be a fixed proportion of liveweight. The proportions used ranged from 1.4 – 3.6% depending on species, type of husbandry system and geographic region (Hendy et al., 1995). To make the CH_4 emissions specific to grasslands, non-grassland derived feed intake was subtracted from total feed intake. The proportions of non-grassland derived feeds (e.g. crop residues, forage crops and concentrates) were taken from Bouwman et al., (2004). The assumed gross energy (GE) content of forages ranged between 18.0 and 18.4 MJ/kg DM (Andrieu et al., 1988). Methane emissions were derived from the forage feed energy intake, assuming that in free ranging animals between 6.5 and 8% of the GE consumed is lost as CH_4 (Johnson & Ward, 1996; McCaughey et al., 1997; Lassey et al., 2002; De Ramus et al., 2003), and that concentrate supplementation below 40% of the diet does not greatly influence CH_4 yield (Vermorel, 1995; Boadi et al., 2002). The global estimate of 44 Tg CH_4/yr from grassland derived feeds implies that compared to IPCC estimates (IPCC, 2001b) approximately 20% of all agricultural CH_4 emissions, and between 40 and 55% of the total ruminant CH_4 emissions, arise from grasslands. For comparison with other IPCC estimates the methods used here estimated enteric CH_4 emissions from feed sources to be 70.5 T g/yr.

Nitrous oxide

Estimating global N_2O emissions from pastoral agricultural soils worldwide is extremely difficult as it requires detailed dietary information (quantity of feed consumed and protein content of feed), detailed information on manure management systems, the quantity of nitrogenous fertiliser used, and information on such things as the quantity of animal dung collected and burnt or used as a building material. A lack of data ruled out a complex methodology and we adopted an IPCC Tier 1 approach. This involved using the grassland feed intake data, calculated from the CH_4 inventory, along with estimates of the N% in the diet (1.6 - 2.4%) and the N% retained in animal products (7 - 20%), to obtain an estimate of excreta N arising from grasslands. Default IPCC emission factors (IPCC, 1996; 2000) were then used to estimate direct and indirect N_2O emissions. This estimate makes no attempt to differentiate between manure deposited directly onto pastures or managed in manure management systems, and uses the IPCC default emission factor of 2% of N deposited. It also does not account for manure removed from pastures and used for other purposes. Nitrous oxide emissions from N fertilisers were estimated using the grassland N fertiliser use estimates of the United Nations Food and Agriculture Organisation (FAO) (FAO, 2001) and IPCC default values.

Estimated N_2O emissions from grasslands for the year 2003 are shown in Table 2. These data indicate that between 16 and 33% of the total estimated agricultural N_2O emissions (IPCC 2001b) arise from grasslands.

Table 2 Estimates of nitrous oxide (N₂O) production by ruminant livestock from grassland forage intake (Tg N$_2$0/year)

| | N input to soil (Tg N/yr) | Direct N$_2$O losses | Indirect N$_2$O losses from: | | Total N$_2$O losses |
			volatislised N	leached N	
			(Tg N$_2$O-N/yr)		
Nitrogen fertiliser	4.331	0.049	0.004	0.029	0.082
Excreta nitrogen	34.812	0.696	0.070	0.261	1.027
Total	39.143	0.745	0.074	0.290	1.109

Mitigating methane and nitrous oxide emissions from grazing ruminants

Mitigating CH_4 and N_2O emissions from grazing ruminants poses a particular challenge since solutions requiring frequent manipulation of the grazing animal, or changes in pasture and soil conditions, are likely to be difficult to implement in many livestock systems. A second, more generic issue is that in the developed world, emissions from agriculture are generally minor compared to total CO_2 equivalent emissions. For example, in the EU in 2000 (and accepting that there are substantial differences between states), enteric CH_4 emissions from the agricultural sector comprised 3.2% of total CO_2 equivalent emissions (UNFCCC, 2004), down from 3.4% in 1990. The situation is similar for N_2O where emissions from agricultural soils in 1990 and 2000 comprised only 4.6% of total CO_2 equivalent emissions (UNFCCC, 2004). There is therefore little incentive to give high priority to the agricultural sector when funding research into GHG mitigation.

Methane mitigation

Improving efficiency of the animal production system

Improving the efficiency of livestock production as a route to reducing CH_4 emissions from livestock systems is an area that has the capability to cause considerable perplexity. Farmers continue to strive for improvements in the efficiency of production in order to survive in a competitive global market and, although improved production efficiency can influence CH_4 output, it is unlikely that efficiency increases by themselves will solve the CH_4 problem. For the purposes of this paper we will define improvements in efficiency as equating to increasing the amount of milk, meat or wool produced per unit of feed ingested. Defined in this way, efficiency is closely related to the partitioning of feed intake between that required for maintenance and that required for production. Viewed simplistically there will be a fixed CH_4 output associated with the maintenance portion of the diet, and a variable CH_4 emission that is associated with the production portion of the diet. As feed intake, and hence production increases, the proportion of total CH_4 output associated with maintenance goes down, and CH_4 output per unit of product declines (Table 3). Thus for a fixed amount of product, it will be beneficial in terms of CH_4 emissions to produce this from a smaller number of high producing animals than a large number of low producing animals.

Unfortunately, although improvements in efficiency will reduce the amount of CH_4 emitted per unit of product, they will not necessarily reduce the amount of CH_4 produced in total. A reduction in the total will only occur if the amount of product produced is static or rises at a slower rate than the rate of decline in CH_4 emitted per unit of product. For example, in New

Zealand the quantity of CH_4 produced per unit of product has declined since 1990 for beef and milk, but CH_4 emissions have increased in both sectors because of increases in the quantity of product produced (Clark & Ulyatt, 2002). Methane emissions from the sheep sector have fallen, principally because sheep number declined by 30% between 1990 and 2000, and increases in the quantity of sheepmeat and wool produced were small (Clark & Ulyatt, 2002).

Table 3 An estimation of the proportion of methane (CH_4) emission attributable to maintenance or milk production, in a 450kg grazing dairy cow at various levels of digestible DM intake (DDMI)

DDMI (kg/d)	Milk yield (kg/d)	CH_4 (kg/d)	% CH_4 associated with		CH_4/milk (g/kg)
			Maintenance	Production	
4.0	0	105	100	0	
7.9	12	206	51	49	17.2
10.5	20	272	39	61	13.6
11.7	24	305	34	66	12.7

Source: O'Hara et al., (2003)

Improving herbage quality

One of the principle aims of grassland management is to increase the quality of the forage ingested by grazing ruminants. Methane production is highly correlated with fibre digestion in the rumen (Kirchgessner et al., 1995), and so it would be logical to assume that decreasing the fibre content of forages would reduce CH_4 emissions. Empirical evidence to support this comes from the work of Blaxter & Wainman (1964), who found with hay based diets fed at twice maintenance intake levels, that CH_4 emissions increased from 3.5 to 7.0% of GE intake, as the crude fibre in the diet increased from 2.2 to 33.8%. In a summary of 339 experiments with sheep and cattle (Blaxter & Clapperton, 1965), it was found that at intakes above twice maintenance, the percentage of GE lost as CH_4 was reduced as digestibility increased. Since fibre content and digestibility of forages are negatively correlated, and are responsive to management manipulation, at first site it appears that increasing the digestibility of forages could be an effective CH_4 mitigation option for grazing livestock. However, this may not be the case in many situations.

Recent work using animals fed fresh, as opposed to dried, forage diets suggests that in C3 grasses at least the percentage of GE lost as CH_4 may be relatively insensitive to forage quality over the range of intakes found in grazing systems. Pinares-Patiño et al. (2003a), working with *Phleum pratense* L. (timothy grass) at four stages of maturity spanning an organic matter digestibility of 56 – 78% and a neutral detergent fibre (NDF) content of 52–76%, could find no relationship between digestibility or NDF and the percentage of GE intake lost as CH_4 in cattle fed at 1 - 1.5 above maintenance. Similarly Molano et al. (2003) working with *Lolium perenne* L. (perennial ryegrass) at two stages of growth and four levels of feeding, found no relationship between CH_4 emissions per unit of DM intake and digestibility (Table 4).

Table 4 Methane (CH₄) emission by sheep at four levels of voluntary feed intake, consuming *Lolium perenne* L. (perennial ryegrass) harvested at the vegetative and reproductive stage of growth

	Reproductive				Vegetative				
Apparent digestibility (%)	61.5	62.5	61.1	65.1	74.5	76.9	74.1	75.9	*P<0.001*
DMI kg/d	0.57	0.73	0.91	1.37	0.78	0.95	1.15	1.54	*P<0.001*
CH₄ g/day	11.5	17.7	24.3	31.9	15.6	22.7	27.4	35.9	*P<0.001*
CH₄ g/kg DMI	20.5	24.2	26.6	23.3	20.1	24.1	24.0	23.5	*NS*

Source: Molano *et al.*, (2003)

A second issue related to forage quality is that even if it does not influence the CH_4 yield, it can indirectly reduce CH_4 emissions since it affects how much feed is needed to achieve a given level of production. Increasing forage quality, could be used to decrease emissions per head simply because less feed is processed in the rumen to achieve a given level of production. However, in practice, if feed quality is increased without any reduction in the quantity of feed available, the intake of individual animals and/or the number of animals kept per unit area will increase. These would both tend to increase CH_4 production either per animal or per unit area. Therefore reductions in CH_4 could only be guaranteed if the number of stock kept, or the amount of product produced was also controlled.

Forage plants with low methane yield

Forage species have been shown to influence CH_4. Waghorn *et al.* (2002) found in sheep, that legumes generally reduced the quantity of CH_4 produced per unit of feed intake compared to C3 grasses. The data of Kurihara *et al.* (1999) and O'Hara *et al.* (2003) suggests that C4 grasses have a higher CH_4 yield than C3 grasses. However, some caution needs to be exercised since, with the exception perhaps of C4 grasses, the differences between forage species in CH_4 emissions may in practice be small. In an experiment with *Trifolium repens* L. (white clover) fed at varying proportions in the diet (Lee *et al.*, 2004), it was reported that even when incorporated at 60% of a grass:clover diet (a quantity only likely to be achieved for short periods in grazed swards), the reduction in CH_4 was only 16%. When *Trifolium repens* was included at 15% of the diet (a more realistic figure in many practical situations), the reduction was only 4%.

The difficulties surrounding plant solutions to CH_4 mitigation are perhaps best exemplified by condensed tannin (CT) containing plants. Plants containing CT have been found to reduce CH_4 emissions in cattle (Woodward *et al.*, 2001), and sheep (Waghorn *et al.*, 2002; Pinares-Patiño *et al.*, 2003b). In addition they have been found to increase liveweight gains and decrease the severity of gastrointestinal worm infestations (Min *et al.*, 2003). The disadvantage of CT containing plants in temperate pastures is that they do not compete well with other temperate species. As pointed out by O'Hara *et al.*, (2003), the benefits of CT containing plants have been recognised for over 30 years but to date we still do not have a competitive CT containing pasture plant.

A number of strategies for influencing CH_4 production by direct manipulation of the rumen ecosystem have been promulgated (for a review see McAllister *et al.*, 1996). Some of them, notably the use of halogenated CH_4 compounds such as chloroform and bromochloromethane, have been shown to be highly effective at suppressing CH_4 production (van Nevel & Demeyer, 1996; McCrabb *et al.*, 1997), but they are in many cases also unstable compounds which are potentially toxic to ruminants (Lanigan *et al.*, 1978). Similarly the control of protozoa (which live in symbiosis with methanogens), which can be responsible for up to 25% of rumen CH_4 emissions (Newbold *et al.*, 1995), can only be achieved by the use of potentially toxic chemicals. Other strategies such as the manipulation of methanogens by bacteriophage and bacteriocins, and the promotion of acetogenesis as an alternative hydrogen sink are at an early stage. Two strategies are much closer to being available and these are discussed in more detail below.

Scientists working for CSIRO in Australia have developed animal vaccines that reduce methanogenesis by stimulating the production of antibodies in the host animal, which restrict the activity of rumen methanogens (Baker, 1999). This work has progressed to the stage where vaccines have been tested *in-vivo*. The limited data available show no clear evidence that the current formulations can consistently reduce CH_4 emissions (Table 5). However, the promising aspect of the Australian trial is that both vaccine formulations were able to boost antibody titres (IGa and IGg) in blood and saliva compared to control animals. Clearly considerably more work is needed to develop a vaccine with proven efficacy but the approach is one that is highly attractive in grazing animals, since it holds out the promise of an effective mitigation technology allied to an infrequent and simple delivery mechanism.

Table 5 Percentage changes in the quantity of methane (CH_4) emitted per unit feed intake, compared to adjuvant only controls following vaccination with three different anti-methanogenic vaccine preparations (AMG-v). All data non-significant except for *, where *P=0.51*

	Post-primary vaccination			Post-booster vaccination		
	AMG-v1	AMG-v2	AMG-v3	AMG-v1	AMG-v2	AMG-v3
Australia[1]	-6	Not used	-1	-7.7*	Not used	+0.8
New Zealand[2]	-4	+2	Not used	+2	+9	Not used

Source: [1]Wright *et al.*, (2004); [2]Clark *et al.*, (2004)

Ionophores, particularly monensin, have been used routinely in animal production systems for many years as growth promoters. There is evidence to suggest that they can reduce CH_4 through a combination of reduced voluntary intake, reduced acetate production and the inhibition of H_2 release from formate (Goodrich *et al.*, 1984; van Nevel & Demeyer, 1996; Tedeschi *et al.*, 2003). Slow release delivery devices are available, thus monensin is potentially suitable for use in grazing animals. There are two principle issues surrounding its use as a CH_4 mitigation tool. First, there are doubts as to the duration of the direct CH_4 suppressing effect (Tedeschi *et al.*, 2003). However O'Kelly & Spiers (1992), working with steers fed Lucerne hay, found that 55% of the reduction in CH_4 was attributed to the anorectic effect (reduced intake) and 45% to the direct effect on rumen fermentation. This implies that even in the absence of a direct effect on rumen methanogenesis, CH_4 production would still be reduced in situations where ionophores reduce herbage intakes. A second issue is that ionophores are classed as antibiotics and there is a strong move to phase out the routine use of antibiotics in livestock production systems. Hence even if the efficacy of monensin as a long-term CH_4 inhibitor could be conclusively demonstrated, its routine use may not be readily acceptable to both consumers and regulatory authorities.

Exploiting animal to animal variation in methane production

Since the development of the SF_6 tracer technique for estimating CH_4 production in unconfined ruminants (Johnson & Johnson, 1995), it has been possible to simultaneously measure emissions from groups of animals consuming the same diet. The vast majority of work with grazing animals fed fresh forage has been carried out in New Zealand. A common finding is that there are large differences in emissions per unit of feed intake (Ulyatt *et al.,* 2002). This phenomenon has been confirmed recently in a single experiment when CH_4 emissions were measured from 302 grazing dairy cows over a four-week period (Table 6). In addition in sheep, differences between individual animals have been found to persist for up to five months (Pinares-Patiño *et al.,* 2003c).

Table 6 Methane (CH_4) emissions from a herd of 302 Friesian x Jersey dairy cows measured between January 12 and February 6, 2003

	Min	Max	Mean	St. Dev	Lower quartile	Upper quartile
CH_4 g/day	213.9	478.8	332.1	38.1	285.6	381.0
CH_4 kg DMI/day^{-1}	11.0	31.1	19.3	2.9	16.1	23.1

Source: C. Pinares-Patiño & H. Clark (*unpublished data*).

Since CH_4 is produced by microbial fermentation in the rumen, the existence of animal-to-animal variation suggests that there is an interaction between the animal and its microbes. This leads onto issues of whether this interaction is genetically based, and if it is a heritable trait? In New Zealand, cows of a US genetic background have been found to have lower CH_4 emissions per unit of dry matter intake than cows of a New Zealand genetic background (O'Hara *et al.,* 2003). Similarly Ferris *et al.* (1999) found that the percentage of GE lost as CH_4, was lower for high genetic merit than for medium genetic merit Holstein cows. These two studies indicate that it may be possible to breed animals that have inherently low CH_4 emissions. Although work on exploiting animal variation in CH_4 emissions is at a preliminary stage, breeding low CH_4 producing animals does offer an extremely attractive solution. It has applicability across all types of production systems, exists for the life of the animal and is open to continuous improvement.

Nitrous oxide mitigation

Improving efficiency of the animal production system

In a similar manner to that already discussed for CH_4, improving the efficiency of production can reduce N_2O emissions in situations where constraints are placed on product output. If the productivity of each animal is increased, less total dry matter intake is needed to produce a given amount of product (see Table 3). This in turn leads to a reduction in the total N being recycled through the grazing animal for a given amount of product. The quantity of N_2O emitted per animal is therefore likely to be higher, but fewer animals are needed to obtain the quantity of product required. Mitigating N_2O (and CH_4) by improving the efficiency of production does however imply that product output is restricted to some extent.

Diet manipulation

Decreasing either the total N content and/or the N degradability of ruminant diets will reduce the amount of N excreted during the grazing process (Kebreab *et al.*, 2001). This solution is most applicable to the dairy sector where there are more opportunities to manipulate the diet. Optimising fertiliser applications has an important role here since the N content of plants is directly related to N supply (Whitehead, 1995). The replacement of high N content grass, with low N content high-energy feeds such as maize silage is a possibility in some circumstances. For example, in New Zealand maize silage is commonly given as a supplement to grazing dairy cows. Modelling studies by de Klein & Ledgard (2005), have shown that substituting fertilised grass with fertilised maize silage can reduce N_2O emissions from the typical New Zealand dairy farm by 27% (Table 7).

Table 7 Estimated nitrogen (N) fertiliser use, N excretion rates and nitrous oxide (N_2O) emissions from an average dairy farm in New Zealand, under a 'business-as usual' scenario and when replacing fertilized grass with maize silage

	Business-as usual	Maize silage supplement
N fertiliser use		
On farm (t N/yr)	7.4	0
Off-farm (t N/yr)	0	1
N excreted (kg N/ha/yr)	345	318
N_2O emissions (t CO_{2equiv}/yr)	218	159 (-27%)

Adapted from de Klein & Ledgard (2005)

The quantity of N voided by grazing ruminants can also be influenced by the protein: carbohydrate ratio of the diet (van Vuuren & Meijs, 1987; Kebreab *et al.*, 2001). Studies at the Institute of Grassland and Environmental Research (IGER) in Wales have shown that feeding beef cattle, silage made from grass cultivars containing elevated concentrations of water soluble carbohydrates, increased the N use efficiency for microbial growth in the rumen from 46 to 68% (Merry *et al.*, 2003). Similarly, studies with dairy cows suggested that high sugar grasses reduced N excretion rates and, under some conditions, increased milk yield and milk protein yield (IGER, 2001). However, recent New Zealand research suggests that the effectiveness of these grasses as a N_2O mitigation strategy might be limited to cooler climates, as a warm-temperate climate may limit grass expression of high sugar content (Parsons *et al.*, 2004).

Plants containing CT's have already been discussed in relation to CH_4 mitigation. They also have the potential to influence N_2O emissions from grazed pastures because of their ability to influence protein breakdown and absorption in ruminants (Min *et al.*, 2003). Unfortunately, as already discussed, the inferior agronomic characteristics of these plants limit their usefulness at present. Similarly, the ionophore monensin, which can reduce CH_4 evolution from ruminants, can also influence N retention (Tedeschi *et al.*, 2003). This can be directly through increased N retention or indirectly through its anorectic effect. However, as discussed previously, the widespread use of monensin to mitigate GHG emissions may not be readily acceptable in practice.

Management of fertilisers and excreta deposited during grazing

In general, practices that increase the efficiency of use of applied N will reduce emissions of N_2O from soils. Likewise, the timing, quantity and type of N fertilisers have all been shown to influence N_2O emissions (for a review see O'Hara *et al.*, 2003). Slow release fertilisers, formulated to achieve a better synchrony between the demand and supply of N, have been shown to be effective at reducing N_2O emissions (Smith *et al.*, 1997). Similarly, fertilisers containing, or applied in conjunction with, nitrification inhibitors such as dicyandiamide (DCD) have also proved to be effective at reducing N_2O emissions by as much as 60% (Belastegui Macadam *et al.*, 2003; Williamson & Jarvis 1997).

Although improved management of synthetic fertilisers can help to reduce N_2O emissions from grasslands, a more pressing problem is that of reducing emissions from animal wastes deposited directly onto pastures by grazing animals. Options here are very limited. A field study by de Klein *et al.* (2005) suggests that the strategic use of a feed pad can reduce total N_2O emission by avoiding urine and dung being deposited during wet conditions when N_2O emissions are likely to be high. Their results suggested that for a typical dairy farm in the southern part of New Zealand, N_2O emissions could be reduced by about 10%. In addition, small reductions can be achieved by altering soil conditions e.g. liming, improving drainage and avoiding soil compaction (Clark *et al.*, 2001). However, the general applicability of these methods is limited. Work with nitrification inhibitors to reduce N_2O emissions from animal urine does however hold some promise. Williamson & Jarvis (1997) reported reductions of over 70% in N_2O emissions from urine applied to pasture in conjunction with DCD, compared to urine alone between 6 – 21 days after application. In lysimeter studies Di & Cameron (2002, 2003) found that DCD reduced N_2O emissions from urine treated grassland by about 80% following spring and/or autumn applications of urine with or without DCD. The addition of a nitrification inhibitor directly into the urine stream from an animal mounted dispenser has also been advocated, although no results are available to attest to the efficacy of this approach (Quin, 2004). The research conducted so far does indicate that applying DCD to grazed pastures could be used as a practical method of reducing N_2O from urine patches, although issues of toxicity (DCD has been shown to be exhibit phytotoxic effects to *Trifolium repens* (Belastegui Macadam *et al.*, 2003), timing and longevity of the effect need to be assessed. Additionally research needs to be conducted at the system level to determine the long-term effects of nitrification inhibitors on N cycling dynamics in the soil, plant, and atmosphere system. For example, Belastegui Macadam *et al.* (2003), showed that DCD increased the N concentration in *Trifolium* plants, and this will influence the quantity of excreta N cycled through the animal.

Conclusions

Grassland ecosystems are major contributors to agricultural emissions of CH_4 and N_2O. Options do exist to mitigate emissions from grazed systems, but in general they do not have universal applicability and for many situations practical methods of reducing emissions do not exist at present. Research into GHG emissions from agriculture is low priority in most developed countries, and this will need to be addressed if more rapid progress is to be made. Priority research areas need to be those that have high efficacy, and cost effective and simple delivery mechanisms. For the long term direct manipulation of the rumen ecosystem provides the best opportunity for large reductions in CH_4 emissions. This is a neglected area of research in most countries. Breeding low CH_4 emitting animals is an exciting prospect but more work is needed on the fundamental basis of animal-to-animal variation before breeding programmes are contemplated. For N_2O emissions the research priorities are two-fold. Firstly, studies should be conducted that focus on providing experimental evidence of the effectiveness of mitigation options. In particular on options which focus on reducing the amount of N excreted by grazing animals, as this source of N_2O constitutes almost 90% of the total global N_2O emissions (Table 2). Secondly, the development of accurate models is important. Due to the high spatial and temporal variability of N_2O emissions, accurate measurements at a whole systems level are near impossible. Therefore, the development of systems models that utilise and link the experimental evidence of component studies to evaluate the effect of mitigation strategies at a systems level is a priority area.

Such models should also have the ability to collectively assess all major GHG emissions. In this paper, as in most others, N_2O and CH_4 have been considered separately. In reality they are both emissions from the same production system and, in the short term at least, manipulations of the system as a whole may offer the best hope of reducing net GHG emissions from pastures. This also means looking at both sources and sinks of GHG. Modelling studies by Lambert & Clark, (2005) have demonstrated that for beef and sheep farms in NZ it is possible to maintain farm incomes and reduce GHG emissions by a combination of the intensification of animal production and the planting of trees. It would be surprising if opportunities for this type of system manipulation didn't exist in other countries.

Finally, although the aim of GHG mitigation technologies is to reduce actual emissions to the atmosphere, international treaty obligations mean that countries also have to be able to demonstrate in their GHG accounts that they have done so. This means that national inventories capable of accounting for mitigation technologies need to be developed alongside measurement systems that can verify claimed emission reductions.

References

Andrieu, J., C. Demarquilly & D. Sauvant (1988). Tables de la valeur nutritive des aliments. In: R. Jarrige (ed.) Alimentation des bovins, ovins et caprins (Feeding of cattle, sheep and goats), INRA, Paris, 351–464.

Baker, S.K. (1999). Biology of rumen methanogens and stimulation of animal immunity. In: P.J. Reyanga & S.M. Howden (eds.) Meeting the Kyoto Target. Implications for the Australian livestock industries. Bureau of Rural Sciences, 31-40.

Belastegui Macadam, X.M.B., A. del Prado, P.Merino, J.M. Estavillo, M.Pinto & C. Gonzalez-Murua (2003). Dicyandiamide and 3,4-dimethyl pyrazole phosphate decrease N_2O emissions from grassland but dicyandiamide produces deleterious effects in clover. *Journal of Plant Physiology*, 160, 1517-1523.

Blaxter, K.L. & J.L. Clapperton (1965). Prediction of the amount of methane produced by ruminants. *British Journal of Nutrition*, 19, 511-522.

Blaxter, K.L. & F.W. Wainman (1964). The utilization of the energy of different rations by sheep and cattle for maintenance and for fattening. *Journal of Agricultural Science, Cambridge*, 63, 113-128.

Boadi, D.A., K.M. Wittenberg & W.P. McCaughey (2002). Effects of grain supplementation on methane production of grazing steers using the sulphur (SF_6) tracer gas technique. *Canadian Journal of Animal Science*, 82, 151-157.

Bouwman, A.F., K.W. Van der Hoek, B. Eickhout & I. Soenario (2004). Exploring changes in world ruminant production systems. *Agricultural Systems,* (In press).

Clark H., C. de Klein & P. Newton (2001). Potential management practices and technologies to reduce nitrous oxide, methane and carbon dioxide emissions from New Zealand agriculture. Report to Ministry of Agriculture and Forestry - Oct 2001, Wellington, New Zealand, 81pp.

Clark, H. & M.J. Ulyatt (2002). A recalculation of enteric methane emissions from New Zealand ruminants 1990-2000 with updated emission predictions for 2010. A report prepared for the Ministry of Agriculture and Forestry, Wellington, New Zealand, 38pp.

Clark, H., A.D. Wright, K. Joblin, G. Molano, A. Cavannagh & J. Peters (2004). Field testing an Australian developed anti-methanogen vaccine in growing ewe lambs. *Proceedings of the workshop on the science of atmospheric trace gases, 2004*. NIWA Technical Report 125, Wellington, New Zealand, 167pp.

de Klein, C.A.M. & S.F. Ledgard (2005). Nitrous oxide emissions from New Zealand agriculture – key sources and mitigation strategies. *Nutrient cycling in AgroEcosystems*. (in press).

de Klein, C.A.M., L.C. Smith & R.M. Monaghan (2004). Restricted autumn grazing to reduce nitrous oxide emissions from dairy pastures in Southland, New Zealand. *Agriculture, Ecosystem and Environmen*, (in press).

DeRamus, H.A., T.C. Clement, D.D. Giampola & P.C. Dickison (2003). Methane emissions of beef cattle on forages: efficiency of grazing management. *Journal of Environmental Quality*, 32, 269-277.

Di, H.J. & K.C. Cameron (2002). The use of a nitrification inhibitor, dicyaniamide (DCD), to decrease nitrate leaching and nitrous oxide emissions in a simulated grazed and irrigated grassland. *Soil Use and Management*, 18, 395-403.

Di, H.J. & K.C. Cameron (2003). Mitigation of nitrous oxide emissions in spay-irrigated grazed grassland by treating the soils with dicyaniamide, a nitrification inhibitor. *Soil Use and Management*, 19, 284-290.

FAO (2001). Global estimates of gaseous emissions of NH_3, NO and N_2O from agricultural land. Report by the international fertilizer industry association and the Food and Agriculture Organisation of the United Nations, Rome, 58pp. Available from <http:www.fao.org/DOCREP/004/Y2780E/y2780e00.htm#P-10>

FAOSTAT (2003). FAO Statistical Databases. Food and Agriculture Organization of the United Nations, Rome. Available from <http://faostat.fao.org/faostat>

Ferris, C.P., F.J. Gordon, D.C. Patterson, M.G. Porter & T. Yan (1999). The effect of genetic merit and concentrate proportion in the diet on nutrient utilisation by lactating dairy cows. *Journal of Agricultural Science, Cambridge*, 132, 483-490.

Firestone, M.K. (1982). Biological denitrification. In: F.J. Stevenson (ed.) Nitrogen in agricultural soils. ASA-CSSA-SSA, Madison, Wisconsin, U.S, 289-326.

GLIPHA (2003). Global Livestock Production and Health Atlas. Food and Agriculture Organization of the United Nations, Rome. Available from<http://www.fao.org/ag/aga/glipha/index.jsp>

Goodrich, R.D., J.E. Garrett, D.R. Ghast, M.A. Kirich, D.A. Larson & J.C. Meiske (1984). Influence of monensin on the performance of cattle. *Journal of Animal Science*, 58, 1484-1498.

Haynes, R.J. & R.R. Sherlock (1986). Gaseous losses of nitrogen. In: R.J. Haynes (ed.) Mineral nitrogen in the plant-soil system. Orlando, Academic Press, 242-302.

Hendy, C.R.C., U. Kleih, R. Crawshaw & M. Phillips (1995). Interactions between livestock production systems and the environment: impact domain: concentrate feed demand. FAO Consultancy Report for Livestock and Environment Study, 147pp.

IGER (2001). <http://www.IGER.bbsrc.ac.uk/igerweb/AnnRep2000/arasciadv.htm#Sweet>

Immig, I. (1996). The rumen and hindgut as source of ruminant methanogenesis. *Environmental monitoring and assessment*, 42, 57–72.

IPCC (1996). Revised IPCC guidelines for national greenhouse gas inventories. IPCC National Greenhouse Gas Inventories Programme. Published for the IPCC by the UK Meteorological Office, Bracknell.

IPCC (2000). Good practice guidance and uncertainty management in national greenhouse gas inventories. IPCC National Greenhouse Gas Inventories Programme. Published for the IPCC by the Institute for Global Environmental Strategies, Japan.

IPCC (2001a). IPCC Third Assessment Report: Climate Change 2001: Synthesis Report, R.T. Watson & the Core Writing Team (eds.), IPCC, Geneva, Switzerland, 34pp.

IPCC (2001b). IPCC third assessment report: climate change 2001: the scientific basis. Contribution of Working Group I to the Third Assessment Report of the Intergovernmental Panel on Climate Change (IPCC) J.T. Houghton, Y. Ding, D.J. Griggs, M. Noguer, P. J. van der Linden and D. Xiaosu (eds.), Cambridge University Press, UK, 881pp.

Johnson, D.E. & G.M. Ward (1996). Estimates of animal methane emissions. *Environmental monitoring and assessment*, 42, 133–141.

Johnson, K.A. & D.E. Johnson (1995). Methane emissions from cattle. *Journal of Animal Science*, 73, 2483-2492.

Kebreab, E., J. France, D.E. Beever & A.R. Castillo (2001). Nitrogen pollution by dairy cows and its mitigation by dietary manipulation. *Nutrient Cycling in AgroEcosytems*, 60:275-285.

Kirchgessner, M., W. Windisch & H.L. Muller (1995). Nutritional factors for the quantification of methane production. In: W. von Engelhardt, S. Leonhard-Marek, G. Breves & D. Gieseke (eds.) Ruminant physiology: digestion, absorption, growth and reproduction. Ferdinand Enke Verlag, Stuttgart, 333-348.

Kurihara, M., T. Magner, R.A. Hunter & G.J. McCrabb (1999). Methane production and energy partition of cattle in the tropics. *British Journal of Nutrition*, 81, 227-234.

Lambert, M.G. & H. Clark (2005). A systems approach to managing greenhouse gases on New Zealand sheep & beef farms. In: *Proceedings of the XX International Grassland Congress*, Dublin, offered paper (in press).

Lanigan, G.W., A.L. Payne & J.E. Peterson (1978). Antimethanogenic drugs and *Heliotropium europaeum* poisoning in penned sheep. *Australian Journal of Agricultural Research*, 29, 1281-1291.

Lassey, K.R., C.S. Pinares-Patiño & M.J. Ulyatt (2002). Methane emission by grazing livestock: some findings on emission determinants. *Proceedings of the third International Symposium on non-CO2 Greenhouse Gases (NCGG-3)*, Maastricht, The Netherlands, 95-100.

Lee, J.M., S.L. Woodward, G.C. Waghorn & D.A. Clark (2004). Methane emissions by dairy cows fed increasing proportions of white clover (*Trifolium repens*) in pasture. *Proceedings of the New Zealand Grassland Association*, 66, 151-155.

McAllister, T.A., E.K. Okine, G.W. Mathison & K.J. Cheng (1996). Dietary, environmental and microbial aspects of methane production in ruminants. *Canadian Journal of Animal Science*, 76, 231-243.

McCaughey, W.P., K. Wittenberg & D. Corrigan (1997). Methane production by steers on pasture. *Canadian Journal of Animal Science*, 77, 519-524.

McCaughey, W.P., K. Wittenberg & D. Corrigan (1999). Impact of pasture type on methane production by lactating beef cows. *Canadian Journal of Animal Science*, 79, 221-226.

McCrabb, G.J., K.T. Berger, T. Magner, C. May & R.A. Hunter (1997). Inhibiting methane production in Brahman cattle by dietary supplementation with a novel compound and the effects on growth. *Australian Journal of Agricultural Science*, 48, 323-329.

Merry R.J., M.R.F. Lee, D.R. Davies, J.M. Moorby, R.J. Dewhurst & N.D. Scollan (2003). Nitrogen and energy use efficiency in the rumen of cattle fed high sugar grass and/or red clover silage. *Aspects of Applied Biology*, 70, 87-92.

Miller, T.L. (1995). Ecology of methane production and hydrogen sinks in the rumen. In: W. von Engelhardt, S. Leonhard-Marek, G. Breves & D. Gieseke (eds.) Ruminant physiology: digestion, metabolism, growth and reproduction. Ferdinand Enke Verlag, Stuttgart, 317-331.

Min, B.R., T.N. Barry, G.T. Attwood & W.C. McNabb (2003). The effect of condensed tannins on the nutrition and health of ruminants fed fresh temperate forages: a review. *Animal Feed Science and Technology*, 106, 3-19.

Molano, G., T. Renard & H. Clark (2003). The effect of level of feeding and forage quality on methane emissions by wether lambs. *Proceedings of the 2nd joint Australia and New Zealand forum on Non-CO2 greenhouse gas emissions from Agriculture*, R. Eckard & B. Slattery (eds.). Cooperative research centre for greenhouse gas accounting. Canberra, Australia, F14.

Mosier, A.R., J.M. Duxbury, J.R. Freney, O. Heinemeyer, K. Minami & R. Merckx (1996). Nitrous oxide emissions from agricultural fields: assessment, measurement and mitigation. *Plant and Soil*, 181, 95-108.

Murray, R.M., A.M. Bryant & R.A. Leng (1976). Rates of production of methane in the rumen and large intestine of sheep. *British Journal of Nutrition*, 36, 1-14.

New Zealand Climate Change Office (2004). National inventory report. Greenhouse gas inventory 1990 - 2002. The national inventory report and common reporting format Tables, April 2004. Wellington, New Zealand, 155pp.

Newbold, C.J., B. Lassalas & J.P. Jouany (1995). The importance of methanogens associated with ciliate protozoa in ruminal methane production *in vitro*. *Letters in Applied Microbiology*, 21, 230-234.

O'Hara, P., J. Freney & M. Ulyatt (2003). Abatement of agricultural non-carbon dioxide greenhouse gas emissions. A study of research requirements. A report prepared for the Ministry of Agricultue and Forestry on behalf of the Convenor, Ministerial Group on Climate Change, the Minister of Agriculture, and the Primary Industries Council. May 2003, New Zealand Ministry of Agriculture and Forestry, Wellington, 171pp. <www.govt.nz/publications>

O'Kelly, J.C. & W.G. Spiers (1992). Effect of monensin on methane and heat production of steers fed lucerne hay either *ad libitum* or at the rate of 250 g/hour. *Australian Journal of Agricultural Research*, 43, 1789-93.

Parsons A.J., S. Rasmussen, X. Hong, C.B. Anderson & G.P. Cosgrove (2004). Some 'high sugar grasses' don't like it hot. *Proceedings of the New Zealand Grassland Association*, 66, (in press).

Pinares-Patiño, C.S., R. Baumont & C. Martin (2003a). Methane emissions by charolais cows grazing a monospecific pasture of timothy at four stages of maturity. *Canadian Journal of Animal Science*, 83, 769-777.

Pinares-Patiño, C.S., M.J. Ulyatt, C.W. Holmes, T.W. Barry & K.R. Lassey (2003c). Persistence of the between-sheep variation in methane emission. *Journal of Agricultural Science, Cambridge*, 140, 227-233.

Pinares-Patiño, C.S., M.J. Ulyatt, G.C. Waghorn, C.W. Holmes, T.N. Barry, K.R. Lassey & D.E. Johnson (2003b). Methane emission by alpaca and sheep fed on lucerne hay or grazed on pastures of perennial ryegrass/white clover or birdsfoot trefoil. *Journal of Agricultural Science, Cambridge*, 140, 215-226.

Quin, B. (2004). Nitrification inhibitors: N-sense or nonsense? FertScience, the science of soil fertility. The Official Newsletter of Summit-Quinphos (NZ) Ltd. (Autumn 2004), 10pp.

Schmidt, E.L. (1982). Nitrification in soil. In: Stevenson, F.J. (ed.) Nitrogen in agricultural soils. ASA-CSSA-SSA, Madison, Wisconsin, U.S, 253-288.

Seré, C. & H. Steinfeld (1996). World livestock production systems: current status, issues and trends. FAO Animal Production and Health Paper 127. Food and Agriculture Organization of the United Nations, Rome, Italy, 92pp.

Smith, K.A., I.P. McTaggart, H. Tsuruta & K. Smith (1997). Emissions of N_2O and NO associated with nitrogen fertilization in intensive agriculture, and the potential for mitigation. *Soil Use and Management*, 13, 296-304.

Tedeschi, L.O., D.G. Fox & T.P. Tylutki (2003). Potential environmental benefits of ionophores in ruminant diets. *Journal of Environmental Quality*, 32, 1591-1602.

Ulyatt, M.J., H. Clark & D.K.R. Lassey (2002). Methane and climate change. *Proceedings of the New Zealand Grassland Association*, 64, 153-157.

UNFCCC (2004). Greenhouse gas inventory database.
<http://ghg.unfccc.int/default1.htf?time=05%3A02%3A54+AM>

USEPA (1995). Global impact domain – methane emissions. Consultant report for the livestock and the environment study. USEPA, Washington D.C. Available from
<http://www.fao.org/WAIRDOCS?LEAD?X6116E/x6116e05.htm#5.%20Exhibits>

van Nevel, C.J. & D.I. Demeyer (1996). Control of rumen methanogenesis. *Environmental Monitoring & Assessment*, 42, 73-97.

van Vuuren, A.M. & J.A.C. Meijs (1987). Effects of herbage composition and supplement feeding on the excretion of nitrogen in dung and urine by grazing dairy cows. In: H.G. van der Meer, T.A. van Dijk & G.C. Ennik (eds.) Animal manure on grassland and fodder crops. Dordrech, The Netherlands, Martinus Nijhof, 17-25.

Vermorel, M. (1995). Productions gazeuses et thermiques résultant des fermentations digestives (Gas and thermal production resulting from digestive fermentation). In: R. Jarrige, Y. Ruckebusch, C. Demarquilly, M.H. Farce & M. Journet (eds), Nutrition des ruminants domestiques, INRA, Paris, 649–670.

Waghorn, G.C., M.H. Tavendale & D.R. Woodfield (2002). Methanogenesis from forages fed to sheep. *Proceedings of the New Zealand Grassland Association*, 64, 167-171.

Wheeler, R.O., G.L. Cramer, K.B. Young & E. Ospina (1981). The world livestock product, feedstuff, and food grain system. Winrock International Livestock Research and Training Center, Arkansas, USA, 85pp.

Whitehead, D.C. (1995). Grassland nitrogen. CAB International, Wallingford, UK, 397pp.

Williamson, J.C. & S.C. Jarvis (1997). Effect of Diccyaniamide on nitrous oxide flux following return of animal excreta to grassland. *Soil Biology and Biochemistry*, 29, 1575-1578.

Wolin, M.J., T.L Miller & C.S. Stewart (1997). Microbe-microbe interactions. In: P.N. Hobson & C.S Stewart (eds.) The rumen microbial ecosystem. Blackie Academic and Professional, London, 467-491.

Woodward, S.L., G.C. Waghorn, M.J. Ulyatt & K.R. Lassey (2001). Early indications that feeding Lotus will reduce methane emissions from ruminants. *Proceedings of the New Zealand Society of Animal Production*, 61, 23-26.

Wright, A.D.G., P. Kennedy, C.J. O'Neill, A.F. Toovey, S. Popovski, S.M. Rea, C.L. Pimm & L. Klein (2004). Reducing methane emissions in sheep by immunization against rumen methanogens. *Vaccine*, (in press).

Relationships between biodiversity and production in grasslands at local and regional scales

A. Hector[1] and M. Loreau[2]

[1]*Institute for Environmental Sciences, University of Zürich, CH-8057, Zürich, Switzerland*
Email: ahector@uwinst.unizh.ch
[2]*Laboratoire d'Ecologie, UMR 7625, Ecole Normale Supérieure, 46 Rue d'Ulm, F-75230 Paris Cedex 05, France*

Key points

1. Experimental manipulations of plant species diversity in unfertilised prairies and meadows has revealed that increasing diversity often leads to increased productivity (range of observed relationships varies from flat to log-linearly positive); driven by a combination of facilitation, niche-partitioning and sampling/selection effects.
2. The longer-term effects of diversity on ecosystem stability are not as clear and in need of further work.
3. Recent applied work, and a new review of the grassland literature, both show the potential for biodiversity to increase productivity under realistic field conditions.
4. The longer-term feedback of grazers on biodiversity gradients is unknown, and grassland biodiversity experiments that incorporate grazers will be needed to test whether patterns differ from those seen in ungrazed prairies and meadows.
5. The relationship between diversity and productivity seen in local experiments is often different from regional-scale correlations, and the scaling-up of experimental results remains a research priority.

Keywords: grazing, ecosystem functioning

Local and regional relationships

One of the main debates in ecology during the last few years has focused on the relationship between diversity and productivity, largely based on data from grasslands. A major factor contributing to this so-called 'biodiversity debate' is a failure to clearly distinguish between patterns and processes that act at local, versus regional scales. In trying to explain large-scale patterns in diversity, many ecologists have focused on productivity as a key factor; generations of ecologists are used to seeing graphs in which diversity is plotted as a function of productivity. While the frequency of different types of patterns (positive, negative, unimodal etc.) and their causes are still open to debate (Grace 1999; Mittelbach *et al.,* 2001), diversity is often correlated with productivity at large-scales (from regional to global). What has been less well researched is the biotic feedback: can local diversity itself influence productivity? The newer question - how does biodiversity influence productivity and ecosystem functioning - was prompted by concern over the ongoing loss of species from ecosystems. The potential for confusion is immediately obvious since now the scale has changed and the graph is reversed: productivity is now the response variable and diversity the explanatory variable. In larger-scale patterns, productivity and diversity are both covarying with environmental factors that change from place to place, while the local influence of diversity on ecosystem functioning will depend on the traits of the species present and their ecological interactions (Loreau 1998; Loreau *et al.,* 2001). The methodologies employed to study these two questions also vary. Larger-scale patterns are usually investigated with the analysis of collected observational data while the local influence of diversity on productivity has been investigated by the controlled experimental manipulation of diversity, both by removal experiments and with experimentally controlled diversity gradients ('biodiversity experiments'). While the relationship between biodiversity and ecosystem functioning has only existed as a focused research area for about a decade (since the book edited by Schulze & Mooney, 1993), the conceptual link can be traced all the way back to Darwin (Hector & Hooper 2002). The correlation between diversity and productivity at larger scales is the subject of several reviews (Ricklefs & Schulter 1993; Grace 1999; Mittelbach et al., 2001), consequently this paper will focus on how changes in local diversity within grasslands can affect productivity and other ecosystem processes and services. This area has been comprehensively reviewed by Loreau et al. (2001); Kinzig et al. (2002); Loreau et al. (2002b); Hooper et al. (in press), and so the focus will be on the most important ideas, experiments and results. Major stages in the development of pure research in this area are reviewed before looking at more applied studies on the potential value of biodiversity in grasslands.

Grassland biodiversity and production

The first controlled experiment deliberately designed to test the relationship between biodiversity and ecosystem functioning was carried out not in grassland, but with experimental communities of annual species grown in an Ecotron controlled environment facility (Naeem et al., 1994). The experiment compared ecosystem processes measured in a hypothetical intact community with those in two depauperate versions that had species omitted at random. Diversity was simultaneously reduced at four trophic levels. The key result from this experiment was that the depauperate communities were less productive. However, since only a single intact community and two increasingly depauperate versions were compared the generality of the result was not clear - the results could have been specific to the particular order of species extinction examined. It was also impossible to separate effects at different trophic levels and so the mechanism generating the patterns was not clear.

In response to these limitations, Tilman et al. (1996) conducted a field experiment on Minnesota prairie grassland at Cedar Creek where they established an experimental diversity gradient, but where each level of diversity (a given number of species) was replicated with different mixtures of species selected at random from the species pool. Productivity was again positively related to diversity, the longer diversity gradient producing an asymptotic curve (on a log scale the pattern proved to be linear, see Figure 1A). While the random selection of species mixtures within diversity levels produced a more general result, the mechanism was still unclear. Interpretations had tended to focus on ecological 'niche differentiation' as the likely underlying mechanism, but a simpler 'sampling' effect had been initially missed. When combined in mixed communities some species in these experiments increased in abundance while others decreased. The sampling effect hypothesis (Huston, 1997) proposed that if relative abundance in a mixture is positively related to productivity in monoculture, then productive species would come to dominate mixtures and this effect will be stronger at higher diversity since there is a greater chance of randomly selecting the most productive species. Simple models demonstrated that with these assumptions, dominance of mixtures by productive species could generate a positive asymptotic curve in the absence of other ecological differences between species.

However Hooper & Vitousek (1997) performed a further biodiversity experiment on Californian grassland at Jasper Ridge, in which they combined different functional groups of species (early season annuals, late season annuals, nitrogen fixing legumes and perennial bunchgrasses). In this instance the results were more complex: there was evidence for resource partitioning and facilitation between the different groups of species, but the overall pattern was flat - no significant effect of diversity (Figure 1C). It subsequently transpired that in this system the experimental communities were dominated not by the high-biomass perennial bunchgrasses, but by low-biomass early season annuals. Competition appeared to be largely for nitrogen and the early season annuals were somehow the best nitrogen competitors despite their small shoot and root systems. Dominance of communities by low-biomass species resulted in the opposite of the early sampling effect models - a negative sampling effect. This negative sampling effect cancelled out the effects of resource partitioning and positive interactions to produce the flat relationship between diversity and productivity on average.

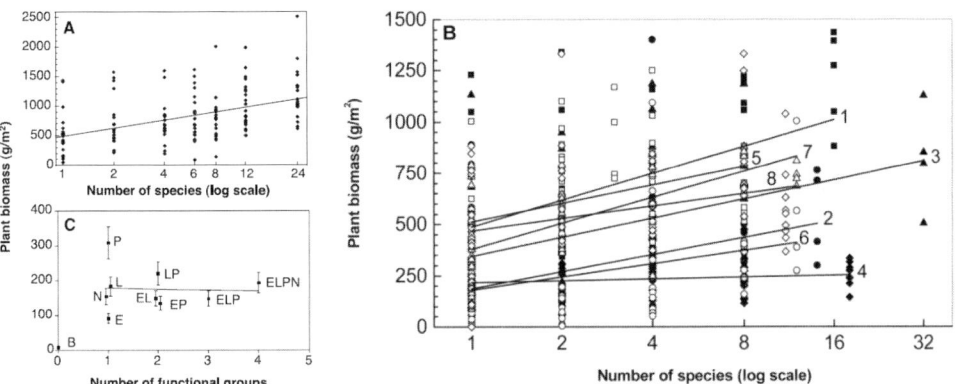

Figure 1 Examples of responses of total (A) or aboveground (B and C) plant biomass (in g/m^2) to experimental manipulations of plant species richness (A, B) or functional-group richness (C) in grasslands in Minnesota (A) (Tilman *et al.*, 1996), across Europe (B) (Hector *et al.*, 1999), and in California (C) (Hooper & Vitousek, 1997). Points in (A) and (B) are data for individual plots. In (B) different regression slopes are shown for the eight sites to focus on between-location differences rather then the general log-linear relationship reported elsewhere (Hector *et al.*, 1999). Closed squares = Germany, line 1; closed circles = Portugal, line 2; closed triangles = Switzerland, line 3; solid diamonds = Greece, line 4; open squares = Ireland, line 5; open circles = Sweden, line 6; open diamonds = Sheffield (UK), line 7; open diamonds = Silwood Park (UK), line 8. Symbols in (C) correspond to functional groups and their combinations: B = bare ground, E = early-season annuals, L = late-season annuals, P = perennial bunchgrasses, N = N fixers. Reproduced from Loreau *et al.*, (2001) with permission.

At this point in time, there were only small numbers of data from single locations available, and it was unclear whether the relationship was highly idiosyncratic or whether some patterns tended to be more common than others. The EC-funded BIODEPTH project (BIODiversity and Ecosystem Processes in Terrestrial Herbaceous communities) deliberately set out to test the generality of relationships between biodiversity and ecosystem processes in European grasslands. The same biodiversity experiment was conducted at eight different grassland sites using standardised methodologies. In the main, ecological studies are carried out by individual research groups at individual locations, and it is interesting to speculate how results would have been interpreted had the experiments been conducted individually? When viewed in this way, patterns varied from site to site between the extremes seen in the Cedar Creek and Jasper Ridge studies (Figure 1B). The Greek BIODEPTH site showed no significant effect of diversity on productivity while other sites showed the positive log-linear curve. At other sites effects were nonlinear. However, when combined in a single analysis the interaction between sites and species richness was not significant, producing a final statistical model with eight parallel log-linear curves. Thus, all else being equal, a progressive loss of diversity from European grasslands can be expected, resulting in accelerating declines in productivity - with variation around the average expectation that includes both stronger and weaker (flat) relationships. No negative effect of diversity on production was found in this study (or from any other study to date).

While the range of different patterns and the typical result was now better understood, the contribution of different underlying mechanisms was still unknown. What was needed was a method that could take the overall pattern and divide it into the portion due to sampling-type effects and the portion due to niche-partitioning and facilitative effects. Loreau & Hector (2001) devised a new methodology to do just this. The methodology extends relative yield approaches previously used in intercropping and plant ecology. It defines a net effect of biodiversity, which is the difference between the observed yield of a mixture and that of the average monoculture. This net effect is then partitioned into two additive components: the selection and complementarity effects. The selection effect is the standard statistical measure of covariance applied to the relationship between yield in monoculture and relative yield in mixture. Selection effects will be positive when species with higher-than-average yield dominate communities and negative when species with lower-than-average yield dominate. The complementarity effect uses relative yields to ask whether increases in the abundance of some species exactly cancel declines in others. When this is the case, resource partitioning is a zero-sum game with some species taking more of a fixed total pool of resources and others taking less. Positive complementarity effects occur when decreases in the abundances of some species do not compensate for the increases in the abundance of other species, and could result from facilitation, resource partitioning or decreased impact of natural enemies in more diverse communities. When applied to the BIODEPTH productivity data, the additive partitioning equation revealed that the complementarity effect drives the relationships more than the selection effect. Selection effects were sometimes positive and sometimes negative. Negative selection effects have since emerged as a widespread result, even though ecologists predicting the results of biodiversity experiments did not anticipate them. One consequence is that complementarity effects are effectively partly 'hidden' by negative selection effects that counter and mask them.

While the patterns and contribution of selection and complementarity effects had now become clearer, the biological mechanisms behind these effects were still largely untested. In particular, complementarity effects could be driven entirely by facilitation between nitrogen-fixing legumes and other species (note that this effect can also be seen as resource partitioning in which nitrogen is accessed from the soil or atmosphere). The existing studies had not been designed to isolate the effects of legumes and were limited in how well they could address this question. The problem was that more diverse plots were also more likely to contain legumes, and to contain more species of legumes. The effects of different components of biodiversity are collinear and whichever is entered first into statistical models often takes the greater portion of the shared variation (Schmid et al., 2002). To better test whether positive biodiversity effects hinged on the presence of legumes, van Ruijven & Berendse (2003), established a new biodiversity experiment using many of the species found in the BIODEPTH project but omitting legumes from all communities. To their surprise the experiment produced a strong positive relationship similar to those seen in the earlier studies. It appears that while legumes play an important role in generating relationships between biodiversity and productivity, they may also sometimes mask the effects of other aspects of diversity. For example, natural enemies play a role in generating the biodiversity effects seen in the work by van Ruijven & Berendse (2003), effects that may have been hidden had legumes been included.

The main Cedar Creek biodiversity experiment (Tilman *et al.*, 2001), the Jasper Ridge experiment (Hooper & Dukes, 2004) and the BIODEPTH project fieldsites (Hector *et al.*, *in press*; Spehn *et al.*, *in press*) have all been monitored in the longer term. In general, the strength of the relationship between biodiversity and ecosystem processes in these studies became stronger, with complementarity effects playing a greater role. A wide range of ecosystem processes in addition to primary production are also affected by diversity.

How many species?

To determine how many species are important in biodiversity effects is difficult both in principle and in practice. When there is a positive relationship between diversity and productivity it is often asymptotic, and so the difficulty then is to decide where on the continuum is the cut-off point at which there are enough species to provide the appropriate level of an ecosystem process. An alternative would be to quantify how many species provide different levels of productivity. Tilman *et al.* (2001, 2002) addressed this question by ranking species from most to least productive in their high-diversity community. They then constructed diversity indices that quantified how many of the 2, 3, etc, most productive species were present in a plot. When above and belowground biomass was analysed as a function of the different diversity indices, they found that indices with between 9 and 13 of the most productive species produced the highest r^2s. Thus, in these analyses, a large proportion of the species present were needed to best explain the productivity of a plot. Additional analyses of species-specific contributions suggest that much of the biodiversity effects can be explained by legumes and C4 grasses, but with additional species coexisting alongside them and contributing to total productivity (Lambers *et al.*, 2004). In contrast, the BioCON experiment at Cedar Creek (a combination of biodiversity gradients, elevated carbon dioxide and nitrogen fertilisation) suggests that the greater range of conditions present in this experiment provides opportunities for a greater range of species to contribute to the biodiversity effect (Reich *et al.*, 2004).

Functional groups

Can species be amalgamated into functional groups that impact ecosystem processes in similar ways? The experiments discussed above used traditional functional group schemes based on traits expected to impact ecosystem processes. Legumes are grouped due to their ability to fix nitrogen from the atmosphere, and C4 and C3 grasses are separated on the basis of their different photosynthetic pathways. Other herbs (forbs) are more difficult to group but can be divided with respect to growth form (rosette versus tall), rooting depth etc. In many cases these groups do capture many of the functionally important differences between species, but frequently additional significant effects of species richness remain after controlling for the effects of functional groups. In particular, this may be the case when there is a high degree of environmental heterogeneity. In the BioCON experiment (which examined the same biodiversity gradient under different conditions of CO_2 and N enrichment), species and functional group richness had largely independent effects across the whole range of conditions, such that species within groups were not functionally redundant but made separate contributions (Reich et al., 2004). In situations like this a more appropriate approach may not be to force species into groups, but to search for continuous measures of functional diversity (Petchey, 2002; Petchey et al., 2004).

Another point of debate is whether the functional traits and groups that distinguish species across large-scale gradients are the same traits that are functionally relevant within a particular ecosystem. For example, specific leaf area seems to provide a distinguishing trait when analysing patterns across different ecosystems, or the same type of ecosystem in different geographic locations. However, it is not clear if this is a particularly useful functional trait when looking at a particular community of species in one location.

Other ecosystem processes

Most of the biodiversity experiments to date, particularly those with plants, have tended to focus on biomass production as a key ecosystem process. While production is of high interest, particularly in grasslands, other ecosystem processes have received less attention than they deserve. Some ecosystem processes can be integrated with biomass production in a logical fashion. For example, in some biodiversity experiments a positive relationship between diversity and biomass is mirrored by a negative relationship with soil nitrogen (Tilman et al., 1996; Scherer-Lorenzen et al., 2003). Lower levels of free resources would be expected if production is higher due to more complete resource capture in high-diversity plots. However, the relationship between plants and soil resources can be complex - particularly when nitrogen fixers are included in communities. Decomposition rates often appear to be less sensitive to changes in diversity, at least in the short-term, than is production (Hector et al., 2000; Knops et al., 2001), which matches with build up of biomass in many studies (Reich et al., 2001). One of the new areas of development in biodiversity experiments is the intention to take a broader whole-system view focusing on element cycles (Roscher et al., 2004).

Diversity and stability

Theory predicts that biodiversity should have a 'portfolio' or 'insurance' effect if asynchronies in the fluctuations of species populations serve to average out fluctuations at the ecosystem level (e.g. total community biomass) (Loreau et al., 2002a). Unfortunately, experimentation has not kept pace with the development of new theory, and to date only a single experimental test exists for grasslands. Pfisterer & Schmid (2002) subjected the Swiss BIODEPTH biodiversity gradient to an artificial drought. Unexpectedly, they found that in absolute terms high diversity communities showed a greater reduction in biomass than did lower diversity communities. However, it must be kept in mind that the relationship between diversity and productivity before drought was already positive. When looked at in relative terms all communities suffered a similar proportional reduction in biomass. Work from the Portuguese BIODEPTH experiment (Caldeira et al., in press) shows similar patterns. Thus, so far there is no strong experimental evidence that biodiversity acts as insurance in natural ecosystems. This lack of evidence, however, may be due to the small number and limited length of studies performed on this issue.

Management strategies, sustainability and ecosystem services in grasslands

All of the experiments reviewed above are from basic ecology. They were intended to perform the first general tests on the influence of biodiversity on ecosystem functioning. However, results from pure science do not always translate into more applied situations in the real world. Bullock *et al.* (2001) set out to perform a more realistic biodiversity experiment. They selected six field sites in the South East of England with similar soil type. The UK Ministry of Agriculture, Fisheries and Food (MAFF) recommended particular seed mixtures as appropriate for these site conditions. At each field site, Bullock and colleagues compared the performance of the recommend seed mixtures with higher-diversity mixtures where they supplemented the recommended mixtures with extra appropriate species. The result was a linear increase in yield with the extra species, a result consistent across the six field sites. The economic assessment obviously depends on the worth of the extra hay produced, how long this effect persists and the price of the seed. Nonetheless, under realistic conditions Bullock and colleagues showed that notwithstanding the recommendations of an expert organisation, increasing biodiversity lead to increased production.

Grazing

All of the grassland biodiversity experiments to date have excluded large grazers. The study systems have been prairies where periodic fires maintain the grassland or meadows where humans act as the main grazer by swathing hay. Sanderson *et al.* (2004) reviewed the grazing literature and suggest that biodiversity experiments incorporating grazers are the next area awaiting good experimentation. Their review of the relevant applied literature produced some old results showing that at least some grassland mixtures designed for pastures exhibit a positive relationship between diversity and production (Figure 2). The untested aspect is what effect grazing would have on diversity; the feasibility of maintaining a diversity gradient once grazers have been added, is not known. While difficult to perform, grassland biodiversity experiments that include grazers will be needed to determine whether relationships differ from those observed in ungrazed prairies and meadows.

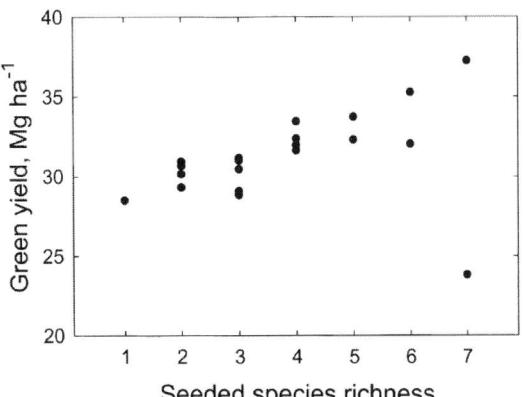

Figure 2 Positive relationship between diversity and yield of fresh green forage in selected grass-legume mixtures grown under grazing and irrigation at Logan, Utah, from 1947 to 1951. Data points are averages of 5 years (Reproduced from Sanderson *et al.*, (2004) with permission).

Scaling-up

One of the major objections to the positive relationships found in experiments is the apparent conflict with correlational patterns found at regional scales (Grace 1999; Mittelbach *et al.*, 2001). A suggested resolution to this apparent conflict is that in experiments the direct effect of biodiversity on ecosystem processes is isolated, whereas in larger-scale surveys both diversity and ecosystem processes are driven by changes in other environmental variables (Loreau, 1998; Loreau *et al.*, 2001). Levine (2000) provides an enlightening example of how local and regional relationships can vary in relation to different ecological processes acting at these different scales.

However, this poses a common problem in ecology: how to scale-up small-scale experimental results to the larger landscape and regional scales? There have been few attempts to do this. Tilman *et al.* (2002) used the empirical species-area approach to predict how many species would be needed to maintain ecosystem processes within the Great Plains prairies (based on work at the Cedar Creek site). This empirical approach is useful as an order-of-magnitude guess about the functional consequences of biodiversity at larger spatial scales. However, it ignores the dynamics of both diversity and ecosystem processes at these scales. Recent theoretical work on metacommunities (sets of communities connected by dispersal of organisms) shows that landscape connectivity can have dramatic influences on diversity and productivity, and predicts that local species diversity, productivity and ecosystem stability will all be highest at intermediate rates of dispersal across communities (Loreau *et al.*, 2003). Thus, investigating processes at landscape and regional scales becomes an urgent need to predict the changes and functional consequences of biodiversity, at the scales where human influences are strongest.

Conclusions

Experimental manipulations of diversity in unfertilised prairies and meadows has revealed that increasing diversity often leads to increased productivity, although the range of observed relationship varies from flat to linearly positive. The effects are driven by a combination of facilitation, niche-partitioning and sampling/selection effects. The longer-term effects of diversity on ecosystem stability are not as clear and in need of further work. The local patterns revealed by biodiversity experiments often differ from regional-scale correlations between diversity and productivity, and the scaling-up of experimental results remains an important challenge.

References

Bullock, J.M., R.F. Pywell, M.J.W. Burke & K.J. Walker (2001). Restoration of biodiversity enhances agricultural production. *Ecology Letters*, 4, 185-189.

Caldeira, M., Hector, A., Loreau, M. & J.S. Pereira (*in press*). Species richness, temporal variability and resistance of biomass production in a mediterranean grassland. *Oikos*.

Grace, J.B. (1999). The factors controlling species density in herbaceous plant communities: an assessment. *Perspectives in Plant Ecology, Evolution and Systematics*, 2, 1–28.

Hector, A., E. Bazeley-White, E.M. Spehn, J. Joshi, M. Scherer-Lorenzen, B. Schmid, C. Beierkuhnlein, M.C. Caldeira, M. Diemer, P.G. Dimitrakopoulos, J. Finn, H. Freitas, P.S. Giller, J. Good, R. Harris, P. Högberg, K. Huss-Danell, A. Jumpponen, C. Körner, P.W. Leadley, M. Loreau, A. Minns, C.P.H. Mulder, G. O'Donovan, S.J. Otway, C. Palmborg, J.S. Pereira, A.B. Pfisterer, A. Prinz, D.J. Read, E.D. Schulze, A.S.D. Siamantziouras, A.C. Terry, A.Y. Troumbis, F.I. Woodward, S. Yachi & J.H. Lawton. Ecosystem effects of the manipulation of plant diversity in European grasslands: data from the BIODEPTH project. *Ecological Archives*, M075-001-S1.

Hector, A., A. Beale, A. Minns, S. Otway & J.H. Lawton (2000). Consequences of loss of plant diversity for litter decomposition: mechanisms of litter quality and microenvironment. *Oikos*, 90, 357-371.

Hector, A. & R.E. Hooper (2002). Darwin and the first ecological experiment. *Science*, 295, 639-640.

Hector, A., B. Schmid, C. Beierkuhnlein, M.C. Caldeira, M. Diemer, P.G. Dimitrakopoulos, J.A. Finn, H. Freitas, P.S. Giller, J. Good, R. Harris, P. Högberg, K. Huss-Danell, J. Joshi, A. Jumpponen, C. Körner, P.W. Leadley, M. Loreau, A. Minns, C.P.H. Mulder, G. O'Donovan, S.J. Otway, J.S. Pereira, A. Prinz, D.J. Read, M. Scherer-Lorenzen, E.D. Schulze, A.S.D. Siamantziouras, E.M. Spehn, A.C. Terry, A.Y. Troumbis, F.I. Woodward, S. Yachi & J.H. Lawton (1999). Plant diversity and productivity experiments in European grasslands. *Science*, 286, 1123-1127.

Hooper, D.U., F.S.I. Chapin, J.J. Ewel, A. Hector, P. Inchausti, S. Lavorel, J.H. Lawton, D.M. Lodge, M. Loreau, S. Naeem, B. Schmid, H. Setälä, A.J. Symstad, J. Vandermeer & D.A. Wardle. Effects of biodiversity on ecosystem functioning: a consensus of current knowledge and needs for future research. *Ecological Monographs*, 75, 3-35.

Hooper, D.U. & J.S. Dukes (2004). Over yielding among plant functional groups in a long-term experiment. *Ecology Letters*, 7, 95-105.

Hooper, D.U. & P.M. Vitousek (1997). The effects of plant composition and diversity on ecosystem processes. *Science*, 277, 1302-1305.

Huston, M.A. (1997). Hidden treatments in ecological experiments: re-evaluating the ecosystem function of biodiversity. *Oecologia*, 110, 449-460.

Kinzig, A., D. Tilman & S. Pacala (2002). The functional consequences of biodiversity: empirical progress and theoretical extensions. Princeton University Press, Princeton, 392pp.

Knops, J.M.H., D. Wedin & D. Tilman (2001). Biodiversity and decomposition in experimental grassland ecosystems. *Oecologia*, 126, 429-433.

Lambers, J.H.R., W.S. Harpole, D. Tilman, J. Knops & P.B. Reich (2004). Mechanisms responsible for the positive diversity-productivity relationship in Minnesota grasslands. *Ecology Letters*, 7, 661-668.

Levine, J.M. (2000). Species diversity and biological invasions: relating local process to community pattern. *Science*, 288, 852-854.

Loreau, M. (1998). Biodiversity and ecosystem functioning: a mechanistic model. *Proceedings of the National Academy of Sciences of the USA*, 95, 5632–5636.

Loreau, M., A. Downing, M. Emmerson, A. Gonzalez, J. Hughes, P. Inchausti, J. Joshi, J. Norberg & O. Sala (2002a). A new look at the relationship between diversity and stability. In: M. Loreau, S. Naeem & P. Inchausti (eds.) Biodiversity and ecosystem functioning: synthesis and perspectives. Oxford University Press, Oxford, 79-91.

Loreau, M. & A. Hector (2001). Partitioning selection and complementarity in biodiversity experiments. *Nature*, 412, 72-76 [erratum: 413, 548pp.]

Loreau, M., N. Mouquet & A. Gonzalez (2003). Biodiversity as spatial insurance in heterogeneous landscapes. *Proceedings of the National Academy of Sciences of the USA*, 100, 12765–12770.

Loreau, M., S. Naeem & P. Inchausti (2002b). Biodiversity and ecosystem functioning: synthesis and perspectives. Oxford University Press, Oxford, 306pp.

Loreau, M., S. Naeem, P. Inchausti, J. Bengtsson, J.P. Grime, A. Hector, D.U. Hooper, M.A. Huston, D. Raffaelli, B. Schmid, D. Tilman & D.A. Wardle (2001). Biodiversity and ecosystem functioning: current knowledge and future challenges. *Science*, 294, 804-809.

Mittelbach, G.G., C.F. Steiner, S.M. Scheiner, K.L. Gross, H.L. Reynolds, R.B. Waide, M.R. Willig, S.I. Dodson & L. Gough (2001). What is the observed relationship between species richness and productivity? *Ecology*, 82, 2381-2396.

Naeem, S., L.J. Thompson, S.P. Lawler, J.H. Lawton & R.M. Woodfin (1994). Declining biodiversity can alter the performance of ecosystems. *Nature*, 368, 734-737.

Petchy, O.L. (2002). Functional diversity (FD), species richness and community composition. *Ecology Letters*, 5, 402-411.

Petchey, O.L., A. Hector & K.J. Gaston (2004). How do different measures of functional diversity perform? *Ecology*, 85, 847-857.

Pfisterer, A.B. & B. Schmid (2002). Diversity-dependent production can decrease the stability of ecosystem functioning. *Nature*, 416, 84-86.

Reich, P.B., J.M.H. Knops, D. Tilman, J. Craine, D. Ellsworth, M. Tjoelker, T. Lee, D. Wedin, S. Naeem, D. Bahaudlin, G. Hendrey, J. Shibu, K. Wrage, J. Goth & W. Bengston (2001). Plant diversity enhances ecosystem responses to elevated CO_2 and nitrogen deposition. *Nature*, 410, 809-812.

Reich, P.B., D. Tilman, S. Naeem, D.S. Ellsworth, J. Knops, J. Craine, D. Wedin & J. Trost (2004). Species and functional group diversity independently influence biomass accumulation and its response to CO_2 and N. *Proceedings of the National Academy of Sciences of the USA*, 101, 10101-10106.

Ricklefs, R.E. & D. Schulter (1993). Species Diversity in Ecological Communities. Chicago University Press, Chicago, 414pp.

Roscher, C., J. Schumacher, J. Baade, W. Wilcke, G. Gleixner, W. Weisser, B. Schmid & E.D. Schulze (2004). The role of biodiversity for element cycling and trophic interactions: an experimental approach in a grassland community. *Basic and Applied Ecology*, 5, 107-121.

Ruijven, J.V. & F. Berendse (2003). Positive effects of plant species diversity on productivity in the absence of legumes. *Ecology Letters*, 6, 170-175.

Sanderson, M.A., R.H. Skinner, D.J. Barker, G.R. Edwards, B.F. Tracy & D.A. Wedin (2004). Plant species diversity and management of temperate forage and grazing land ecosystems. *Crop Science*, 44, 1132-1144.

Scherer-Lorenzen, M., C. Palmborg, A. Prinz & E.D. Schulze (2003). The role of plant diversity and composition for nitrate leaching in grasslands. *Ecology*, 84, 1539-1552.

Schmid, B., A. Hector, M.A. Huston, P. Inchausti, I. Nijs, P.W. Leadley & D. Tilman (2002). The design and analysis of biodiversity experiments. In: M. Loreau, S. Naeem & P. Inchausti (eds.) Biodiversity and ecosystem functioning: synthesis and perspectives. Oxford University Press, Oxford, 61-78.

Schulze, E.D. & H.A. Mooney (1993). Biodiversity and ecosystem function. Springer-Verlag, Berlin, 528pp.

Spehn, E.M., A. Hector, J. Joshi, M. Scherer-Lorenzen, B. Schmid, E. Bazeley-White, C. Beierkuhnlein, M.C. Caldeira, M. Diemer, P.G. Dimitrakopoulos, J. Finn, H. Freitas, P.S. Giller, J. Good, R. Harris, P. Högberg, K. Huss-Danell, A. Jumpponen, P.W. Leadley, M. Loreau, A. Minns, C.P.H. Mulder, G. O'Donovan, S.J. Otway, C. Palmborg, J.S. Pereira, A.B. Pfisterer, A. Prinz, D.J. Read., E.D. Schulze, A.S.D. Siamantziouras, A.C. Terry, A.Y. Troumbis, F.I. Woodward, S. Yachi & J.H. Lawton. Ecosystem effects of the manipulation of plant diversity in European grasslands. *Ecological Monographs*, (in press).

Tilman, D., J.M.H. Knops, D. Wedin & P.B. Reich (2002). Experimental and observational studies of diversity, productivity and stability. In: A. Kinzig, D. Tilman & S. Pacala (eds.) Functional consequences of biodiversity: experimental progress and theoretical extensions. Princeton University Press, Princeton, 42-70.

Tilman, D., P.B. Reich, J.M.H. Knops, D. Wedin, T. Mielke & C. Lehman (2001). Diversity and productivity in a long-term grassland experiment. *Science*, 294, 843-845.

Tilman, D., D. Wedin & J. Knops (1996). Productivity and sustainability influenced by biodiversity in grassland ecosystems. *Nature*, 379, 718-720.

Enhancing grassland biodiversity and its consequences for grassland management and utilisation

J. Isselstein
Institute of Agronomy and Plant Breeding, University of Goettingen, Von-Siebold-Str. 8, 37075 Goettingen, Germany
Email: jissels@gwdg.de

Key points

1. Grasslands make an important contribution to the biodiversity of rural landscapes.
2. Biodiversity has the potential to support the production function of grassland. The conditions for this are largely unexplored.
3. The enhancement of biodiversity on agriculturally improved, species-poor grasslands is difficult to achieve due to seed limitation and high residual soil fertility.
4. Measures to overcome constraints are addition of propagules, depletion of soil nutrients, and the use of specific sward treatments.
5. Agri-environmental schemes will play an increasing role in achieving the biodiversity targets in the future, however, the efficiency of such schemes needs improvement.

Keywords: species richness, production, agri-environment schemes, seedling recruitment

Introduction

Grassland biodiversity has received a great deal of attention by society and scientists in recent years. As a definition, biodiversity describes the whole variability found among living organisms, and encompasses the diversity within species, between species, and of whole ecosystems. Agriculture represents a dominant form of land use all over the world, and rural landscapes make up a significant part of the earth's total biodiversity. Two reasons are responsible for the increased attention to biodiversity. Firstly, biodiversity is generally perceived as a heritage that mankind needs to preserve for future generations. The convention on biodiversity is a major expression of this perception. Secondly, biodiversity and ecosystem functioning are related. For grassland systems, it has been shown that biodiversity is positively related to the primary production, stability and resilience of a system, as well as the nutrient retention and efficiency (Tilman *et al.*, 1996; Loreau, 2000). This raises the hope that contrasting functions of grassland, e.g. production and biodiversity, may be brought together so that management systems can be developed to the benefit of both. However, recent findings from ecological research have not been seriously considered for being implemented in farming practice (Sanderson *et al.*, 2004a).

Agriculture and farming activities are closely related to the biodiversity at a particular site or of whole landscapes. Biodiversity has co-evolved with land-use practices since the onset of crop and livestock husbandry in prehistoric times. Agriculture has been responsible for both an increase and loss of biodiversity in rural areas; its actual effect being dependent on the intensity of land-use.

The challenges of current grassland farming with regard to biodiversity are: 1) traditionally managed species-rich grasslands are at risk of being abandoned with an attendant loss of species diversity. How can biodiversity be maintained through agricultural management? 2) Can the diversity of agriculturally improved grasslands with a low species number be enhanced, and which management measures will be effective in this respect (Marriott et al., 2004)? Although the decrease in plant species diversity in grassland through the intensification of grassland use has been well investigated and the processes involved are generally understood, the situation is much less clear for other organisms and the appropriate measures for maintaining or enhancing their diversity have also not been elucidated.

The objectives of this paper are to investigate the potential benefits of biodiversity on the production function of grassland, and to present ways in which biodiversity can be increased on sites that have experienced a history of intensive management leading to a concomitant deterioration of diversity. Finally, consideration will be given to those management measures appropriate to achieving the goal of increased grassland biodiversity and to the consequences for grassland farming.

The potential benefits of utilising biodiverse grasslands

The biodiversity of semi-natural species-rich grasslands has developed over centuries under the influence of traditional human management (Ellenberg, 1996). It is the result of a broad range of environmental conditions interacting with a large variety of grassland utilisation systems for livestock production (Nösberger & Rodriguez, 1996). In temperate climates, there is only a relatively small amount of 'natural' grassland still in existence. Most of the grasslands require regular utilisation if their botanical composition and plant species diversity are to be maintained. Grassland that is abandoned from agricultural use shows successional change with an increasing dominance of tall species and the invasion of woody plants. In many situations, this will cause a decrease in plant species diversity. Across Europe, semi-natural grasslands are becoming increasingly at risk of being abandoned from agricultural use as these grasslands produce low herbage yields compared to intensively managed grasslands, and the feeding value of the herbage often fails to fulfil the energy requirements of livestock of high genetic merit (Peeters & Janssens, 1998; Tallowin & Jefferson, 1999). However, these grasslands are highly valued by society, since they have a high biodiversity and are attractive for tourists and urban populations seeking recreation (Nösberger & Rodriguez, 1996; Schüpbach et al., 2004). In addition, species-rich grasslands are important for agriculture as they provide a reservoir of plant genotypes, which can be, used in future plant improvement schemes (Nösberger & Rodriguez, 1996).

Grassland that is utilised for intensive livestock husbandry is usually characterised by low species diversity with only a few valuable forage species dominating the grass swards. These species are introduced by sowing, they have a high yield potential and feeding value. High doses of mineral or organic fertilisers are required to maintain the dominance of these species. The herbage is defoliated frequently and stocking rates are high. In addition, grasslands are regularly renovated by reseeding or oversowing. Given this situation, farmers perceive species-rich grassland mainly as a constraint to productive livestock husbandry as traditionally, species-rich semi-natural grasslands have a low herbage production and quality, leading to the common perception of farmers that species-rich grasslands are not a valuable resource for profitable livestock production.

In contrast, recent ecological research on the relationship between biodiversity and ecosystem services suggests that preserving and restoring biodiversity has the potential to support the agricultural function of grasslands (Minns et al., 2001). Three main aspects of this function will be considered: 1) herbage production, 2) the stability and resilience of swards, including susceptibility to weed invasion, and 3) nutrient cycling.

Herbage production

In recent years, grasslands have been the focus of ecological research on the relationship between plant species diversity and productivity (Tilman *et al.*, 1996). A prominent experiment was performed on European grasslands at eight different sites. Its results showed an advantage of increasing plant species numbers on the aboveground productivity (Hector *et al.*, 1999). Plant communities were established on sterilised soils by random selection from a species pool and were maintained by weeding. A controversy developed over the mechanisms behind this relationship (Loreau *et al.*, 2001). It was proposed that facilitation among plant species and niche-differentiation were the major processes involved. However, a 'sampling effect' was also suggested, which may have arisen due to the increased probability of selecting highly productive species with an increasing number of species in the mixture (Wardle, 1999).

The number of functional traits that are represented by the plant species in the mixture is also positively related to production. The inclusion of nitrogen-fixing legumes usually gives a leap in the productivity of unfertilised swards. The addition of common non-legume species has also been shown to support the relationship between diversity and productivity. Such species may have other special traits, i.e. many forbs with deep root systems or the grasses with a shallow, but intensive root system (van Ruijvens & Berendse, 2003).

Ecologists have suggested that the relationship between diversity and productivity should be better exploited to the benefit of grassland farming (Bullock *et al.*, 2001; Minns *et al.*, 2001). There is, however, little evidence that this can be readily done by standard farming practices. Most of the experiments done in this field were not designed to produce data that could be directly applied to farming nor could it be used in current livestock production systems. When summarising the currently available knowledge, a number of constraints can be identified.

Previous investigations on the diversity-productivity relationship were based on sown grasslands ('experimental grasslands') and yields were followed over a period of up to five years. This is a ley-farming situation where grasslands are established on arable land for a couple of years and then returned to an arable crop rotation (Hopkins, 2000). In the first year in particular after these grasslands are ploughed up or re-established by sowing, but also in the following years, the nutrient flows and organic matter turnover are highly dynamic and in a state of imbalance (Whitehead, 1995). Thus, the sward response to varying species diversity may not reflect the situation of permanent grasslands that in many cases have not experienced any soil disturbance by tillage or seed addition for decades. However in many countries, permanent and semi-natural grassland covers a larger part of the total area farmed than short-term leys.

The botanical composition of newly established swards also shows a dynamic development. The occurrence and cover of sown species as well as of those not sown entering the sward, may change considerably in the early years after sowing. The direction and amount of change is dependent on the site and the management conditions. Deduced from long-term studies, the dynamic of change of the sward's botanical composition after sowing will decrease with time (Arens, 1971). This experience has been indirectly confirmed by recent research with 'experimental grasslands', where weeding of plots were pivotal in maintaining the intended species number and composition of the plots. If weeding was stopped, the differences between plots with a varying number of plant species diminished rapidly (Pfisterer *et al.*, 2004).

In order to avoid such artificial systems when studying the effect of species number on productivity, species removal from existing permanent swards and oversowing treatments have been proposed. This would leave the existing swards largely undisturbed, and little vegetation change due to disturbance would occur. However, only a few data from such experiments are available. In one multi-site experiment, Hopkins et al. (1999) found little establishment of oversown seed and thus little effect on herbage yield, compared to a control was not oversown. Hofmann & Isselstein (2005) achieved a successful establishment of oversown seeds of wildflower species into an existing species-poor hay-meadow. Four years after oversowing, the herbage yield of the oversown plots was 23% higher compared to the unsown control.

So far, ecological research with 'experimental grasslands' has focused on the primary production; grazing and livestock performance have not been considered. As farmers use the grasslands to feed their livestock and to produce milk, meat and fibre, the benefits of biodiversity for grassland farming should be measured in terms of animal performance (Sanderson et al., 2004a). This should include a determination of the feeding value of the herbage, the feed intake, and the production of milk, meat and fibre. The majority of grasslands worldwide are utilised by grazing. In addition, grazing behaviour and herbage intake are dependent on species diversity and sward heterogeneity (Rook et al., 2004a). Thus, the response of grazers and their performance to varying plant species diversity needs to be investigated. A pilot experiment in this respect was performed by Sanderson et al. (2004b), who sowed forage species mixtures with a varying number of species, and then measured the primary production as well as the performance of grazing animals. A limited response of livestock performance to varying levels of forage species diversity was found.

A further drawback of many biodiversity experiments with regard to their relevance for current grassland farming is the low defoliation frequency. In the BIODEPTH experiment (Hector et al., 1999), swards were cut once or twice a year. This is considerably less than the usual frequency on many intensive livestock farms. Hence, systems with a more frequent defoliation need to be included to study the effect of species number on production. There is some indication that the defoliation frequency or - more generally - the disturbance level has an impact on the biodiversity/productivity relationship. Sala (2001) suggested that the species complementarity that is assumed to be mainly responsible for a positive relationship is higher in communities that have a history of a long co-evolution of species. Accordingly, systems with a higher disturbance level should have a less strong species complementarity effect. However experimental verification for this is still missing.

Various experiments with sown grassland showed that a significant effect of plant species number on productivity occurred at a log-linear scale (Hector et al., 1999). This means that the lower the species number, the higher is the productivity response per added species. This result could be applied in grassland farming and research. For profitable grassland farming, only a limited number of valuable forage species are usually considered for use in seed mixtures. These are mainly a few productive grasses and legumes. The inclusion of other selected forage species like grassland forbs could lead to a significant increase in production without compromising forage quality. The necessary research in this field could be based on the huge experience in grass clover-systems that has been gathered during the last two decades. Another application in research could arise from the prospect of studying biodiversity effects on production in swards with a comparatively low species number. This would allow not only the investigation of the physiological processes involved in the response of production to biodiversity, but also the use of modelling in order to simulate yield formation, competition and facilitation among species, as well as niche-differentiation in the swards.

Stability and resilience of swards

Ecologists have hypothesised that species-rich grasslands maintain their ecosystem function and production more effectively if they face disturbance or stress through defoliation, disease infection or weed infestation (Knops *et al.*, 1999; Pywell *et al.*, 2002). For seed mixtures with an increasing number of selected forage species, it was found that weed infestation was unrelated to the number of sown species. However, weeds played a decreasing role with increased evenness of the sown plant community (Tracy *et al.*, 2004). It was concluded that maintaining an even distribution of competitive forage species in the mixture rather than the forage species number would be important in reducing weed invasion (Tracy *et al.*, 2004).

An important agronomic characteristic of grass swards is the variability of herbage production during the growing season. Most of the experiments on grassland biodiversity and ecosystem functioning include infrequent defoliation by cutting, so that so far, there is no experimental proof that diverse systems lead to a lower variability.

Nutrient cycling

Increasing plant species diversity may increase the nutrient uptake and nutrient efficiency of the vegetation. Plant species with different functional traits exploit different resources for their nutrients; examples being nitrogen fixation by legumes or the exploitation of soil nutrients in the lower horizons by deep rooting forbs. Nutrient retention was also found to be related to plant species diversity. Nitrate leaching in grasslands was lower with an increasing number of species, as excess nitrate in the soil occurred less frequently with diverse systems (Reich *et al.*, 2001; Scherer-Lorenzen *et al.*, 2003).

Enhancing the biodiversity of grasslands with a history of agricultural improvement

Maintaining species-rich semi-natural grasslands that have not experienced a history of agricultural improvement is a primary target of nature conservation activities in the rural landscape (Söderstrom *et al.*, 2001; Jeangros & Thomet, 2004). In countries where most of the grasslands were 'improved' during a period of intensification of livestock husbandry and thus, are species-poor, priority has also been given to the restoration of biodiversity (Marriott *et al.*, 2004). Today, a trend towards extensification of grassland use can be seen in many countries where dairy cows and beef cattle feeding ratios are more and more based on highly digestible forages from arable land rather than on grass (Feehan *et al.*, 2005). Therefore, developing extensive grassland systems primarily for biodiversity rather than for livestock production is a new target in a changing agri-environment, with a need to manage a relatively large grassland area with relatively few stock (Rook *et al.*, 2004b). Restoration activities on species-poor agriculturally improved grassland are mainly focussed on vegetation. It is commonly accepted that increased plant species diversity will also benefit the diversity of other living organisms, such as mammals, birds, invertebrates, or soil organisms (Duelli & Obrist, 1998). However, there is also evidence from research on semi-natural grassland, which shows that single taxa are not necessarily good indicators for total biodiversity (Vessby *et al.*, 2002). Therefore, restoration success should generally be evaluated by measuring a group of organisms rather than single taxa.

In order to enhance biodiversity, extensification objectives for grassland management are generally embedded in the many agri-environment schemes. Extensification means reduction or cessation of fertilisation, reduction of stocking rates and cutting frequencies, and the abandonment of sward improvement measures, like the application of herbicides or reseeding (Barthram *et al.*, 2002; Smith *et al.*, 2003, Marriott *et al.*, 2004). The efficiency of methods for the enhancement of biodiversity is discussed in separate sections for grassland vegetation and grassland fauna.

Plant species diversity

Many attempts at enhancing plant species richness on species poor grasslands by switching from intensive to extensive management have failed, at least in the short term (Dyckmans *et al.*, 1999; Marriott *et al.*, 2004). If at all, species number recovered only after several years of extensive management (Jeangros & Bertola, 2002; Smith *et al.*, 2003). According to van Diggelen & Marrs (2003), several reasons have to be considered for this result. They identified the necessary steps for a successful restoration of species-rich plant communities on agriculturally improved land: 1) the abiotic conditions have to be within the tolerance limits of the target vegetation; 2) propagules from species of the target community must be available, they must be viable and able to establish seedlings on the sites; and 3) grassland management needs to be adapted to the requirements of the introduced species.

(1)
Unimproved species-rich grasslands are characterised by a low nutrient availability in the soils. For species-rich grasslands on mineral soils, the phosphorus concentration is usually low and this is a key factor for the diversity of the swards (Janssens *et al.*, 1998; Critchley *et al.*, 2002). Various investigations have demonstrated that a moderate to high soil nutrient concentration is a major constraint to the restoration of diverse swards (Peeters & Janssens, 1998; Bakker & Berendse, 1999). After the onset of extensive management, the soil nutrient concentration may remain at a high level for some years, and productive forage species will maintain their vigorous growth and competitiveness (Koch & Masé, 2001). This will reduce the availability of regeneration niches for seedling recruitment (Grubb, 1977), and emerging seedlings will be eliminated through competition and shading by the existing sward (Muller *et al.*, 1998; Isselstein *et al.*, 2002). Peeters & Janssens (1998) have pointed out that unless the residual soil fertility and the soil phosphorus concentration have markedly decreased, restoration will not be successful. From earlier ecological research, it was concluded that on mesotrophic grassland, an annual herbage production above 6 t DM/ha is incompatible with a high plant species number (Oomes, 1992). High soil fertility has an adverse effect on the restoration success even if species-rich grasslands are created on ex-arable land and no existing vegetation suppresses seedling growth. Emerging arable weeds and ruderals will limit the establishment of the desired species (Hopkins *et al.*, 1999; Pywell *et al.*, 2002). In order to reduce the soil fertility and the nutrient availability, various depletion measures have been studied. Deturfing and topsoil removal is a most drastic but effective technique for the reduction of soil nutrients and the limitation of crop growth. Using this method, species-rich swards could be restored within a comparatively short time span (Hopkins *et al.*, 1999). Another technique employing deep cultivation to dilute the nutrient concentration was shown to reduce topsoil P and K concentrations, and to facilitate the establishment of sown forbs (Pywell *et al.*, 2002). Deturfing and deep cultivation are techniques which are not particularly favoured if sites are to enter agri-environment schemes (Walker *et al.*, 2004). Therefore, beyond the depletion of soil nutrients, an urgent need exists for the development of restoration techniques that are effective in the presence of elevated nutrient concentrations (Critchley *et al.*, 2002).

In addition to the overriding importance of soil nutrients for the restoration of plant species diversity, soil biotic diversity also plays a role (Wardle *et al.*, 2004). High input grassland systems tend to have a lower diversity of soil fauna, and favour bacterial-pathways of decomposition. In comparison, low-input systems have higher soil fauna diversity and favour fungal-pathways. Obviously, low-input management has a potential to increase soil biodiversity and therewith ecosystem self-regulation (Bardgett & Cook, 1998). The fungal:bacteria biomass ratio was shown to be useful in the indication of the long-term intensity of grassland management with higher values in low-input grasslands (Bardgett & McAlister, 1999). However, the conversion of intensive grassland management to low-input management did not result in a rapid change of the fungal:bacterial biomass ratio (Bardgett & McAlister, 1999), indicating that the potential for a rapid enhancement of plant species diversity would also be low.

(2)
Several experiments aimed at restoring extensive grasslands have shown a decrease in herbage production ranging from 0.5 to 0.8 of the value before the extensification (Oomes, 1990; Berendse *et al.*, 1992; Dyckmans *et al.*, 1999). Such loss of herbage yield is likely to be accompanied by a decreased percentage yield of competitive plants like *Lolium* spp. (Hofmann & Isselstein, 2005). In principle, the reduced herbage growth should facilitate the establishment of emerging seedlings, and yet, species number often remains low. The reason for this is that no propagules of the desired species are available, neither in the soil seed bank nor from the adjacent vegetation. Accordingly, the addition of propagules is necessary to induce an increase in diversity (Berendse *et al.*, 1992; Bakker & Berendse, 1999; Pywell *et al.*, 2002; Matejkova, *et al.*, 2003).

Grasslands have transient soil seed banks. After species have disappeared from the vegetation, they will stay in the soil seed bank only for a limited time, with the duration depending on the vegetation type and the site conditions (Matus *et al.*, 2003). Intensively used landscapes are fragmented, many semi-natural habitats are lost and the adjacent vegetation of intensive grassland is often poor in species. Agricultural techniques that had formerly allowed grassland species to set seed, and supported the exchange of propagules between different sites (e.g. traditional haymaking, communal grazing or shepherding, etc.), have been abandoned and this has further limited seed availability and seedling recruitment (Poschlod *et al.*, 1998). Thus, restoration of species diversity is often seed-limited (Bakker & Berendse, 1999).

(3)
If plant species are reintroduced into grasslands, there is a need to implement management measures that support the establishment success such as appropriate cutting dates, grazing regimes and other sward treatments (Smith *et al.*, 2000; Marriott *et al.*, 2004). It is necessary to apply such measures before the seeds from the desired species are over-sown into the existing swards. These measures comprise the lowering of the sward density by infrequent cutting and taking a hay crop immediately before over-sowing or by mechanical disturbance of the sward, e.g. through harrowing. Both these measures reduce the tiller density and increase the availability of gaps in the canopy, which is a prerequisite for germination and emergence (Hopkins *et al.*, 1999; Isselstein *et al.*, 2002; Losvik & Austad, 2002; Hofmann & Isselstein, 2004a).

Even more important is the management after seeds have been sown. Various experiments have shown that defoliation of the existing sward following seedling emergence is pivotal for the survival of the seedlings (Marriott *et al.*, 2004). This is particularly important when the production of the existing sward is still moderate to high and seedlings suffer from shading. Defoliation increases the amount of light available for the seedlings and thereby enhances seedling survival (Hofmann & Isselstein, 2004a). On a site with high residual fertility, it was demonstrated that frequent defoliation, irrespective of whether it was done by continuous sheep grazing or cutting, significantly increased seedling survival and establishment compared to infrequent defoliation (Isselstein *et al.*, 2005). An alternative approach for successful restoration on a moderately fertile site was developed by Pywell *et al.* (2004). Sowing of the parasitic plant *Rhinanthus* spp., significantly increased germination and establishment of a blend of over-sown wildflower species. This was obtained by a reduction in competitiveness of the existing sward by *Rhinanthus* and by the dead *Rhinanthus* shoots leaving gaps for seedling colonisation.

It has to be emphasised that during seedling recruitment, the requirements of plants with regard to suitable microhabitats may change. Germination and emergence are mainly enhanced by the availability of gaps with bare soil. Seedling survival and establishment are dependent on defoliation of the existing vegetation. Moreover, adult plants may again have different requirements as has been shown for species typical for traditional hay meadows. These species are susceptible to frequent defoliation, and yet, seedling establishment was enhanced by frequent cutting (Hofmann & Isselstein, 2004a).

Restoration attempts by introducing propagules from desired species have to consider species-specific requirements for successful establishment. Species vary in their response to competition by the existing sward, to drought and other types of environmental stress and this has implications for seedling recruitment (Hofmann & Isselstein 2004b). More research is needed to identify the specific requirements of the desired species with regard to seedling recruitment so that management measures on the grass sward can be more successfully targeted at obtaining the establishment of a high number of seedlings.

Seed provenance
The availability and origin of wildflower seeds to be used in grassland restoration have led to concerns among ecologists and agronomists. It is commonly accepted that the introduced species should be adapted to the site and management conditions (Walker *et al.*, 2004) and should originate from a local source. Collecting seeds from species-rich grasslands that are similar to the target vegetation of the restoration site, and that are located in the same region would undoubtedly be the best technique. The introduction of ecotypes from other regions bears the risk of failure with regard to establishment and persistence. It may also pose a threat to indigenous populations of the species through hybridisation and competitive exclusion (Keller *et al.*, 2000).

Germination under controlled conditions of several grassland and arable species from different provenances has been investigated by Keller & Kollmann (1999). There was considerable variability among provenances for the investigated species, and a relationship with the climatic conditions of the sites where the seeds had been collected was found. These adaptations were seen as an argument against the introduction of foreign species for the enhancement of biodiversity (Keller & Kollmann, 1999). However, the collection of seeds from semi-natural grasslands is laborious and increases the cost of restoring biodiversity. In Switzerland, where the establishment of species-rich hay meadows is encouraged by agri-environment schemes, a commercial system of seed multiplication has been set up in order to provide wildflower seeds at a reasonable cost, however a problem may arise with this technique as the genetic variation within a species may be reduced by repeated seed multiplication. This potential drawback has not been investigated to any great depth (Rüegger & Zanetti, 2001).

The selection of appropriate genotypes for restoration is mainly seen in relation to their potential to restore specific species assemblages or communities. However, as Gray (2002) has pointed out, restoration should also consider the functioning of ecosystems, which would include the process of evolutionary change. Thus, in a long-term perspective, the genetic diversity of introduced species must be such that it enables the species to cope with future abiotic and biotic change and thereby to escape extinction. There is no general answer to the question on the necessary extent of genetic diversity. On isolated restoration sites where genetic exchange with neighbouring habitats is not likely to occur, the genetic diversity of introduced species needs to be higher. Similarly, the size and degree of disturbance of the site at the beginning of restoration is of importance as more disturbed sites are more likely to change in habitat conditions than less disturbed ones (Gray, 2002).

Grazing as a management tool

Utilising semi-natural grasslands by grazing livestock has been shown to be effective in the maintenance of biodiversity (Bakker, 1994; Spatz, 1994; Stammel et al., 2003). Evidence also shows the potential of grazing for restoring biodiversity on semi-natural unimproved grassland that had been abandoned from agricultural use for some years, and therefore had lost species (Pykala, 2003; Lindborg & Eriksson, 2004; Hellström et al., 2005). Grazing is also considered to have potential for the restoration of biodiversity on agriculturally improved species-poor grasslands. On a local scale, grazing increases sward structural heterogeneity by selective defoliation due to dietary choice, treading, nutrient cycling and propagule dispersal (Rook et al., 2004a,b). Grazing at low stocking rates compared to moderate or high stocking, markedly increased sward heterogeneity (Correll et al., 2003; Isselstein et al., 2003).

Within a growing season, patches of tall and short grass develop with time and remain relatively constant (Correll et al., 2003). Therefore, at any given time, a range of microhabitats exists within a paddock providing a range of microsites for the different phases of the development of a population. Patches with short grass would enhance seedling recruitment, whereas patches of tall grass would allow adult plants to produce and distribute diaspores. Although such a situation seems likely to occur on extensively grazed grassland, the experimental proof is still scarce. In an ongoing Europe-wide, multi-site experiment on improved, mesic grasslands, extensive grazing showed an immediate effect on the sward structure compared to moderate stocking, but such a rapid effect was not seen on either the botanical composition of the sward or the species number (Rook et al., 2004b; Scimone et al., 2004).

The effect of livestock grazing on biodiversity is not only visible on a local scale, but also at a landscape scale. Traditional grazing systems, such as communal grazing or extensive set stocking, are characterised by relatively unrestricted grazer access to the whole pasture land. Thus, 'natural' grazing behaviour will be expressed and preferences for grazing areas, resting areas, etc., will be seen in the development of a mosaic of different plant communities. Today, examples of such conditions of high community diversity can still be observed in traditionally managed alpine grasslands with summer grazing of dairy cows for hard cheese production (Spatz, 1994). In a recent investigation on abandoned alpine grassland, it was shown that abandonment from grazing led to a significant long-term decline in species number. This effect occurred markedly at the landscape scale, and it was suggested that the reason was mainly community displacement rather than competitive exclusion at the local scale (Dullinger *et al.*, 2003). This result emphasises the need to consider biodiversity at a landscape scale (Eriksson *et al.*, 2002), and to maintain and re-establish large scale stocking systems. Recent nature conservation concepts have taken up this consideration and in many countries, large reserves with mega-herbivore grazing have been established (Redecker *et al.*, 2002). Grazers such as Heck cattle, Konik horses, Red and Roe deer are kept outdoor all year round. These animals are not utilised for livestock production purposes, though in theory they could be.

Employing extensive grazing to restore biodiversity within an agricultural system could offer economic advantages over non-agricultural mega-herbivore grazing, as the sale of the livestock products can be used to balance the animal husbandry costs. Despite such income, additional support by agri-environment schemes will still be indispensable to maintain such grazing systems, but the need for support by public budgets is likely to be higher for the non-agricultural systems. Within agricultural production, when grazing is compared with cutting, grazing seems to be more advantageous. Hay or silage from extensively managed improved grassland often has poor feeding quality, and their use is not advised in intensive beef, dairy or sheep farming (Dyckmans *et al.*, 1999; Isselstein *et al.*, 2001). However, when the herbage is grazed, livestock may achieve an individual performance that is comparative to the standards in intensive systems (Fothergill *et al.*, 2001, Hofmann *et al.*, 2001; Isselstein *et al.*, 2004), which gives some room for the integration of such grassland in modern livestock farms.

Fauna diversity

As with the restoration of plant species diversity on agriculturally improved species-poor grassland, the reversal from intensive to extensive management does not immediately lead to an increase in the diversity of the fauna. The mechanisms involved in the exclusion of many animal species during the intensification of grassland management are obviously not simply reversed and, in addition, are far from being fully understood (Robinson & Sutherland, 2002). The decline of animal species has been particularly marked amongst habitat specialists, and many of the taxa still common on farmland are habitat generalists. During the intensification of agriculture, landscape diversity showed a dramatic reduction, farm number and farm labour declined, and agricultural yields markedly increased (Robinson & Sutherland, 2002).

For birds, it has been shown that with the intensification of grassland management, the suitability of habitats for breeding and feeding declined (Vickery *et al.*, 2001). However, the relative significance of the individual factors associated with grassland intensification, whether it is the chemistry or hydrology of the site, the defoliation frequency, or the application of fertilisers and agrochemicals, etc., remains unclear (Vickery *et al.*, 2001). Multivariate analyses of long-term field data by Benton *et al.* (2002) suggest that the effect of management on bird populations is mainly an indirect one through changes in the invertebrate fauna, i.e. the quality and quantity of the food available for birds. In addition, it seems obvious that processes operating at the landscape or regional level have a strong effect on bird species diversity and abundance as birds easily move over larger distances.

Vickery *et al.* (2001) proposed that low input livestock systems should be established with low amounts of organic fertilisers and moderate levels of grazing pressure, in order to increase sward heterogeneity for the enhancement of bird population and species diversity. Monitoring of a large-scale grassland extensification project within an agri-environment scheme in Switzerland, revealed variable effects on grassland birds (Walter *et al.*, 2004). Holistic approaches that consider the various factors affecting bird populations in grasslands at the local and the landscape scale need to be investigated and developed for the benefit of farmland birds (Chamberlain *et al.*, 2000).

Other groups of grassland fauna showed a slightly clearer response to a switch from intensive to extensive management. The abundance of grasshoppers and butterflies usually increases within a relatively short time. The number of species, however, increases much slower (Walter *et al.*, 2004; WallisdeVries *et al.*, 2005). In a comparison of intensively and extensively utilised grasslands in Germany and Switzerland, Kruess & Tscharntke (2002) and Jeaneret *et al.* (2004) found on extensive grasslands, higher species numbers for solitary bees, wasps, their natural enemies and for carabids. The mechanism through which extensification enhances the diversity of these creatures is mainly by an increasing plant species number and in particular an increased sward height and standing crop at pasture (Söderstrom *et al.*, 2001; Dennis *et al.*, 2004; WallisdeVries, 2005).

If species-rich grasslands were established by sowing diverse mixtures on ex-arable land, a diverse fauna (butterflies, grasshoppers) was obtained in a relatively short time. However, no rare species occurred (Bosshard, 2001), indicating that the restoration of the animal diversity comparable to unimproved semi-natural grassland is difficult.

The role of agri-environmental schemes

In many countries, agri-environment schemes (AES) play a key role in attempts to maintain and restore grassland biodiversity (Kleijn & Sutherland, 2003). It is estimated that in the European Union, at least 20% of the farmland is managed under an AES (Rounsevell *et al.*, 2005). Such schemes have two objectives. Firstly, they are designed to reverse the deteriorating effect of intensive grassland farming on the biotic and abiotic resources, and to strengthen the multi-function of grassland (Jeangros & Thomet, 2004). Secondly, they have gained increasing importance in the profitability of grassland farming, where a significant percentage of the farm income is now obtained through these schemes.

Agri-environmental schemes differ markedly with regard to their targets. There are schemes that have precise prescriptions in order to achieve a precisely set target, and there are schemes that follow a broader approach with a range of targets, and support extensive farming in more general ways. An example of the former type of AES are the various schemes to protect meadow birds; an example for the latter, is the support of organic farming or the set aside of farmed land. Most of the schemes are measure-orientated rather than result-orientated (Bertke *et al.*, 2005). This means that farmers, when they join a scheme, have to follow detailed management prescriptions with regard to the input of fertilisers and agrochemicals, the date and frequency of cutting, or the stocking rate, etc. As long as they farm is managed according to the prescription, the farmers receive the payments. In comparison, result-orientated schemes would set a precise target and would leave it up to the farmer as to how he/she achieves the target. Thus, the performance in terms of its environmental or biodiversity effect rather than the measure itself would be honoured by the payments (Gerowitt *et al.*, 2003).

In recent years, several monitoring studies and meta-studies have been set up in order to investigate the efficiency of AES in grasslands. The results show that overall the AES have only had a small effect on biodiversity (Rounsevell *et al.*, 2005). However, taxa differed in their response to AES (Kleijn & Sutherland 2003); the effects on the number of plant species were often small, whereas invertebrates showed a stronger response (Kleijn & Sutherland 2003, Walter *et al.*, 2004).

Why have AES been of relatively minor efficiency? Several reasons have been advanced: (i) the measures have little or no effect on biodiversity or on the desired species, thus the wrong measures have been chosen; (ii) single measures have little effect, so complex measures are necessary to achieve the desired result. However, in many cases the effects of complex measures and the interactions between measures are not sufficiently understood (Chamberlain *et al.*, 2000; Robinson & Sutherland, 2002). This is particularly true for the effect of factors operating at the landscape scale that are likely to interact with factors operating at the local scale (Söderstrom *et al.*, 2001; Eriksson *et al.*, 2002). (iii) Measures that are known to be responsible for the decline of species number during the intensification of grassland use are often simply reversed in order to restore biodiversity. However, due to a hysteresis effect, the reversal of such factors does not give the desired benefit. Thus, more complex measures are required to overcome this constraint.

In order to improve the efficiency of AES, there is a strong need to increase scientific knowledge on how biodiversity targets can be achieved, and which measures have to be taken on species-poor grasslands that have seen a switch from production to biodiversity-orientation. This knowledge is urgently required in the farming community, so that farmers can take the appropriate action. In addition, the design and administration of schemes should rely more on the accomplishment of results. This would include the farmers' knowledge and experience on their particular site, which would improve the efficiency compared to that generally attained with standardised prescriptions for management.

References

Arens, R. (1971). Neuanlage von Dauergrünland (Establishment of permanent grassland). In: E. Klapp (ed.) Wiesen und Weiden (Meadows and pastures). Verlag Paul Parey, 4th edition, 326-369.

Bakker, J.P. (1994). Nature management in Dutch grasslands. In: R.J. Haggar & S. Peel (eds.) Grassland management and nature conservation. BGS Occasional Symposium No 28, Reading, UK, 115-124.

Bakker, J.P. & F. Berendse (1999). Constraints in the restoration of ecological diversity in grassland and heathland communities. *Trends in Ecology and Evolution*, 14, 63-68.

Bardgett, R.D. & R. Cook (1998). Functional aspects of soil animal diversity in agricultural grasslands. *Applied Soil Ecology*, 10, 263-276.

Bardgett, R.D. & E. McAlister (1999). The measurement of soil fungal : bacterial biomass ratios as an indicator of ecosystem self-regulation in temperate meadow grasslands. *Biology and Fertility of Soils*, 29, 282-290.

Barthram, G.T., C.A. Marriott, T.G. Common & G.R. Bolton (2002). The long-term effects on upland sheep production in the UK of a change to extensive management. *Grass and Forage Science*, 57, 124-136.

Benton, T.G., D.M. Bryant, L. Cole & H. Crick (2002). Linking agricultural practice to insect and bird populations: a historical study over three decades. *Journal of Applied Ecology*, 39, 673-687.

Berendse, F., M.J.M. Oomes, H.J. Altena & T. Elberse (1992). Experiments on the restoration of species-rich meadows in The Netherlands. *Biological Conservation*, 62, 59-65.

Bertke, E., J. Isselstein & R. Marggraf (2005). Species-rich grassland as an ecological good in an outcome-based payment scheme. In: *Proceedings of the XX International Grassland Congress, Dublin*, offered paper (in press).

Bosshard, A. (2001). Wie erfolgreich ist die Ansaat artenreicher Wiesen in der Praxis? (How successful is the establishment of species-rich meadows by sowing). *Schriftenreihe der FAL*, 39, 34-44.

Bullock, J.M., R.F. Pywell, M.J.W. Burke & K.J. Walker (2001). Restoration of biodiversity enhances agricultural production. *Ecology Letters*, 4, 185-189.

Chamberlain, D.E., R.J. Fuller, R.G.H. Bunce, J.C. Duckworth & M. Shrubb (2000). Changes in the abundance of farmland birds in relation to the timing of agricultural intensification in England and Wales. *Journal of Applied Ecology*, 37, 771-788.

Correll, O., J. Isselstein & V. Pavlu (2003). Studying spatial and temporal dynamics of sward structure at low stocking densities: the use of an extended rising-plate-meter method. *Grass and Forage Science*, 58, 450-454.

Critchley, C.N.R., B.J. Chambers, J.A. Fowbert, A. Bogha, S.C. Rose & R.A. Sanderson (2002). Plant species richness, functional type and soil properties of grasslands and allied vegetation in English Environmentally Sensitive Areas. *Grass and Forage Science*, 57, 82-92.

Dennis, P., J. Doering, J.A. Stockan, J.R. Jones, M.E. Rees, J.E. Vale & A.R Sibbald (2004). Consequences for biodiversity of reducing inputs to upland temperate pastures: effects on beetles (*Coleoptera*) of cessation of nitrogen fertilizer application and reductions in stocking rates of sheep. *Grass and Forage Science*, 59, 121-135.

Duelli, P. & M.K. Obrist (1998). In search of the best correlates for local organismal biodiversity in cultivated areas. *Biodiversity and Conservation*, 7, 297-309.

Dullinger, S., T. Dirnbock, J. Greimler & G. Grabherr (2003). A resampling approach for evaluating effects of pasture abandonment on subalpine plant species diversity. *Journal of Vegetation Science*, 14, 243-252.

Dyckmans, A., H. Mack & F. Weissbach (1999). The effect of grassland extensification on yield, forage quality and botanical composition at different grassland locations. *Landbauforschung Völkenrode*, SH 206, 125-139.

Ellenberg, H. (1996). Die Vegetation Mitteleuropas mit den Alpen (The vegetation of Central Europe including the Alps). 5th Edition. Ulmer Verlag, Stuttgart, Germany, 1095pp.

Eriksson, O., S.A.O. Cousins & H.H. Bruun (2002). Land use history and fragmentation of traditionally managed grasslands in Scandinavia. *Journal of Vegetation Science*, 13, 743-748.

Feehan, J., D.A. Gillmor & N. Culleton (2005). Effects of an agri-environment scheme on farmland biodiversity in Ireland. *Agriculture, Ecosystems & Environment*, (in press).

Fothergill, M., D.A. Davies & C.T. Morgan (2001). Extensification of grassland use in Welsh uplands: sheep performance in years 1-6. *Grass and Forage Science*, 56, 105-117.

Gerowitt, B., J. Isselstein & R. Marggraf (2003). Rewards for ecological goods – requirements and perspectives for agricultural land use. *Agriculture, Ecosystems & Environment*, 98. 541-547.

Gray, A.J. (2002). The evolutionary context: a species perspective. In: M.R. Perrow & A.J. Davy (eds.) Handbook of ecological restoration, Volume 1, Principles of Restoration. Cambridge University Press, 66-80.

Grubb, P.J. (1977). The maintenance of species-richness in plant communities: the importance of the regeneration niche. *Biological Review*, 52, 107-145

Hector, A., B. Schmid, C. Beierkuhnlein & the BIODEPTH project (1999). Plant diversity and productivity experiments in European grasslands. *Science*, 286, 1123-1127.

Hellström, K., A.P. Huhta, P. Rautio, J. Tuomi, J. Oksanen & K. Laine (2005). Use of sheep grazing in the restoration of semi-natural meadows in northern Finland. *Applied Vegetation Science*, (in press).

Hofmann, M. & J. Isselstein (2004a). Seedling recruitment on agriculturally improved mesic grassland: the influence of disturbance and management. *Applied Vegetation Science*, 7, 193-200.

Hofmann, M. & J. Isselstein (2004b). Effects of drought and competition by a rygrass sward on the seedling growth of a range of grassland species. *Journal of Agronomy and Crop Science*, 190, 277-286.

Hofmann, M. & J. Isselstein (2005). Species enrichment in an agricultural improved grassland and its effects on botanical composition, yield and forage quality. *Grass and Forage Science,* (in press).

Hofmann, M., N. Kowarsch, S. Bonn & J. Isselstein (2001). Management for biodiversity and consequences for grassland productivity. *Grassland Science in Europe*, 6, 113-116.

Hopkins, A. (ed.) (2000). Grass – its production and utilization. Blackwell Science, 3rd edition, 440pp.

Hopkins, A., R.F. Pywell, S. Peel, R.H. Johnson & P.J. Bowling (1999). Enhancement of botanical diversity of permanent grassland and impact on hay production in Environmentally Sensitive Areas in the UK. *Grass and Forage Science*, 54, 163-173.

Isselstein, J., M. Benke & C. Hermanspahn (2001). Bewirtschaftung von Niedermoorgrünland unter Naturschutzauflagen (Management of peat grasslands under nature conservation schemes). *Landnutzung und Landentwicklung*, 6, 264-267.

Isselstein, J., J.R.B. Tallowin & R.E.N. Smith (2002). Factors affecting seed germination and seedling establishment of fen-meadow species. *Restoration Ecology*, 10, 173-184.

Isselstein, J., O. Correll, J. Strodthoff, G. Zhao & M. Hofmann (2003). Variability of sward structure and plant species composition of pastures at low stocking rates. *Grassland Science in Europe*, 8, 606-609.

Isselstein, J., B.A. Griffith, P. Pradel & S. Venerus (2004). Effects of livestock breed and stocking rate on sustainable grazing systems: 3. agronomic potential. *Grassland Science in Europe*, 9, 620-622.

Isselstein, J., N. Kowarsch, S. Bonn & M. Hofmann (2005). The effect of sheep grazing at two stocking rates on the seedling recruitment of grassland forbs. In: *Proceedings of the XX International Grassland Congress, Dublin*, offered paper (in press).

Janssens, F., A. Peeters, J.R.B. Tallowin, J.P. Bakker, R.M. Bekke, F. Fillat & M.J.M. Oomes (1998). Relationship between soil chemical factors and grassland diversity. *Plant and Soil*, 202, 69-78.

Jeangros, B. & C. Bertola (2002). Long-term evolution of an intensively managed meadow after cessation of fertilisation and reduction of cutting frequency. *Grassland Science in Europe*, 7, 794-795.

Jeangros, B. & P. Thomet (2004). Multi-functionality of grassland systems in Switzerland. *Grassland Science in Europe*, 9, 11-23.

Jeanneret, Ph., L. Pfiffner, S. Pozzi & T. Walter (2004). Impact of low input meadows on arthropod diversity at habitat and landscape scale. *Grassland Science in Europe*, 9, 237-239.

Keller, M. & J. Kollmann (1999). Effects of seed provenance on germination of herbs for agricultural compensation sites. *Agriculture, Ecosystems and Environment*, 72, 87-99.

Keller, M., J. Kollmann, & P.J. Edwards (2000). Genetic introgression from distant provenances reduces fitness in local weed populations. *Journal of Applied Ecology*, 37, 647-659.

Kleijn, D. & W.J. Sutherland (2003). How effective are European agri-environmental schemes in conserving and promoting biodiversity? *Journal of Applied Ecology*, 40, 947-969.

Knops, J.M.H., D. Tilman, N.M. Haddad, S. Naeem, C.E. Mitchell, J. Haarstad, M.E. Ritchie, K.M. Howe, P.B. Reich, E. Siemann & J. Groth J (1999). Effects of plant species richness on invasion dynamics, disease outbreaks, insect abundances and diversity. *Ecology Letters*, 2, 286-293.

Koch, B. & G. Masé (2001). Extensivierung von intensiv bewirtschaftetem Grasland (Extensification of formerly intensively managed grasslands). *Schriftenreihe der FAL*, 39, 61-68.

Kruess, A. & T. Tscharntke (2002). Grazing intensity and the diversity of grasshoppers, butterflies, and trap-nesting bees and wasps. *Conservation Biology*, 16, 1570-1580.

Lindborg, R. & O. Eriksson (2004). Effects of restoration on plant species richness and composition in Scandinavian semi-natural grasslands. *Restoration Ecology*, 12, 318-326.

Loreau, M. (2000). Biodiversity and ecosystem functioning: recent theoretical advances. *Oikos*, 91, 3-17.

Loreau, M., S. Naeem, P. Inchausti, J. Bengtsson, J.P. Grime, A. Hector, D.U. Hooper, M.A. Huston, D. Raffaelli, B. Schmid, D. Tilman & D.A. Wardle (2001). Biodiversity and ecosystem functioning: current knowledge and future challenges. *Science*, 294, 804-808.

Losvik, M.H. & I. Austad (2002). Species introduction through seeds from an old, species-rich hay meadow: effects of management. *Applied Vegetation Science*, 5, 185-194.

Marriott, C.A., M. Fothergill, B. Jeangros, M. Scotton & F. Louault (2004). Long-term impacts of extensification of grassland management on biodiversity and productivity in upland areas. A review. *Agronomie*, 24, 447-461.

Matejkova, I., R. van Diggelen & K. Prach (2003). An attempt to restore a central European species-rich mountain grassland through grazing. *Applied Vegetation Science*, 6, 161-168.

Matus, G., R. Verhagen, R.M. Bekker & A.P. Grootjans (2003). Restoration of the *Cirsio dissecti-Molinietum* in The Netherlands: can we rely on soil seed banks? *Applied Vegetation Science*, 6, 73-84.

Minns, A., J. Finn, A. Hector, M. Caldeira, J. Joshi, C. Palmborg, B. Schmid, M. Scherer-Lorenzen, E. Spehn & A. Troumbis (2001). The functioning of European grassland ecosystems: potential benefits of biodiversity to agriculture. *Outlook on Agriculture*, 30, 179-185.

Muller, S., T. Dutoit, D. Alard & F. Grevilliot (1998). Restoration and rehabilitation of species-rich grassland ecosystems in France: a review. *Restoration Ecology*, 6, 94-101.

Nösberger, J. & M. Rodriguez (1996). Increasing biodiversity through management. *Grassland Science in Europe*, 1, 949-956.

Oomes, M.J.M. (1990). Changes in dry matter and nutrient yields during the restoration of species-rich grassland. *Journal of Vegetation Science*, 1, 333-338.

Oomes, M.J.M. (1992). Yield and species density of grasslands during restoration management. *Journal of Vegetation Science*, 3, 271-274.

Peeters, A. & F. Janssens (1998). Species-rich grasslands: diagnostic, restoration and use in intensive livestock production systems. *Grassland Science in Europe*, 3, 375-393.

Pfisterer, A.B., J. Joshi, B. Schmid & M. Fischer (2004). Rapid decay of diversity-productivity relationships after invasion of experimental plant communities. *Basic and Applied Ecology*, 5, 5-14.

Poschlod, P., S. Kiefer, U. Tränkle, S. Fischer & S. Bonn (1998). Plant species richness in calcareous grasslands as affected by dispersability in space and time. *Applied Vegetation Science*, 1, 75-91.

Pykala, J. (2003). Effects of restoration with cattle grazing on plant species composition and richness of semi-natural grasslands. *Biodiversity and Conservation*, 12, 2211-2226.

Pywell, R.F., J.M. Bullock, A. Hopkins, K.J. Walker, T.H. Sparks, M.J.W. Burke & S. Peel (2002). Restoration of species-rich grassland on arable land; assessing the limiting processes using a multi-site experiment. *Journal of Applied Ecology*, 39, 294-309.

Pywell, R.F., J.M. Bullock, K.J. Walker, S.J. Coulson, S.J. Gregory & M.J. Stevenson (2004). Facilitating grassland diversification using the hemiparasitic plant *Rhinanthus minor*. *Journal of Applied Ecology*, 41, 880-887.

Redecker, B., P. Finck, W. Härdtle, U. Riecken & E. Schröder (eds.) (2002). Pasture landscapes and nature conservation. Springer Verlag Berlin-Heidelberg, 435pp.

Reich, P.B., J. Knops, D. Tilman, J. Craine, D. Ellsworth, M. Tjoelker, T. Lee, S. Naeem, D. Wedin, D. Bahauddin, G. Hendrey, S. Jose, K. Wrage, J. Goth & W. Bengston (2001). Plant diversity enhances ecosystem responses to elevated CO_2 and nitrogen deposition. *Nature*, 410, 809-812.

Robinson, R.A. & W.J. Sutherland (2002). Post-war changes in arable farming in Great Britain. *Journal of Applied Ecology*, 39, 157-176.

Rook, A.J., B. Dumont, J. Isselstein, K. Osoro, M.F. Wallis deVries, G. Parente & J. Mills (2004a). Matching type of livestock to desired biodiversity outcomes in pastures – a review. *Biological Conservation*, 119, 137-150.

Rook, A.J., M. Petit, J. Isselstein, K. Osoro, M.F. WallisdeVries, G. Parente & J. Mills (2004b). Effects of livestock breed and stocking rate on sustainable grazing systems: 1. Project description and synthesis of results. *Grassland Science in Europe*, 9, 572-574.

Rounsevell, M.D.A., F. Ewert, I. Reginster, R. Leemans & T.R. Carter (2005). Future scenarios of European agricultural land use - II. Projecting changes in cropland and grassland. *Agriculture, Ecosystems & Environment*, (in press).

Rüegger, A. & S. Zanetti (2001). Wiesenblumensaatgut: Vermehrung, Kontrolle und Qualität (Hay meadow seed: multiplication, control and seed quality). *Schriftenreihe der FAL*, 39, 87-92.

Sala, O.E. (2001). Price put on biodiversity. *Nature*, 412, 34-36.

Sanderson, M.A., R.H. Skinner, D.J. Barker, G.R. Edwards, B.F. Tracy & D.A. Wedin (2004a). Plant species diversity and management of temperate forage and grazing land ecosystems. *Crop Science*, 44, 1132-1144.

Sanderson, M.A., K.J. Soder, N. Brzezinski, L.D. Muller, R.H. Skinner, M. Wachendorf, F. Taube & S.C. Goslee (2004b). Plant species diversity influences on forage production and performance of dairy cattle on pasture. *Grassland Science in Europe*, 9, 632-634.

Scherer-Lorenzen, M., C. Palmborg, A. Prinz & E.D. Schulze (2003). The role of plant diversity and composition for nitrate leaching in grasslands. *Ecology*, 86, 1539-1552.

Schüpbach, B., A. Grünig & T. Walter (2004). Grassland and landscape aethetics. *Grassland Science in Europe*, 9, 186-188.

Scimone, M., R.E.N. Smith, J.P. Garel & N. Sahin (2004). Effects of livestock breed and stocking rate on sustainable grazing systems: 4. short-term effects on vegetation. *Grassland Science in Europe*, 9, 623-625.

Smith, R.S., R.S. Shiel, D. Millward & P. Corkhill (2000). The interactive effects of management on the productivity and plant community structure of an upland meadow: an 8-year field trial. *Journal of Applied Ecology*, 37, 1029-1043.

Smith, R.S., R.S. Shiel, R.D. Bardgett, D. Millwards, P. Corkhill, G. Rolph, P.H. Hobbs S. Peacock (2003). Soil microbial community, fertility, vegetation and diversity as targets in the restoration management of a meadow grassland. *Journal of Applied Ecology*, 40, 51-64.

Söderstrom, B., B. Svensson, K. Vessby & A. Glimskar (2001). Plants, insects and birds in semi-natural pastures in relation to local habitat and landscape factors. *Biodiversity and Conservation*, 10, 1839-1863.

Spatz, G. (1994). Freiflächenpflege (Management of open landscapes). Ulmer Verlag, Stuttgart, Germany, 296pp.

Stammel, B., K. Kiehl & J. Pfadenhauer (2003). Alternative management on fens: response of vegetation to grazing and mowing. *Applied Vegetation Science*, 6, 245-254.

Tallowin, J.R.B. & R.G. Jefferson (1999). Hay production from lowland semi-natural grasslands: a review of implications for livestock systems. *Grass and Forage Science*, 54, 99-115.

Tilman, D., D. Wedin & J. Knops (1996). Productivity and sustainability influenced by biodiversity in grassland ecosystems. *Nature*, 379, 718-720.

Tracy, B.F., I.J. Renne, J. Gerrish & M.A. Sanderson (2004). Effects of plant diversity on the invasion of weed species in experimental pasture communities. *Basic and Applied Ecology*, 5, 543-550.

van Diggelen, R. & R.H. Marrs (2003). Restoring plant communities - introduction. *Applied Vegetation Science*, 6, 106-110.

van Ruijven, J. & F. Berendse (2003). Positive effects of plant species diversity on productivity in the absence of legumes. *Ecology Letters*, 6, 170-175.

Vessby, K., B. Söderstrom, A. Glimskär & B. Svensson (2002). Species-richness correlations of six different taxa in Swedish seminatural grasslands. *Conservation Biology*, 16, 430-439.

Vickery, J.A., J.R.B. Tallowin, R.E. Febers, E.J. Asteraki, P.W. Atkinson, R.J. Fuller & V.K. Brown (2001). The management of lowland neutral grasslands in Britain: effects of agricultural practices on birds and their food resources. *Journal of Applied Ecology*, 38, 647-664.

Walker, K.J., P.A. Stevens, D.P. Stevens, J.O. Mountford, S.J. Manchester & R.F. Pywell (2004). The restoration and re-creation of species-rich lowland grassland on land formerly managed for intensive agriculture in the UK. *Biological Conservation*, 119, 1-18.

WallisdeVries, M.F., J.R.B. Tallowin, J.P. Dulphy, M. Sayer & E. Diana (2005). Effects of livestock breed and stocking rate on sustainable grazing systems: butterfly diversity and abundance. *Grassland Science in Europe*, 10, (in press).

Walter, T., F. Herzog, S. Birrer, S. Dreier, M. Hunziker, Ph. Jeanneret, A. Lüscher, B. Peter, L. Pfiffner & M. Spiess (2004). Effects of ecological compensation areas on species diversity in the Swiss grassland – an overview. *Grassland Science in Europe*, 9, 171-173.

Wardle, D.A. (1999). Is 'sampling effect' a problem for experiments investigating biodiversity-ecosystem function relationships. *Oikos*, 87, 403-407.

Wardle, D.A., R.D. Bardgett, J.N. Klironomos, H. Setala, W.H. van der Putten, D.H. Wall (2004). Ecological linkages between aboveground and belowground biota. *Science,* 304, 1629-1633.

Whitehead, D.C. (1995). Grassland nitrogen. CAB International, Wallingfort, UK, 416pp.

Section 3

Delivering the benefits from grassland

Grassland and forage to improve livelihoods and reduce poverty

S.G. Reynolds, C. Batello, S. Baas and S. Mack
Food and Agriculture Organization of the United Nations, Viale delle Terme di Caracalla, 00100 Rome, Italy
Email: Stephen.Reynolds@fao.org

Key points

1. Grasslands contribute to the livelihoods of over 800 million people including many poor smallholders where government focus will have the biggest impact on poverty and livelihoods, but development must be done within local communities with their full participation.
2. Rapid urbanization has increased demand for forages for peri-urban smallholder dairies. In temperate areas provision of reliable winter fodder or feed remains the biggest need.
3. Greater use can be made of forages under tree crops and agroforestry systems. The food potential of grassland plants in semi-arid areas is neglected.
4. Management of extensive grasslands must consider the public good as well as production; policies are needed to remunerate pastoralists who manage grasslands to safeguard watersheds and provide public goods and services.
5. In some areas such as Eastern Europe there is a move away from intensive systems to extensive grassland use combining grazing and wildlife management, which provides income while preserving biodiversity.

Keywords: pastoralists, development, smallholders, policy

Introduction

The world's population has doubled to over six billion in forty years (FAO, 2002a) and is expected to reach 8 billion by 2025. Unsustainable and environmentally unsound policies and practices have caused widespread degradation of the environment and increased people's vulnerability to food shortages (UNEP, 2002). The crucial agricultural dilemma is how to satisfy the demand for food and at the same time sustain the natural resource base. In 1999 some 23% of the population of the developing world (1.2 billion people), had incomes or consumption levels below one US dollar a day; 2.8 billion lived on less than two dollars. Three quarters of these live in rural areas with nearly half in Southeast Asia and a quarter in sub-Saharan Africa, where the problem is growing (IFAD, 2001). In developing countries half the population is involved in agriculture compared to 5.5% in the developed world: it is highest in sub-Saharan Africa with 58.3%, 53.9% in east and southeast Asia and 49% in south Asia. The agricultural population is 87% of the rural population, although income may be diversified with other activities. FAO estimated that in 2003 some 840 million people lacked enough food to lead healthy, active lives. The share of international development assistance going to agriculture has fallen from about 20% in the late 1980s, to only 12% in 2000 (IFAD, 2001). Bilateral development assistance to agriculture from OECD countries fell from US$4.1 billion in 2001 to US$3.8 billion in 2002. Poverty is widespread and the United Nations Millennium Development Goals (UN, 2000) include eradication of extreme poverty and hunger (Goal 1. Eradicate extreme poverty and hunger with two targets, between 1990 and 2015, of halving the proportion of people whose income is less than US$1 a day and halving the proportion of people who suffer from hunger). What role can grasslands and forages play in improving this situation?

The context

Estimates of the proportion of the earth covered by grassland vary between 20 and 40%, depending on definition. Natural pastures cover more than twice the area of the world's cropped land. Grasslands are a source of food, forage and energy, a wildlife habitat, a place for *in situ* conservation of genetic resources, for carbon and water storage, and recreation, while protecting the catchments of many major river systems. Grasslands contribute to the livelihoods of over 800 million people (White *et al.*, 2000) and for almost 200 million people in the more marginal areas (particularly arid and semi-arid regions) grazing livestock is the main source of livelihood. Much of the best land of the world's major grassland areas is now under crops. Many of the world's poor depend on mixed farming in humid and sub-humid regions where climates permit cropping and livestock. This contributes to intensification, nutrient recycling and diversification of income thus providing a buffer against risk (Thornton *et al.*, 2002; Wright, 2005). Livestock feed is derived from fodder crops, crop residues, fodder trees and by-products and from marginal land, fallows, and roadsides.

Dixon *et al.* (2001) emphasise that small farmers produce most of the developing world's food but are much poorer than the rest of the population. Dealing with poverty and hunger means tackling the problems of small farmers in their daily struggle for livelihood security. FAO (2003) notes "no sustainable reduction in poverty is possible without improving livelihoods in rural areas". As well as listing five main strategies (income diversification; extension of land area; intensification of land use; increased non-farm income; and leaving farming altogether), Dixon *el al.* (2001) propose elements for reducing hunger and poverty: refocusing of institutions, policies and public goods; trade liberalisation and market development; enhancing agricultural infrastructure and human capital, and improving techniques and management of natural resources. The Oxford English Dictionary defines livelihood as "a means to a living". Ellis (2000) suggests that "a livelihood comprises the activities, assets and the access to these that together determine the living gained by an individual or household", and stresses that they alter as they adapt to change. In the context of grasslands, livelihoods mean not only cash incomes or higher livestock production, but livelihood outcomes may also be the provision of food for home consumption, or power to cultivate crops, as grasslands are multifunctional (Hervieu, 2002) providing goods and services as a common good.

This paper examines some opportunities for increased or alternative use of grasslands and forages to improve livelihood security and reduce poverty.

Some major opportunities

The most obvious indicator of opportunities is the likely rise in demand for livestock products: per capita consumption in developed countries will be relatively constant over the next 20 years but will increase in developing ones (Delgado *et al.*, 1999). Much of the demand is for white meat (Delgado *et al.*, 1999; Upton, 2004) but the projection is for a doubling of beef consumption and a 2.3-fold increase in milk consumption between 1993 and 2020 in the developing world (Table 1), especially in China and Southeast Asia where there is rapid economic growth.

Table 1 Projected trends in production and consumption of selected livestock products (million tonnes) (Delgado *et al.*, 1999)

	Developed world		Developing world	
	1993	2020	1993	2020
Beef	35 (32)[1]	38 (36)	22 (22)	44 (47)
Milk	348 (245)	371 (263)	164 (168)	401 (391)

[1]Consumption figures in brackets

As indicated above, grasslands support a range of livestock systems in addition to other functions and therefore the opportunities (as well as the constraints) are numerous:

- the demand for livestock products will present opportunities for both:
  extensive systems: including the introduction and wider use of supplementary winter and dry-season feeding and the sustainable use of irrigation for localised forage production, and,
  intensive systems: particularly sustainable crop-pasture mixed farming and cut and carry systems for smallholders;
- with increased urbanisation there will be a major focus on peri-urban systems, especially in terms of dairy production, which provides an opportunity for increased fodder production;
- an opportunity frequently mentioned but so far under-exploited is the forage available under various tree crops and in agroforestry and especially sylvopastoral systems;
- in semi-arid areas various grassland species have considerable potential for food production (as well as for medicinal purposes, energy and as a genetic resource for much greater future exploitation). To date (on a world scale) only 100-150 forage species have undergone selection or been cultivated out of a total of some 10,000 species of grasses (Peeters, 2004);
- mountain grasslands and other areas with a strong consumer brand image have a potential in terms of the development of niche products, and for many natural grassland areas there is a considerable potential to add value to livestock production in terms of 'natural' products, organic meat/milk/wool, and from the harvesting of wildlife meat and products. With increasing public interest in the methods employed in food production (Younie & Baars, 2005) and concerns about food safety, there is considerable scope for focus on the nutritional quality, hygiene issues and food safety aspects of products from 'natural' systems, which in the public's view is often identified with grasslands. For example, recent FAO assistance to the mountainous and remote farming communities of Babine and Brodarevo in Prijepolje Municipality in the Sandzak region of Serbia and Montenegro, has identified a strong demand and well-developed marketing outlets for the distinctive cheeses from this area (FAO, 2004a).
- a major aspect that will have increasing significance in future is the issue of carbon sequestration, and whether farmers and pastoralists should be compensated for providing carbon sequestration services (Lipper & Cavatassi, 2003);
- use of grassland areas for watershed protection as reserves and protected areas and habitats for wildlife. Likewise should pastoralists or mixed farmers be compensated for safeguarding watersheds (Pagiola *et al.*, 2004) by adopting appropriate grazing management policies?
- there is also the ethical dimension and the importance particularly, but not exclusively, of some of the arid and semi-arid grasslands in terms both of cultural heritage (Wright, 2005) and religious significance for pastoralists, such as in Tibet and also in Dolpo, Nepal where the pasture of Kunasa is considered a 'beyul' (sacred, hidden land) by the Bonpo communities of Phoksundo (Aumeeruddy-Thomas *et al.*, 2004).

While there are major opportunities for increased or alternative uses of forage and grassland to improve livelihoods in developing countries, there are constraints, and a need for successful, applicable models wherein the factors for success and the actions needed can be identified.

Case studies and projects

This paper focuses on a number of the above opportunities and in particular stresses five areas for increased or alternative uses of forage and grassland to improve livelihoods in developing countries:
1. fodder production and livestock in small mixed farming systems;
2. grazing under trees in agroforestry systems;
3. grazing systems with winter fodder in temperate areas;
4. extensive grazing systems, and protected areas;
5. grasslands as a source of traditional food in the semi-arid/arid tropics.

Examples are given of successful FAO projects and other case studies to illustrate possibilities for livelihood improvement through grassland and forage development and some of the problems faced. These are drawn from various countries and regions: Chad, China, Czech Republic, Kenya, Mongolia, Nepal, Pakistan, Patagonia, Southeast Asia, Sudan and Syria.

1. Fodder production and livestock in small mixed farming systems

Mixed farming supplies over 90% of the world's milk, 70% of sheep and goat meat, and 35% of the beef (de Haan et al., 1997; Wright, 2005). Most of the world's poor are associated with small mixed farms, so this area has the greatest scope for reducing poverty. In populous developing countries natural pastures are being converted to crops. Rapid urbanization is increasing the demand for cheap food - by 2005 over half the world's population will live in cities (FAO, 2002b). In the cities of the developing world (with poverty rates often over 50%), urban and suburban farms supply food to about 700 million – a quarter of the world's urban population. Dairying in and around cities is important in many countries where there is a demand for fresh milk but no large-scale dairy sector. The greatest development of smallholder milk for urban consumption has been in India where the "White Revolution" has improved supply and quality (NDDB, 2004). Horne et al. (2005) estimate that demand for red meat per capita will increase by 40–60% by 2010 in Southeast Asia compared with 2000. The major constraint to satisfying regional demand for livestock products is limited feed supply and quality.

Peri-urban dairying in Pakistan

In Pakistan, demand for milk has led to the development of specialised fodder production and long-distance transport of fodder to city dairies, creating a great deal of employment. Commercial fodder growing and dairying, with good money flows and increased incomes, represents a major opportunity for smallholders. Urban dairying has been important for a very long time, but in the late nineteen-eighties it began to grow rapidly (Dost, 2003); milk production rose by 81% in the past decade. Farmers took up commercial forage production and thousands of jobs have been created in milk marketing, forage production, harvesting, chopping and transport. Peri-urban supply has increased greatly, but costs are lower in the countryside so huge quantities of milk and forage are transported to urban areas. Forty percent of dairy animals, mainly buffalo, are peri-urban and 60% rural; most are stall-fed in the irrigated plains. Lahore with 250,000 animals has the largest urban herd and finds fodder within 100 kilometres; Faisalabad (100,000) and Gujranwala (150,000) obtain fodder close by. In rainfed areas fodder comes from afar: Rawalpindi (160,000) from up to 350 km; Karachi (175,000) 350 km and Quetta (60,000) over 400 km (Suttie, 2000).

The winter fodders are *Trifolium alexandrinum* (berseem), and *Avena sativa* (oats); summer fodders are: (*Sorghum bicolor*) sorghum, (*Sorghum x drummondii*) Hybrid Sweet Sorghum, (*Pennisetum glaucum*) millet and *Zea mays* (maize). Forage growing is popular and more profitable than cash crops such as *Triticum aestivum* (wheat), *Gossypium* spp. (cotton), *Saccharum* spp. (sugar cane) and *Oryza sativa* (rice); cultivation is easier and cheaper as it requires less weeding, sprays, and fewer inputs, and gives quick, regular income (Dost, 2003). Straw and dried stover forms the bulk of dairy rations. Drying coarse cereals is widespread in rainfed districts; they are bought by big stockists, stored where they are produced then supplied to peri-urban dairies. High forage feeding with low concentrate use is the best system for maximum profit. Net income per hectare was highest for *Sorghum x drummondii* (Rs 146 266) followed by *Medicago sativa* (lucerne) (Rs 139 990) and *T. alexandrinum* (Rs 138 925), and was lowest for *Z. mays* and *Sorghum bicolor* (Dost, 2003). Farmers involved only in forage production and sale were satisfied with the business, earning around Rs 100,000 to Rs 150,000 (April 2004, 1 US$ = 57.6 Pak. Rs) from 0.8 - 1.23 ha of forage. Crop production needs more investment and inputs, and little income is available for 4-6 months. A comparison of economic returns showed that benefits from raising seventeen dairy animals on a hectare (buying some feeds), and selling milk gave maximum profit (Rs 732,113) followed by forage production for sale to urban dairies (Rs 96,250) and mixed farming i.e. animal + crops (Rs 62,550) respectively, compared with all the other agriculture related businesses.

Women farmers and milk production from Napier grass in Kenya

In Kenya dairying is developing around urban areas with a change to stall-feeding. Units are small and local; fodder is traded by farmers who may not own stock, so employment and income are spread throughout the community. In 1996 more than 70% of income on smallholder mixed farms was from dairying; smallholders contribute 56 and 70% of total and marketed milk respectively, from about 2.5 million cattle on 625,000 small farms (Karanja, 1999). Deficits in milk supply are evident, and farmers, especially close to urban centres have responded by intensifying milk production. Supportive policies on milk liberalization allow smallholders to realize the benefits of dairying: regular cash generation, employment creation and improved nutrition.

A typical farmer keeps 2-3 cows with followers on a hectare, and also grows crops (Orodho, 2005). Cattle are mixed exotic breeds or crosses with zebu. Stall-feeding of residues and grass is widespread and increasing. Purchase of fodders or hay is common (Staal *et al.*, 1998). There has been a rapid adoption of *Pennisetum purpureum* (Napier grass); other fodders include *Ipomoea batatas* vines, *Vicia* spp., *Desmodium uncinatum*, *D. intortum* and trees such as *Calliandra calothyrsus* and *Leucaena leucocephala*. Farmers buy grain, concentrates and by-products. Increased farm gate prices have encouraged feed conservation for the dry season when the milk price is highest, and hence an increase in forage production on small dairies. Smallholders with zero-grazing units take the dung straight to the *P. purpureum* plots. Many households with intensive dairying hire labour, usually from surrounding areas. Increased milk production increases home consumption, and it was noted that the incidence of chronic malnutrition was significantly lower among children of households with dairy cattle than those without. Dairying is evenly spread across income categories (Nicholson *et al.*, 1999). Women are particularly involved in dairying and decision-making in dairy households and do most of the work (except spraying and herding). The contribution to dairy work on small farms was: women – 30%; children – 26%; men – 25%; and hired labour – 19%. Women control income from milk while men deal with cash crops and decisions regarding animal sales and purchases (Staal *et al.*, 1998). A survey of 202 households showed that intensive dairying substantially increased household income, generated secondary employment and improved the nutrition of pre-school children (Nicholson *et al.*, 1999). Profits obtained by farmers growing forages for intensive dairy production are high. However, the currently used clones of *P. purpureum* are over fifty years old, and Orodho (2005), has indicated an increased incidence of diseases and pests, in particular head smut and stunting. The pressing need is to develop disease resistant *P. purpureum* cultivars and identify alternative fodders.

2. Grazing under trees in agroforestry systems

There is considerable opportunity for increasing livestock production through grazing under tree crops in the millions of hectares of agroforestry systems worldwide (Reynolds, 1995). Systems include *Elaeis guineensis* (oil palm), *Hevea brasiliensis* (rubber), *Cocos nucifera* (coconut), *Syzygium aromaticum* (clove), *Anacardium occidentale* (cashew), *Juglans regia* (walnut), citrus and various other fruit trees as well as sylvopastoral systems in both temperate and tropical areas. However, it is in Southeast Asia that livestock-tree crop integration is likely to provide a considerable increase in production for improved livelihoods and poverty reduction, as the large tree crop areas produce as much forage as the region's pasturelands (Wong *et al.*, 2005). *Elaeis guineensis* covers 5.2 million hectares,), *Cocos nucifera* 6.6 and), *Hevea brasiliensis* 5.7 million hectares; young *Tectona grandis* (teak) plantations are also suitable for grazing and vast areas are presently ungrazed. Over 90% of *Cocos nucifera* are grown by smallholders. Stock raising in plantations can be by cut-and-carry or grazing. Grazing under tree crops and in plantations is not new, and is often a supplementary enterprise to diversify and stabilise income (Reynolds, 1995). Traditionally, grazing cattle in plantations has often raised tree crop yields through weed removal and dung returns to improve soil fertility. Also, low copra and oil prices are a good reason for integration, and in Indonesia cattle may contribute from 50-75% of income in cattle-*Cocos nucifera* systems. For over one hundred years cover crops like *Calopogonium mucunoides* and *Centrosema pubescens* have been sown in young plantations for weed control and soil improvement (Horne *et al.*, 2005).

However, to date the potential of livestock-tree crop integration has been under-exploited and assessment is needed of the key impediments, both technical and institutional. Light levels in plantations vary from almost full daylight at planting to below 10% when the canopy closes and improved forages are replaced by native vegetation. With *Cocos nucifera* light transmission increases again with more mature palms. Introduced forages grow well when plantations are young but poor persistence limits their usefulness. Finding shade-tolerant forages is as elusive as it was in the nineteen-seventies. Indigenous grasses like *Axonopus compressus* and *Paspalum conjugatum* perform well, and *Desmodium ovalifolium*, *Calopogonium caeruleum* and *C. mucunoides* persist well, but appropriate management is necessary to ensure persistence and productivity. Alternative tree crop spatial arrangements could provide for a considerable increase in livestock production (Reynolds, 1995). Tree by-products for livestock include oil-palm fronds and press fibre as roughage and rubber seed meal, copra cake and palm kernel cake. Grazing for weed control gives low returns so emphasis is likely to be on high-yielding trees with livestock to increase output per unit area. In Malaysia, estates using cattle rose from 120 in 1998 (with 56,000 cattle), to 167 with 115,400 in 2000. Positive results in Malaysian plantations have led to a national policy whereby the sector was targeted to contribute 240,000 tons of red meat by 2010 (Wong *et al.,* 2005).

3. Grazing systems with winter fodder in temperate areas

Thirteen percent of the world's pastures are temperate grazing systems on natural vegetation (de Haan *et al.*, 1997), mainly in China, USA and South America. Much of these pastures are in relatively good condition, but there are areas where grazing pressure is high such as in China and Mongolia, and grasslands have degraded. However, studies in the eastern parts of temperate Asia (Suttie & Reynolds, 2003) indicate a lack of data to ascertain the extent of poor pasture condition and the evolution of degradation. FAO work in the Altai region of China has demonstrated the complementarity of mobile pastoralism and agro-pastoral development and the benefits of irrigated fodder for winter feed, and shown that technological development should go in parallel with social, environmental and ecological considerations. In the Northern Areas of Pakistan, with the introduction of appropriate fodders and by working closely with farmers it has been possible to achieve year-round fodder availability. In Nepal, where *Avena sativa* fits the cropping system and provide green feed in scarcity periods, farmers are adopting new technologies such as small-bag silage, which suit their system.

Winter-feed and transhumant herding by Kazakhs in Altai, Xinjiang, China

Altai prefecture has an extreme continental climate. Transhumant herding is the main land use, and lack of winter-feed is a major problem (Li-Menglin *et al.*, 1996). Several countries have tried to settle herders, often with undesirable social results; the Altai project has maintained herd mobility while providing settled bases for herders, where fodder is grown and education and social services for their families are available. Over half the population of Altai follow transhumant herding (cattle, sheep, horses, goats and camels). Traditional herders have good summer grazing on land above 1,300m only between late June and September. In spring and autumn the transition routes are heavily grazed; winter grazing (December to the end of March) is on desert plains.

Winter feed problems have been solved by developing irrigated land. From 1988 to 1997 over 32,000 hectares were developed for *M. sativa* hay rotated with crops; the work was assisted by World Food Programme and UNDP/FAO. By 1997 some 6,100 households had been voluntarily settled; a further 26,700 hectares of *M. sativa* have been established bringing the total to more than 50,000 hectares, and another 20,000 were foreseen in the Ninth Five Year Plan. The project, which transformed desert into farms and herders into herder-farmers, is accepted as a model for further voluntary settlement schemes. Food security of the region and the target population has been dramatically improved without dismantling the socio-economic basis of transhumance. Family incomes have increased steadily as has access to education and medical facilities. In 1999 and 2000, transhumance patterns and socio-economic conditions of project herders and nomads following the 'old ways' were studied (Suttie & Reynolds, 2003) and compared. Settled groups were able to over-winter more stock and sold more animals. There are clear income differences between the groups as shown in Table 2.

Table 2 Income (RMB)1 of study groups: traditional nomads v project herder-farmers, in Altai, Xinjiang

	Nomadic		Herder-farmer	
	1999	2000	1999	2000
Per capita income	2,583	3,303	6,005	8,198
Income per household	17,511	21,658	39,369	55,569

[1]in 1999/2000 US$1 was equivalent to approx. 8.28 Renminbi (RMB)

The herder-farmer group's herds are increasing; they have good access to services and are investing in production equipment and household goods. Although *M. sativa* yields are far below their potential because herders are loath to provide inputs for 'grass', winter weight loss in sheep has been converted to gain; flocks are mated earlier and lambs slaughtered in their first year. The nomadic group had lower incomes, little access to services and own what they carry in their baggage train; their sheep have to be kept through a second summer.

New forages, winter feed and benefits for smallholders in Pakistan's Northern Areas

The Northern Areas[5] of Pakistan cover 72,500 km^2 of steep, broken land with some of the highest mountains in the world. Cultivable land is very scarce, but there are three million livestock, which spend summer on alpine pastures but rely on crop residues in winter (Dost, 2001). Cereals and orchards are the mainstay of cropping systems and fodder legumes such as *M. sativa* and *Trifolium resupinatum* (shaftal) are intercropped with trees. After harvest there is a potential fodder-growing period of three months in double-cropping areas (1,200-2,000 m) and two months above 2,200 m. Traditionally all lands are open to grazing between autumn harvest and sowing the winter cereal, a serious constraint to fodder growing which has to be addressed by communal agreement.

FAO fodder crop activity began in 1994, and integrated with work of the Aga Khan Rural Support Programme (AKRSP). Traditional fodders were *Z. mays* in summer, a winter-dormant landrace of *M. sativa* which gives only three cuts, and *T. resupinatum* which produces only in late spring and gives 2-3 cuts by late April; green *T. aestivum* was cut in time of scarcity.

Introduced forages and cultivars have had a striking impact. Non-winter-dormant *M. sativas* are well adapted to much of the area and cv 'Sundar' provides 7-9 cuts in double-cropping areas and 4-6 in single crop areas. *T. alexandrinum*, which was not grown in the region before the project, produces from November-January to late April, and gives 5-6 cuts and 2-3 times higher yields than *T. resupinatum* which gives 2-3 cuts in March-April. 'Agaiti' *T. alexandrinum* yields 135 tonnes/ha green matter compared with 80 tonnes for *T. resupinatum*. A mixture of *Avena sativa* (oats) with *T. alexandrinum* improves yields and provides feed at a time of great scarcity; improved *Avena* grow much earlier and are more productive than other winter-green cereals. Multi-cut *Sorghum x drummondii* produce three or four times as much as local *Z. mays*. Adaptive research continues along with demonstration, and fodders for the highest areas are still sought. With green feed and hay it is now possible to ensure year round fodder for dairying. By 1997 some 7,000 farms were using improved forage and cereal

[5] A territory administered by Pakistan comprising the disputed territories other than Azad Jammu and Kashmir - the old Gilgit Agency

varieties and these numbers continue to climb. Local merchants stock seed on project advice. Depending on the area under improved fodder, farmers saved between 6,000 and 10,000 Rs per annum on winter-feed; many also earned 3,000-5,000 Rs from sale of hay. Daily milk yield increased from 1.4 to 4 litres and, in addition to improved family nutrition most farmers earn 5,000-8,000 Rs annually from sale of milk and butter (exchange rate 1US$=Rs 64.25 - September 2001).

Nepal – fodder oats and small-bag silage

Nepalese farming ranges from the tropical Terai to the High Hills where only one crop a year is possible; farms are small. Urban demand for milk has risen, as has the numbers of improved and crossbred cattle and buffaloes. In Brahmin areas not only milk but cow dung and urine are essential in domestic rituals. Straw and crop residues are fed from September/October but run out in January. *Avena* spp. are grown on *Oryza* land but have to be off in time to get the land ready for the new *Oryza* crop; a few single-cut cultivars are used, but seed quality maintenance is poor (Pariyar, 2002). An *Avena* improvement project has focused on poor farmers in the Terai, Low and Middle Hills. *Avena* is well accepted by farmers and new multi-cut cultivars being screened and demonstrated are high yielding. The project assures seed production, works with farmers to assure uptake and is introducing fodder conservation. At high altitudes, where crop choice is very limited, *Avena* could be grown for hay. *Avena* provides green feed in the winter lean period, from the Terai to the Middle Hills. Seed supply is through local production with maintenance of stock by local institutions. Hay cannot be made in wet conditions and ensilage is a better method at lower altitudes. To avoid wastage associated with small silos a 'small bag' system (Lane, 2000) has been adopted and is rapidly becoming popular. Bags are of 6 or 12 kg (a daily ration for one or two buffaloes). Silage can be made in summer when fodder is available but conditions are very wet; a few bags can be made at a time using slack periods in the farming day. A number of farmers have reported significant increases in milk production and income with little additional production cost. This technology is applicable to like situations in the Himalayas.

4. Extensive grazing systems and protected areas

Extensive grazing systems are the only feasible form of agriculture in much of the arid and semi-arid regions of Africa, Latin America, the Near East and Central Asia where the climate is harsh, rainfall unreliable and the risk factor is high (Upton, 2004; Suttie & Reynolds, 2003). On communal grasslands there are often problems of grazing rights and pastoralists have difficulty in managing grazing resources sustainably when there is continual likelihood of trespass and conflict. Each grassland system has its characteristics, cultures and problems but the practical problems associated with production often depend on their method of ownership and management; the problems of commercial stock-raising on privately-owned land require a very different approach from those of pastoralists on communally owned pastures (Riveros, 1993). In some Near East countries, such as Syria, where the land is State-owned, there has been a breakdown of traditional systems of range management. Rising population and livestock numbers, and cultivation, have increased pressure and resulted in deterioration of large areas of the rangelands with consequent loss of biodiversity. In areas of Eastern Europe the break up of collectives has resulted in the abandonment of land, and extensive land-use systems combining grazing and wildlife management offer an alternative for additional income and preservation of natural resources and biodiversity.

Participatory range rehabilitation, wildlife and biodiversity awareness in Syria

Rising population and livestock numbers, and a high degree of mechanisation has caused severe deterioration of large areas of the Al-Badia steppe. Many recognize a dramatic degradation of the environment over the past 30-40 years, but show little awareness of the causes and the possibility of remedial action. A project for the rehabilitation of degraded range, re-introduction of wildlife and establishment of a wildlife reserve, biodiversity awareness raising and the re-establishment of sustainable grazing management based on community participation began in 1996. Activities included re-seeding and shrub planting, closure to grazing, preparation of management plans, assessment of alternative energy sources and income generating activities. Training and capacity building among the stakeholders was stressed, and they participated in planning, maintenance and management of the Talila Wildlife Reserve to which gazelle and oryx have been re-introduced. A holistic approach to resource management and involvement of all stakeholders is crucial for the long-term sustainability of the reserve and its surrounding rangelands; community participation and Bedouin extension were emphasised. The Palmyra project has changed people's attitudes to biodiversity from indifference to awareness (Serra *et al.*, 2003). This has occurred at a local level through formal training, involving hunters and nomads in the work and through the results of the project, in the form of the reserve, the gazelles and the oryx, and education. At national and international level the spectacular rediscovery of the *Geronticus eremita* (Northern Bald Ibis) created widespread awareness of the value and significance of Syria's natural heritage.

Grasslands in Patagonia: desertification, depopulation and diversification

On Patagonia's cool semi-arid grazing lands (approximately 750,000 km^2) extensive sheep farming is a monoculture (Borrelli &Cibils, 2005). Supplementary feed is not used and there are mortalities during big snowfalls. Over half the sheep farmers are poor and own less than 10% of the sheep, for the rest, farm income no longer satisfies their economic expectations. About 400 sheep farmers are economically viable and own approximately 40% of all sheep. A drought in 1983, followed by a harsh winter when 1.5 million sheep died showed that an ecological threshold had been crossed; grasslands could no longer provide enough forage for sheep survival. Recommendations for grazing management were released, but few farmers followed them. Scientists realised that there was a serious problem, politicians and farmers did not. Displaced persons moved to the cities looking for aid, not drawn by economic activity. The government tried to keep the farmers and their personnel in rural areas. However, because farmers were paid in relation to labour use, not sustainable land use, this attenuated the signal and maintained the *status quo*; so more farms were abandoned and rural depopulation accelerated. In Santa Cruz, 100,000 km^2 have been abandoned in the last decade; half could re-start farming if they combined into large operations or included other economic activities, but the rest are too degraded to support sheep.

Carefully managed sheep do not damage the environment, and can be compatible with range conservation and alternative uses if proper grazing is imposed. Some Government policies address this: Argentina established a Fiduciary Fund to support five Sustainable Sheep Development Programmes for farmers' proposals. Access is conditional on the adoption of some basic technologies. Livelihoods could be improved for sheep farmers in Patagonia by focusing on the natural aspect of the native rangelands and aiming to obtain organic or other quality certification so that quality and marketing strategies become the key issues. The aim should be to supply traceable, differentiated lamb meat; increase lamb prices and farm profit, and use market incentives to promote sound grazing management (Borelli & Cibils, 2005).

Risk management for herding in Mongolia

A good example of a livelihood-based approach to grazing land development is the World Bank Sustainable Livelihoods Project in Mongolia (World Bank, 2004) designed with the technical support of FAO. The programme has four complementary components: pastoral risk management; micro finance outreach to herders; a local initiatives fund to improve social infrastructures in grassland areas; and management and policy support. These support local asset creation and institutional capacity and create response mechanisms to satisfy locally identified demands relevant to herders' production systems. Work began in 2002. The most innovative part of the project is the Pastoral Risk Management Strategy which pilots and adapts options for 143 districts for managing covariant risk, with emphasis on *preparedness*. It aims to reduce the vulnerability of herders and enhance their resilience to weather events and other shocks through four activity areas: (a) risk forecasting and contingency planning; (b) grazing and pasture management, supporting community-based pasture management, and mechanisms for conflict resolution; rehabilitation of emergency grazing reserves in areas of high quality under-utilized pasture (FAO, 2003b); (c) herder self-help initiatives, organisational strengthening with matching loans to assist groups to establish revolving funds for livestock investments; (d) hay and fodder development. Herders' participation and the interest-driven replication rate from 16 districts in 2002 to 143 in 2004 are strong indicators of the high potential of such innovative approaches.

Grazing and wildlife - multiple land use in the Czech Republic

After 50 years of collective organisation, land is being returned to its original owners or given to new owners; as a consequence of this change some agricultural holdings will be small and not economically viable, and in marginal areas may be abandoned. Marginal areas often favour biodiversity and form the basis of an attractive landscape, which are assets for eco-tourism and suitable for extensive grazing. A project to provide training and introduce pilot techniques on this 'abandoned' land provides a model for converting this resource into sustainable grasslands, and promoting and enhancing biodiversity (FAO, 2004b). Breeding game in reserves has a long tradition in the Czech Republic; currently there is a move to change game reserves to zones of extensive ecological agriculture, which would protect the landscape on the one hand and produce sport and develop agro-tourism on the other. Livestock may be managed to co-exist with the wildlife. Grassland in the forest provides fodder and shelter for game; temporary grassland is species poor and regular bush control is necessary. Maintaining a balance between the forest and the food requirements of the wildlife requires careful management. Feed for game, grown near to forests, will be managed as crops or permanent grassland, and the needs of specific game considered, including cover. Where storms damage trees, and gaps in the canopy result, this will provide increased herbaceous cover with beneficial effects on wildlife. Various institutes, which have strengths in the biophysical and socio-economic attributes of land-use systems need to collaborate to provide farmers with the advice they need to manage these areas.

5. Grasslands as a source of traditional food in the semi-arid/arid tropics

A wide range of plants is used as emergency and seasonal food in the Sahel, including Sudan and Chad, and contributes to nutrition and rural incomes. Some are collected every year, others in emergencies. Leaves are eaten fresh or cooked while fruits are consumed fresh, dry or processed. Seeds are ground to produce porridge or bread. Many plants are important items of traditional medicine. Some have entered local trading activities and are marketed:

for instance *Adansonia digitata*, fruits of *Ziziphus, Tamarindus indica, Grewia tenax, Hyphaene* and *Balanites*. These form an important source of income for rural households particularly in the event of crop failure. Gathering should be guided and encouraged when compared with less desirable activities like wood collection, tree cutting and cultivation of marginal lands. Grasses have significant food potential and the seed of several are harvested, known collectively as *kreb* or *kereib;* prominent grasses are species of *Brachiaria, Cenchrus, Dactyloctenium, Echinochloa, Oryza* (ruz al wadi – *Oryza barthii* - is much sought after and is more expensive than *Pennisetum glaucum* and *Rottboellia*. In 2003, FAO undertook various field studies on the extent to which harvested foods contribute to food security; their function in the ecosystem; existing species and varieties and *in situ* conservation practices, with a view to estimate market opportunities and income generation possibilities - especially for women who collect, process, store and sell traditional food from grasslands. Markets were chosen for study, as they are end points for the products. Leafy plants are sold locally in small quantities and at low prices while dried fruits are marketed at more regular prices. In Chad *kreb* is mainly harvested for domestic consumption, although some is sold; its price varies from half the price of *Triticum* spp. to the same or double the price of *Pennisetum glaucum*. A sample of *kreb* from N'djamena comprised *Echinochloa colona* (79%), *Panicum coloratum* (11%) and traces of *Panicum antidotale, Cyperus* spp. and *Eragrostis* spp. with 9% impurities (Batello *et al.*, 2004). It contained 8.8% protein and 1.4% lipids on a dry matter basis. In monetary terms a 100 kg sack will fetch between 15,000 and 80,000 CFAF (Communauté Financière Africaine franc, with 655.957 CFA francs/€). Thus there is a genuine market for *kreb*, and potential to develop it as a replacement for other cereals, and as a delicacy to be sold as a 'niche' product to the people who use it in traditional recipes.

Key factors for success in improving grassland livelihoods and reducing poverty

In the past, failure of development projects was usually blamed on the reluctance of farmers or pastoralists to take up the proposed techniques, on problems of land tenure, or on the inadequacies of administrations; it was rarely attributed to project formulators who did not consult the beneficiaries and failed to understand their needs and expectations. While the situation has improved, the importance of understanding the role of local institutions and community regulatory mechanisms is still insufficiently stressed. A review of 800 projects (LID, 1999) found that most had failed to bring about significant sustainable improvements in the livelihoods of the poor, and concluded that "the key lesson is the importance of institutions in defining the success of pro-poor measures", and the need to respect property rights which provide incentives for land conservation and improvement. IFAD (2004) recently focused on the failure of present-day policies and practices to consider the needs of rural smallholders, and proposed strategies to help provide rural livestock keepers with the tools they need to overcome their poverty. Whilst different lessons were learned in the various case studies included in this paper, a number of key features emerged:

1. Project duration, government commitment and farmer/herder participation
   Large-scale development programmes should be preceded by small pilot projects and impact assessment studies. Projects are usually of short duration but address long-term problems; longer (or linked) projects stand a better chance of success. Governments must be fully committed to providing the resources needed over the long-term and be willing to take the necessary political and economic decisions as demonstrated in on-going developments in Xinjiang, China and the Al-Badia in Syria. Crucially, the local population must have a major stake in a project; dialogue may be protracted but long-term success depends on their participation. Participation requires training of technicians and local

people as well as feedback mechanisms. Projects working with pastoralists, have successfully demonstrated that they can be provided with a settled base, without destroying their pattern of life, as long as herd mobility is maintained so that flexibility remains within the system.

2. Technologies, market access and opportunities

Interventions should be targeted at real farmer problems such as critical shortage periods in the feed calendar. Technologies should be appropriate and relatively inexpensive, and should result in a real impact on incomes and livelihoods. Access to markets will encourage farmers to consider adopting new technologies; rapid urbanisation has created a considerable demand for peri-urban dairying and a market for milk produced by smallholders based on intensive fodder production. Greater focus is needed on under-exploited forage opportunity areas, e.g. the large areas of tree crop plantations and agroforestry systems, and also the many un- or under-utilized food plants from grasslands. These need to be assessed for their potential as cereal replacements, 'niche' products, and their role in traditional food security systems. Given the importance of traditional foods and of genetic diversity, greater focus is needed on conserving genetic diversity, as it is the poor who suffer most from its steady erosion. New systems such as extensive land use combining grazing and wildlife management may offer prospects for additional income and the preservation of natural resources and biodiversity.

3. Government policies

Technical, socio-economic, institutional and environmental factors must be considered together. Particular focus is needed on the breakdown of traditional grazing systems, land tenure, grazing rights, management of grasslands on critical watershed areas, access to grazing lands and state ownership of pasture. Clear policies on food security, alternative employment, long term sustainability and risk management are necessary, as well as good governance. Small producers may need protection from changes in international market access. Crucially, government policies should recognize that smallholders produce most of the developing world's food, but are much poorer than the rest of the population, so the biggest impact on poverty and livelihoods is likely to be made by focusing on this category.

The way forward

This paper has identified a number of opportunities in developing countries and presented various case studies. A vast amount of research has been carried out on pastures and range management, but uptake by small farmers and herders has been limited. The present challenge is how to put this *corpus of science* to better use, integrating it into the users' knowledge and needs, working with the farmers and herders in collaboration with other disciplines to improve livelihoods and reduce poverty. Attention is needed in particular in three areas:

Development needs

Increased awareness is required of the need for public sector and international funding, where public goods such as environmental services on water catchment areas are provided by pastoralists grazing communal lands. Long-term public sector support for delivery systems is essential if development is to succeed; capacity building of decentralised bodies will contribute to the sustainability of results during and after project execution. Development programmes should focus more on areas with greatest poverty in smallholder systems. Emphasis should be on the ability of systems to produce goods and services, not on maximising production that often has negative long-term consequences. More attention should be given to valorisation of the multiple functions of grasslands and to new mechanisms of technology transfer that combine traditional and recent systems.

Research needs

Grassland research should focus more on the needs of the poor, and difficult environments dominated by low fertility, saline soils and arid areas, and on ensuring that research effort is reflected through extension in successful outcomes. More emphasis is needed on fodder conservation, particularly in tropical areas. A critical need is for work on developing and making use of the many under and unutilized species from grassland areas, to enhance the number adapted to different soils, climates and farming systems. Ensuring seed availability is another critical need. For parts of Africa a priority is to identify disease resistant *Pennisetum purpureum* clones.

Policy needs

Economic policies must ensure farmers an adequate return while encouraging environmental sustainability. Policies should be based on scientific knowledge with recognition of traditional practices and social acceptance. Only continuous dialogue and consultation among all stakeholders can lead to good governance and sustainable agricultural development. International commitments such as the *Convention on Biological Diversity* and the *Treaty on Genetic Resources* should provide a fair framework for the development of regional, national and local policies. Use of subsidies should focus on raising the livelihood of poor farmers and promoting maintenance of natural resources. Support to production of quality food for all should be given through local production, more promotion of local varieties and breeds, controlled food processing and fair international trade agreements.

An initial target is the Development Goal of the 2000 UN Millennium Declaration, which seeks to "eradicate extreme poverty and hunger" with the aim of halving, between 1990 and 2015, the proportion of people whose income is less than one dollar a day, and those who suffer from hunger. Making progress towards this target depends on many factors, not least of which are international political will and the mobilization of additional resources; to this can be added a greater appreciation of the role that grasslands and forages can play in improving livelihoods and reducing poverty. While some of the answers may be contained in the UN Millennium Project's *Investing in Development* report (UN, 2005), this operational framework will need technical inputs from many areas including those from grassland scientists.

Acknowledgements

The contribution of Mr Jim Suttie, former FAO officer, to the preparation of this paper is much appreciated.

References

Aumeeruddy-Thomas, Y., Y.C. Lama & S. Ghimire (2004). Medicinal plants within the context of pastoral life in the village of Pungmo, Dolpo, Nepal. In: C. Richard & K. Hoffman (eds.) Strategic innovations for improving livelihoods in the Hindu Kush-Himalayan Highlands Vol. II, ICIMOD, Kathmandu, Nepal, 107-128.
Batello, C., M. Marzot & A.H. Touré (2004). The future is an ancient lake. Traditional knowledge, biodiversity and genetic resources for food and agriculture in Lake Chad Basin ecosystems. FAO Interdept. Working Group on Biological Diversity for Food and Agriculture, 307p.

Borelli, P. & A. Cibils (2005). Rural depopulation and grassland management in Patagonia. In: S.G. Reynolds & J. Frame (eds.) Grasslands: developments, opportunities, perspectives. FAO and Science Publishers Inc., Rome & Enfield, USA, 461-487.

de Haan, C., H. Steinfeld & H. Blackburn (1997). Livestock and the environment. Finding a balance. European Commission Directorate General for Development. (Wrenmedia, Eye, UK, 115pp.

Delgado, C., M. Rosegrant, H. Steinfeld, S. Ehui & C. Courbois (1999). Livestock to 2020: the next food revolution. IFPRI Food, Agriculture and the Environment Discussion Paper, No. 28. International Food Policy Research Institute, Washington DC, 72pp.

Dixon, J., A. Gulliver & D. Gibbon (2001). Farming systems and poverty. Improving farmers' livelihoods in a changing world. FAO and World Bank, Rome & Washington DC, 412pp.

Dost, M. (2001). Fodder success story: improved fodder crop production in the northern areas of Pakistan. *Integrated Crop Management*, FAO, Rome, Italy, 4, 23pp.

Dost, M. (2003). Fodder production for peri-urban dairies in Pakistan. <http://www.fao.org/ag/AGP/AGPC/doc/pasture/dost/fodderdost.htm>.

Ellis, F. (2000). Rural livelihoods and diversity in developing countries. Oxford University Press, Oxford, UK, 273pp.

FAO. (2002a). Mobilizing the political will and resources to banish world food hunger. Foreword by Director-General, FAO, to: The World Food Summit – five years later. Technical Background Document, 105pp.

FAO. (2002b). Focus on the issues: feeding an increasingly urban world. <http://www.fao.org/worldfoodsummit/english/newsroom/focus/focus2.htm>

FAO. (2003a). The state of food insecurity in the world 2003. FAO, Rome, Italy, 36pp.

FAO (2003b). TCP/MON/0066 - pastoral risk management strategy project. <http://www.fao.org/ag/AGP/AGPC/doc/publicat/field2/TCP0066.htm>

FAO. (2004a). GCP/FRY/001/NET - development assistance to livestock farmers in the mountainous areas of Prijepolje, Sjenica and Tutin in the Sandzak region. <http://www.fao.org/ag/AGP/AGPC/doc/publicat/field2/GCP001.htm>

FAO. (2004b). TCP/CEH/2902 - sustainable utilization of agricultural 'abandoned' land in Czech Republic. <http://www.fao.org/ag/AGP/AGPC/doc/publicat/field2/TCP2902.htm>

Hervieu, B. (2002). Multi-functionality: a conceptual framework for a new organisation of research and development on grasslands and livestock systems. In: J.L. Durand, J.C. Emile, C. Huyghe & G. Lemaire (eds.) Multi-function grasslands. Quality forages, animal products and landscapes. EGF Grassland Science in Europe, 7, 1-2.

Horne, P., W. Stur, P. Phengsavanh, F. Gabunada & R. Roothaert (2005). New forages for smallholder livestock systems in southeast Asia: recent developments, impacts and opportunities. In: S.G. Reynolds & J. Frame (eds.) Grasslands: developments, opportunities, perspectives. FAO and Science Publishers Inc., Rome & Enfield, USA, 357-382.

IFAD (International Fund for Agricultural Development) (2001). Rural poverty report: the challenge of ending rural poverty. Oxford University Press, UK for IFAD. 266pp.

IFAD. (2004). Livestock services and the poor. a global initiative. Collecting, coordinating and sharing experiences. IFAD, Rome, Italy, 132pp.

Karanja, G.M. (1999). Nutrition requirements of the dairy cow. In: G.M. Karanja & E.N. Sabiiti (eds.) *Proceedings of the workshop on integrated smallholder dairy farming systems in peri-urban areas with emphasis on forages and fodder tree utilization*. Embu, Kenya (June 13-17), 27-36.

Lane, I. (2000). Little bag silage. In: L. t'Mannetje (ed.) Silage making in the tropics with particular emphasis on smallholders. *FAO Plant Production and Protection Paper* No. 161, 70-83.

LID. (1999). Livestock in poverty-focused development. Crewkerne, Somerset: Livestock in Development. OutHouse Publishing Services, Wiltshire, UK, 94pp.

Li-Menglin, Yuan Bo-Hua & J.M. Suttie 1996. Winter feed for transhumant livestock in China: the Altai experience. *World Animal Review*, 87 (1996/2), 38-44.

Lipper, L. & R. Cavatassi (2003). Land use change, carbon sequestration and poverty alleviation. *ESA Working Paper*, Agricultural and Development Economics Division, FAO No. 03-13, 22pp.

NDDB (National Dairy Development Board) (2004). <http://www.nddb.org/perspective/perspective-more.html#goal>

Nicholson, C.F., P.K Thornton, L. Mohammed, R.W. Muinga, D.M. Mwamachi, E.H. Elbasha, S.J. Staal & W. Thorpe (1999). Smallholder dairy technology in coastal Kenya. An adoption and impact study. *ILRI Impact Ass. Series*, No. 5. 59pp.

Orodho, A.B. (2005). Intensive forage production for smallholder dairying in eastern Africa. In: S.G. Reynolds & J. Frame (eds.) Grasslands: developments, opportunities, perspectives. FAO and Science Publishers Inc., Rome & Enfield, USA, 433-459.

Pagiola, S., P. Agostini, J. Gobbi, C. de Haan, M. Ibrahim, E. Murgueitio, E. Ramirez, M. Rosales & J.P. Ruiz (2004). Paying for biodiversity conservation services in agricultural landscapes. *Environment Dept. Papers* No. 96, The World Bank, 27pp.

Pariyar, D. (2002). Fodder oats in Nepal. <http://www.fao.org/ag/AGP/AGPC/doc/pasture/spectopics/oatsnepal.html>

Peeters, A. (2004). Wild and sown grasses. Profiles of a temperate species selection: ecology, biodiversity and use. FAO and Blackwell Publishing, Rome, 314pp.

Reynolds, S.G. (1995). Pasture-cattle-coconut systems. FAO-RAPA Publication, Bangkok, Thailand, 668pp.

Riveros, F. (1993). Grasslands for our world. In: M.J. Baker (ed.) Grasslands for our world. SIR Publishing, Wellington, NZ, 6-11.

Serra, G., D. Williamson & C. Batello (2003). From indifference to awareness. Encountering biodiversity in the semi-arid rangelands of the Syrian Arab Republic. FAO Interdept. Working Group on Biological Diversity for Food and Agriculture, FAO, Rome, 47pp.

Staal, S.J., L. Chege, M. Kinyanjui, A. Kimani, B. Lukuyu, D. Njubi, M. Owango, J. Tanner, W. Thorpe & M. Wambugu (1998). Characterization of dairy systems supplying the Nairobi milk market. A pilot survey in Kiambu District for the identification of target groups of producers. Smallholder Dairy (R&D) Project. KARI, ILRI and Livestock Prod. Dep. (Min. of Agric.), Nairobi, Kenya.

Suttie, J.M. (2000). Hay and straw conservation for small scale farming and pastoral conditions. *Plant Production and Protection Series* No. 29, FAO Rome, 244-248.

Suttie J.M. & S.G. Reynolds (2003). Transhumant grazing systems in temperate Asia. *Plant Production and Protection Series* No 31, FAO Rome, 331pp.

Thornton, P.K., R.L. Kruska, N. Henninger, P.M. Kristjanson, R.S. Reid, F. Atieno, A.N. Odero & T. Ndegwa (2002). Mapping poverty and livestock in the developing world. International Livestock Research Institute (ILRI), Nairobi, Kenya, 117pp.

UN (2000). <http://www.unmillenniumproject.org/html/dev_goals1.shtm>

UN (2005). Investing in development. <http://www.unmillenniumproject.org/reports/index.htm>

UNEP (2002). Global environment outlook 3: past, present and future perspectives. Earthscan Publications Ltd., London, 416pp.

Upton, M. (2004). The role of livestock in economic development and poverty reduction. Pro-Poor Livestock Policy Initiative (PPLPI) Working Paper No.10. FAO, Rome, 59pp. <http://www.fao.org/ag/againfo/projects/en/pplpi/project_docs.html>

White, R., S. Murray & M. Rohweder (2000). Pilot analysis of global ecosystems (PAGE): grassland ecosystems. World Resources Institute (WRI), Washington DC, USA, 100pp.

Wong, C.C., F. Moog & C.P. Chen (2005). Forage and ruminant livestock integration in tree crop plantations of southeast Asia. In: S.G. Reynolds & J. Frame (eds.) Grasslands: developments, opportunities, perspectives. FAO and Science Publishers Inc., Rome & Enfield, USA, 403-431.

World Bank (2004). Sustainable Livelihoods Project <http://web.worldbank.org/external/projects/main?pagePK=104231&piPK=73230&theSitePK=40941&menuPK=228424&Projectid=P067770>

Wright, I.A. (2005). Future prospects for meat and milk from grass-based systems. In: S.G. Reynolds & J. Frame (eds.) Grasslands: developments, opportunities, perspectives. FAO and Science Publishers Inc., Rome & Enfield, USA, 161-179.

Younie, D. & T. Baars (2005). Organic grassland: principles, practices and potential. In: S.G. Reynolds & J. Frame (eds.) Grasslands: developments, opportunities, perspectives. FAO and Science Publishers Inc., Rome & Enfield, USA, 207-232.

Participatory approach to common use grazing management in dry area developing countries

J.A. Tiedeman, A. Larbi, F. Ghassali and N. Battikha
International Center for Agricultural Research in the Dry Areas (ICARDA), P.O. Box 5466, Aleppo, Syria
Email: j.tiedeman@cgiar.org

Key points

1. Range restoration technology is available but useless when not followed by management.
2. Institutional mechanisms for grazing management are needed for communal range.
3. Community participatory approaches help pastoralists better manage rangeland grazing.

Keywords: communal range, rotational grazing, herders

Introduction

Most rangelands in developing countries are grazed in common, usually overgrazed and severely degraded, producing much less forage than potential. Use rights vary depending upon the region: from completely free access (open to everyone) to exclusive use by tribes, communities or extended families. Livestock numbers are usually not controlled unless by government intervention. Indigenous management systems that were once sustainable such as the 'Hema' system common to the Middle East have broken down for various reasons, but mostly because of increased population pressure or inappropriate government policy and land tenure. Range research and development has focused on range restoration, either by direct seeding or transplanting shrubs, but most fail because of inadequate management once the area is improved. Overgrazing and continuous grazing are usually the causes of degradation, but other factors include over-harvesting for fuelwood and temporary cultivation. These problems are being addressed by ICARDA, which serves all the dry-area developing countries for rehabilitation and management of rangelands.

Participatory approach to grazing management used for rangelands by ICARDA

The International Center for Agricultural Research in the Dry Areas (ICARDA), in collaboration with national agricultural research organizations, have developed and used a participatory approach since 1989 in range research and development. This continues as a central theme within the overall core project 'Rehabilitation and Improved Management of Rangelands in Dry Areas.' The goal of this project is to halt desertification, restore the productivity of rangelands and improve pastoralist income. Most of our experience with the participatory approach was through two agropastoral projects funded by the Swiss Agency for Development and Cooperation (SDC), and the Mashreq/Maghreb (M & M) project funded by the International Fund for Agricultural Development (IFAD) and the Arab Fund for Economic and Social Development (AFESD). From these projects, rangeland restoration technology was developed. However, it soon became evident that management, not restoration technology, was most needed. Restored land rapidly degraded without proper grazing management. For management to change, stakeholder participation was essential. Involving farmers and communities fully in the development and research process leads to success. Government institutions must also participate as partners in the process to empower the herders, and provide the community with assurance that their efforts and investments are rewarded. Policy reform is often needed. The participatory approach used in range management by these projects can be outlined in the following steps:

- Representative pilot areas are selected based upon existing or new socio-ecological survey data. This is done so that tested technology can be transferred to other areas.
- Education on grazing management is provided to the community via scientists and officials. Traditional management is documented, and potential options that merge science and indigenous knowledge are discussed.
- Existing relevant biological and socio-economic community data is accumulated and analysed. Missing information is gathered using rapid rural appraisals, community or household surveys, ecological inventory, mapping, and range condition assessment.
- Community leaders draw a map of their rangeland area including different types of range and use patterns. Researchers go into the field with the pastoralists to characterise, evaluate condition, and map rangeland types as defined by the community. Data are assembled in GIS layers of both social and biological data to serve as a base map for the community grazing management plan.
- Analysis is done and potential management options tested. A community grazing management plan is developed, implemented and monitored.

The approach was successfully used in Morocco to establish a process of community participation that has led to better planning and grazing management (Bounejmate & El Mourid, 2001). An assessment of the social-ecological conditions was used to develop management plans, and to test alternative approaches and policies to management. The agro-pastoral project at Aïn Béni Mathour (Morocco) is an example of how farmers participated in the ecological assessment and research, and where scientific ecological assessment was combined with farmer interviews. Farmers provided their knowledge of the changes in vegetation and conditions as detected and quantified by scientific studies, remote sensing image analysis and ecological data sampling at ground truth sites. This blend of indigenous knowledge and scientific research has made it possible to assess and characterize the socio-ecological resources used in priority setting and the development of community management plans.

Substantiated change in ecological condition (trend) requires intensive sampling over long periods of time. There is no scientific proof of reduced degradation, but there are visual and measurable increases in vegetative cover on cooperative land – which is a step towards recovery. Morocco (El Harizi, 1998) and Algeria have both recorded 5 times as much biomass on managed, compared to free access rangeland. This is not a change in productivity, but an indication of dramatic changes in vegetative cover that could halt desertification.

The observed change in human behaviour is of much greater significance because such change is necessary to halt desertification. There is now dialogue between government officials and the herders at Aïn Béni Mathour, resulting in the development of joint plans - something that had not occurred previously. This approach has spread and is now applied by the national program throughout Morocco, Algeria (26 communities) and in neighbouring countries.

In eastern Morocco, the vegetation cover of a grazing cooperative improved where members had to pay a fee to use the 'improved' rangeland. During high rainfall years they choose not to use the cooperative land because free grazing was found elsewhere. As a result, the cooperative land is rested from grazing during the growing season, which is the critical time for plants to rest and regain vigour. The cooperative's land and livestock improved (but what happens to the adjoining land is another story!). Livestock populations increase, and eventually peak at an equilibrium-stocking rate that is too excessive to be profitable. As the land becomes degraded, the point of equilibrium shifts downward in catastrophic events that occur when ecological thresholds are exceeded. Overgrazing problems are often aggravated by subsidy feeding programs, which discourage migration and allow livestock numbers to increase well beyond natural balances, e.g. an emergency feeding program in Tunisia backfired when stock growers imported more livestock in order to increase the amount of subsidy they could claim!

There are only two basic options for grazing management of communal range: controlled stocking rate and rotational grazing. Year-long range rest is an option often applied but is drastic. Partial season rest during the critical growing season can provide the same improvement without wasting the forage - it would increase seasonal forage production, and can be applied every year with rotational grazing. Excluding grazing from one area may increase pressure on nearby areas or alternatively, producing more on-farm fodder may reduce the pressure on adjacent rangeland. This concept must be re-evaluated under free-access grazing, where maximum livestock pressure may have already been reached, and flocks are forced out of the area or fed concentrates once the 'free' forage is gone. Subsidising feed, range restoration, or growing more fodder will not eliminate rangeland degradation. This can only be achieved by a change in management strategy.

By dividing rangeland into paddocks, rotational grazing provides the plants a short rest period to build carbohydrate reserves, produce seed, and grow more forage. In developed countries, private ranches install fences to apply grazing management strategies. On public land, government agencies enforce stocking rate and rotations. Developing countries have a greater opportunity to apply intensive rotations because livestock are herded, so fences are not needed. The problem is that herders must cooperate to apply the rotation. This seldom occurs even in villages where a strong leader maintains control, and is much less likely in remote seasonal pastures. If the means for cooperation in herding are developed, desertification could be halted and rangeland production substantially increased.

Conclusions

The participatory approach fails to achieve its objectives if all stakeholders are not involved in the decision-making process, or if the community is not granted the authority to manage the resource. There is no point to rangeland restoration if the poor management that degraded the range is not changed. Rotational grazing and stocking rate control are the only known options to sustain the system and stop degradation. Community participatory approaches can help pastoralists and governments find a way to enable them to better manage the land. Herders who cooperate in rotational grazing or stock control can reverse the downward spiral of degradation of common-use range.

References

Bounejmate, M. & M. El Mourid (2001). Gestion durable des ressources agropastorales. (Sustainable management of the agro-pastoral resource base). *Proceedings regional workshop*, February 20-22, Oujda, Marocco, ICARDA, Alep, Syrie. Iv + 217p.
El Harizi, K. (1998). Morocco case study: blending old with new institutions – an innovative approach to sustainable range land development and management in a traditional society. Case studies, IFAD, WBI's CBNRM Initiative, Suatainable rural development information systems (SRDIS) web site, CIESIN and World Bank <http://srdis.ciesin.org/cases/morocco-001.html>

Adoption of *Brachiaria* grasses in Mexico and Central America: a successful story

F. Holmann[1], P.J. Argel[2] and C.E. Lascano[2]
[1]*International Livestock Research Center (ILRI) and Centro Internacional de Agricultura Tropical (CIAT), A.A. 6713, Cali, Colombia*
Email: F.Holmann@cgiar.org
[2]*Centro Internacional de Agricultura Tropical (CIAT), A.A. 6713, Cali, Colombia*

Key points

1 Of all cultivars released, grasses from the *Brachiaria* genus currently dominate the market.
2 The main beneficiary from the adoption of *Brachiaria* pastures has been Costa Rica.
3 The main adopters of *Brachiaria* pastures have been dual-purpose farmers.

Keywords: tropical forages, adoption, milk and beef production

Background and strategy

In Latin America and the Caribbean (LAC) there has been a major effort to develop new pastures technologies, to increase livestock productivity for the extensive systems prevailing in the tropical lowlands. This multi-national and inter-institutional effort was initiated through the International Network for the Evaluation of Tropical Pastures (RIEPT, by its name in Spanish), which operated from 1976 to 1996 under CIAT leadership. This network became a platform for institutions to train technicians, share forage material from existing gene banks, study the behaviour of new germplasm under different environments, and established the exchange of scientific information to extrapolate research results (Toledo, 1982). Six hundred and forty five agronomists from 24 countries in LAC were trained by RIEPT, in subjects related to forage agronomy and pasture evaluation. Training was key for the success of RIEPT, because these professionals carried out evaluations of new and improved forages under contrasting ecosystems and provided feedback. In addition, during this period participating institutions in RIEPT released 11 selected grasses as commercial cultivars, most of them from the *Brachiaria* genus, as well as 16 forage legume cultivars (CIAT, 2003). In Central America and Mexico these cultivars were released between 1990 and 1996. Forage evaluation activities in this region continues at present through a joint research agenda between CIAT and ILRI, as well as between CIAT and the private seed sector. Of all pasture cultivars released; grasses from the *Brachiaria* genus currently dominate the market – accounting for approximately 84% of all grass seed sales in Mexico and Honduras, 90% in Nicaragua, 85% in Costa Rica, and 97% in Panama during the last 5 years (Holmann *et al.*, 2004). The objective of this paper is to estimate the impact of the adoption of *Brachiaria* grasses released through RIEPT during the period 1990-2003 on milk and beef production and to describe how this was achieved.

Impact

The estimated marginal increase in milk and beef production attributed to *Brachiaria* cultivars was determined by converting seed sales to areas sown (Sáez & Andrade, 1990), and comparing productivity between the traditional and the improved technology (Holmann *et al.*, 2004).

Adoption

In each country the pattern of adoption has followed the theoretical lag – rapid adoption – plateau schedule (Figure 1). Producers found out about new options first through the public sector, and later from private seed companies marketing the new products. Producer's interest in adoption was stimulated by an increased domestic demand for milk and beef, and from a decline in productivity from less-well adapted species and degraded pasturelands.

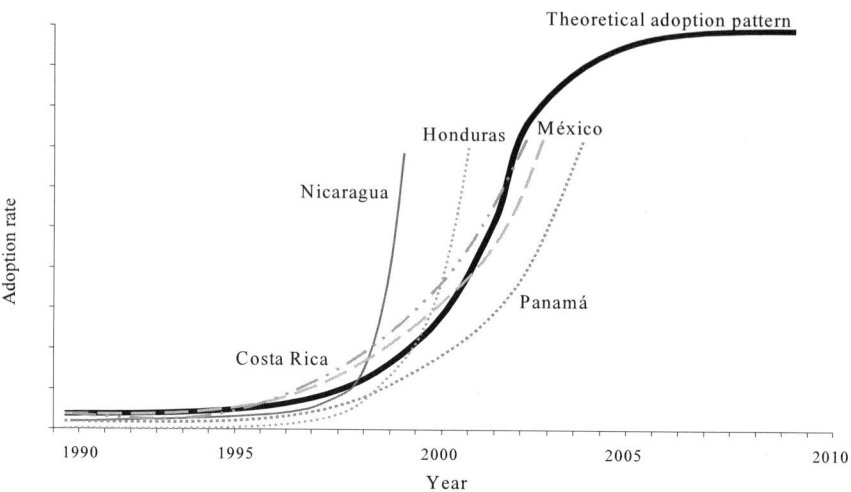

Figure 1 Theoretical adoption pattern of new forage technologies and adoption curves in México and Central America based on *Brachiaria* seed sales (Holmann *et al.*, 2004).

Areas planted with Brachiaria grasses

It is assumed that most seed was allocated to renovate pastures in advanced stages of degradation, or naturalised pastures with low productivity. The largest volumes of seed sales and planted areas correspond to Mexico (9,100 million t (Mt) of seed with 2,616,130 ha planted). Costa Rica is the country in Central America with the largest seed sales and planted areas (1,692 Mt of seed and 437,516 ha planted), followed by Honduras (671 Mt and 186,788 ha), Nicaragua (134 Mt and 35,822 ha) and Panama (40 Mt and 10,952 ha). During this period the annual rate of increase in seed sales was respectively 32% in Mexico, 62% in Honduras, 45% in Nicaragua, 39% in Costa Rica, and 54% in Panama. Total area planted with *Brachiaria* cultivars during this period amounts to 6.5% of the total area of permanent pastures in Mexico, 12.5% in Honduras, 1% in Nicaragua, 18.7% in Costa Rica, and 0.1% in Panama.

Additional milk and beef production

In relative terms, the main beneficiary from the adoption of *Brachiaria* cultivars has been Costa Rica because the additional production resulting from the adoption of these grasses is equivalent to 55% of the national milk production (437,000 Mt fluid milk) and 18% of the beef (26,000 Mt) produced in 2003 (Holmann *et al.*, 2004). These benefits are followed by Mexico where the increase in productivity from the adoption of *Brachiaria* cultivars is equivalent to 24% of national milk production and 5% of beef production. In Honduras the marginal increase in milk and beef production is equivalent to 25% and 12% of national production, respectively. In Nicaragua and Panama the adoption of *Brachiaria* grasses has been the lowest in the region. As a result, the additional increase in milk in Nicaragua and Panama amount to 11% and 5% of national production, respectively. In the case of beef production, the additional increase in Nicaragua and Honduras has been 2% and 1% of domestic production, respectively. These figures suggest that adopters of *Brachiaria* grasses are producers oriented toward dairying and to a lesser extent beef. Dual purpose production systems are common in the region (i.e. where approximately 70% of gross sales come from milk and the remainder from the sale of weaned male calves), and it can be argued that the main adopters of these grasses have been dual-purpose farmers.

Success factors

- *Availability of locally adapted cultivars.* The underlying hypothesis was that lack of adaptation of commercial cultivars selected in other continents could be overcome by selection of locally adapted forage germplasm. This in turn required a large effort on multi-locational screening for adaptation to prevailing biotic and abiotic constraints.
- *Involvement of private seed sector.* The region dos not have comparative advantages for *Brachiaria* seed production, particularly in terms of soil and climatic conditions. However, grass seed production is a large commercial activity in Brazil, and much of the *Brachiaria* cultivars were marketed by Brazilian companies and sold regionally through local seed companies.
- *Increased forage seed demand.* The process of adoption of new *Brachiaria* cultivars has been stimulated by the availability at reasonable cost of commercial seed produced in Brazil, and a rapidly expanding regional livestock sector geared towards meeting increased consumer demand for milk and beef.

What could have been done better?

The results presented in this paper indicate that investments of public funds in Central America and Mexico to support a forage evaluation R&D network, paid of in terms of adoption of improved grasses and increased supplies of beef and milk, and staple food commodities for consumers across income levels in the region. In spite of this, RIEPT should have worked more closely with the private sector (i.e. local seed companies, universities, and livestock producer associations), both in terms of people trained and exchange of germplasn, to achieve a faster and greater diffusion of selected cultivars. Most of the effort concerning seed multiplication and dissemination were concentrated on the public sector, which was significantly reduced in size during the 90's which had a negative effect.

References

CIAT (2003). Annual Report of the Tropical Forages Project. CIAT, Cali, (218pp).

Holmann, F., L. Rivas, P. Argel, & E. Pérez (2004). El impacto de la adopción de gramíneas del género *Brachiaria* sobre la productividad e ingreso de los productores: Un estudio de caso de Centroamérica y México (The impact of the adoption of *Brachiaria* grasses on livestock productivity and income of producers: a case study of Mexico and central America). CIAT. Working Document 197, 31pp.

Sáez, R.R. & R.P. Andrade (1990). Impactos técnico-económicos de *Andropogon gayanus* en los Cerrados de Brasil (Technical and economic impacts of *Andropogon gayanus* in the Cerrados of Brazil). EMBRAPA-CPAC, Brasilia, 38p.

Toledo, J.M. (1982). Objetivos y Organización de la Red Internacional de Evaluación de Pasturas Tropicales (Objectives and organisation of the international network for the evaluation of tropical pastures). In: J.M. Toledo (ed.) Manual para la Evaluación Agronómica. Red Internacional de Evaluación de Pastos Tropicales (RIEPT). CIAT, Cali, 13-21.

Improved livelihoods from grasslands; the case of Napier grass in smallholder dairy farms in Kenya

D.M. Mwangi[1], D. Romney[2], S. Staal[2], I. Baltenweck[2] and S.W. Mwendia[1]
[1]*Kenya Agricultural Research Institute (KARI), Animal Production Research Programme P.O. Box 58711-00200, Nairobi, Kenya*
Email: DMMwangi@kari.org
[2]*International Livestock Research Institute (ILRI), P.O.Box 30709-00100, Nairobi, Kenya*

Key points

1. Many in Kenya consider smallholder dairying as the path out of poverty.
2. The geographic distribution of the smallholder dairy industry is influenced by a combination of socio-economic market access and biophysical constraints on forage production.
3. Napier grass has established itself as the forage crop of choice in intensive smallholder dairy farms.
4. Combining bio-physical constraints to the adoption of planted fodder with farmer's socio-economic situation can help to better target forage technologies.
5. Over dependency on one crop, especially where the genetic diversity is low, can be dangerous.

Keywords: market access, forage adoption, socio-economic factors, forage technology

Introduction

In Kenya, smallholder farmers produce about 80% of the marketed milk. The farming systems vary from mixed farms with up to 10 ha of land and <10 dairy cows (Gitau *et al.,* 1994; Anon., 1985), to intensive smallholder dairy producers in the high human population central Kenya region with 0.9 to 2 ha of land and 3-4 dairy cows (Staal *et al.,* 2001a). Milk production depends heavily on the cultivation of forages, with *Pennisetum purpureum* (Napier grass) by far the most important. An estimated 350,000 of the 600,000 smallholder farms in Kenya grow and utilize *P. purpureum* on their farms.

There are various published data indicating the level of dependence on sown forages. In a survey of 21 smallholder dairy farmers in the highlands of Kenya, Romney *et al.* (2004) found that *P. purpureum* supplied approximately 40 and 60% of the feed offered to dairy cows in the dry and wet months respectively, with the remaining feed provided by concentrates, crop residues (mainly *Zea mays* (maize) stover) and other cut and carry fodder such as roadside grass. In the more intensive cut and carry systems of production practiced in central Kenya, McLeod *et al.* (2003) found that *P. purpureum* was grown by over 70% of the smallholder farmers in their study area. In farm level characterisation surveys of over 3300 households conducted between 1996 and 2000 in central Kenya, 62% kept livestock and more than 50% were growing *P. purpureum*. Farmers were also growing fodder legumes such as *Sesbania grandiflora* (Sesbania), *Leucaena leucocephala* (Leucaena), *Calliandra calothyrsus* (Callindra), *Desmodium intortum/uncinatum* (Desmodium) and *Medicago sativa* (Lucerne), but the frequency did not exceed 7.5% (Staal *et al.,* 2001b).

What factors lead to the adoption of *Pennisetum purpureum*?

It is argued that the adoption of a forage crop is not influenced by fodder characteristics alone, but by a complex set of factors including farmer and farm resource base, agro-climatic factors and market access. Staal *et al.* (2002) showed the likelihood of adopting *P. purpureum* increased as i) the number of years of farming experience and education of the household head increased, ii) distance from urban centres decreased, and iii) rainfall increased. Thus, while the adoption of *P. purpureum* as the feed of choice in Kenya is principally due to its forage characteristics in comparison with alternatives, the adoption of a cultivated forage *per se* is likely due to the other socio-economic factors outlined above.

Forage characteristics

In order to establish a basis for further discussion, the forage characteristics of *P. purpureum* are briefly outlined prior to addressing the more important factors that led to its adoption.
- The ease of establishment from stem cuttings or root splits overcomes a major constraint to the adoption of a forage crop (Thomas & Sumberg, 1995; Mwangi & Wambugu, 2003).
- High biomass yield is an essential characteristic for smallholder dairy systems where land is a major limiting factor. Dry matter production of 10-20 t/ha per annum have been reported with minimal fertilizer application (Aninda & Potter, 1986), with increases to about 30 t/ha per annum with manure application or intercropping with forage legumes (Mwangi & Wambugu, 2003; Mwangi *et al.*, 2004). These yields compare with 6-8 t/ha per annum from *Chloris gayana* (Rhodes grass) under similar conditions, making *P. purpureum* attractive to smallholder farmers facing land shortages.
- *Pennisetum purpureum* will grow in areas receiving over 600 mm of rainfall per annum, making it suited to most of the east African highlands. Figure 1 maps the potential natural distribution of *P. purpureum*, taking bio-physical factors into account.

The widespread use of *P. purpureum* in Kenya is being threatened by the spread of Napier headsmut, caused by *Ustilago kamerunesis*, which can reduce annual yield by 50-90%. Fortunately there is some variation for resistance to the disease among *P. purpureum* genotypes, and a conventional plant breeding program is under way to produce smut resistant cultivars.

Socio-economic factors

A study by Staal *et al.*, (2002, 2001a) has shown that the adoption of a cultivated forage (in this case mainly *P. purpureum*) was heavily influenced by farmer characteristics and access to milk markets. Data was collected from a household survey involving 3,311 households between 1996 and 2000, and GIS maps were produced using population density, the distance to the nearest urban centre and the condition of the road as a proxy for market access. Potential adoption of *P. purpureum* was predicted using these market access data and then combined with the bio-physical limits to determine areas where adoption of the grass was expected to be high (Figure 2). The findings of the survey were compared with the prediction made using GIS layers on market access and bio-physical limits (rainfall, temperature and soils) and the fit was more that 80% (Figure 3).

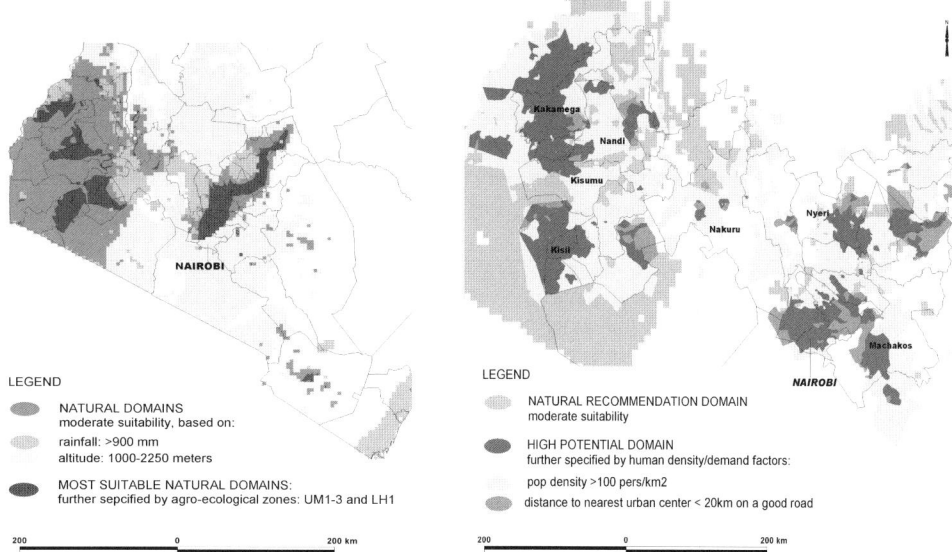

Figure 1 Natural recommendation domains for Napier grass based on bio-physical limits. (Source, Staal *et al.,* ILRI, 2001a)

Figure 2 Recommendation domain after combining the bio-physical limits and socio-economic factors. (Source, Staal *et al.,* ILRI, 2001a)

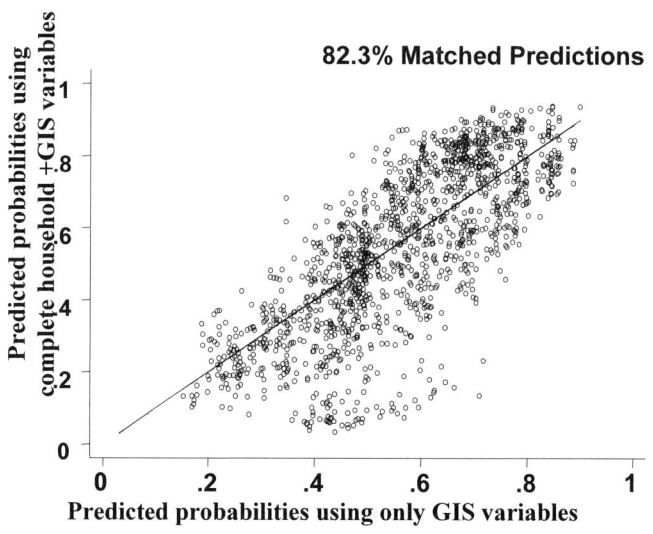

Figure 3 Comparison of planted forage predicted using GIS tools and household survey

This milk industry case study indicates that where data are available, it is possible to use GIS and socio-economic statistics to define recommendation domains for planted forages. This would enhance adoption as dissemination efforts could be better targeted.

Lessons learnt

1 The adoption of planted forages is not only determined by forage characteristics but also by the socio-economic situation of the farmer. Market access for livestock products and land availability are important factors in the adoption of *P. purpureum*. It is contended that policies that improve market access by the poor will result in uptake of planted forages.
2 GIS predictions can be used to fine-tune recommendation domains for forage technology so that extension activities can be targeted at areas where adoption is most likely.
3 The smallholder dairy sector in central Kenya has become reliant on *P. purpureum,* which provides 40-80% of the feed in the system. This high dependency on one fodder crop has a potential downside with the incidence of plant disease. This is particularly the case with *P. purpureum,* which is propagated vegetatively, meaning that the population will have a narrow genetic base. For this reason it is important to maintain the biodiversity of the forage plant and/or to develop alternative forages.

References

Anindo, D.O. & H.L. Potter (1986). Seasonal variation in productivity and nutritive value of Napier grass at Muguga, Kenya. *East Africa Agricultural and Forestry Journal*, 59, 177-185.
Anonymous (1985). Ministry of Livestock Development, Annual Report, Ministry of Livestock Development, Nirobi, Kenya, 167pp.
DANIDA/MoLD. (1991). Kenya dairy master plan. Study report. Danish International Development Agency (DANIDA) and Ministry of Livestock Development (MoLD), Nairobi, Kenya, 289pp.
Gitau, G.K., C.J. O'Callaghan, D. McDermott, A.O. Omore, P.A. Odima , C.M. Mulei & J.K. Kilungo (1994). Description of smallholder dairy farms in Kiambu district, Kenya. *Preventative Veterinary Medicine,* 21,155-166.
Mcleod, A., J.M.G. Njuguna, F. Musembi, J. Maina & D.M. Mwangi (2003). Farmer Strategies for maize growing, maize streak virus control and feeding of smallholder dairy cattle in Kiambu district, Kenya. First Technical Report, Dairy and Maize Project (R7955/ZCO180), 33pp.
Mwangi, D.M., G. Cadisch, W. Thorpe & K.E. Giller (2004). Harvesting management options for legumes intercropped in napier grass in the central highlands of Kenya. *Tropical Grasslands*, 38,234-244
Mwangi, D.M. & C. Wambugu (2003). Adoption of forage legumes: the case of *Desmodium intortum* and *Calliandra calothyrsus* in central Kenya. *Tropical Grasslands,* 37, 227-238.
Romney, D., C. Utiger, R. Kaitho, P. Thorne, A. Wokabi, L. Njoroge, L. Chege, J. Kirui, D. Kamotho &S. Staal (2004). Effect of intensification on feed management of dairy cows in the central highlands of Kenya. In: E. Owen, T. Smith, M.A. Steele, S. Anderson, A.J. Duncan, M. Herrero, J.D. Leaver, C.K. Reynolds, J.L. Richards & J.C. Ku-Vera (eds.) Responding to the livestock revolution – the role of globalisation and implications for poverty alleviation. British Society of Animal Science Publication No. 33, Nottingham University Press, Nottingham, UK, 167-178.
Staal, S.J., I. Baltenweck, M.M. Waithaka, T. deWolff, & L. Njoroge (2002). Location and uptake: integrated household and GIS analysis of technology adoption and land use, with application to smallholder dairy farms in Kenya. *Agricultural Economics*, Special Issue on Spatial Analysis, 27, 295-315.
Staal, S., L. Njoroge, T. de Wolff, M. Waithaka, D. Romney, A. Wokabi, P. Wanjohi, J. Nyangaga, & I. Baltenweck (2001a). Integrated household and GIS analysis of planted fodder adoption on smallholder farms in the Kenyan highlands. Presented at International Workshop on forage demand and adoption by smallholder livestock keepers, ILRI-Addis Ababa, June 18-20, (proceedings not available).
Staal, S., M. Owango, H. Muriuki, M. Kenyanjui, B. Lukuyu, L. Njoroge, D. Njubi, I. Baltenweck, F. Musembi, O. Bwana, K. Muriuki, G. Gichungu, A. Omore, & W. Thorpe (2001b). Dairy system characterisation of the greater Nairobi milk shed. SDP Collaborative Research Report. Smallholder Dairy (R&D) Project, MoARD/KARI/ILRI, Nairobi, Kenya, 69pp.
Thomas, D. & J.E. Sumberg (1995). A review of the evaluation and use of tropical forage legumes in sub-Saharan Africa. *Agriculture, Ecosystems and Environment,* 54,151-163.

Role of information and information providers in technology transfer

D.J. Undersander

University of Wisconsin, 1575 Linden Drive, Madison, WI 53717 USA
Email: djunders@wisc.edu

Key points

1 Technology transfer requires the following three steps:
   a) Information must be provided;
   b) The target audience must become aware of the information;
   c) The target audience must implement the new technology.
2. Successful technology transfer considers the economic, social and environmental concerns of the audience.
3. Differences in learning methods among and within audiences should be considered.
4. Marketing principles using specific audience data from public and private sources should be used to develop technology transfer plans.

Keywords: knowledge transfer, technology transfer models, information dissemination

Introduction

Technology transfer is an essential component of economic change in society. Developers of new technology often fail to realize that there is a science to technology transfer. This lack of appreciation for the skills involved in information dissemination and in activities necessary to affect a change of action in an audience often severely limits the rate and amount of technology transfer that occurs. Significant differences exist between doing a news release or writing a publication and causing audience acceptance of a new technology. The old standards of expecting adoption of a new technology simply because it "will profit farmers" or because it is tested and recommended by "revered" public researchers has not been, and will not be sufficient to cause acceptance. Instead it is necessary to consider the audience and how its concerns relate to technology adoption, how audiences learn (which varies with different audiences), and match the presentation methods to the learning preferences of the specific audience.

Model of technology acceptance

The increased use of computers and associated software has stimulated significant research into technology transfer and acceptance. Many of the concepts can be applied to other technology transfer scenarios, such as for agriculture. The Technology Acceptance Model (Davis 1989; Davis *et al.*, 1989), adapted from the Theory of Reasoned Action (Ajzen & Fishbein, 1980; Fishbein & Ajzen, 1975), offers a useful explanation for user acceptance and usage behaviour. The basic model suggests that user acceptance is determined by two key beliefs: perceived usefulness and perceived ease of use. Perceived usefulness is the extent to which a person believes that using a particular technology will enhance her/his job performance, while perceived ease of use is the degree to which a person believes that using a technology will be free from effort (Davis, 1989). Further research has suggested that perceived quality of life and subjective norms also have significant influence on technology acceptance (Venkatesh & Morris, 2000). Subjective norms include factors such as individualism/collectivism, masculinity/femininity, power distance (decentralised vs. centralised power) and uncertainty avoidance.

There are several practical lessons that can be learned from studying these models. Some key concepts that apply to agriculture are:

- That perceived not actual benefits are the inputs to the models.
- Gender differences can be significant. Feminine culture individuals tend to place more importance on perceived ease of use and quality of life, while male culture individuals place more emphasis on perceived usefulness.
- Some groups place more importance on receiving and considering input from peers/superiors than others.
- Uncertainty avoidance may be a strong factor in certain socio-economic audiences and may significantly limit willingness to make changes.
- Age of target audience may be a significant factor as younger audiences may place more emphasis on extrinsic rewards, and older audiences may be more influenced by social concerns.
- Social influences are more important for a mandated effect (e.g. environmental concern) than a voluntary effect (e.g. selecting a new variety).

Thus understanding the target audience is crucial to success in technology transfer. Do technology transfer specialists emphasize productivity benefits, or do they emphasize process/usability issues and social factors? Do specialists emphasize primarily one factor or do they emphasize usefulness issues for men, while offering women a more balanced analysis that includes productivity aspects, process issues, and testimonials from peers or superiors?

The following considers the above target audience differences to consider technology transfer efforts.

Making information available to a target audience

Truly making information available is not a simple task. The first challenge is gathering information. This is obviously much simpler if the information is an improved variety than if a management system change is to be transferred. However, even for the variety information transfer, the information specialist must have the information the target audience needs, in order to accept the change. Information providers must be closely associated with the applied researchers. Often extension is a separate federal agency that does not have close ties to the federal research branch or to the universities. Such distancing hurts information transfer in several ways: (1) research information may not get to extension specialists in an expedient manner, (2) extension personal may not fully understand the conditions associated with the research information, and (3) feedback from extension to researchers about needed research or applicability of research is often lacking (Murray, 1999). Further, when extension is part of an agency that has regulatory function, acceptance of technology is reduced because of the inability of the audience to distinguish between the regulatory and educational functions. Information must be made available in a form and from a source that is respected by the intended audience.

The second step is for the audience to become aware of the information. This can be a major problem if information is provided in a form that is not received by the target audience (e.g. a publication that is not well distributed or a non-user friendly website). Considering learning differences and preferences, it is imperative that information is made available by via several different methods. For example, the successful release of audiotapes in Australia allowed farmers to listen to them while driving, which allowed information to flow even when individuals didn't have time to read a publication (Hartley & Hayman, 1992). Multiplicity of release formats increased the audience that was exposed to the information. In this regard it is important to consider the age effect, on the form information is most readily received by the target audience. As shown in Figure 1, reading is declining among younger generations.

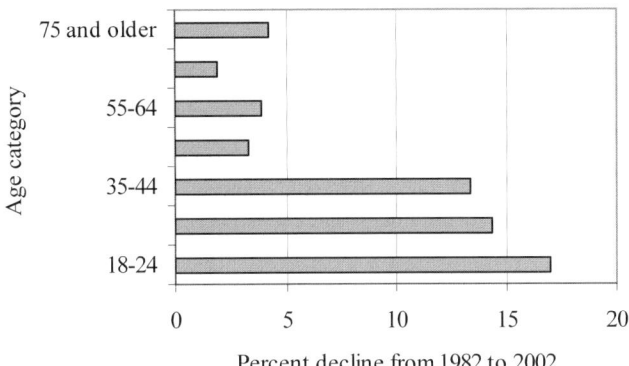

Percent decline from 1982 to 2002

Figure 1 Young adults read less

It is generally accepted by marketing agencies, that the average person must be exposed to information a minimum of five times before he or she fully becomes aware of it. Too often public agencies distribute information once and think everyone should know what is being taught. Information disseminators need to be aware of where the target audience is receiving information and attempt to place information accordingly. Table 1 is a survey of information sources indicated by farmers in different regions of the U.S by a private company. Only the top three sources are listed. While University/extension was high, it was not always first. Thus it is necessary to be working with multiple groups and sources in order to make information available, and that the importance of difference sources varies with different audiences.

Table 1 Ranked sources of information of farmers in regions of the USA

Region	Rank of importance				
	University/ extension	Seed dealer	Farm publication	Crop consultant	Other farmers
Midwest	2	1	-	3	-
Plains	1	2	-	-	3
Northwest	2	1	-	-	3
California	-	-	2	-	1

Marketing plans divide potential audiences many ways. A common method is to consider at least three groups: leaders, followers and nonresponders. In the first instance the target group has to be identified, and then characterised. The size of each group varies slightly depending on the type of trait being transferred and the specific farm audience. Leaders often comprise about 10 to 15% of most farm audiences, followers comprise 40 to 50% of farm audiences and nonresponders are about 40 to 45%. Leaders are identified in marketing efforts by factors such as farm size, participation in farm organizations, and certification/training. For some types of information, such as environmental management issues, the audience may truly be the entire agricultural sector, but for most improved technology and/or management concepts, the audience is likely to be only the first two groups. While public agencies may feel some need to address all groups, causing change in the nonresponder group is very time consuming and expensive. Change in this group is often only accomplished by a combination of information, support (both financial and intellectual) and regulations.

The third step occurs when the target audience implements the new technology. This step is really the completion of the information transfer and depends on the information itself, its value (both real and perceived) to the audience and the method of presentation. From an evaluation process, this should be the final evaluated result. It can be as simple as how much of a particular variety was sold or more complex considerations such as the number of individuals implemented specific environmentally sound practices, or feed budgets?

Social economic and environmental concerns of the audience

Methods of delivery and content for effective technology transfer differs amoung audience groups. For example, leader will tend to be more fact and data oriented (as are scientists) but other groups may be more influenced by factors such as status or images of green fields or healthy animals.

Demographics of the intended audiences are critical in technology transfer efforts and key to success. For example, what is the make-up in terms of age, gender, income level, education, beliefs and values and what activities is the target group currently involved in? If a large portion of the audience is at the subsistence level, attempting to transfer technology that has a relatively high cost may not be feasible, even if the economic returns are good. Another example is a situation where additional profit is not the first objective of farmers who may be satisfied with their situation, work effort and income, and have no desire to increase cattle herd size. Some farmers may desire self reliance more than wanting to purchase feed and so will use stocking rates below carrying capacity to avoid purchases in periods of low production, even though the latter may actually increase profit.

It is also worth considering characterising audiences by specific knowledge and other shared similarities. In audiences where certain educational efforts have been exceptionally strong, it may be possible to build on the general knowledge level. The technology transfer specialist should be aware of such effort to be able to build on them by using the same terminology and phrases. For example, companies marketing *Medicago sativa* (lucerne) in the U.S. have learned that farmers are more knowledgeable of diseases and disease resistance in some regions than others. This knowledge leads to different approaches in selling product.

Who are the decision makers within the farm unit? All too often educators/extension workers etc. have targeted the senior male when others may be involved. Results of surveys in Wisconsin (Zepeda *et al.*, 1997) revealed that individual goals and contributions to decision making within families were related to the time allocated to tasks. That is, day to day decisions tended to be made by the person performing the task, whereas, long-term decisions tended to be made jointly. The most male-dominated activity was crop production where 90% of both men and women say that men or mostly men make these decisions. Feeding decisions (82%), decisions regarding pasture (66%), capital purchases (60%) and those regarding the dairy herd (58%) are all predominately made by men. Longer term decisions however, e.g. whether or not to switch to grazing from confinement feeding and beginning of off-farm work, tended to be joint decisions. Regional and age differences also appeared to play a role in decision-making. Thus effective technology transfer should consider who in the family unit is targeted.

The same study found that a greater proportion of men than women (52 vs. 32.2%) were likely to seek out information. The preferred information source for almost half of both men and women was reading (87% of men and 79% of women preferred farm periodicals, 5% of both men and women preferred newsletters, while extension publications were preferred by about 3% of women and less than 2% of men). Open discussion was the second most preferred option, cited by nearly a third of women and a quarter of the men. 'Seeing' was the third most preferred method (20% of men vs. 12% women), and 'hearing' preferred by fewer than 10% of both men and women. (Note the difference between 'discussion' and 'hearing'.

In addition to demographics, the social concerns of the audience must also be embraced. This has become particularly true as many farmers are motivated by factors such as time spent with the family (does new technology take more time, or save time?), religious beliefs and social concerns e.g. Genetically Modified Organisms (GMOs). The GMO issue is both a social issue (do I want to grow/consume a crop with an incorporated gene?) and a market issue (will I be able to sell the product?). In addition, many farmers are concerned about risks associated with a new technology or practice; they may not be able to afford failure.

The history of many technologies is that potential benefits are less at farm level than was originally promised by research. This occurs because benefits are determined from controlled studies, and numerous factors (environment, inputs, management etc.) are often very different at farm level, where things are not done exactly right or at the right time due to knowledge of the user, time constraints and conflicting priorities. Also, only early adopters tend to benefit economically from new technology while others adopt to remain economically viable.

Technology is growing rapidly and the issue of information overload is real. This affects implementation of new technology because farming is a complicated system that involves many facets. Many farmers simply do not have enough time to absorb all the information presented to them. Additionally, adoption of a technology often involves other changes in the farming operation: can the technology be incorporated into an existing system as a package rather than be put out as separate technology where the user must figure out how to integrate into his or her system? Incorporating the technology into a system also avoids some of the problems of mixed signals given to farmers e.g. harvest early for quality but later for higher yield and persistence. Side effects of too much information include anxiety, poor decision-making, difficulties in memorising and remembering, and reduced attention span (F. Heylighen, 1999). These effects add to the stress caused by the need to constantly adapt to a changing situation.

Specific audience characteristics that impact learning

Understanding adult learners and adult learning is a critical component of providing information for technology transfer. The different ways in which learning occurs must be recognised. For technology transfer to succeed, the learner must be understood and provided with meaningful information rather than simply providing information in ways familiar or easily available to those disseminating the technology.

Table 2 illustrates this point where; data are from a private company survey of its customers. The best-received activity was a field day where people could see a technology or product first hand, and talk with experts about it. Newsletters ranked higher in the previously reported study of the general public (Zepeda *et al.*, 1997). The major difference between general customers and those characterised as price conscious was the enhanced use of the web by price conscious individuals.

Table 2 Communication strategies considered valuable

	% responding	
	General public	Price conscious
Plot tours/field days	87	87
Newsletters	80	83
Magazine articles and advertisements	79	77
Toll-free number with technical support	75	82
Brochures or other mailings	74	74
Interactive website	33	62
Email message or online newsletter	24	47
Radio advertisements	23	25

It is clear (data not shown) that, while field days ranked highest, a number of those who did not attend gained the same information from reading either print or electronic material.

Experience with web-based courses also stresses the need and value of allowing for variation in the time of learning and methods among audiences. A web-based course allows for 'on-demand learning' where the course is available whenever the learner wants rather than available at a specific time and place. This leads to students taking the course when interested rather than when available, and increases their retention of learned material. By tracking when individuals sign on the web to take lessons, clear learning differences emerge. Some individuals take lessons over lunch, some right after work and some in the late evening (after the children are in bed). A web-based course allows the learner to pick and choose what parts he or she wants rather than having to sit through familiar material, and get to their specific areas of interest. In this context, teachers must allow some flexibility and concentrate on objectives (i.e. to cause learning, not necessarily to have student exposed to all lessons).

Lastly it is appropriate to consider new ways of using technology to facilitate learning. Technology does not present a new way of learning (Leamnson, 2001). All learning is biological brain change and all teaching is an attempt to encourage and stimulate students to do what it takes to make those changes, that is, to focus their attention and to practice. However, computers and associated technology, and the access they afford, can constitute a new way of studying. It is important to recognise that expected learning outcomes remain unchanged when new technology is utilised. What changes is the ability to reach new and old audiences in different ways. What humans think *about* changes almost daily, but the *way* we think has not changed in many thousands of years.

Technology should be considered from the aspects of:
- Does it improve the users' problem-solving abilities?
- Does it encourage concentration?
- Does it enhance engagement with content?
- Does it build facility with language?

The last decade has seen great increases in the use of the web, email, and interactive TV as well as CDs and DVDs. Each of these technologies must be developed independently, but if used properly can have a significant place in information transfer (Unknown, 2002). The leader group of individuals is particularly more adaptive of modern learning technology.

A major mistake with use of the web has been institutions, organizations, or companies thinking organisationally (e.g. by department) rather than to think how the user would access the information. This has resulted in some wonderful and rich websites that are severely limited by the ability of clients to access the desired information. Another common mistake is to change medium without redesigning the message format – it is not appropriate to switch from print to video without reformatting, but it is often the case taat the printed work is placed on the web without any consideration of the audience. If the material was designed for print medium, it is unlikely to be as effective as it could be for the web. Material should not be transferred from one media to another without specifically redesigning for the new medium.

Lastly, it is necessary to consider what types of material work well for the different media. Every medium (print, web, email, radio, video and interactive TV) can be useful in technology transfer, but some are stronger for certain types of messages than others. For example print media can be useful for everything from promotional flyers to reprints of refereed journal articles. However, as all of us know, the space needed to keep the items is significant and expensive, and locating a specific fact or detail may be difficult. The web can easily handle different levels of material through layering, and can find specific points with search engines, but may not be available to everyone. Also, the audience may not know the web address. Email and radio are good for short messages or alerts. Interactive TV will be useful for providing ancillary data to a main message (on the video), e.g. information on identification or control characteristics of an insect, disease or weed being discussed, or containing a web address for ordering or obtaining more information.

In summary, technology transfer is key to the economic development of any society, but lags far behind technology development. Typically many agronomic researchers and educators, whilst well versed in their subject matter, often lack expertise in how best to transfer the information (Murray, 1999). Information dissemination is a critical component of technology transfer. The goal is not just to distribute information, but also to cause awareness and change within the target audience. To be effective information must be presented with the audience in mind. Therefore it is important to know and understand the target audience. The message must be based on the social and economic concerns of the audience, and multiple media chosen with the audience in mind. Learning differences of individuals and of audiences must be taken on board. New technology should be used where appropriate, but only after considering the different design considerations of each medium. Information should only be distributed with appropriate marketing principles to be effective in causing change.

References

Ajzen, I., & M. Fishbein (1980). Understanding attitudes and predicting social behaviour. Prentice-Hall, Englewood Cliffs, NJ, 278pp.

Davis, F.D. (1989). Perceived usefulness, perceived ease of use, and user acceptance of information technology. *Management Information Systems Quarterly*, 13 (3), 319-339

Davis, F.D., R.P. Bagozzi & P.R. Warshaw (1989). User acceptance of computer technology: a comparison of two theoretical models. *Management Science*, 35, 982-1002.

Fishbein, M. & I. Ajzen (1975). Belief, attitude, intention and behaviour: an introduction to theory and research. Addison-Wesley, Reading, MA, 578pp.

Hartley, R. & P. Hayman (1992). Information without the transfer - a common problem? *Journal of Extension*, 30 (1), <http://www.joe.org/>

Heylighen, F. (1999). Change and information overload: negative effects. In: F. Heylighen, C. Joslyn & V. Turchin (eds.) Principia Cybernetica Web, Principia Cybernetica, Brussels, <http://pespmc1.vub.ac.be/CHINNEG.html>

Leamnson, R.N. (2001). Does technology present a new way of learning? Educational Technology & Society, vol. 1. <http://ifets.ieee.org/periodical/vol_1_2001/leamnson.html>

Murray, M. (1999). A Contrast of the Australian and California extension and technology transfer processes. *Journal of Extension*, 37 (2). <http://joe.org/joe/1999april/a1.html>

Unknown (Apr 13, 2002). Power in your hand. *The Economist*, 363, Issue 8268, 2-3 <www.economist.com.>

Venkatesh, V. & M.G. Morris (2000). Why don't men ever stop to ask for directions? Gender, social influence, and their role in technology acceptance and usage behaviour. *Management Information Systems Quarterly*, 24 (1), 0276-7783.

Zepeda, L., M. Goodale, C. Lay, K. McSweeney & D. Undersander (1997). Results of four Wisconsin focus groups: roles of husbands and wives in farm decision. *Review of Agricultural Economics*, 19, 291-307.

Participatory research for smallholder livestock systems: applying common sense to complex problems

P.M. Horne[1] and W.W. Stür[2]
[1]CIAT, P.O. Box 783, Vientiane, Lao PDR
Email: p.horne@cgiar.org
[2]CIAT, 22 Seventh Avenue, Windsor, Qld, 4030, Australia

Key points

1. Participatory approaches to research (PAR) bring researchers closer to farmers, the intended users of research outputs.
2. Active, functional participation of farmers in the evaluation and development of new technologies requires researchers to make an important commitment: respecting the knowledge, skills and opinions of farmers while maintaining confidence in their own scientific knowledge.
3. Farmer experimentation is not usually suitable to provide quantitative biophysical data (this can be achieved more effectively in researcher controlled experiments), but to provide qualitative information and improve understanding. This type of information can be collected systematically to enable rigorous analysis.
4. While participatory approaches are likely to lose some of their current 'favoured status', the principles of farmer participation will remain an essential component of agricultural research.

Keywords: farmer participation, on farm experimentation

Introduction

'Research' – as a distinct activity separate from the everyday life of farming – is a relatively new phenomenon. Only in the last century has agricultural science become sufficiently complicated and large-scale to be separated from the realities of farming. Agricultural research stations were built and became a reliable, predictable comfort zone for researchers. Did agricultural research then became less participatory? Perhaps, but the consequences were not immediately obvious.

In the 1960's, there existed an optimistic mood that modern scientific approaches in agriculture (especially plant breeding and selection) could solve the perceived threat to the food supply of developing countries. The underlying assumption was that modern science had answers and technologies that farmers needed to produce more food. The Green Revolution of the 1950's and 60's, based on new crop varieties and extensive use of fertilisers, irrigation, herbicides, pesticides and machinery, resulted in spectacular yield increases. By the 1990s, almost 75% of the area of paddy rice in Asia and half the area of wheat in Africa was sown using Green Revolution varieties and methods (Rosset et al., 2000). Most of the gains were made in relatively uniform production environments where farmers had the means and motivation to aspire to the production levels achieved on research stations. The net increases in food production came at the cost of a greater dependence on fossil fuels and agricultural chemicals, reduced agro-biodiversity and an increased disparity between those that had, and hadn't access to food and food producing resources.

The fact that so many of the world's poorest farmers were either left behind or further marginalised by these new technologies was a stimulus for many to reassess the way research for agriculture was conducted in the developing world. There was a growing demand that research must deliver benefits for the poor. The farming systems which support most of the world's rural poor are complex, risk-prone and often marginal for agriculture. This paper discusses the contribution participatory approaches have made to the needs and opportunities of smallholder farmers. It identifies (i) situations in which it is particularly useful to adopt a participatory approach and (ii) the enabling factors for the successful implementation of participatory approaches to research.

The promise of participatory approaches to research

In the 1970's and early 1980's, Farming Systems Research (FSR) attempted to make technology development more relevant to the realities of smallholder livelihood systems, through a better understanding of the complex interaction of factors that govern the success or failure of new technologies. Farming Systems Research introduced into the discourse an appreciation of (i) the importance of non-biophysical factors on the success of promising new technologies and (ii) the complexity of the farming systems in which they would be tested. However, FSR did not fundamentally change the way that new technologies were generated and 'delivered' (Sumberg & Okali, 1997).

Participatory rural appraisal

In the mid '80's, the concept of 'participation' in agricultural R&D took hold and spread rapidly through research institutions, NGO's and bilateral/multilateral agencies involved in rural development. Many individuals and institutions recognised the need for a broader participation of stakeholders, and contributed to developing practical approaches for implementation (Chambers, 1997). Participatory Rural Appraisal (PRA) emerged as both an approach and a diverse set of relatively simple tools that enabled development-oriented organisations to work with farmers in a more collaborative way, to define and prioritise the key issues in their livelihood systems. Development specialists, adopted PRA widely because:
- Many of the tools were intuitive, providing development workers with a process that was easily followed, and helped them break communication barriers with farmers.
- Increasingly donor agencies, seeking ways to make their projects more effective in delivering impacts, adopted the principles of 'participation' in their projects.
- Experienced development workers found that using these tools, information and insight could be gained relatively quickly.
- The tools were fun – gone were the dull interviews that produced masses of data that rarely got analysed.

Subsequently however, the concepts and directions of 'participation in development' seemed to become disconnected from the realities of much of the implementation on the ground. While a handful of influential groups, networks and individuals (notably among them, the International Institute for Environment and Development (IIED), the University of East Anglia and the Natural Resources Institute (NRI)) continued to stimulate the debate, encouraging a focus on the quality and action learning aspects of participatory processes, the field application in many cases became bogged down by a fascination with the tools. There are instances of development projects which invested enormous effort into trying to understand farmers' realities using PRA tools (in some cases for up to two years) but when it finally came down to answering the

question "So what can we offer these farmers?", the process reverted to 'transfer of technology'. While PRA had made a major contribution in altering the way development specialists approached rural development, it had not substantially changed the way in which new ideas or technologies were generated to address farmers' problems.

Participatory research approaches

At the same time as the concepts of PRA were spreading, a new set of participatory research approaches, tools and terms emerged (Jiggins, 1989; Sumberg & Okali, 1997, Veldhuizen *et al.*, 1997; Stür *et al.*, 2002). The arguments driving the development of these more participatory research approaches were:

- Normal modes of agricultural research and extension had failed to make significant contributions to resource-poor farmers in complex and risk prone environments.
- The detailed knowledge which local farmers have of their environment and farming systems was not being utilised in the normal modes of R&D. In the 'transfer of technology' approaches, 'finished' technologies were developed with the expectation that farmers didn't need to adapt the technologies; they needed to change their farming practices to take advantage of the promise the technologies offered.
- Farmers in risk-prone livelihood systems are invariably hungry for ideas and 'raw technologies' to evaluate and adapt to their local opportunities and constraints. They are looking for quick action from researchers.
- Participatory approaches to research advocate farmers' active involvement as decision makers at all stages of the research process, including the early stages of problem identification and setting of research priorities. The promise of participatory approaches to research was that a) inappropriate or poorly-adapted technologies would be rejected early in the process b) researchers would gain a better understanding of the factors that contributed to particular technologies being integrated into smallholder livelihood systems c) technologies would have a greater chance of subsequent adoption because farmers were involved in screening and developing the technologies over a wider range of conditions than would happen on research stations and d) new problems and opportunities would arise that need new strategic research.

The two main approaches that emerged were Farmer Participatory Research (FPR) and Participatory Technology Development (PTD). The most common distinction made between these two approaches is that FPR is somehow more rigorous and technically-focused while PTD places a greater emphasis on community empowerment. In practice, however, there is little to separate them and this paper will refer only to Participatory Approaches to Research (PAR). The main difference between PAR methods is in the emphasis they place on technology development and adaptation. Conroy (2005), refers to "a process in which local people and outside facilitators work together purposefully and creatively to identify, experiment with, and validate technologies that effectively address important problems or opportunities, while simultaneously strengthening the capacity of local communities to address other related problems and opportunities in the future". This definition places emphasis on *development of impacts from technologies.* By contrast, Braun & Hocdé (2000) define "a process whereby a group or a community identifies a problem or question of interest, reviews what is known about it, conducts research on it, analyses the information generated, draws conclusions and implements solutions". This definition places emphasis on *empowerment of communities* to resolve problems through whatever means. The main characteristics of these modes of PAR and comparison with strategic research are summarised in Figure 1. In reality approaches to PAR fall somewhere in between these broad categories.

	Strategic research	Participatory approaches to research	
Main focus	Biophysical Information	Development of Impacts from Technologies	Empowerment of Communities
Objectives	• Define the biophysical adaptation and potential of technologies on research stations; • Deliver the most promising technologies for adoption.	• improve the effectiveness of research in delivering impacts to farmers; • help research focus more on issues of importance to farmers.	• empower communities through building the capacity of farmers' groups to conduct collaborative research; • enhance self reliance and ability to resolve broader community issues.
Where	• Research stations	• Research stations and farms	• Farms and the broader community

	Contractual	Consultative	Collaborative	Collegiate
Types of participation	Farmers' land and services are hired or borrowed	There is a doctor-patient relationship. Researchers consult farmers, diagnose their problems and try to find solutions	Researchers and farmers are roughly equal partners in the research process, continuously collaborating in activities	Researchers actively encourage and support farmers' own research and experiments

	Strategic research	Participatory approaches to research	
Stakeholders	Researchers and 'key' farmers.	Researchers, extension workers and farmers.	Many internal and external stakeholders (including farmers, researchers, NGO's, public and private sector).
Typical activities	Detailed, controlled biophysical research leading to either (i) technology packages being identified for extension or (ii) 'raw technologies' being identified as options for PAR.	Farmers identify, test and evaluate technology options, adapting the most promising to their local conditions, opportunities and constraints.	A community group identifies a problem or question of interest, reviews what is known about it, conducts research on it, analyses the information generated, draws conclusions and implements solutions.
Types of information gathered	Quantitative, biophysical	Quantitative and qualitative	Largely qualitative
Local relevance	Low	Medium – high	High

Figure 1   Comparison of strategic research and participatory approaches to research (modified after Probst & Hagmann, 2003)

Participatory approaches to research in practice

The three modes of research summarised in Figure 1 have strengths at different stages of a research agenda. Government research organisations typically aim to contribute to all three modes of research but are active only in the first. NGO's and development projects are typically focused on the second and third modes of research, but rely heavily on the first mode as a source of raw technologies. International agricultural research centres are increasingly becoming involved in all three modes of research.

To date in international agricultural research, there are fewer examples of PAR being implemented to empower communities than to deliver impacts from technologies. Approaches involving empowerment of communities (such as 'learning alliances' that aim to strengthen the ability of communities to conduct sustained participatory learning and action) challenge the mandate and responsibilities of research institutions, and are generally more difficult to implement, are less bounded and require longer-term commitment with intangible outcomes. Despite this, there have been some notable examples of PAR aimed at developing local research capacity in the fields of; natural resource management (Pound et al., 2003), management of communal grazing lands (Waters-Bayer & Bayer, 1994; Bayer & Waters-Bayer, 2002), formation or strengthening of community research groups (Ashby et al., 2000; Braun & Hocdé, 2000), and the Landcare movement (Garrity et al., 2000). There are few examples where attempts have been made to strengthen the capacity of communities of livestock-keepers to conduct their own research.

The emphasis of PAR in international agricultural research has been largely on the 'development of impacts from technologies' (Stür et al., 2000; Conroy et al., 2002; Stür et al., 2002; Franzel et al., 2003; Peters & Lascano, 2003; Pengelly et al., 2004; Conroy, 2005). Specific methods have ranged from formal experimentation managed by farmers with assistance from researchers, through to completely informal testing of raw technologies and ideas, with researchers encouraging changes and innovation. There are three main types of experimentation focused on development of impacts from technologies:

- Type 1 - conventional research trials designed and managed by researchers (either on-station trials or the same trial but conducted on farms);
- Type 2 - trials using PAR designed by researchers and managed by farmers;
- Type 3 - trials using PAR designed and managed by farmers.

There are many variations within this typology, especially in the extent of collaboration between farmers and researchers in Type 2 trials, and the extent of scientific rigour applied (controls, replication and precision of data collection). Despite this, the typology is useful in highlighting that the most appropriate type of trial to implement depends very much on 1) the objectives of the researcher in conducting PAR and 2) the degree of understanding about the potential of the raw technologies being tested to deliver impacts (see Table 1).

The typology does not imply that one approach is better than another, or that greater levels of participation somehow lead to better outcomes. Strategic research and the two typologies of PAR trials have different strengths that complement each other. The strategic research of plant breeders, for example, is a main source of raw technologies for PAR. It is important, however, that 'researchers select more thoughtfully and consciously between the different options at hand to explore the most appropriate strategy towards impact' (Probst & Hagmann, 2003).

Table 1 Situations in which it is particularly useful to adopt Participatory Approaches to Research (PAR) (after Franzel & Coe, 2002)

Objective	Likelihood of outputs from each trial type[1]		
	Type 1[2]	Type 2	Type 3
Precise and accurate biophysical data	H	M – L	L
Farmers' preferences about:			
• New raw technologies (eg new forage varieties; New anthelmintics)	L	H	M
• New management practices	L	M	H
Information about impacts and adoption:			
• Likelihood of wider adaptation and adoption	L	M	H
• Identifying farmer innovations	0	L	H
• Determining biophysical and socioeconomic boundary conditions for technologies	H	H	H

[1]0 = none, L=low, M=medium, H=high; [2]Type 1 = conventional trials designed and managed by researchers; Type 2 = trials using PAR designed by researchers and managed by farmers; Type 3: trials using PAR designed and managed by farmers.

It is worth noting from Table 1 that:
- Problems commonly arise when researchers expect to get both rigorous biophysical information and sound farmer evaluation from the same trial. In most cases this requires two different kinds of trial. Some PAR approaches have been developed to maximize both biophysical rigour and quality of farmers' participation (Snapp, 2002).
- If the goal of the research is to encourage farmer-innovation, it is important to allow farmers to evaluate the raw technologies to see how they work, discover how they fit into their existing farming system and, most importantly, identify new opportunities that a raw technology or practice may offer (not just focus on the solution to an existing problem). An example of the significance of the unexpected outcomes from farmer innovation comes from recent development of smallholder forage systems in northern Vietnam (Stür et al., 2000). Farmers in Tuyen Quang province identified the lack of feed resources for their buffalo and cattle as a high priority problem and requested support from researchers to find a solution. The scarcity of feed meant that the labour input from farmers to finding sufficient feed for their animals was making their livestock systems unviable. In 1997, researchers introduced a range of broadly adapted forage varieties for farmers to evaluate, and encouraged them to find innovative ways to integrate the varieties on their farms. They grew the forages in very small areas near their animal pens to test the feasibility of growing feed for livestock rather than relying fully on communal feed resources. After two years, the 53 farmers who had been testing the varieties concluded there was not enough available land in this region of intensive agriculture to grow sufficient feed for buffalo and cattle. Two farmers, however, found by chance that several of the varieties could be fed to their ponded-fish (grass carp and common carp). This reduced the time needed to feed fish from 2-3 hours per day for collecting soft native grasses to less than 30 minutes per day for cutting planted grasses grown in small plots near the fish ponds. This innovation generated considerable interest within the community and other farmers started to feed planted forages to fish (Yen & Binh, 2000) (Table 2). In this region fish are one of the most productive forms of farm activity with an average farm pond of 600–900 m^2 producing 240–360 kg of fish, worth US$250–300. This is equivalent to the income from two high yielding crops of irrigated rice from 0.25 hectares of land. This development was

totally unexpected to the researchers who only understood the role of forages for ruminants. Through networking, this innovation has been introduced and adopted by farmers in other provinces of Vietnam and other countries in Southeast Asia. In Ea Kar district of central Vietnam, for example, >100 farmers have built substantial fish ponds to take advantage of this production system, and irrigate forages from the ponds during the dry season to maintain production. They have also introduced cattle fattening to the system to utilise the excess feed available when the fish are still just fingerlings. Based on experiences in smallholder forage systems in other parts of the world, the impacts are expected to expand exponentially. These unexpected outcomes from farmer innovation and would not have been possible using conventional research approaches.

Table 2 Expansion of planted forages in Yen Son district, Tuyen Quang, Vietnam (Vu Hai Yen, *pers. comm.*)

	1997	1998	1999	2000	2001	2002	2003
Farmers growing forages (no.)	9	53	138	158	312	529	629
Average size of forage area (m^2)	75	75	200	500	500	500	700

- A criticism of Type 2 trials is that because of a lack of control, they do not provide sufficiently reliable data to be useful for researchers. In the context of PAR, researchers need to question the value of seeking reliable averages. The farming systems in which most of the rural poor live are highly diverse, risk-prone and marginal. Any new technologies emerging from PAR will be further modified through site-specific adaptation. Unless the raw technologies are particularly broadly adapted and robust (as in the case of many forage varieties or treatments for Helminths in cattle and pigs), then developing reliable average results from Type 2 trials will not address farmers' imperative for site-specific adaptation. Ceccarelli *et al.* (1994) argue that for crop varieties in highly variable environments it is more sensible to aim for specific, as opposed to general adaptation. This highlights the importance of understanding the variability of results from PAR trials rather than searching for the average result.

Situations that are particularly suited to these kinds of PAR investigations include:
- *Defining problems.* The nature of a particular problem of importance to farmers may be poorly defined (for example, farmers reporting deaths of buffalo calves but with non-specific symptoms or causes). Researchers can work with farmers to investigate the causes of the problem and with farmers identify options worth testing.
- *Understanding farmers' criteria.* The characteristics of a technology option that are important to farmers may not be well understood. Researchers tend to think of production-oriented characteristics whereas farmers very often have other criteria that are more important in selecting one technology option over another (for example, the green revolution rice varieties were high yielding but not preferred by subsistence farmers because of the poor taste).
- *Encouraging impacts and providing feedback to research.* Often the motivation for conducting PAR has been to better understand which of a range of technology options farmers prefer and why. In this case, researchers need to play less of a role in designing trials and place more effort on monitoring impacts. The lessons for research can be

insightful as the impacts that farmers gain from technologies are sometimes unrelated to the resolution of the initial problem. Sometimes they come from farmers changing their management practices to take advantage of a new opportunity identified during their experimentation. For example, smallholder livestock systems in Laos are mostly extensive (low input and low output systems) where the role of livestock is as a 'livelihood safety net'. Under these circumstances, few farmers are able to test new technologies that would allow them to move from being livestock keepers to producers. Forage researchers needed to find 'entry points' that would provide early and substantial benefits to interest farmers in making a substantial change in their livestock systems. In PAR trials that commenced in 2001, the most common entry point for >1300 farmers was using plots of forages as a source of cut feed to save labour for farmers at particular times of year when they needed to keep animals closer to home. By 2005, around 25% of these farmers had generated significant livelihood impacts by changing their livestock systems to take advantage of the forage resource available, keeping their cattle and buffalo closer to home and fattening them for sale. Thus the impacts came not from resolving the initial problems but from farmers changing their livestock systems to take advantage of new opportunities. These impacts are now expanding rapidly to neighbouring farmers, villages and districts. By 2005, 950 farmers had started feeding the legume, *Stylosanthes guianensis* CIAT 184 to their pigs, both as fresh feed and dried leaf meal (Horne, unpublished data). The impacts that emerged on pig productivity have been significant, novel and surprising to researchers, stimulating a new research effort into the potential role of legumes for improving productivity of village pig systems.

- *Encouraging adoption of complex technologies such as soil conservation practices.* Howeler *et al.* (2005) found that farmers participating in PAR trials of a range of system improvements for *Manihot* spp. (cassava) on sloping agricultural land were more likely to adopt soil conservation practices such as contour hedgerows of *Vetiveria zizionoides* (vetiver grass) and *Paspalum atratum* than non-participating farmers (Table 3). In contrast, 'simple' technologies such as new varieties were adopted more readily by non-participating farmers.

Table 3 Adoption of new technologies by farmers participating in Type 2 trials and non-participating farmers in nearby areas in *Manihot* spp. (cassava) systems in Thailand[1] (after Howeler *et al.*, 2005)

	Participating farmers (% adoption)	Non-participating farmers (% adoption)
Varieties	100	86.6
Soil conservation practices	79.5	29.2
Intercropping	28.2	9.6
Fertilization	100	87.6

[1]Data based on a survey of 439 households

Situations where PAR may not be appropriate include:
- Trials requiring good biophysical data to better understand the environmental boundaries of an untried, raw technology (for example, understanding the seed yield potential of a new crop before testing it with farmers as a possible seed cash crop).
- Trials where farmers have limited knowledge about the potential benefits of a technology option and where the technology option requires substantial effort to develop (for example, forage tree legumes can be a long term source of high-quality feed but require up to two

years of careful management to establish successfully). In these cases, conventional demonstrations on rented farmers' fields may be more appropriate.
- Trials that impose significant risk to farmers' livelihoods. In such cases (such as testing new treatments for livestock disease or evaluating seed production of a new crop with farmers), researchers may either choose to conduct more on-station research to better understand the risk, or agree to compensate farmers in case of losses due to problems with the technology option.

There are many documented examples of PAR with a focus on delivering impacts from technologies. Despite the specific differences in methodological detail between them, they generally share a common sequence of activities, facilitated by outside organisations and conducted either by individuals or groups (Veldhuizen *et al.*, 1997; Horne & Stür, 2003; Conroy, 2005);
- identify key problems and opportunities using tools such as mapping, ranking, calendars and problem trees.
- identify, test and evaluate new ideas or raw technologies to address these problems and opportunities.
- evaluate the outcomes of the trials using tools such as ranking and scoring as well as collecting relevant conventional research data.
- decide what steps to take next (including the need for further experimentation or new options).

While these activities and tools are conducive to a participatory mode of research, they are not inherently participatory (Conroy 2005). Active, functional participation of farmers in the evaluation and development of new technologies requires researchers to make an important commitment: respecting the knowledge, skills and opinions of farmers while maintaining confidence in their own scientific knowledge. It is a key factor for successful PAR that researchers demonstrate to farmers, through words and actions, that they are respected as equal stakeholders in the PAR process.

When PAR is carried out with this kind of commitment and using genuinely promising raw technologies or ideas, impacts (both direct impacts and impacts on livelihoods) are likely to materialise. The process can rapidly move from 'identifying, experimenting with, and validating new technologies' to 'expanding the benefits to more people, more quickly and equitably'. That is, the process can quickly move from research to extension. A broad coalition of stakeholders is needed to take the outcomes of participatory research into extension, and the formation of such coalitions has to happen earlier rather than later in the process to foster ownership and commitment by development partners (e.g. government extension service). There is no abrupt end to participatory research or 'handing over' of results to development partners; research and development overlap for considerable periods with a continuing need for researchers to support the innovation process, and for development partners to find ways of short-cutting the time needed for other farmers to learn about the results and adapt the outcomes to their situations. Researchers alone usually do not have the skills and mandate to follow-through with extension opportunities.

Participatory approaches to research in livestock systems

Smallholder livestock keepers in the tropics have generally been poorly served by research. The tendency has been for research to focus on technical aspects to improving productivity without fully understanding the constraints facing the livestock keepers. Despite this, there are a handful of well-known examples where smallholder livestock keepers have benefited from changes brought about by;

- *Regulation* - some large-scale beneficial impacts in farming systems have came about through enforcement rather than participation. For example in Amarasi district of West Timor, traditional regulations ('adat') were imposed in the 1930's and 1940's compelling shifting cultivators to plant the tree legume *Leucaena leucocephala* in hedgerows. This provided sufficient feed for livestock and soil improvement for cropping to eliminate seasonal famine by the 1960's. By the 1980's, 500km^2 of Amarasi was covered with *Leucaena* based crop-livestock systems (Shelton *et al.*, 2000).
- *Strong government support programs* - these can support development of widespread impacts that would not have emerged on their own. Examples include the development of a network of >4000 smallholder forage seed producers in Thailand with government support over 25 years (Hare & Horne, 2004) and the spread of the green manure legume *Astragalus sinicus*, over more than 8 million hectares of paddy rice in southern China (Wen *et al.*, 2000).

There are however, many more examples of large failures of these two approaches than successes. In Southeast Asia, there have been numerous unsuccessful attempts to 'photocopy' locally successful systems (e.g. forage hedgerows for erosion control and simple agro-livestock technologies) from one location to another. These failures came about from not recognising that often the technology (e.g. the hedgerows) is just the manifestation of 'a complex conjunction of people, technologies, events and luck...often with unanticipated outcomes' (Rob Cramb, *pers. comm.*). Some technologies have been actively promoted despite well-understood reasons for their repeated failure to deliver impacts. The benefit of treating rice straw with urea, for example, comes from higher digestibility of the straw increasing feed intake. This is only potentially useful in areas where there is a surplus of straw, yet the technology is often promoted in areas where all the rice straw is already utilised by animals in the dry season. Similar stories exist for promotion of molasses-urea-blocks, cross breeding with exotics to produce 'better' (i.e. larger) animals and delivery of vaccines to eradicate infectious diseases that are of greater global than local importance. Little research has focussed on the priority issues of the poor in relation to livestock (LDG, 2004). Most of these issues (see Table 4) are researchable and potential options already exist that may be able to overcome the problems.

As PAR were largely developed and implemented with crop technologies; some argue that it is more difficult to use PAR in livestock systems research because of:

- The greater time scales involved in livestock research, especially breeding and production research.
- Problems in sampling procedures and replication because of small numbers of animals available on farms. This makes it difficult to deal with between-farm and between-animal variability.
- The difficulties for farmers of managing individuals or groups of animals differently on the same farm, especially in feeding trials.
- Large variation in basal diets against which treatments are compared between and within farms (Morton *et al.*, 2002).
- Potentially greater risk to farmers from research on farm animals (e.g., testing new vaccination procedures with pigs). The loss of one animal (whether it was a direct result of the experimentation or not) can be significant to the farmer and hamper the relationship with researchers.
- The mobility of livestock, especially in extensive systems.

Table 4 Typical constraints in smallholder livestock systems and potential options for Participatory Approaches to Research (PAR) (after Conroy, 2005)

Constraint	Impacts on		Example Options for PAR
	Numbers	Production	
Seasonal feed shortages		✓	Forages and other feed resources (e.g.
Insufficient feed all year	✓		sweet potato, cassava and maize for pigs)
Not enough labour to tend animals or cut feed	✓	✓	Having a managed feed resource allowing animals to be housed closer to
Wandering animals damage crops, get injured or stolen		✓	home, enabling better management
Fatal diseases	✓		Strategic use of veterinary medicines
Productivity-limiting diseases		✓	combined with better housing,
Poor management	✓	✓	herd management and feeding
Scarce water	✓	✓	Options will be site specific
Poor access to markets	✓	✓	Livestock producers groups

In reality many of these problems either do not arise, or there are ways to overcome them, moreover these issues are part of the nature of smallholder livestock systems, and represent the context in which any new technology option must be tested. So, while it may be true that PAR in livestock research is more difficult to implement than in cropping systems, the justification for, and potential benefits from PAR in livestock research are as great as they are in crop research. Given this potential, what are the main factors needed for successful implementation of participatory approaches to livestock research?

Individual commitment

The attitudes, facilitation skills and empathy of the researchers for PAR are key factors that will determine the outcomes of the PAR. Central to this is respect for farmers' views and their role as equal partners in the process.

Institutional commitment

Research institutions need to have a long-term commitment to PAR, especially if the work is targeted at developing impacts and encouraging innovation. Inherent in this commitment is the need for a broad skills base. It is desirable to have at least one 'process specialist' (not just biophysical researchers in the team), but PAR should not be implemented just by process specialists – the interaction between farmers and researchers is both insightful and necessary if there are technical issues or opportunities to overcome. It is also vital to have access to good technical advice and the raw materials (e.g. seed) for any technical options that are being tested. Finally, the organisation will need adequate funding. The financial resources required for PAR are often underestimated. Monitoring and analysis may be more time-consuming than they are in more conventional research modes (Conroy, 2005).

A researchable issue

It is necessary to have an issue that farmers consider important enough for them to commit time and resources to finding a solution. It should be an issue that faces many farmers in the area and for which researchers have something to offer. Researchers have to bring technical options and ideas that will be adapted by farmers in a participatory approach. If farmers could have developed innovations to solve their problems with what they already have in hand, they would have done it long ago! There are many instances of organisations that are strong on participatory processes but lack access to the best available technical options. The result is a strong community process without any options to offer farmers or worse, inappropriate options, such as fruit tree varieties that are poorly adapted to a region or have a limited market. Key technical information associated with the options can be just as important but does not flow as easily as 'hard' technologies.

A clear research process

All of the stakeholders need to have a clear idea of the different stages in the process – diagnosis of issues, identification of options, testing, evaluation and planning – and the specific activities that will be conducted. This process needs to be flexible and alert, partly because the inherent nature of on-farm work, but also because PAR creates a moving research target. Technical developments that lead to impacts often do not come from solving the immediate problems (which are usually the entry points), but from farmers changing their production systems to take advantage of a new opportunity. This was clearly demonstrated in the example of feeding forages to fish in Tuyen Quang, described earlier in this paper.

Challenges for wider acceptance of participatory approaches to research

One of the main benefits of the spread of PAR concepts has been that more researchers are listening to, and working with farmers. This increased interaction with the 'end-user' has been an important shift in the research process and, in a sense, is a return to a common sense approach to complex problems. Few would now disagree that farmers need to be involved in agricultural research, and few would agree that researchers can continue to develop agricultural technologies assuming they will be disseminated by extension processes. The concepts and practice of farmer participation in research have become part of the comfortable norm of many researchers involved in adaptive research. However, even among researchers who have empathy for the principles of PAR, there are criticisms about its practice that present challenges to the wider acceptance of PAR.

Breaking down differences between rhetoric and reality

The discourse about participatory approaches in agricultural research and natural resource management is way ahead of the realities of implementation on the ground. The issues facing field implementation are very practical; (i) developing the basic skills, technical abilities and experience of field staff to do collaborative research with farmers; (ii) creating a common understanding of the PAR process among field staff so that they can continue to implement activities as a sequence of events; building on the last and preparing for the next, and (iii) engendering a problem solving and systems-oriented (as opposed to discipline-oriented) approach to PAR. A challenge for wider acceptance of PAR approaches is to demonstrate that these basic issues of implementation can be achieved, replicated and institutionalised in a cost-effective way.

Overcoming institutional inertia

International and national organisations engaged in agricultural research need to redefine their roles and mandates to accept PAR and multi-stakeholder approaches to research as a normative mode of action. This includes accepting a mandate and responsibility for research to engage in a wider spectrum of activities - from strategic research to achieving impacts on farmers livelihoods.

Applying PAR systematically

LDG (2004) give one example where the priorities of livestock experts influenced farmers to identify Foot and Mouth disease as a major issue for research, even though 41% of households did not own animals that could be affected by FMD. The danger is that poor implementation of PAR and false perceptions drive agendas, and these contribute to the perception of PAR as lacking rigour. While the data collected from PAR trials may not be as rigorous as on-station research, it can be collected in systematic ways to improve its reliability. This can include triangulation of methods to avoid researcher bias and check on the repeatability of farmers' preferences (Conroy, 2005). A challenge for the wider acceptance of PAR, is the development and application of more-rigorous approaches to analysing data from trials that are not specifically aiming at encouraging farmer innovation (Bellon & Reeves, 2000).

Moving beyond appraisal

For many, the perception of 'participation in research' is Participatory Rural Appraisal. Organisations involved in PAR need to move beyond this, providing farmers with action and access to ideas and technologies that are addressing their problems.

Sustaining PAR beyond projects

What happens when the donor funds run out and the 'experts' go home? How much of the success can be attributed to their time and effort? Is it possible to reach a stage towards the end where not only can the immediate groups with whom they have worked continue to develop, but where the messages can spread beyond the point of contact? Sustainability of a farming system is not a static endpoint with a checkbox to be ticked, but an ongoing and dynamic response to changing markets, environments and policies. New problems and opportunities are continually arising. Sustainability will be better achieved by local people having access to a broad range of technical information and raw materials of technologies along with the ethos of problem solving, so that they can respond to those changing markets, environments and policies. Given the practical realities of PAR described above, it is easier to identify these needs than to implement them. Learning alliances, to assist communities to conduct sustained learning and action through research, are a promising way to address the issue of sustainability of PAR.

Avoiding tokenism

Many of the criticisms of PAR refer to the perception that 'participation' has become just another necessary component of funding proposals. This 'tokenism' results in cynicism about 'participation'. The challenge for advocates of PAR is to demonstrate clearly that these approaches provide substantial benefits to research that cannot be gained in other ways.

Similarly, a common criticism of PAR by biophysical researchers is that 'participation' has become an end rather than a means!

It is not necessary to fundamentally change the way agricultural research is done to make it somehow more participatory. Rather, there is a need to conduct agricultural research that includes farmer participation (of varying types and levels), rather than conducting farmer participatory research as a distinct and separate activity (Okali & Sumberg, 1997). Participatory approaches to research will probably lose some of their current favoured status, but the principles will remain and should be institutionalised. The trend to increased interaction with and involvement of 'end-users' in agricultural research is here to stay.

Acknowledgements

In preparing this paper, we recognized that we have been enthusiastic advocates of PAR to the point where it might have become a 'blind spot'. We are grateful to colleague's and development practitioners for their opinions about PAR. Special thanks are due to Douglas Gray (ILRI) and John Connell (CIAT) for their ongoing inputs to the discussion.

References

Ashby, J., A. Braun, T. Gracia, M. Guererro, C.A. Quirós & J.I. Roa (2000). Investing in farmer researchers: experience with local agricultural research committees in Latin America. CIAT Publication 318, Centro International de Agricultura Tropical, Cali, Colombia, 199p.

Bayer, W. & A. Waters-Bayer (2002). M&E with pastoralists: a review of experiences and annotated bibliography. ETC Press, Eschborn, 56p.

Bellon, M.R. & J. Reeves (2000). Quantitative analysis of data from participatory methods in plant breeding. The International Maize and Wheat Improvement Centre (CIMMYT), Mexico, 143p.

Braun, A.R. & H. Hocdé (2000). Farmer participatory research in Latin America: four cases. In: W.W. Stür, P.M. Horne, J.B. Hacker & P.C. Kerridge (eds.) Working with farmers: the key to adoption of forage technologies. Proceedings of an International Workshop held in Cagayan de Oro, Mindanao, Philippines, October 12-15, 1999. The Australian Centre for International Agricultural Research (ACIAR) Proceedings 95, Canberra, 32-53.

Ceccarelli, S., M. Erskine, S. Grando & J. Hamblin (1994). Genotype by environment interaction and international breeding programmes. *Experimental Agriculture,* 30, 177-187.

Chambers, R. (1997). Whose reality counts? Intermediate Technology Publications, London, 297p.

Conroy, C. (2005). Participatory livestock research: a guide. Natural Resources Institute, London, 295p.

Conroy, C., Y Thakur & M. Vadher (2002). The efficacy of participatory development of technologies: experiences with resource-poor goat-keepers in India. *Livestock Research for Rural Development,* 14 (3).

Diamond, J. (1998). Guns, germs and steel. Random House, NY, 480p. (ISBN: 0-393-31755-2).

Franzel, S. & R. Coe (2002). Participatory on-farm technology testing: the suitability of different types of trials for different objectives. In: M.R. Bellon & J. Reeves (eds.) Quantitative analysis of data from participatory methods in plant breeding. The International Maize and Wheat Improvement Centre (CIMMYT), Mexico, 1-8.

Franzel, S., C. Wambagu & P. Tuwei (2003). The adoption and dissemination of fodder shrubs in Central Kenya. *Agriculture Research and Extension Network (AgREN) Paper 131,* Overseas Development Institute, London, 10p.

Garrity, D., A. Mercado, R. Howeler & S. Fujisaka (2000). Participatory methods in research and extension for using forages in conservation farming systems: managing the trade-offs between productivity and resource conservation. In: W.W. Stür, P.M. Horne, J.B. Hacker & P.C. Kerridge (eds.) Working with farmers: the key to adoption of forage technologies. Proceedings of an International Workshop held in Cagayan de Oro, Mindanao, Philippines, October 12-15, 1999. The Australian Centre for International Agricultural Research (ACIAR) Proceedings No. 95, Canberra, 254-272.

Hare, M.D. & P.M. Horne (2004). Forage seeds for promoting animal production in Asia. *Proceedings of the 10th Asian Seed Congress, Seoul, Korea,* September 13-17, Asia Pacific Seed Association, Bangkok, (in press).

Horne, P.M. & W.W. Stür (2003). Developing agricultural solutions with smallholder farmers - how to get started with participatory approaches. The Australian Centre for International Agricultural Research (ACIAR) Monograph No. 99, CIAT and ACIAR, 119p.

Howeler, R.H., W. Watananonta & T. Ngoc Ngoan (2005). Working with farmers: the key to achieving adoption of more sustainable cassava production practices on sloping land in Asia. Paper presented at the International Potato Centre, UPWARD Network Meeting, held in Hanoi, Vietnam, January 19-21, International Potato Centre, (in press).

Jiggins, J. (1989). Farmer participatory research and technology development. *Occasional Papers in Rural Extension 5. Department of Rural Extension Studies,* University of Guelph, Canada, 12p.

LDG (2004). Receptors, end users and providers: the deconstruction of demand-led processes and knowledge transfer in animal health research. Livestock Development Group, School of Agriculture Policy and Development, University of Reading, 101p.

Morton, J., B. Adolph, S. Ashley & D. Romney (2002). Conceptual, methodological and institutional issues in participatory livestock production research. *Livestock Research for Rural Development,* 14 (4).

Okali, C., J. Sumberg & J. Farrington (1994). Farmer participatory research: rhetoric and reality. IT Publications, London, 156p.

Pengelly, B.C., A. Whitbread, P.R. Mazaiwana & N. Mukombe (2004). Tropical forage research for the future – better use of research resources to deliver adoption benefits to farmers. In: A.M. Whitbread & B.C. Pengelly (eds.) Tropical legumes for sustainable farming systems in southern Africa and Australia. The Australian Centre for International Agricultural Research (ACIAR) Proceedings No. 115, ACIAR, Canberra, 28-37.

Peters, M. & C.E. Lascano (2003). Forage technology adoption: linking on-station research with participatory methods. *Tropical Grasslands,* 37, 197-203.

Pound, B., S. Snapp, C. McDougall & A. Braun (2003). (eds.) Managing natural resources for sustainable livelihoods. Uniting science and participation. Earthscan Publications, London, 252p.

Probst, K & J. Hagmann (2003). Understanding participatory research in the context of natural resource management – paradigms, approaches and typologies. *Agriculture Research and Extension Network (AgREN) Paper 130,* ODI, London, 16p.

Rosset, P, J Collins & F. Moore Lappé (2000). Lessons from the Green Revolution. Report of the Institute for Food and Development Policy, Oakland, CA, USA, *Tikkun,* 15 (2), 52-56.

Shelton, H.M., C.M. Piggin, R. Acasio, A. Castillo, B.F. Mullen, I.K. Rika, J. Nulik & R.C. Gutteridge (2000). Case studies of locally-successful forage tree systems. In: W.W. Stür, P.M. Horne, J.B. Hacker & P.C. Kerridge (eds.) Working with farmers: the key to adoption of forage technologies. Proceedings of an International Workshop held in Cagayan de Oro, Mindanao, Philippines, October 12-15, 1999. The Australian Centre for International Agricultural Research (ACIAR) Proceedings No. 95, Canberra, 120-131.

Snapp, S. (2002). Quantifying farmer evaluation of technologies: the mother and baby trial design. In: M.R. Bellon, & J. Reeves (eds.) Quantitative analysis of data from participatory methods in plant breeding. The International Maize and Wheat Improvement Centre (CIMMYT), Mexico, 9-17.

Stür, W.W., P.M. Horne, F. Gabunada Jr., P. Phengsavanh & P.C. Kerridge (2002). Forage options for smallholder crop-animal systems in Southeast Asia – working with farmers to find solutions. *Agricultural Systems,* 71, 75-98.

Stür, W.W., P.M. Horne, J.B. Hacker & P.C. Kerridge (2000). Working with farmers: the key to adoption of forage technologies. In: W.W. Stür, P.M. Horne, J.B. Hacker & P.C. Kerridge (eds.) Proceedings of an International Workshop held in Cagayan de Oro, Mindanao, Philippines, October 12-15, (1999). The Australian Centre for International Agricultural Research (ACIAR) Proceedings No. 95, Canberra, 325p.

Sumberg, J. & C. Okali (1997). Farmers' experiments. Creating local knowledge. Lynne Riener Publishers, London, 186p.

Veldhuizen, L. van, A. Waters-Bayer & H. de Zeeuw (1997). Developing technology with farmers: a trainer's guide. Zed Books, London, 230p.

Waters-Bayer, A. & W. Bayer (1994) Planning with pastoralists - a review of methods focused on Africa. Working Paper. GTZ, Eschborn, Germany, 90pp.

Wen, Shi Lin, Xu Minggang & Qin Daozhu (2000). *Astragalus sinicus* L. in rice farming systems in southern China. In W.W. Stür, P.M. Horne, J.B. Hacker & P.C. Kerridge (eds.) Working with farmers: the key to adoption of forage technologies. Proceedings of an International Workshop held in Cagayan de Oro, Mindanao, Philippines, October 12-15, (1999). The Australian Centre for International Agricultural Research (ACIAR) Proceedings No. 95, Canberra, 181-183.

Yen, Vu Hai & Le Hoa Binh (2000). Forage options for fish and pigs in Vietnam. In: W.W. Stür, P.M. Horne, J.B. Hacker & P.C. Kerridge (eds.) Working with farmers: the key to adoption of forage technologies. Proceedings of an International Workshop held in Cagayan de Oro, Mindanao, Philippines, October 12-15, (1999). The Australian Centre for International Agricultural Research (ACIAR) Proceedings No. 95, Canberra, 281-283.

The contribution of participatory research: on-farm research

P.F. Fennessy[1], N.J. Daniels[1], S.A. Chadwick[1] and P.A. Speck[2]
[1]*Abacus Biotech Limited, PO Box 5585, Dunedin, New Zealand*
Email: pfennessy@abacusbio.co.nz
[2]*Infineon Technologies, Melbourne, Australia (PAS)*

Key points

1. The issue of farmer participation throughout the research – extension process continues to challenge research scientists, but the benefits evident at the individual farm enterprise level are such that researchers cannot afford to ignore the opportunity.
2. Researchers need to develop processes and methodologies to ensure that their core research competencies can be deployed in on-farm participatory settings.
3. The application of new electronic technologies, especially Radio Frequency Identification for individual animal data capture, has the potential to revolutionise on-farm research.

Keywords: extension work, systems development, radio frequency identification, RFID, sheep

Introduction

Participatory research in agriculture may range from research and technology development (R&D), carried out on a research station with some involvement of farmers, through to genuine participatory research involving researchers and farmers working together. The latter involves the end-user in actually carrying out aspects of the research and/or in the development and evaluation of technology that is appropriate to commercial enterprises. Researchers often question the validity of the 'findings of on-farm participatory research' as they are more comfortable with the 'controlled' environment of the research station. However if research is to be applied appropriately on farms, it must go through a period of evaluation on-farm.

This paper summarises perspectives relating to participatory on-farm research, highlighting some opportunities that new technology is providing by considering five key areas as follows:
- Participatory research within the agricultural research enterprise (current situation);
- On-farm research compared with in-station research (options);
- A New Zealand example of on-farm research;
- New opportunities in on-farm research;
- Future models for participatory on-farm research.

Participatory research within the agricultural research enterprise

From research to best practice

The major driver for research in grassland and associated livestock enterprises is the development or evaluation of technologies or systems that will ensure adequate profitability of the enterprise, coupled with long-term sustainability of the grazing-livestock ecosystem. The general approach involves definition of key factors that impact on the performance and product output of the system, with the overall objective of assisting farmers to adopt and follow 'best practice' in managing their farming operations profitably. Thus participatory or on-farm research must be considered in the context of the overall objective of achieving best practice in farm and grassland management. The definition of 'best practice' may differ between different stakeholders and observers. However, regardless of the target, rural research is invariably focused on innovation and improvement within farm systems. They are the desired outcomes irrespective of whether the research approach is conventional (researchers operating external to the rural system to which new technologies are applied), or participatory (where farmers or their beneficiaries are involved in the research process). Irrespective of the manner in which research is conducted, process improvement is fundamental to the ability of rural ecosystems to adapt and change to meet new challenges.

Farming systems research, extension and learning, and the value of participatory research

The key elements of farming systems research (FSR) include a holistic approach, orientation towards the needs of defined target groups, and high levels of farmer participation and hence co-learning by farmers and specialists (Petheram & Clark, 1998). Recently, a combined Australia-New Zealand group (Barlow *et al.*, 2002) outlined an approach to Farming Systems Research and Learning (FSRL) that incorporates on-farm research as a major component. They highlighted the fact that FSRL plays a role in ensuring that innovative technologies and practices will add value to farming businesses, but that it is the addition of the human element to traditional biophysical enquiry that makes FSRL a unique part of the innovation cycle.

The value of on-farm participatory research can be realised at several different levels. For example, it can enhance both the diffusion of technologies and the capacity of farmers to engage in a continuous pattern of innovation (Hildebrand, 1993; Johnson *et al.*, 2003). The involvement of farmers at the formulation and design stage helps define overall priorities and identify factors and variables to be considered. It also helps provide a sense of ownership for the participants, and through involvement in trials, highlight issues that may impact on, or even compromise, the application of the technology in the wider practical on-farm situation.

Participatory on-farm research is especially relevant in grassland research, where the system involves the interaction of the grazing animal and the grassland environment. The system is, by definition, extraordinarily complex, with numerous 'inexplicable' interactions. In considering R&D that is relevant to grassland, it is critical to consider the purpose of the research. From a practical perspective, the different phases of research may be regarded as:
- *Component research*: defining the impact of changing a specific input component on the output (such as the effect of level of feed on milk production);
- *Novel technology or input*: evaluation of the impact of a specific technological input (such as a new pasture cultivar);
- *Whole system research*: defining the impact of a number of components on productivity within the context of the whole system (see section on the New Zealand example).

Figure 1 outlines a scenario for the research-innovation process through the phases from the research station to adoption on-farm. Early-stage, component or investigatory research is generally best managed on a research station where unforeseen difficulties can be managed, whereas the evaluation of the impact of a new technology on a system is best evaluated on farm. At all stages there are major benefits in involving farmers and other end-users in the planning and assessment.

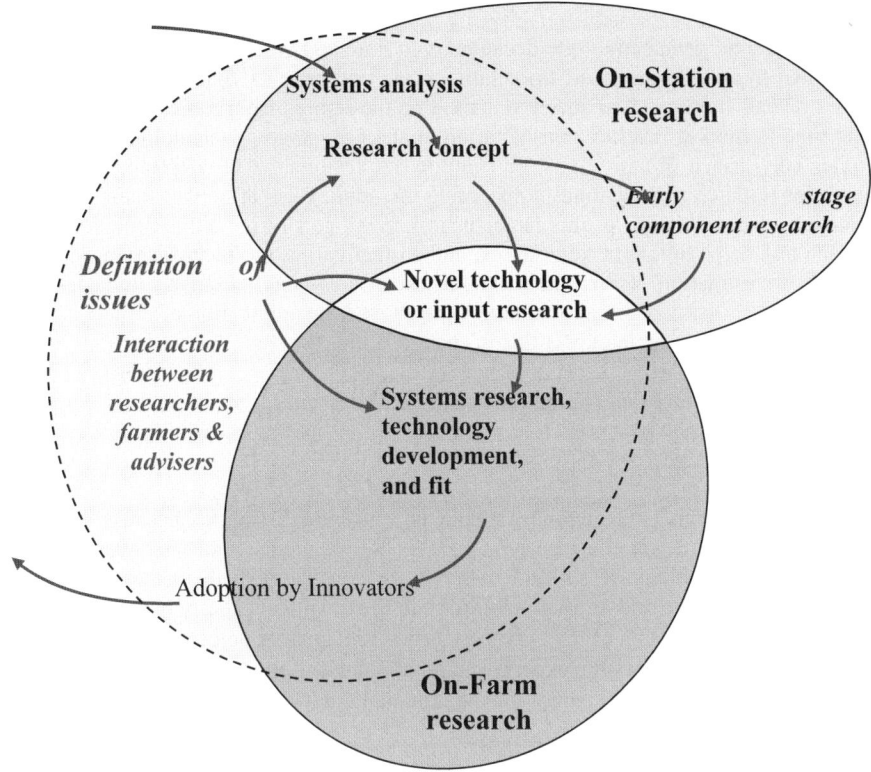

Figure 1 An outline of the research-innovation process (in part after Barlow *et al.*, 2002)

On-farm research in the context of the system

On-farm research is particularly appropriate when designing experiments to challenge hypotheses about systems. Researchers commonly refer to this phase as validation, but the term is inappropriate. While the distinction between falsification and validation should be fundamental in all research, it is particularly pertinent in systems research where, in the absence of clearly defined hypotheses, the sheer number of observed variables can cloud the interpretation of the data. It is essential to develop well-defined hypotheses for on-farm research and establish designs and analyses to investigate the proposed key factor(s); otherwise the sheer magnitude of the observed variance and a lack of appropriate data could well obscure the effect of the factor under investigation. At another level, some investigators may regard it as simpler to collect large amounts of data via a type of 'population census' approach and then search for relationships. Irrespective of the manner in which an experiment or trial[6] is conducted, a second set of trials may well be required to test for specific effects.

---

[6] We distinguish between an experiment and a trial: a trial is essentially an investigation of a known technology, while an experiment implies a greater level of rigour involving investigation of unknowns.

A cost-benefit perspective

From a pragmatic perspective, the decision as to how and where to conduct research should come down to an assessment of the relative costs and benefits. Cost-benefit or investment analysis reveals the impact of speed to market on the returns from any investment. The issue of the time to market is a key consideration at an early stage of technology development in commercially-driven organisations, but time to market or time to adoption, is not usually a priority for research organisations considering the application of new systems in agricultural production. Consideration of this factor is an integral part of an active project management approach and helps highlight key factors, and so facilitating the identification of appropriate targets for investigation. Given the increased use of cost-benefit analysis in decision-making with regard to investments in agricultural research, and the application of project management approaches to R&D, we can expect this to change, albeit slowly.

On-farm or in-station research: challenges and limitations

A researcher perspective

Conventional research is (preferably) conducted on research stations because of the greater control scientists have on the research process, both at conceptual and operational levels. Therefore, the limitations of on-farm research are considered here in the context of those factors that scientists regard as likely to impact on the scientific validity of research, namely:
- The clarity of expression of the hypotheses and experimental objectives;
- The quality of the design, the experimental methodology and control of the variables;
- The quality of the management of the experiment (and the management rules);
- The quality of the data collected;
- The quality of the analyses;
- The quality of the interpretation.

The quality of the experimental design and of the methodology are especially important in on-farm research, as the ability of the researcher to control the environmental variables (including variations in management between farms) is limited, compared with that on a research station. However the limitations of data collection, the analysis of the data and the interpretation are also factors. Failures at any stage of this process can lead to inappropriate conclusions and a failure of the R&D process. This in turn can constrain the process of innovation, and at its worst discredit the role of R&D in the innovation process, and as a result, a formal R&D process may well be abandoned.

The end-user perspective

The potential for a direct impact of the findings of on-farm research on the farmer or end-user is a key benefit to be gained from on-farm participatory research. There is considerable literature around this issue that discusses the merits of various levels of farmer participation in research, particularly in respect of developing technologies that are more appropriate for intended users (and therefore more likely to be adopted), and/or improving the innovation process itself (Johnson *et al.*, 2003). As noted above, key factors that scientists regard as impacting on the scientific validity of research are associated with the quality of the research. In contrast, the key factors that influence the value of research for the farmer or end-user are:
- The perceived importance of the function under investigation;
- The perceived reliability of the research;

- The perceived applicability of the research;
- The perceived cost-benefit relationship.

The perceptions of farmers are fundamental to achieving uptake and successful adoption of a technology or system. Thus a coherent value proposition is critical to convincing farmers of the potential value of a technology. In this respect, 'real world' on-farm research provides the opportunity for researchers to generate sound value propositions that will appeal to farmers. Consequently while there are major opportunities to generate real value from on-farm R&D where commercial farmers are directly involved, the scientist and farmer generally have contrasting value perceptions, with the farmer interpreting research findings in relation to one's own experience, particularly around integration of the technology into farm practices.

The research process

The following discussion of the research process allows the identification of the critical parts of the process that are amenable to on-farm, as distinct from on-station, research. The issues are highlighted mainly from the scientist perspective, which is focused largely on what is regarded as valid scientifically supportable data and subsequent analysis and interpretation.

Hypotheses and objectives
Clarity of thinking around both hypotheses and objectives of an experiment are fundamental to good research, but this is especially important in systems-based research, where the quantity and quality of data can greatly magnify the difficulty of objective analysis. The problems arise mainly from the complexity of the system and inherent interactions. This is a feature of systems research *per se*, rather than the location of the research (i.e. on-farm versus on-station). However in the experience of the authors, carrying out systems R&D on-farm in the 'real world' offers major advantages, and in this respect, investigators have shown that participatory models offer advantages over conventional research models in the determination of research objectives (Johnson *et al.*, 2003).

Design, methodology and control of the variables
Why are researchers traditionally wary of on-farm research? The basic problem is the issue of control, in that it is very difficult to control the numerous variables introduced by operating in a commercial on-farm situation. This problem is well illustrated in the following expression from quantitative genetics:

$$\sigma^2 \text{ observed} = \sigma^2 \text{ environment} + \sigma^2 \text{ genetic} + \sigma^2 \text{ (genetics x environment)},$$

where σ^2 environment $= \sigma^2$ (fixed and random).

Adapting this more specifically to the current situation, the following expression is appropriate:

$$\sigma^2 \text{ observed} = \sigma^2 \text{ treatments} + \sigma^2 \text{ environment (fixed + random)} + \sigma^2 \text{ (treatment x environment)},$$

where   the fixed and random effects are considered separately.

Considering on-farm research in the context of these expressions, a key issue is the variance structure which is likely to be different to that encountered in on-station research. In the case of the latter, it can be expected that a greater proportion of the environmental variation will be allocated to the fixed component than in the situation with on-farm trials.

The above considerations also highlight the issue of replication. While there are several variations on the theme, replication of treatments in researcher-directed, on-farm R&D is often derived from replication across farms (including different managements), which virtually by definition will be more variable than sites within or even across research stations. This will result in a greater observed variance for on-farm trials and a greater random variance, as unassigned variance (that not due to treatments or fixed effects or those interactions) is allocated to the error or random environmental term. One solution to this dilemma may be to exert greater control over the site of the experiment by running the trials on a research station. However this may reduce the impact of the work in the eyes of farmers by restricting the perception of its applicability due to the control of variables that farmers may not be able to achieve in a commercial setting. An alternative is on-farm research with active involvement of farmers coupled with the collection of more data to compensate for the inherently greater variability around farm environments. In addition, a focus on the experimental design will be rewarding, by seeking to allocate variance to known fixed effects or seeking to define the factors that are important in generating the variation between farms.

Management and management rules

Participation of farmers in a study often results in variation in trial methodology due to managerial decisions that need to be made 'on-the-day'. For a study to produce conclusive or indicative outcomes, the consistency of farmer decisions across farms needs to be managed. Trials can be managed through the imposition of defined experimental rules at the outset. However, it is desirable to allow farmers some flexibility (within defined parameters), and accept variation resulting from commercial decision-making, and/or incorporate likely parameters resulting from commercial decisions into the experimental hypotheses and/or analyses. Accounting for such real-life variables is a challenge in rural systems research. This point is emphasised by Collins *et al.* (2001) who describe one of the constraints to on-farm research as the need to ensure that the co-operators -(farmer researchers), project facilitation staff and technicians, all understand the balance between the practical operation of the system and the need to collect accurate data. However, a pragmatic approach to dealing with the practical realities ensures that: 1) research is conducted in concert with the commercial realities of the farming operation; and hence that 2) the experimental results will have greater commercial relevance to farmers; and thus 3) can be expected to generate a greater buy-in and acceptance by farmers.

Data collection

The ability to collect adequate data (with respect to both quantity and quality) is a fundamental issue when comparing on-farm to on-station research. As noted previously, researchers often take the view that the results and interpretation of on-farm research are compromised due to the difficulty of controlling or managing the problem of the variation across farms. This has the effect of greatly increasing variability (variance in the statistical model), thus reducing the chance of detecting a statistically significant result. The normal response to such a problem is to propose an increase in the size of the experiment and/or a better definition or control of the environmental variables. In this context, the limitation is generally the ability to collect and manage additional data. Of all the factors that limit the use of on-farm trials, data collection is by far the most important as it impacts on the cost of collecting sufficient data and the accuracy of those data. However, new technologies can greatly enhance the process of data collection providing new opportunities in on-farm research. This is discussed in more detail in the section on new opportunities.

Data analysis and interpretation

The challenges of data analysis and interpretation in on-farm research arise from two sources, namely 1) the variation introduced by the range in environments (including the participation of the farmers in managing trials), and 2) the complexity of the interactions. However, where the process of data collection can be simplified, more data can be collected. Given the availability of such additional data, the feasibility of developing and testing more complex hypotheses is greatly enhanced. In particular, trials can be designed to ensure that factors that may vary across farms (including management) become amenable to investigation.

A New Zealand example of on-farm research: the CF2000 Scheme

On-farm research is ideally suited to investigations where the objective is to define the factors that influence outputs of the whole system, and then to use these findings to help evaluate new technologies and drive innovation at the farm level. A good example is the CF2000 scheme in the Clutha region of New Zealand. This case study provides an example of an approach to on-farm research where data were collected to allow the testing of a number of hypotheses. For example on New Zealand sheep farms, the lambing percentage (lambs weaned per ewe put to the ram) is a key factor that impacts on the profitability of sheep farming. While this may be 'widely known', the CF2000 study has helped reveal the impact, and also has highlighted other factors that are important in ensuring that an increase in the numbers of lambs born results in a greater profit. That is, on-farm research firstly highlighted the impact, and then secondly helped bring about real change in the industry in this region.

In the CF2000 study, a group of farmers, researchers, farm consultants, accountants and veterinarians lead by one enthusiastic farmer developed a package of measures to help define the factors that might be expected to impact on profitability on sheep farms. The concept was then sold to a wider group of farmers as an opportunity to develop benchmarking parameters that would enable them to compare their performance with that of others in their region. About 60 farmers took up the opportunity. Two highly respected advisers (a farm consultant and an accountant) had strong involvement, and assisted their clients with the interpretation of analyses. The involvement of the advisers, along with twice-yearly meetings and workshops, helped propagate findings among the group. Over time, several farmers shared experiences, and this free exchange of information became a further driver for change and innovation.

CF2000 parameters and analyses

In developing the scheme, several parameters considered likely to be important were identified. Farmers were responsible for their own data compilation although they had assistance from their farm accountant (who would have required much of the physical and financial data to complete the financial accounts for taxation purposes anyway). The approach involved the collection and compilation of base data and additional performance data (*) as in Table 1.

Over the few years that CF2000 has been operating, several types of analyses have been performed. Regression analysis has been used frequently (including analysis of covariance), to present an overall regression line for a relationship (as below), and highlighting the performance of individual farms has provided a potent stimulus to discussion.

Table 1 Data recorded by farms participating in the CF2000 scheme

Farm classification data:	Area & class of land, soil type, sub-division, water supply etc.,
Physical input data:	Breed, numbers, and age classes of sheep (& other stock) *Fertiliser use: timing & type *Pasture renovation: timing & type *Feed conservation: silage, balage or hay *Feed purchases: timing & type *Use of anthelmintics, vaccination & animal health treatments
Physical output data:	Numbers & weights of products sold (meat, wool, livestock) *Timing of sales *Quality parameters, including sex of lambs
Financial data:	Inputs: Maintenance, fertiliser, seed, labour, contractors etc., Outputs: Returns as for physical outputs
Performance parameters:	*Weight & condition of ewes at key times (mating & weaning) *Pregnancy rate & litter size (at ultrasound scanning by ewe age etc.), lamb deaths, weaning rate *Ewe deaths (number & timing) *Live weights & weight gains of lambs (monitor mobs)

Generally statistical analyses were performed using farms as replicates, with farms grouped as 'treatments' as appropriate to test the effects of different factors (e.g. breed or strain of sheep, ratio of sheep:cattle, stocking rate within land class etc.) on performance parameters and the relationship between physical and financial outputs and profitability on performance parameters and so on. For the farmers, the most interesting findings around the performance parameters included the:

- impact of breed or strain of sheep on litter size at ultrasonic scanning during pregnancy;
- relationship between litter size and live weight at mating;
- relationship between litter size and the date by which all surplus lambs are sold;
- seasonal pattern of ewe deaths;
- variation between (and within) farms in the growth rate of lambs.

In many cases these findings influenced the behaviour of farmers. The involvement of an experienced farm consultant to assist with interpretation and to provide the opportunity for the farmer to discuss ideas was also important. For example, the realisation of the:

- impact of breed or strain on litter size (by ultrasonic scanning) resulted in some farmers evaluating (and in many cases) changing to other breeds or strains of sheep;
- effect of liveweight at mating changed the way farmers managed ewes after weaning;
- impact of feed supply available for ewes in the month or so before mating (as affected by the competition from surplus lambs being grown to higher weights) lead to farmers becoming more focused on lamb growth rates and selling lambs by defined dates;
- seasonal pattern of ewe deaths resulted in farmers sorting ewes by litter size to enable preferential management of ewes carrying triplets in the period prior to lambing;
- variation between farms in lamb growth rate resulted in some farmers changing grazing management of lambs after weaning (including sowing newer pasture cultivars).

In combination with analysis of performance, analysis of profitability (gross margin/hectare) proved very helpful to farmers, particularly in relation to gross margin as a function of four factors, namely 1) the average weight of meat sold per hectare; 2) average weight of lambs sold per hectare; 3) pattern of sales of surplus lambs, and 4) the average litter size of the flock.

As a consequence of their involvement in CF2000, and the identification of some of the key factors impacting on profitability, the farmers involved became more confident about evaluating new technologies and looking at new ways of managing their businesses.

New opportunities in on-farm research

Constraints to the increased use of on-farm research

Background
The CF2000 experience highlighted the value of participatory R&D, on-farm analysis and benchmarking. Having shown some of the benefits and identified the critical components of the research process, the challenge now is to target the key constraints to the increased use of on-farm R&D. The premise is that a greater use of on-farm research will increase levels of participation, and hence can be expected to facilitate the process of innovation and the adoption of improved practices by farmers.

The development of research objectives
The joint development of the research agenda is a critical part of ensuring that all parties are engaged in tackling the same objective. A shared goal that all stakeholders have 'bought into' provides the commitment required to overcome the difficulties and constraints encountered during the research process.

Farmer appreciation of issues in research design
The formulation and design of research in agricultural systems is an advanced skill. Even on research stations, these skills may be under-developed and research compromised, hence the involvement of professional biometricians in the design process is essential. An understanding of design issues is particularly pertinent in participatory research. Thus researchers must ensure that farmers are aware of the nature of the research process, and the necessity for some constraints on the research design. When farmers are aware of the factors involved in robust experimental design, they can be expected to have a greater appreciation of the constraints. Two good examples are the necessity for appropriate replication in experiments, and the classical trade-off in progeny tests between slaughter of animals at an appropriate time versus slaughter at the same weight.

Perspective
A key constraint in the use of on-farm research is not a lack of researchers who are willing to be involved, but rather a shortage of researchers who can:
• develop processes and experimental designs that can leverage on-farm research to provide scale and ensure faster uptake of research findings;
• design research processes compatible with commercial farming practices;
• communicate research processes, and obtain buy-in from farmer stakeholders;
• interpret the results in ways that are both meaningful and relevant to the farmer.

Data capture technologies

The above discussion focused on areas where the skills of researchers can be effectively deployed in on-farm participatory research programmes. While researchers must develop processes and methodologies to engage farmers in such programmes, new electronic data capture technologies will play a critical role in expanding the opportunities for on-farm research. Many people see a paradigm shift that will transform on-farm livestock and associated grassland research. Key technologies include Radio Frequency Identification (RFID) systems, Geographical Information Systems (GIS), associated electronic data collection, and other related technologies (e.g. novel technologies that will enable cost-effective collection of data on soil and plant conditions and pasture growth). However, the following discussion is confined with respect to RFID systems.

Radio Frequency Identification systems as a tool for improved data collection
RFID systems are being rapidly deployed in a range of logistics and product tracking situations (Want, 2004), including livestock identification and traceability. Specific benefits of the application of RFID systems will be an enhanced ability to rapidly capture data from large numbers of individual animals, and to retain records in an easily retrievable form for subsequent analyses (e.g. intensive recording of individual milk production, live weight, health and reproductive data etc.). Thus RFID will provide the means to radically change the scale of on-farm research.

Why are new data capture technologies are so important?
The use of such tools will allow for the collection of data of a quality required for formal analysis through conventional research processes, within a commercial setting. Given appropriate guidance by researchers, electronic capture of data effectively raises the ability of on-farm managers, to record data of a quality previously captured by highly trained technicians and research staff.

In summary, the application of RFID systems will greatly facilitate the collection of animal data. The impact of this technology is considered to be much greater for on-farm research than for on-station research. Table 2 presents a summary of the comparative importance of issues, advantages and disadvantages of on-farm R&D compared with on-station R&D, highlighting the impact of the application of RFID systems.

The impact of new data capture technologies on participatory research

The major limitation to research is money. Hence regardless of whether research is on-station or on-farm, the balance of the cost-benefit equation is dependent on the comparative benefits of the two alternatives. As discussed previously, researchers often take the view that the results and interpretation of on-farm research are compromised by the difficulty of controlling or managing variation on commercial farms, i.e. that on-farm research involves new variables and management challenges, such that there is a risk that the research will lack the scale required to produce statistically valid and useful research outcomes. Given the use of RFID technology and associated systems, the key question in considering any R&D project is whether the benefits of on-farm research are outweighed by the risks. In this respect, the key benefits of on-farm research (across many farms) include the following:
- opportunity to look at whole systems, various environments and technology interactions;
- the 'sell' opportunity to create fast and significant adoption (in this sense, word-of-mouth will always be the most powerful selling tool);
- farmer participation helps drive support and understanding of the contribution of research.

In most cases it is the data limitation that inhibits or compromises on-farm research. Hence many of the negatives around the on-farm research approach can be reduced by the collection (and analysis) of more (quality) data. In this context, RFID is an enabling technology. By utilising RFID systems, the potential benefits of participatory research for whole systems research over that of research based on research stations can be realised.

Table 2 Issues, and advantages/disadvantages of on-farm compared with on-station R&D, with particular respect to the impact of the application of Radio Frequency Identification (RFID) systems to collect animal data

Design & management		
Clarity around hypotheses and design	Greater importance for on-farm research, due to the need to communicate & engage farmers	
Control over experimental management and process	Less control with on-farm R&D, but RFID reduces the disadvantage of on-farm research	
Opportunity to investigate many factors determining performance of system	Greater opportunity through on-farm than on-station R&D	Greater ability to capture data means more factors can be considered in on-farm research with RFID than without
Opportunity to integrate technologies into whole systems		
Data analyses & interpretation		
Number of parameters (complexity) to be accounted for (or not accounted for) in the statistical analysis	More complexity on-farm means that there is a lesser chance of producing statistically significant results	RFID facilitates data collection so that the quality and quantity of on-farm data are enhanced
Chance of producing statistically significant results		
Farmer involvement & long-term benefit		
Farmer participation and support	Much greater through actual involvement of farmers in the R&D	Enhanced impact by application of RFID as more meaningful R&D is achievable
Opportunity for interactions with farmers around the technology		
Expected overall level of adoption & rate of adoption		

Potential future models for participatory on-farm research

In considering the opportunity for the further development of on-farm participatory research within livestock-grassland systems, it is interesting to outline aspects of the innovation process in other industries, and then compare these with those within the grassland sector.

On-site research within the manufacturing sector
Most ground-breaking research in manufacturing is conducted and funded by large multi-national corporations, with some basic research conducted within Universities. However, in most manufacturing sites, process engineers typically assume responsibility for on-going process improvement and innovation. While ground-breaking technologies or new systems are frequently sourced from outside the company (e.g. from the multi-nationals, or specialist suppliers), process engineers possess the detailed knowledge of company processes, constraints and targets, and have access to the statistical tools to research and implement process improvement and innovation.

On-site research in seed and vegetable production
In a similar manner to the manufacturing sector, large-scale vegetable production often relies on external sources for significant innovation in processes and in the development of new cultivars. However, scientific skills in the evaluation of new varieties and methodologies are

typically found only on the largest of commercial vegetable production operations. The suppliers of key inputs, such as seeds and fertiliser, often undertake the role of process improvement. In this respect, a service commonly provided by the seed companies is that of the establishment and management of research trials of new cultivars and varieties for growers. A New Zealand company (Xenacom) has developed a trial management and traceability software application for the seed industry to accelerate the process of development and evaluation of new plant varieties. The package provides a system to record commercial trial data using standardised evaluations, thus facilitating the selection process and enhancing relationships with breeders and growers.

Developing new 'on-site' participatory models for grassland farming

Background
In most parts of the world, commercial grassland farming is dominated by family or owner-operator businesses. The characteristics of such businesses include:
- production systems that are too complex for simple trials to yield useful information;
- few significant external inputs, which could be used to subsidise the provision of research skills (as in the vegetable grower example above);
- the skills required to operate the business do not rely on formal research skills, so that innovation conducted at the farm level is not of a formal research standard (in contrast to the process engineer example above), and hence on-farm stakeholders have difficulty engaging in participatory research programmes;
- detailed performance data (e.g. individual animal data) are unlikely to be required to operate the farm, but such data are vitally important in the more formal R&D process.

The implications of the above are such that development of participatory models in on-farm livestock-grassland research is not a simple extension of current systems. The availability of new tools that will greatly increase the capacity to collect data, coupled with the opportunity to devise much more powerful experimental designs, can greatly increase both the power and scale of on-farm research. This requires the involvement of skilled biometricians with an understanding of biological variation, who can design, analyse and help interpret experiments in association with the end-users (Johnstone, 1998). New technologies will allow for far greater levels of end-user participation in research than has previously been the case.

On-farm as an extension of on-station research
Several on-farm sites can be used as replicates for on-station research programmes, using technology to facilitate data capture, and with formal experimental design and analysis of results supplied by on-station experts. Data can also be transferred regularly back to the researcher for monitoring. The obvious potential benefits include a more powerful experiment, with increased scale at less cost, plus the enhanced participation of farmers and their likely acceptance of the relevance of the research.

Collective on-farm
A development of the research model above would be where farmers themselves initiate the research process, engaging researchers to provide key skills (experimental design and data analysis). Thus, farmers would outline the objectives of the research and have the key responsibilities of management and data collection.

'Distributed' research model

The widespread use of new RFID technologies will facilitate the development of a potentially much more powerful research model. Where several farms routinely capture data via automated RFID equipment, farms will build their own databases of farm performance information. These distributed databases can then provide the data source for a 'distributed' research programme. Data capture is a by-product of the day-to-day farming operations and is not an added burden or requirement of the production system. The potential for such a model is exciting, as it provides minimal constraints on the farming operation, but allows for a wide range of on-going research programmes to be conducted. However, special care needs to be taken in the formulation of hypotheses and design of experiments based on a distributed model. There are situations where the large quantity of data will not compensate for biases that may exist in the data capture process, and systematic bias is a danger for such distributed models. Thus without full participation of both scientists and farmers at all stages of the process, the research outcomes may well be compromised.

Acknowledgements

Thanks to our colleagues, Neville Jopson, Sharl Liebergreen and Peter Amer for their helpful contributions to this paper.

References

Barlow, R., D. Clark, A. Crawford, M. Paine, G. Sheath & J. Weatherley (2002). Guidelines for farming systems research & learning. Dairy Research and Development Corporation, Melbourne, Australia. 18pp.

Collins, R., S. Buck, K. McCosker, G. Lambert & D. Sparkes (2001). Challenges of on-farm research – insights from the central Queensland farming systems experience. *Proceedings of the Australian Agronomy Conference*, <http://www.regional.org.au/au/asa/2001/4/b/collins2.htm>

Hildebrand, P.E. (1993). Targeted technology diffusion through coordinated on-farm research. *Proceedings of Association for Farming Systems Research-Extension*. <http://breeze.ifas.ufl.edu/readings/PEHildebrand.PDF>

Johnson, N.L., N. Lilja & J.A. Ashby (2003). Measuring the impact of user participation in agricultural and natural resource management research. *Agricultural Systems,* 78, 287-306.

Johnstone, P. (1998). Planning and managing agricultural and ecological experiments. Stanley Thornes Ltd, Cheltenham, England. 118pp.

Petheram, R.J., R.A. Clark (1998). Farming systems research: relevance to Australia. *Australian Journal of Experimental Agriculture,* 38, 101-115.

Want, R (2004). RFID: A key to automating everything. *Scientific American* 290 (January), 46-55.

Xenacom. www.xenacom.com

Computer-based forage management tools: historical, current, and future applications

D.H. Hannaway[1], C.P.Q. Daly[2], D.F Chapman[3], B.B. Baker[4,5] and A.S. Cooper[6]
[1]Oregon State University, Department of Crop & Soil Science, Corvallis, OR 97331, USA
Email: david.hannaway@oregonstate.edu
[2]Oregon State University, Department of Geosciences, Corvallis, OR 97331, USA, [3]University of Melbourne, School of Agriculture and Food Systems, Melbourne, Victoria, Australia 3127, [4]The Nature Conservancy, Global Climate Change Initiative, Boulder, CO 80302, USA, [5]Department of Animal Sciences, Colorado State University, Ft Collins, CO 80523, USA, [6]ANE-Asia Associates LLC, Bangkok, Thailand

Key points

1. Historical approach has been traditional, point-based, individual experiments combined with extension education to convey best management practices.
2. Budget cuts have forced efficient new approaches to field experiments.
3. Computer tools provide a means for improving understanding, creativity, and cost-effective visualization and integration of statistical and spatial data and expert knowledge into management decision-aids and improved priority-setting and resource allocation.
4. Vision is for better integration of the various disciplines and technologies to improve global collaboration and management decision-making.

Keywords: models, GIS, remote sensing, DSS, geospatial, quantitative eco-physiology

Introduction

Forage management has been an important human activity since the beginning of civilization. By comparison, the personal computer has been available only in the immediate past. The software developed to deal with the complexity of climate, soil, plant, animal, and socio-economic factors has seen huge changes in a few decades. Mainframe computers facilitated numerical calculations for exploring relationships among dozens of variables. Personal computers opened the door for more individual scientist creativity and routine communication. Web-based communication globalised the option for multidisciplinary teams to tackle problems. Forage-related computer applications abound, allowing farmers, ranchers, and others to more effectively manage the land. This paper describes historical, current, and future computer-based applications that improve understanding and efficiency leading to more economically and environmentally sustainable forage-livestock systems.

Historical review

Agricultural computer applications have come a long way in the past 20 years (Hannaway *et al.*, 1997). In the mid- 1980s, relatively few individuals had access to computers on a daily basis. However, there were already early developments in the use of agricultural software for least-cost ration analysis for livestock, animal weight-gain analysis for various systems, crop simulation, and grazing models (Rasmussen *et al.*, 1985). By the early 1990s, Expert System and Decision Support System (DSS) approaches were well-developed (Bolte *et al.*, 1991; Hannaway *et al.*, 1992). Some of the earliest DSS software included Grazing Land Application (GLA), Rangepack, Grazplan, Stockpol, and Beefman (Stuth *et al.*, 1993).

Computer modelling allowed developers to explore the interactions in a farm system and test new technologies and ideas.

Crop and rangeland models

Crop modelling and systems analysis is still a relatively new discipline within agricultural sciences. However, significant progress has been made in the acceptance of modelling as a tool for research, teaching, and extension (Hoogenboom, 2000a). A few examples are SPUR, DAFOSYM, GRASIM, GrassGro, and the IBSNAT/DSSAT crop models.

SPUR

Originally released in 1987, Simulation Production and Utilization of Rangelands (SPUR) is a deterministic, mechanistic process model that operates on a daily time step. It was developed to design and analyse management scenarios (irrigation, fertilisation, seeding, and grazing systems) that affect rangeland sustainability (Wight & Skiles, 1987; Hanson *et al.*, 1992). Specifically, SPUR simulates growth initiation, germination, carbon assimilation, translocation between roots and shoots, N mineralisation, and nitrogen uptake. In 1987 SPUR was unique, in that it was a multipoint model designed to allow for direct competition between several species of plants for water and nitrogen as well as incorporating the impacts of grazing by wild and domestic herbivores. Over the past seventeen years, modifications have been made to SPUR that reflect ongoing efforts to incorporate improved understanding of the interactions among ecosystem processes, climate dynamics, and management decisions (Baker *et al.*, 1992; Foy *et al.*, 1999; Pierson *et al.*, 2001; Teague & Foy, 2002; Shafer, 2003; Skirvin & Moran, 2003).

DAFOSYM

The Dairy Forage System Model (DAFOSYM) simulates the performance, environmental impact and economics of a dairy farm over multiple years of weather (Rotz, 2001). The simulation includes the growth, harvest, handling and storage of *Medicago sativa* (lucerne), grass, *Zea mays* (corn), small grain, and *Glycine max* (soybean) crops. Farm produced feeds are supplemented with purchased feeds to meet a given level of production for a dairy herd. Manure is returned back to the land where nutrients are lost, accumulated in the soil or used in crop production. Costs of feed production and manure handling are compared to milk, animal, and feed sales to determine a net return. Other farm costs are then included to estimate the net return or profitability of the whole farm. Major submodels include crop growth, harvest, feed storage, feed use, manure handling, tillage, and economic analysis. Simulation of crop production and harvest is based on a daily time step using historical weather data for the location. Crop growth models determine the accumulation of dry matter and the change in quality (nutrient content) based on daily weather and available soil moisture. Total costs of production are subtracted from the various farm incomes to obtain the net return above feed and manure costs and the overall return to management and unpaid factors. DAFOSYM has been used extensively as a research tool and continues to be improved as part of the USDA-ARS Dairy-Forage research team's activities (http://www.dfrc.ars.usda.gov/).

GRASIM

GRASIM is a comprehensive grazing model designed to simulate intensive rotational grazing management linking all components of the pasture system (Mohtar, 2004). It predicts standing biomass, herbage nutritional quality, and nutrient leaching under pasture. It can be used to obtain a better understanding of the pasture system and determine management strategies to improve pasture utilisation. It generates information suitable for estimating the financial and environmental consequences of alternative dairy management strategies including partial mechanical harvest in the context of the year-round feed needs of the dairy herd, storage/harvest needs, year-to-year variability, and stocking rate impacts on supplementation and need for harvested feed. GRASIM operates on a daily time step. Data requirements include minimum and maximum daily temperatures, daily rainfall, average daily solar radiation, soil physical properties, grass growth parameters, soil nitrogen transformation coefficients, and initial levels of soil water and soil nitrogen. GRASIM models multiple paddocks, with each paddock budgeted separately. Paddocks share the same weather conditions but can have different crop and management practices. The grass component contains carbon compartments for storage and structure. The model accounts for root growth and maintenance, shoot growth (partitioned into leaf and stem), shoot respiration, senescence, and recycling. GRASIM models nitrification, mineralisation, uptake, volatilisation, denitrification, and leaching. Soil water is accounted for using a simplified water balance that considers runoff, evapotranspiration, soil matrix water dynamics, and leaching. The harvest component controls the grazing events and handles all pasture management input data. GRASIM is available on-line at: http://danpatch.ecn.purdue.edu/~grasim/grasim.html.

GrassGro

GrassGro is a decision support tool developed by CSIRO Plant Industry's GRAZPLAN group to examine whole enterprise production risk for grazing enterprises in high rainfall temperate zones of Australia (Moore *et al.*, 1997). GrassGro simulates pasture growth and predicts the intake of herbage by ruminants and their productivity using daily weather inputs and user-specified descriptions of soil type, pasture species, and livestock. For any specified site, users can analyse grazing management systems in terms of pasture and animal production, gross margins, and year-to-year variability. Salmon & Moore (2001) reported on an assessment of the adoption of GrassGro based on a survey and training evaluation. Results demonstrated GrassGro's value to clients who had been trained to use the program. The range of problems analysed reflected the diversity of client occupations. The value of these applications was indicated by extrapolation of the results of an individual farm analysis to a regional level.

IBSNAT/DSSAT/CERES

The International Benchmark Sites Network for Agrotechnology Transfer (IBSNAT) project has helped improve agricultural production in developing countries, using a systems analysis approach. The project was headquartered at the University of Hawaii and funded from 1982-1993 by a grant from the United States Agency for International Development (Hoogenboom, 2000a). The Decision Support System for Agrotechnology Transfer (DSSAT) is a software system that facilitates the application of crop simulation models in research, teaching, extension, outreach, and policy and decision-making (Hoogenboom *et al.*, 1999; 2000b). The core of DSSAT consists of 17 crop simulation models, with supporting software in the form of utilities and tools for data handling and analysis programs for applications. DSSAT provides standards for data file formats and naming conventions, and a standard protocol for communication between components and modules. The DSSAT crop models include the generic cereal model CERES for *Triticum* spp. (wheat), *Z. mays*, *Oryza* spp. (rice), *Sorghum bicolour* (sorghum), *Panicum miliaceum* (millet) and *Hordeum* spp. (barley) (Ritchie *et al.*, 1998) and the generic grain legume model CROPGRO for *Glycine max*, *Phaseolus vulgaris* (common bean), *Arachis hypogea* (peanut) and *Cicer arietinum* (chickpea) (Boote *et al.*, 1997, 1998). Other models include the SUBSTOR model for *Solanum tuberosum* (potato), the CROPSIM model for *Manihot* spp. (cassava), the OILCROP model for *Helianthus annuus* (sunflower), and the CROPGRO model for *Solanum lycopersicum* (tomato), *Paspalum notatum* (bahia grass) and a fallow crop.

Web information systems

One of the major computer developments in the past 10 years has been the World Wide Web. Web-based systems provide a means for governmental and university research and education programs to remain viable in spite of budget and personnel cuts. Working together provides many professional development benefits, as well as educational, informational and organization and integration benefits (Green & Hannaway, 1996).

Forage information system

The first comprehensive, forage-related web-based information system was the Forage Information System (FIS; http://forages.oregonstate.edu/). The FIS has been continuously hosted at Oregon State University since it became available to the world in November of 1994. The goal of the FIS is to provide comprehensive information about forage-related topics with contributions from forage scientists and educators worldwide. The FIS is a database-driven system using a combination of proprietary and open-source software to create and maintain segments. Segments of the system include a national forage and grasslands curriculum (Hannaway *et al.*, 1999), and forage-related Classes, Organizations, People, Projects, Topics, and Resources. Segments for *M. sativa*, *Dactylis glomerata* (cocksfoot), *Lolium arundinaceum* (Schreb.) Darbysh. (tall fescue), and other species are under development by groups of experts working together (Hannaway *et al.*, 2001). Recent progress has been made in the area of creating white papers for important forage species in cooperation with the American Seed Trade Association (http://www.amseed.com/). These include quantitative climate and soil tolerances and GIS-based suitability maps (http://forages.oregonstate.edu/main.cfm?Page ID=321). Work by Australian colleagues has led to a more extensive listing of international forage organizations (http://forages.oregonstate.edu/main.cfm?Page ID=246).

Other web system examples

During the late 1990s, many universities, agencies, and organizations in the USA and around the world developed web segments to serve their clients. In addition to Oregon State University, early US university leaders in these efforts have included:
- Cornell (http://www.css.cornell.edu/forage/forage.html);
- Penn State (http://www.forages.psu.edu/);
- Purdue (http://www.agry.purdue.edu/ext/forages/);
- Texas A&M University (http://stephenville.tamu.edu/~butler/foragesoftexas/);
- UC-Davis (http://alfalfa.ucdavis.edu/);
- University of Wisconsin (http://www.uwex.edu/ces/forage/).

The US National Dairy Database was an early leader in assembling information for dairy production (Eastwood *et al.*, 1996). Today, a web segment represents nearly every agricultural content area in each university and agency. The current challenge is using database management techniques to make adding and updating information a simple process for non-computer experts.

Extensive forage-related information is now available in most countries. The following is a list of a few of these sites. A more detailed list is available in the FIS at the following URL (http://forages.oregonstate.edu/main.cfm?PageID=31):

- Australia: (New South Wales) http://www.agric.nsw.gov.au/reader/forage-fodder
  (Pasture Species Database) http://www.meu.unimelb.edu.au/grasslands/
- China: http://www.chinainfowww.com/caw/caw_e.htm
- Ireland: http://www.teagasc.ie/
- Sweden: http://www.ngb.se/Forage/
- UK: http://www.royagcol.ac.uk/flg/

GIS/RS-based geospatial/soil/landuse and landcover information

One of the most significant developments of the last 5 years has been the development and use of geospatial information for topography, climate, soils, land use and cover, and species suitability. This has been made possible by modelling and display software, and the capability of GIS software to display multiple layers of information. An early leader in this work was Engle *et al.* (1997), applying crop models over the landscape. The continuing challenge is availability and sharing of data among scientists, both within and between countries (Hannaway *et al.*, 2000). Also important is the integration of the various layers of information, requiring the availability of similar resolutions for the various types of data needed.

Climate

Excellent progress has been made on climate mapping based on a state-of-the-science model known as PRISM. The Parameter-elevation Regressions on Independent Slopes Model (PRISM) (Daly *et al.*, 1994, 2002, 2003) is especially suited to mapping climate in complex landscapes. The regression-based PRISM uses point data, a digital elevation model (DEM), other spatial data sets, a spatial climate knowledge base, and expert evaluation to generate repeatable estimates of annual, monthly, daily, and event-based climatic elements. These estimates are interpolated to a regular grid, making them GIS-compatible.

The model adopts the assumption that for a localised region, elevation is the most important factor in the distribution of temperature and precipitation. PRISM calculates a linear climate-elevation relationship for each DEM grid cell, but the slope of this line changes locally with elevation as dictated by the data points. Beyond the lowest or highest station, the function can be extrapolated linearly as far as needed. A simple, rather than multiple, regression model was chosen because controlling and interpreting the complex relationships between multiple independent variables and climate is difficult and can produce misleading results. Instead, weighting the data points controls the effects of variables other than elevation.

PRISM is being actively developed and applied at Oregon State University's Spatial Climate Analysis Service (SCAS). Recent mapping efforts include peer-reviewed, official USDA precipitation and temperature maps for all 50 states and Pacific Islands (Daly *et al.*, 2001); a new official climate atlas for the United States (Plantico *et al.*, 2000); a continuing series of monthly temperature and precipitation maps for the conterminous 48 states beginning in 1895; precipitation and temperature maps for Canada, China, Mongolia, and Taiwan (Daly &

Hannaway, 2004), and the first comprehensive precipitation maps for the European Alps region. In progress are updates for the United States and possessions at better than 1-km grid resolution. Additional information can be accessed from the SCAS Web site at http://www.ocs.oregonstate.edu/prism/.

Soils

Digital soil resources are now available for most countries, though the classification systems are many and the scale/resolution varies widely. The availability, cost, and coarse resolution of these resources continue to be an impediment to widespread use in forage-based tools.

USA
The USDA–NRCS, through the National Cooperative Soil Survey (NCSS), has been developing soil geographic databases at three scales: local, regional, and national (Miller & White, 1998). At the regional level, the State Soil Geographic (STATSGO) database was released in 1992 for use in river basin, multicounty, multistate, and state resource planning. This database was created by generalising available soil survey maps (including published and unpublished detailed soil surveys), county general soil maps, state general soil maps, state major land resource area maps, and, where no soil survey information was available, Landsat imagery (Reybold & Tesselle, 1989). STATSGO consists of georeferenced digital map data and associated digital tables of attribute data. The compiled soil maps were created using the U.S. Geological Survey (USGS) 1° by 2° topographic quadrangles (1:250,000 scale, Albers equal area projection) as base maps, which were then merged on a state basis. The NRCS provides web-based access to STATSGO data as part of their web site: (http://www.ncgc.nrcs.usda.gov/products/datasets/statsgo/index.html). Penn State University's Soil Information for Environmental Modeling and Ecosystem Management unit also provides downloadable files for each state as Arc/Info EXPORT format files (http://www.essc.psu.edu/soil_info/index.cgi?soil_data&statsgo). SSURGO is another primary national product that is available in the USA. It is the county survey in digital form at a resolution of 1:24,000. Access to SSURGO data sets in various formats is at: http://www.ncgc.nrcs.usda.gov/products/datasets/ssurgo/data/. Both STATSGO and SSURGO are excellent soil information resources but require significant expertise for appropriate interpretation and use in integrated applications.

China
The Institute of Soil Science, Chinese Academy of Sciences, has successfully produced a 1:1,000,000 soil database of China, which is composed of 3 parts, 1) soil spatial data – the 1:1,000,000 digital soil map of the country, 2) soil attribute data, and 3) Chinese soil reference system (Shi *et al.*, 2004a, 2004b). The soil spatial data are the most detailed countrywide digital soil map available and the Chinese soil reference system has been completed recently.

Canada
The Canadian Soil Information System (CanSIS) has supported the research activities of Agriculture and Agri-Food Canada by building the National Soil DataBase (http://sis.agr.gc.ca/cansis/nsdb/intro.html) and producing GIS products (Schut, 2000). Soils Landscapes of Canada (SLCs) were compiled at a scale of 1:1 million, and information was organised according to a uniform national set of soil and landscape criteria based on permanent natural attributes. Each polygon is described by a standard set of attributes, including surface form, slope, water table depth, permafrost, and lakes. SLCs provide a standardised database of major attributes important to plant growth, land management, and

soil degradation. These data are now used as a framework to support other databases, including Environment Canada's Ecological Land Classification System.

Australia - Australian Soil Resources Information System (ASRIS) is a national database of soil information, suitable for use at national to large regional scale. It was designed to provide nationally conformable information from the extensive soil data that was collected by the State and Territory agencies and CSIRO. It contains both primary data (soil point and soil survey map data) and modelled estimates of soil properties. Until the development of ASRIS, the best available digital coverage of soil information for all of Australia was the Atlas of Australian Soils. This was compiled in the late 1960's at a scale of 1:2,000,000 which provides only the broadest national overview of soil attributes. More detailed soil data have been collected over the last 10 years through national programs such as the National Landcare Program and the Natural Heritage Trust (Australian Natural Resources Atlas v2.0, 2004).

Europe

In collaboration with FAO, IIASA-LUC (Land Use Change program) has created a digitised soil map for North and Central EURASIA covering the countries of the Former Soviet Union, Mongolia and China (Enzlberger, 2001). The database consists of 4942 polygons. For each polygon, information is available on the dominant soil type, associated soils and inclusions, soil texture, slope and soil phases. The 1:5 million scale map, uses the legend of the FAO-UNESCO Revised Legend of the Soil Map of the World (http://www.fao.org/ag/agl/agll/dsmw.stm).

Land use and land cover

The previously described climate and soils modelling and mapping efforts are based on information collected using ground-based weather stations and soil surveys. Land use and land cover projects typically involve satellite-based sensors.

LUTEA project

The growing concern over the impact of changes in land use and cover on environmental conditions, and the increasing human impact on the natural resources in the Temperate East Asian region, led to the formation of the Land Use in Temperate East Asia (LUTEA) steering committee and subsequent project (http://www.nrel.colostate.edu/projects/lutea). The overall objectives of LUTEA were; 1) to better understand the role and consequences of changes in climate, ecosystem dynamics, human demography, and socio-economic transitions on land use and land cover in temperate East Asia during the past 100 years, and into the next decade; and 2) develop a mechanism to assess the short-term and long-term changes in food security and environmental conservation in the region (Ojima, 2000). The region of study included China, Korea, Japan, Mongolia, and Eastern Russia. Project activities included workshops, regional database development for climates, soils, vegetation, and socio-economic factors, conferences, publications, modelling efforts, and analysis tools to better understand how changes in land use and cover in the region will interact with further global environmental changes. Factors affecting changes in pastoral systems were identified as an important regional research issue.

Australian example

Land use and land cover applications specific to forage-livestock systems have been studied in Australia since the mid 1980s. Patterns associated with soil fertility (Vickery & Hedges, 1987) and pasture botanical composition (Vickery *et al.*, 1997) have been mapped in research projects aimed at better understanding the factors influencing pasture availability at the paddock scale. More recently, the use of satellite imagery for practical decision-making has been piloted on commercial farms in Western Australia with promising results. The 'Pastures from Space' project used 30-m resolution imagery to derive estimates of feed on offer and predictions of pasture growth rate for up to 7 days into the future (Henry, 2002). Producers in 'precision sheep production groups' were supplied with this information for each paddock on their property in two successive growing seasons. At the end of the pilot project, 82% of the 63 farmers who were surveyed indicated that the satellite-based information increased their confidence in making pasture and stock management decisions (Gheradi & Oldham, 2003). Sixty-one percent of producers reported that the information helped them to manage risk better, and 59% responded positively when asked if the information had increased the profitability of their farm business.

Current research is seeking to extract better information on forage chemical composition from hypospectral sensors (Held & Hill, 2003), which would add feed quality information to data on feed quantity, and potentially improve animal nutrition decisions. The addition of new satellite platforms with greater spectral imaging capacity and more-frequent coverage of the Australian continent provides the opportunity to monitor pasture mass and composition at spatial resolutions of 250 m at daily intervals.

GPS-based fertiliser and lime applications

Precision agriculture is a management strategy that uses data and information technologies from multiple sources to optimise crop production. Soil type, soil organic matter, plant nutrient levels, topography, water availability, weed pressure, and insect pressure are recognized as parameters that affect crop growth (NESPAL, 2004). Precision farming has developed rapidly with global positioning system (GPS) sensors and intensive soil sampling and yield sensors to guide liming, fertiliser, and pesticide applications for optimal yield. Significant investment is required in equipment and software, but dividends for large-scale producers include reduced production costs and reduced negative environmental impacts. There are three main components of precision agriculture: 1) capturing data at an appropriate scale and frequency, 2) interpretation and analysis of that data, and 3) implementation of a management response at an appropriate scale and time (Sonka et al., 1997). A key difference between conventional management and precision agriculture is the application of high-resolution spatial and temporal data for crop production decision-making. A basic premise is that a larger quantity of better quality information can reduce the uncertainty producers face in decision-making and the unmeasured variability in agronomic conditions. Although there are few examples of cost-effective applications of precision agriculture in forage-livestock systems, these are likely to develop as high-resolution data becomes more generally available.

Integrated expert systems/decision support systems/knowledge based systems

Spatial and temporal variability pose a significant challenge for the management of forage production and utilization systems. In grazing systems, strategic decisions about stocking rate, flock or herd structure, animal management policies, and the grazing program have great bearing on the financial outcome and sustainability of farm businesses. These decisions often pivot around the seasonal cycle of feed supply/demand balance, with the common aim of maximising the proportion of total feed requirement met from home-grown pasture or fodder crop, and minimising dependence on purchased feeds. Many ES/DSS/KBS tools have been developed to assist decision-making in this area, mostly based on the principles and concepts outlined by Milligan et al. (1987), Sheath & Clark (1993), and Holmes et al. (2002). In all cases, these tools use measured or expected average pasture or crop growth rates to determine the supply side of the equation, which is then solved for some combination of animal management policies that make best use of available paddock-grown feed. The decision reached is a good decision (provided the technical data are sound), if it is based on the best information available at the time. However, whether it is the *right* decision depends on the intra- and inter-annual variability in climatic conditions and pasture growth outcomes subsequently experienced.

Climatic variability across many countries is substantial; both in space and time (Nicholls *et al.*, 1997), and therefore production risk looms large in the management decision-making. Not surprisingly, the development of simulation models of pasture growth and animal production that are capable of analysing variability and risk in grazing systems has been a major focus of research activity over 2 decades, leading to several tools that have been applied to both practical and theoretical problems (White *et al.*, 1983; McKeon *et al.*, 1990; Stafford-Smith & McKeon, 1998; McCall *et al.*, 1991; Finlayson *et al.*, 1995; Cacho *et al.*, 1999; Donnelly *et al.*, 1997; Wastney *et al.*, 2002; Johnson *et al.*, 2003a, 2003b). When integrated with location-specific climate data (Jeffrey *et al.*, 2001), these tools allow detailed description of the range of pasture growth outcomes for any particular site (Clark *et al.*, 2003). Given the importance of the pasture base for cost-effective feeding, this can underpin the analysis of risk associated with major management decisions, such as changing the mix of pasture types and fodder crops used to supply feed to animals (Kenny *et al.*, 2005), or changing the animal enterprise (Salmon *et al.*, 2004).

Simulation tools can be integrated with spatial information layers from ground or remote sources to describe spatial variability in pasture growth outcomes at the paddock and farm scale (Hill *et al.*, 1999). Used in this mode, these analytical tools move into the domain of precision agriculture technology for pasture-based systems, but the power of the information technology currently exceeds our capacity to interpret the data it can generate, and our capacity to apply it for the purposes of improved forage management decision-making. While this remains the case, this route seems unlikely to lead to greater on-farm use. However, integrated tools of this sort are being applied to improve farmer learning, and support risk management decision making for sustainable environmental and agricultural planning purposes in Australia (Webby, 2002; Laughlin *et al.*, 2003; Brinkley *et al.*, 2005). Here, simple simulators of pasture growth are used to analyse the reliability of seasonal growth using historical weather records for broad regions. The processing power of super-computers offers a future where pasture simulation models are distributed in virtual space across selected regions and connected to historical and projected weather data (from improved climate models) to predict pasture growth outcomes several weeks ahead. This information would be invaluable for forage management decision-making in pasture- and rangeland-based grazing systems and the down-stream industries that handle their products. Farmers could foresee trends in vegetation cover in response to their grazing management policies, and judge whether that cover will meet targets for environmental protection and/or herbage availability for animal feeding. They could be better positioned to know, for instance, whether or not to apply fertiliser, and if so, when and how much to use, so that the desired outcome (extra feed for animals) occurs at the time it is most needed. Processors may be able to forecast rates of supply for products coming largely from pasture feeding within their procurement regions, and adjust processing capacity and forward orders accordingly. Outcomes like these can add significantly and directly to economic and environmental outcomes, and it is tempting to argue that this should ensure the widespread adoption of such tools. However, history shows this is not a safe assumption, and our knowledge of the causes of persistent poor implementation of DSS products in farm businesses (McCown, 2002) will need to be applied to the solution of the problem for these applications, just as for others.

Optimal species selection

Oregon State University, Penn State University, and Cornell University are developing forage species selection tools. Cornell's forage selection tool (http://www.forages.org/) is made up of several programs which access numerous databases to provide forage species recommendations for New York State, taking into consideration both the available soil type and the intended forage use. Soil type can be selected from a list, or the program can estimate soil type based on zipcode, county, and basic soil characteristics. Similar to Cornell's product, Penn State's selection tool (http://www.forages.psu.edu/selection_tool/index.html) is based primarily on soil information and intended use. The Species Selection Information System at OSU (http://forages.oregonstate.edu/is/ssis/) is a collaborative project to create a web-based, comprehensive knowledge resource to assist land managers and other decision makers in choosing forage and conservation species that are optimally matched with their environment. OSU has integrated climate and soil information and a quantitative tolerances approach to matching species, including intended use and management level. All of these systems will improve as the underlying geospatial information improves and our knowledge of the quantitative ecology of the various species is improved.

Developing country challenges

Developing country challenges are often referred to as 'the digital divide' and include the entire spectrum of needs for computer infrastructure, up-to-date equipment, software, and trained professionals. One important issue is that of proprietary vs. 'open source' software. Developing countries often cannot afford the high price of proprietary software and license fees imposed by the major software companies. Since wages are much lower than in developed countries, time is often invested to develop similar function software. Although this requires less cash expenditure, it is an inefficient use of workers compared to joint development and improvement of 'open source' software.

Development of integrated solutions is often impeded by the cost of data or the regulations preventing the sharing or sale of base layer information (topography, climate, soils, rivers, roads, boundary files, etc.). Significant time is often spent in efforts to assemble data, quality checking, and putting the data in usable form. National and international geospatial data archives are needed in which information can be freely obtained or purchased for a reasonable price.

Well-trained scientists are needed for the development and use of geospatial approaches to agriculture and natural resource management. Strong university programs are a long-term approach that is needed, but international training programs and workshops of weeks to months in length are necessary to meet the immediate demand for trained professionals. On going needs for infrastructure and equipment upgrades need to be built into project budgets. International donors can be helpful in providing these infrastructure needs and training requirements.

Future developments and applications

Historical applications have included crop and rangeland simulation modelling, traditional field trials, and extension outreach provided by field days and extension education programs. Current GIS and RS-based technologies allow users to: 1) integrate information from multiple disciplines, 2) cost-effectively extend traditional point-based trial data, and 3) visualise spatial information across the landscape. Future applications will involve better integration of the various land-based and space-based technologies with the many disciplines that affect management (physiography, physical processes, eco-physiology, socio-economic, and cultural anthropological factors).

A probable development over the next 5 years will be the more routine use of geospatial approaches to agricultural and natural resource management. Refinements in spatial information for topography, climate, soils, and vegetation will allow better selection of crops, simulation of yields and quality, more precise production management, and recommendations for policy and economically and environmentally sustainable systems. Socio-economic elements that are today too often ignored in computer application product development will be more widely included as techniques are developed to routinely analyse, model, and map these factors. Both the 'bottom line' economics and the decision-making process will be included in integrated products.

Current difficulties of obtaining data, quality checking, and spatially modelling data will be more readily resolved due to extensive use of low-cost ground-based and space-based sensors. Basic and applied science fields will work together more seamlessly as cooperating scientists from around the world grasp the mutual benefit of such collaboration. 'Market analysis' approaches will be used and partnerships developed with end user groups (including agricultural lenders) to ensure that the product meets the user's needs. Widespread availability of high-speed networks will eliminate the need to limit products to currently available equipment or transmission speed problems. Focus will be on developing creative solutions to important natural resource management issues.

References

Australian Natural Resources Atlas v2.0 (2004).
   <http://audit.ea.gov.au/anra/land/land_frame.cfm?region_type=AUS&region_code=AUS&info=soil_asris>
Baker, B.B., R.M. Bourdon & J.D. Hanson (1992). FORAGE: a simulation model of grazing behaviour for beef cattle. *Ecological Modelling*, 60, 257-279.
Bolte, J.P., D.B. Hannaway, P.E. Shuler & P.J. Ballerstedt (1991). An intelligent frame system for cultivar selection. *AI Applications in Natural Research Management*, 5, 21-31.
Boote, K.J., J.W. Jones & G. Hoogenboom (1997). Simulation of crop growth: CROPGRO model. In: R.M. Peart & R.B. Curry (eds.) Agricultural Systems Modelling. Marcel Dekker, New York, 651-692.
Boote, K.J., J.W. Jones, G. Hoogenboom & N.B. Pickering (1998). The CROPGRO model for grain legumes. In: G.Y. Tsuji, G. Hoogenboom & P.K. Thornton (eds.) Understanding options for agricultural production. Systems approaches for sustainable agricultural development. Kluwer Academic Publishers, Dordrecht, The Netherlands, 99-128.
Brinkley, T.R., G.P. Laughlin, M.F. Hutchinson & K. Ranatunga (2005). GROWEST PLUS – a tool for rapid assessment of seasonal growth for environmental planning and assessment. *Environmental Modelling & Software*, 20, (in press).
Cacho, O.J., A.C. Bywater & J.L. Dillon (1999). Assessment of production risk in grazing models. *Agricultural Systems*, 60, 87-98.
Clark, S.C., E.A. Austen, T. Prance & P.D. Ball (2003). Climate variability effects on simulated pasture and animal production in the perennial pasture zone of south-eastern Australia. Between year variability in pasture and animal production. *Australian Journal of Experimental Agriculture*, 43, 1211-1219.

Daly, C., W.P. Gibson, G.H. Taylor, G.L. Johnson & P. Pasteris (2002). A knowledge-based approach to the statistical mapping of climate. *Climate Research*, 22, 99-113.

Daly, C. & D.B. Hannway (2004). Suitability zone mapping for forage species using GIS technologies. In: *Proceedings of the 4th International Conference on the Tibetan Plateau*, Lhasa, Tibet, 187-188.

Daly, C., E.H. Helmer & M. Quinones (2003). Mapping the climate of Puerto Rico, Vieques, and Culebra. *International Journal of Climatology*, 23, 1359-1381.

Daly, C., R.P. Neilson & D.L. Phillips (1994). A statistical-topographic model for mapping climatological precipitation over mountainous terrain. *Journal of Applied Meteorology*, 33, 140-158.

Daly, C., G.H. Taylor, W.P. Gibson, T.W. Parzybok, G.L. Johnson & P. Pasteris (2001). High-quality spatial climate data sets for the United States and beyond. *Transactions of the American Society of Agricultural Engineers*, 43, 1957-1962.

Donnelly, J.R., A.D. Moore & M. Freer (1997). Decision support systems for Australian grazing enterprises. Overview of the GRAZPLAN project and a description of the MetAccess and LambAlive DSS. *Agricultural Systems*, 54, 57-76.

Eastwood, B. (1996). Agricultural databases for decision support.
<http://www.reeusda.gov/agsys/adds/addshome.htm>

Engel, T., G. Hoogenboom, J.W. Jones & P.W. Wilkens (1997). AEGIS/WIN - a computer program for the application of crop simulation models across geographic areas. *Agronomy Journal*, 89 (6), 919-928.

Enzlberger, C. (2001). Soil types databases. IIASA-LUC GIS Database.
<http://www.iiasa.ac.at/Research/LUC/GIS/soil_types.htm>

Finlayson, J.D., O.J. Cacho & A.C. Bywater (1995). A simulation model of grazing sheep. Animal growth and intake. *Agricultural Systems*, 48, 1-25.

Foy, J.K., W.R. Teague & J.D. Hanson (1999). Evaluation of the upgraded SPUR model (SPUR2.4). *Ecological Modelling*, 118, 149-65.

Gherardi, S. & C. Oldham (2003). The value proposition for remotely sensed estimates of feed on offer and pasture growth rate. In: *Proceedings of the 1st Joint Conference of the Grassland Society of Victoria, and the Grassland Society of New South Wales*, Grassland Society of Victoria, Mornington, Australia, 19-21.

Green, J.L. & D.B. Hannaway (1996). USDA CSREES global information systems for decision support: using technology to work smarter. <http://forages.oregonstate.edu/projects/gisds/GISDS_Concept_Paper.html>

Hannaway, D.B., S. Griffith, P. Hoagland, M. Runyon & R. Fisher (1997). Forage information systems on the world wide web. In: *Proceedings of the 18th International Grasslands Congress*, Saskatoon, Canada, June 8-19, 24-5,6.

Hannaway, D.B., J.P. Bolte, P.E. Shuler & P.J. Ballerstedt (1992). ACE: Alfalfa cultivar expert. *Journal of Production Agriculture*, 5, 85-88.

Hannaway, D.B., C. Daly, W. Gibson, G. Taylor, D. Johnson, H. Wan, W. Luo, L. Liu, X. Han, L. Feng, Y. Wei, Y. Hu, L. Gao, X. Li, W. Zhang, X. Yang, Z. Xu & A. Gu (2000). GIS-based forage species adaptation mapping for China. International Symposium on Intelligent Agricultural Information Technology, December 1-4, Beijing, China, 257-259.

Hannaway, D.B., K.J. Hannaway, P. Sohn, S. Griffith, B.E. Avery, E. Nowick, W.F. Wedin, L.R. Vough, S.C. Bosworth, G.D. Lacefield, G.E. Bates, D. Undersander, N.P. Martin, J. Caddel, G.L. Kilgore, S.B. Orloff, A.M. Gray, R. Ditterline, T. Griggs, D.H. Putnam & M.J. Ottman (2001). Developing a national alfalfa information system. In: *Proceedings XIX International Grassland Congress*, São Pedro, Brazil, 1069-1070.

Hannaway, K.J., D.B. Hannaway, P.E. Shuler, M.L. Niess, S. Griffith, G.W. Fick & V.G. Allen (1999). World wide web curriculum design using national collaboration. *Journal of Natural Resources and Life Sciences Education*, 28, 59-62.

Hanson, J.D., B.B. Baker & R.M. Bourbon (1992). SPUR II model description and user guide: GPSR technical report No. 1. USDA-ARS, Great Plains Systems Research Unit, Ft. Collins, Colorado, 24pp. (1 diskette).

Held, A. & M. Hill (2003). Satellites and Australia. In: *Proceedings of the 1st Joint Conference of the Grassland Society of Victoria, and the Grassland Society of New South Wales*, Grassland Society of Victoria, Mornington, Australia, 7-11.

Henry, D. (2002). Measuring pastures from space. In: *Proceedings of 43rd Grassland Society of Victoria Conference*, Morwell, Victoria, June 13-14, 91-94.

Hill, M.J., G.E. Donald, P.J. Vickery, A.D. Moore & J.R. Donelly (1999). Combining satellite data with a simulation model to describe spatial variability in pasture growth at a farm scale. *Australian Journal of Experimental Agriculture*, 39, 285-300.

Holmes, C.W., I.M. Brookes, D.J. Graaick, D.D.S. MacKenzie, T.J. Parkinson & G.F. Wilson (2002). Milk production from pasture: principles and practices. Massey University, NZ, 602pp.

Hoogenboom, G. (2000a). Future of canegro in the DSSAT suite of models. In: *Proceedings of International CANEGRO Workshop*, Mount Edgecombe, South Africa, August 4-7, 9-12.

Hoogenboom, G. (2000b). Contribution of agrometeorology to the simulation of crop production and its applications. *Agricultural and Forest Meteorology*, 103 (1-2), 137-157.

Hoogenboom, G., P.W. Wilkens, P.K. Thornton, J.W. Jones, L.A. Hunt & D.T. Imamura (1999). Decision support system for agrotechnology transfer, Ver. 3.5, 1-36. In: G. Hoogenboom, P.W. Wilkens & G.Y. Tsuji (eds.) DSSAT Version 3, Volume 4. University of Hawaii, Honolulu, Hawaii. <http://www.icasa.net/dssat>

Jeffrey, S.J., J.O. Carter, K.B. Moodie & A.R. Beswick (2001). Using spatial interpolation to construct a comprehensive archive of Australian climate data. *Environmental Modelling & Software*, 16, 309-330.

Johnson, I.R., D.F. Chapman, A.J. Parsons, R.J. Eckard & W.J. Fulkerson (2003a). DairyMod: a biophysical model of the Australian dairy system. In: *Proceedings of the Australian Farming Systems Conference*, Toowoomba, Australia. <http://afsa.asn.au/>

Johnson, I.R., G.M. Lodge & R.E. White (2003b). The sustainable grazing systems pasture model: description, philosophy, and application to the SGS national experiment. *Australian Journal of Experimental Agriculture*, 43, 711-728.

Kenny, S., D.F. Chapman & D. Beca (2005). Alternative feedbase systems for southern Australia dairy farms. 2. Seasonal variability. In: *Proceedings of the XX International Grassland Congress*, Dublin, offered paper (in press).

Laughlin, G.P., H. Zuo, J. Walcott & A. Bugg (2003). The rainfall reliability wizard – a new tool for assessing rainfall reliability with examples for the Australian wheat belt. *Environmental Modelling & Software*, 18, 49-57.

McCall, D.G., P.R. Marshall & K.L. Johns (1991). An introduction to Stockpol: a decision support model for livestock farms. In: *Proceedings of the International Conference on Decision Support Systems for Resource Management*, Texas A&M University, 27-30.

McCown, R.L. (2002). Changing systems for supporting farmers' decisions: problems, paradigms and prospects. *Agricultural Systems*, 74, 179-200.

McKeon, G.M., K.A. Day, S.M. Howden, J.J. Mott, D.M. Orr, W.J. Scattini & E.J. Weston (1990). Northern Australian savannas: management for pastoral production. *Journal of Biogeography*, 17, 355-372.

Miller, D.A. & R.A. White (1998). A conterminous United States multilayer soil characteristics dataset for regional climate and hydrology modeling. *Journal of Earth Interactions*, 2 (2), 1-26.

Milligan, K.L., I.M. Brookes & K.F. Thompson (1987). Feed planning on pasture. In: A.M. Nicol (ed.) Livestock feeding on pasture. New Zealand Society of Animal Production, Occasional Publication No. 10, 75-88.

Mohtar, R.H. (2004). Grazing simulation model. Penn State University. University Park, PA. <http://danpatch.ecn.purdue.edu/~grasim/grasim_intro.html>

Moore, A.D., J.R. Donnelly & M. Freer (1997). GRAZPLAN: Decision support systems for Australian grazing enterprises. III. Pasture growth and soil moisture submodels, and the GrassGro DSS. *Agricultural Systems*, 55, 535-582.

National Environmentally Sound Production Agriculture Laboratory (NESPAL) (2004). Precision Agriculture. University of Georgia. Tifton, GA. <http://nespal.cpes.peachnet.edu/PrecAg>

Nicholls, N., W. Drosdowsky, B. Lavery (1997). Australian rainfall variability and change. *Weather*, 52, 67-72.

Ojima, D.S. (2000). Integrated analysis of climate change impacts on land use in temperate East Asia (LUTEA): Integration of ecosystem and economic factors determining land use under climate change. In: *Proceedings of the Land Use in Temperate East Asia Workshop: Current Status and Future Trends*, 1-18.

Pierson, F.B., D.H. Carlson & K.E. Spaeth (2001). A process-based hydrology submodel dynamically linked to the plant component of the simulation of production and utilization on rangelands SPUR model. *Ecological Modelling*, 141, 241-60.

Plantico, M.S., L.A. Goss & C. Daly (2000). A new U.S. climate atlas. In: *Proceedings of the 12th AMS Conference on Applied Climatology*, American Meteorological Society, Asheville, NC, 247-248.

Rasmussen, W.O., C.T.K. Ching, L.A. Linden, P.A. Myer, V.P. Rasmussen Jr., R.S. Rauschkolb & C.B. Travieso (1985). Computer applications in agriculture. Westview Press. Boulder, CO, 143pp.

Reybold, W.U. & G.W. Tesselle (1989). Soil geographic databases. *Journal of Soil and Water Conservation*, 43, 226-229.

Ritchie, J.T., U. Singh, D.C. Godwin & W.T. Bowen (1998). Cereal growth, development and yield. In: G.Y. Tsuji, G. Hoogenboom & P.K. Thornton (eds.) Understanding Options for Agricultural Production. Kluwer Academic Publishers, Dordrecht, The Netherlands, 79-98.

Rotz, C.A. (2001). The dairy forage system model (DAFOSYM). In: D. Fox & C. Rasmussen (eds.) Developing and applying next generation tools for farm and watershed nutrient management to protect water quality. Animal Science Department Mimeo 220, Cornell University, Ithaca, NY, 30-42. <http://pswmru.arsup.psu.edu/software/dafosym.htm>

Salmon, L., J.R. Donnelly, A.D. Moore, M. Freer & R.J. Simpson (2004). Evaluation of options for production of large lean lambs in south-eastern Australia. *Animal Feed Science and Technology*, 112, 195-209.

Salmon, E.M & A.D. Moore (2001). Adoption and influence: industry evaluation of the grassgro decision support tool. In: *Proceedings of the 19th International Grassland Congress*, Sao Paulo, Brazil, 1073-1075.

Schut, P. (2000). Canadian soil information system. Agriculture and Agri-Food Canada. <http://sis.agr.gc.ca/cansis/>

Shafer, W. (2003). The Colorado beef cattle production model: effects of simulation with realistic levels of variability and extreme within-herd diversity. Ph.D. Dissertation. Colorado State University, Fort Collins, CO, 281pp.

Sheath, G.W. & D.A. Clark (1993). Management of grazing systems: temperate pastures. In: J. Hodgson & A.W. Illius (eds.) Ecology and management of grazing systems. CAB International, Wallingford, UK, 301-324.

Shi, X.Z., D.S. Yu, E.D. Warner, X.Z. Pan, G.W. Petersen, Z.G. Gong & D.C. Weindorf (2004a). Soil database of 1:1,000,000 digital soil survey and reference system of the Chinese genetic soil classification system. *Soil Survey Horizons*, 45 (4), 129-136.

Shi, X.Z., D.S. Yu, W.X. Sun, H.J. Wang, Q.G. Zhao & Z.T. Gong (2004b). Reference benchmarks relating to great groups of genetic soil classification of China with soil taxonomy. *Chinese Science Bulletin*, 49 (14), 1507-1511.

Skirvin, S.M. & M.S. Moran (2003). Rangeland ecological and physical modelling in a spatial context. In: K.G. Renard, S.A. McElroy, W.J. Gburek, H.E. Canfield & R.L. Scott (eds.) *Proceedings of the First Interagency Conference on Research in the Watersheds*, U.S. Department of Agriculture, Agricultural Research Service, 451-454.

Sonka, S.T., M.E. Bauer, E.T. Cherry, J.W. Colburn, R.E. Heimlich, D.A. Joseph, J.B. Leboeuf, E. Lichtenberg, D.A. Mortensen, S.W. Searcy, S.L. Ustin & S.J. Ventura (1997). Precision agriculture in the 21st Century. National Academy Press. Washington, D.C. <http://books.nap.edu/html/agriculture/>

Stafford-Smith, D.M. & G.M. McKeon (1998). Assessing the historical frequency of drought events on grazing properties in Australian rangelands. *Agricultural Systems*, 57, 271- 299.

Stuth, J.W., W.T. Hamilton, J.C. Conner & D.P. Sheehy (1993). Decision support systems in the transfer of grassland technology. In: M.J. Baker (ed.) Grasslands for our world. SIR Publishing, Wellington, NZ, 234-242.

Teague, W.R. & J.K. Foy (2002). Validation of SPUR2.4 rangeland simulation model using a cow-calf field experiment. *Agricultural Systems*, 74, 287-302.

Vickery, P.J. & D.A. Hedges (1987). Use of Landsat MSS data to determine the fertiliser status of improved grasslands. In: D. Brare (ed.) *Proceedings of the 4th Australian Remote Sensing Conference*, Remote Sensing and Photogrammetry Association of Australia, Perth, 287-296.

Vickery, P.J., M.J. Hill & G.E. Donald (1997). Landsat derived maps for pasture growth status: association of classification with botanical composition. *Australian Journal of Experimental Agriculture*, 37, 547-562.

Wastney, M.E., C.C. Palliser, J.A. Lile, K.A. Macdonald, J.W. Penno & K.P. Bright (2002). A whole-farm model applied to a dairy system. In: *Proceedings of the New Zealand Society of Animal Production*, 62, 120-123.

Webby, R.W. (2002). The value of decision support models for farmer learning. *Proceedings of the New Zealand Grassland Association*, 64, 45-47. <http://www.grassland.org.nz/Volume64.aspx>

White, D.H., P.J. Bowman, F.H.W. Morley, W.R. McManus & S.J. Filan (1983). A simulation model of a breeding ewe flock. *Agricultural Systems*, 10, 149-189.

Wight, J.R. & J.W. Skiles (1987). SPUR: simulation of production and utilization of rangelands. Documentation and user guide, ARS 63. U.S. Department of Agriculture, ARS.

Paying for our keep: grasslands decision support in more-developed countries

A.D. Moore

CSIRO Plant Industry, GPO Box 1600, Canberra 2601, Australia
Email: Andrew.Moore@csiro.au

Key points

1. A survey of decision support (DS) tools in grassland agriculture illustrates the diversity of decisions supported and of delivery technologies that are used. Larger, whole-enterprise planning tools are undergoing a period where their user interfaces are being adapted to better reflect the requirements and practice of advisory users.
2. The history of use of GrassGro, a 'versatile simulator', is used to illustrate how versatile tools attract a diverse range of users and uses. Lessons learnt by the GrassGro team are discussed.
3. Uptake rates of DS tools in grasslands are generally lower than was expected a decade ago. Nevertheless, if return on investment is used as the criterion then some DS tools – especially smaller ones – are clearly successful. Uptake for educational use can be much more rapid.

Keywords: GrassGro, planning tools, versatile simulators

Introduction

"Models in grassland science have come to stay, and so they must be made to pay for their keep. To do that, it will be necessary to improve their scientific sophistication and their management relevance..." (Seligman, 1993).

Decision support (DS) tools in grassland agriculture have also come to stay. This review will examine the ways and means by which DS tools are being made to pay for their keep, with a focus upon DS efforts from the last decade and on the lessons that arise for those who develop DS tools for grasslands. A survey of the present state of grassland DS is presented, and a case made that there are grounds for modest optimism. As a more detailed case study, the history and impact of GrassGro (a simulation-based DS tool that has been in use since 1997; Moore *et al.*, 1997) will be revisited. Lastly, some trends in technology and practice that will determine the effectiveness of DS tools in the coming decade will be examined. Throughout the review, the focus will be on those parts of the world where agriculture is well developed; consequently the term 'grasslands' should be read with this geographic caveat. A companion paper (Donnelly *et al.*, 2005) addresses a number of technical issues in DS tool development and expands some points that are only touched on here.

Current trends in grassland decision support: a survey

The following survey of grassland DS tools is intended to be illustrative rather than complete. The selected tools show the diversity of adoption strategies and technologies that have been utilised in recent years, and general lessons for the practice of DS tool development. However, the survey is restricted to tools that have been released at the time of writing, and consequently a number of problem domains – especially provision of advice to policy makers – receive little attention.

'Niches that remain promising': McCown's critique of agricultural decision support

In an important paper concluding a special issue of *Agricultural Systems*, McCown (2002) applied theory from the wider discipline of management science to a series of case studies of agricultural DS efforts. He proceeded from a position that DS in agriculture has fallen far short of expectations. His analysis posited that the future of agricultural DS lies in only four directions:
- small tools ('decision calculi') that assist a decision-maker with a self-contained and highly structured task;
- use of models by consultants to aid farmers with both tactical and strategic decision-making;
- use of models as the basis of a facilitated learning process (a 'flight simulator'); and
- provision of formal frameworks that support regulatory objectives in constraining and documenting agricultural practice.

The second and third of these uses depend upon scientific knowledge being embodied in what McCown calls a 'versatile simulator'. A simulator is differentiated from a traditional research model as follows: "the primary aim in making a simulator is not to mimic system *process* but rather system *function* and *performance*, and to do so cost-effectively. Flexibility and ease of specification are key simulator attributes."

This author does not entirely agree with McCown's assessment of agricultural DS nor with his view of the feasible paths to success. The case studies on which he bases this conclusion are mainly concerned with cropping systems, and the authors of the two grazing-lands case studies to which he refers (Donnelly *et al.*, 2002; Stuth *et al.*, 2002) are conspicuously more optimistic. Nevertheless, McCown's categories provide a useful starting point for examining, and in many cases understanding, recent developments amongst grassland DS tools.

The decision calculus in grasslands: feeding, fertilizer and forage choice

There are a wide variety of small DS tools available for graziers in the more-developed countries. These tools are engineering rather than scientific artefacts, so literature citations are uncommon; this section is therefore mainly based on a survey of the World Wide Web and on personal communication with developers and distributors. Internet addresses for the tools discussed here are given in Table 1.

Tactical animal nutrition - Full accounts of the GrazFeed and NUTBAL animal nutrition tools can be found in Donnelly *et al.* (2002), and Stuth *et al.* (2002); these are well-established products that are now in a period of incremental change and accruing benefit to industry. They have been joined by a number of other tools, in particular a DS implementation of the Cornell nutritional model (Fox *et al.*, 2004).

Fertiliser and manure - It is not surprising that choice of fertiliser and application rate is a decision area with considerable activity. The fertiliser tools all follow a common operating method: specification of soil attributes, pasture type and time of year is followed by the computation of a static yield model, with regulatory constraints taken into account where necessary.

Table 1 Some decision calculi for grazing lands, and their location on the World Wide Web

Tool	Purpose	Website (or citation if no site)
GrazFeed	Nutrition of sheep & cattle	www.hzn.com.au/grazfeed.htm
NUTBAL	Nutrition of sheep & cattle	cnrit.tamu.edu/ganlab
CNCPS	Nutrition of sheep & cattle	www.cncps.cornell.edu/cncps/main.htm
Diet Check	Nutrition of dairy cattle	Heard *et al.* (2004)
N Decision Tools	N fertilizer rates	www.nitrogen.unimelb.edu.au
EMA Fertilizer	Fertilizer & manure rates[1]	www.herts.ac.uk/natsci/Env/aeru/ema/
Cayley & Kearney	P fertilizer rates	Cayley & Kearney (2000)
fencepost.com.nz	Fertilizer rates[1]	www.fencepost.com/fertiliser/ft_main.jhtml
Phosphorus for sheep & beef pastures		www.dpi.vic.gov.au/dpi/nrenavh.nsf
M-CLONE4	Manure management	farmcentre.com/english/farmsoftware/start.htm
NMAN	Manure management	www.gov.on.ca/OMAFRA/english/nm/nman/default.htm
OVERSEER	Farm nutrient budgeting	www.agresearch.co.nz/overseerweb/
Forage Species Selection		www.forages.org
Greenhouse Accounting Tools		www.greenhouse.crc.org.au/calculators
StockPlan	Drought management	McPhee *et al.* (2003)
Pl@nteInfo	Yield prediction[†]	www.planteinfo.dk

[1]Part of a larger tool or suite

They are notable, however, for the variety of technologies and distribution channels that they use. Stand-alone software, Web applets, spreadsheets and paper-based systems are all employed, as are distribution via commercial CD (EMA), Internet download (N Decision Tools) and mass mailing (the tool of Cayley & Kearney, 2000). The Nitrogen Decision Tools product (Eckard *et al.*, 2001) uses three different means of deployment, and so provides an interesting example of the way that different user groups exhibit preferences for different means of dissemination. Dairy farmers – the ultimate decision makers – prefer lookup tables, while their advisors tend to prefer a downloadable spreadsheet. For this small tool, online calculation is not favoured by users (R.J. Eckard, *pers. comm.*).

The other side of nutrient management – managing excess nutrients (especially in manure) has also seen considerable DS effort. Tools from New Zealand (OVERSEER, Wheeler *et al.*, 2003), and from the United Kingdom (the nutrient budgeting module of the EMA product, Lewis & Tzilivakis, 2000) are decision calculi: they seek to inform producers about the nutrient balance of their enterprise, not to control it. In Ontario, on the other hand, the decision calculus approach has been overtaken by the fourth category identified by McCown (2002): use of DS tools to satisfy regulatory demand. The MCLONE4 software, developed in Ontario, is a decision calculus for use by advisors. Uptake of MCLONE4 has virtually ceased since 2002, and been replaced by NMAN, a tool specifically designed for the preparation of mandated nutrient management plans. Some of the science used in MCLONE4 has been re-used in NMAN (J. Ogilvie, *pers. comm.*).

Forage species choice - While it is not widely used, the Forage Species Selection Tool developed at Cornell University demonstrates an important technological opportunity for decision calculi. By employing a database (in this case a spatial data set of soil attributes linked to maps and aerial photography), the selection tool tailors its information to specific circumstances (a particular field); because it is deployed across the Internet, the application can proceed despite the large size of the underlying database.

In a previous review, Donnelly & Moore (1999) suggested that many small tools would be oriented to learning and designed to have a limited lifespan. There is some evidence of the former (e.g. StockPlan, McPhee *et al.*, 2003), but the latter does not seem to have happened. Instead, decision calculi tend to be archived on the Internet, awaiting occasional future users at minimal cost. An example of this phenomenon is the greenhouse gas calculator tool of R.J. Eckard and colleagues (Table 1).

Information bases for decision support

A significant number of structured information bases for grassland agriculture have emerged in recent years. Examples include the Forage Information System (forages.oregonstate.edu), Agricultural Databases for Decision Support (www.adds.org) and the Agricultural Document Library module of EMA (Lewis & Tzilivakis 2000; www.herts.ac.uk/natsci/Env/aeru/ema/). These tools require a capacity to distribute and access large quantities of information; the compact disc seems to have been the initial enabling technology, but a clear shift toward Internet delivery is now under way. These tools are vehicles for self-directed, rather than participative, learning. Knowledge is expressed in the form of documents rather than models, which may be why McCown (2002) did not take them into account. The EMA example is of particular interest as it once again shows the value of placing decision calculi (fertiliser and pesticide application tools) into a learning context (the information base).

Tactical grazing management: the problem child of decision support

Allocating resources to optimise production was one of the earliest problems addressed by operations researchers. The classic problem of this kind in grassland agriculture is feed budgeting: the tactical allocation of grazing livestock to fields and other sources of feed. Given the obvious potential for benefit to graziers, it is no surprise that a number of DS efforts have been devoted to this decision area.

The success of these efforts has been mixed at best. Rickert (1998) reports the FEEDMAN project for Queensland graziers as a failure: despite a genuinely participative approach to development, uptake was low and FEEDMAN now appears to be in Internet archive. Rickert (1998) explains this failure in terms of insufficient ongoing commitment by the developing organizations. The technically very similar PRO PLUS tool (McPhee *et al.*, 2000) has been successful within a limited market: an evaluation showed that 60% of graziers provided with PRO PLUS continued to use the software (A.K. Bell, *pers. comm.*). Release of PRO PLUS has been deliberately restricted to participants in a grazing management-training course. This has allowed the developers to find a niche market of users, whose attitude and aspirations predispose them to make use of the technology. The contrasting history of these two tools illustrates two lessons of general importance in DS tool construction: that participative development processes are a necessary but not a sufficient condition for success, and that embedding a tool in a learning process increases the likelihood of success.

The evaluation of PRO PLUS made it clear that a major reason for non-adoption was the time cost involved. Feed budgeting and feed planning have two major costs to a grazier: time spent in first describing the enterprise, and ongoing time spent monitoring pasture and livestock. Below a certain economic enterprise size, the setup cost makes these tools impractical; above a certain geographic size, the cost of monitoring is prohibitive. This explains why New Zealand, with its high ratio of economic size to geographic size, is where feed budgets are most popular.

Whole-enterprise planning: supporting the consultant

The four strategic planning tools in Table 2 share two important features. All are the outcome of DS efforts continuing over more than a decade; and they have all undergone extensive user interface changes in recent years. StockPol (Marshall *et al.*, 1991) has transformed into a new service called FARMAX; the GPFARM tool (Shaffer *et al.*, 2000) is being upgraded to accommodate risk analysis under the name iFARM (L. Ma, *pers. comm.*); GLA has been ported to a Web-server application, webGLA (Stuth *et al.*, 2002). There is a common thread to these changes: a closer alignment of the tools with the requirements and practice of advisors.

In the FARMAX project, the simulator underlying StockPol has been implemented within two distinct interfaces: a strategic-planning interface for consultancy use, and a simplified interface that is intended for use by the grazier. Both interfaces share common data files, which are set up in an advisor-mediated strategic planning process. Grazier use of FARMAX is therefore critically dependent upon the prior involvement of an advisor; this connection underpins the FARMAX business model. From a technological point of view, this re-use of a model within different interfaces parallels the approach of the GRAZPLAN group (Donnelly *et al.*, 2002).

Table 2 Some whole-enterprise planning tools for grasslands

Tool	Origin	Purpose and website
FARMAX	New Zealand	Farm planning service: flock and herd structure, seasonal decision-making (paddock allocation, livestock trading). www.farmax.co.nz
GrassGro	Southern Australia	Evaluation of various strategic and tactical management options for sheep and cattle enterprises (see below) www.hzn.com.au/grassgro.htm
iFARM	Great Plains, USA	Analysing 10-50 year production plans for mixed cropping & livestock enterprises: water, nutrient and pest management. Gpsr.ars.usda.gov
webGLA	Texas	Grazing Lands Application: conservation planning, inventory of forage resources and design of grazing plans.

The webGLA application differs in two important ways from the rest of the tools in Table 2. First, despite its evident viability and although it is intended for consultancy use, it is not centred on a simulation model; it therefore does not fit the typology of potentially successful DS tools identified by McCown (2002). Second, it is unique in being implemented as a Web application. Both the unique structure and implementation arise because GLA exploits the information in a set of extensive databases of plant and land attributes.

GrassGro: a versatile simulator encounters versatile users

GrassGro (Moore *et al.,* 1997) is a DS tool designed to evaluate various strategic and tactical management options for sheep and cattle enterprises. It incorporates simulation models of the grazing ruminant (Freer *et al.*, 1997), soil moisture, growth of multiple pasture species and enterprise management (Moore *et al.*, 1997). Users describe a grazing system in terms of the models and then conduct simulations, from which they draw their own conclusions as to the best course of action to follow. GrassGro is part of a larger effort that included the unequivocally successful GrazFeed DS tool for tactical nutrition of sheep and cattle (Freer *et al.*, 1997; Donnelly *et al.*, 2002).

GrassGro fits squarely within the definition of a versatile simulator (McCown, 2002). Its original intended use was as a tool for consultants, and during development of the software a great deal of effort was expended in building an interface that would make GrassGro 'easy' for consultants to use. Since its release the GRAZPLAN group has trained about 200 people to use GrassGro, most from the high-rainfall zone of southern Australia or the wetter parts of the cereal-livestock zone. Figure 1 shows the growth in user numbers; the distinct pulses are associated with the release of new versions of the software. Practically all users belong in one of three groups: advisors, university educators and researchers who use GrassGro for scientific or policy applications. Different lessons have emerged from our interaction with these distinct user groups.

Advisors

When GrassGro was released, it was intended that advisors would be the primary group of users. Figure 1 shows that this expectation has not been borne out; only 40% of those trained have been advisors. About half the advisors trained in the last two years have been from the private sector, compared with one in four during the first two years.

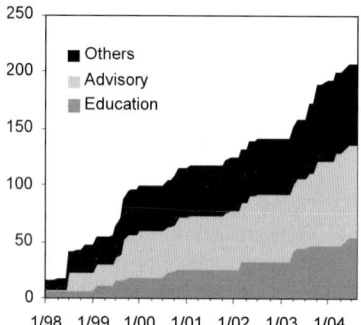

Figure 1 Growth in the number of trained GrassGro users since release

It was recognized that GrassGro users would require significant training, and so a workshop-based process was adopted in which small groups of users gathered twice. At the first workshop, users were exposed to the underlying models – especially the limits of their applicability – and to the software interface. During the interval between workshops, each new user conducted an analysis with GrassGro that was subsequently shared with the group during the second workshop. The second workshop also provided an opportunity to clear up any technical problems that users had encountered. A dedicated GrassGro specialist conducted all workshops, and over time a 'resource kit' for GrassGro users has evolved containing printed help and technical information.

A detailed account of the process by which GrassGro was evaluated is provided by Donnelly *et al*., (2002). These evaluations facilitated the identification of key needs of advisory users that were not met by the GrassGro package:

- Simulations needed to match the clients' situation more closely. Users requested that more management options be available, and that more pasture species be included for use in simulations. Successive releases of GrassGro have incrementally added management options such as more flexible descriptions of grazing management policies, and new parameter sets have been added to the underlying library of pastures (although fewer than demanded). The lesson learnt is that if a tool is versatile, users will always want more from it.
- An imbalance in the training was identified: its focus needed to shift away from technical issues and toward imparting the general technique of using simulation analysis to solve problems. With hindsight, it can be seen that this problem arose because experience gained with GrazFeed (a decision calculus) led the development team to assume that advisors would see for themselves how to use GrassGro (a versatile simulator). To remedy this problem, the training workshops and the resource kit now concentrate much more on the steps of specifying, analysing and reporting on a problem.
- While the 'easy-to-use' software interface was consistently well rated by users, drawbacks emerged when it was put into practice. Those who resumed working with GrassGro after a break often found that re-establishing their skills was difficult. Much of the work of converting simulation results into reports for clients had to be repeated manually, using up time that is an advisor's most valuable resource. These drawbacks have not stopped advisors from using GrassGro, but they have made its use less efficient.

Market research prior to release would not have identified these issues. Given the fragmented nature of the market and the fact that advisors have to learn how to exploit GrassGro, it is clear that the strategy to release the product, get the simulator into use and then to proceed by iterative prototyping (as described by Stuth *et al*., 1993) was the correct one to follow. The GrassGro interface is currently being completely re-developed to focus on the problem-solving process. The main element of the interface no longer represents a single simulation; instead it becomes an 'analysis', which is a *set* of simulations coupled with pre-defined reports (Figure 2). Locally specific combinations of initial conditions are stored as 'scenarios' that can be added to an analysis in a single action. Analyses can therefore be re-used quickly across different circumstances. A user can design an analysis and then transmit it to others, further increasing the value gained from the time spent designing it.

Educators

Tertiary educators form a quarter of the GrassGro user group, and are probably the most active users. This unexpected but welcome outcome is largely due to Jim Scott of the University of New England (UNE), who saw the opportunity and started developing a project to introduce GrassGro to UNE's undergraduate programme within months of GrassGro's release (Daily *et al*., 2000). GrassGro is now used at all levels of the UNE agricultural science degree and in both agronomy and animal production. Teaching exercises with GrassGro have proven to be a unique way of conveying a systems perspective to students. A second project (Daily *et al*., 2005) has disseminated the expertise developed at UNE to other institutions and deployed GrassGro through an innovative internet-based distribution channel. By 2004, over 500 students were learning from GrassGro at 8 of the 11 Australian institutions with tertiary courses in agriculture. The public investment in educational use of GrassGro now approaches the size of the investment in supporting its use by advisors. At the time of writing, the first GrassGro-aware graduates are beginning to appear amongst advisors and their clients.

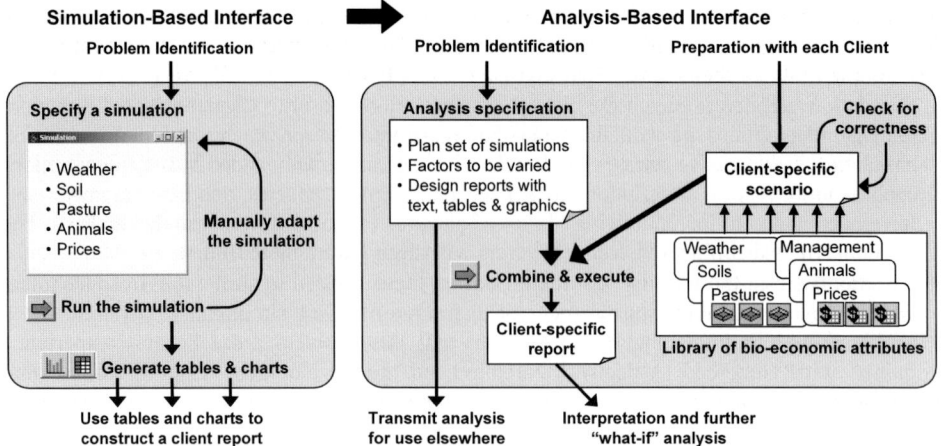

Figure 2 Changing from a simulation-based to an analysis based user interface for the GrassGro simulator. The shaded area denotes actions that are carried out by the GrassGro software

From a technical viewpoint, the GrassGro software has turned out to be well adapted to teaching use. Despite its drawbacks for advisors, a simulation-oriented user interface is a virtue when teaching higher-level undergraduates, because it means that they must learn the problem-solving process as part of the teaching exercise. The very detailed output options that GrassGro provides enable students to explore the basis of a simulated production outcome in plant or animal physiology. From a sociological viewpoint, the success of GrassGro in tertiary education provides an interesting twist on McCown's (2002) category of a versatile simulator used for facilitated learning: 'mutual learning' in a university setting involves students learning about the agricultural system, and lecturers learning how to teach it. Our experience in teaching use has parallels in the experience of the webGLA developers (Stuth *et al.*, 2002).

Scientific and policy users

GrassGro has been used in a variety of ways by scientists, including extrapolation of experimental results (Cayley *et al.*, 1998; Cohen *et al.*, 2004), analysis of climatic variability (Clark *et al.*, 2003; Donnelly *et al.*, 2005) and studies of potential new production systems (Donnelly *et al.*, 2002; Moore *et al.*, 2004a).

Two of these research applications are of particular interest: drought assessment and the assessment of deep drainage from farming systems. Both issues are of pressing concern to Australian agricultural policymakers, since the cost of relief measures during drought is high, and excess deep drainage can lead to dryland salinity in much of the Australian agricultural landscape (Lambers, 2003). The GRAZPLAN development group applied GrassGro to both these problems in relatively small, point-based studies (Donnelly *et al.*, 1998; Cresswell *et al.*, 2002); it was then taken up by others and applied across entire landscapes (Crichton, 2001; Beverly *et al.*, 2003). Applying a point-based model at the landscape scale presented challenges, both in managing the weather, soils, and management inputs required to create the simulations, and in executing multiple simulations with a user interface designed to execute small numbers of runs. In the drought assessment work of Crichton (2001), the spatial resolution was quite coarse (~40 unique areas) and it was feasible to manage the volume of inputs, simulations and outputs manually; however, GrassGro was adapted to accept up-to-date weather data automatically. The landscape water balance study of Beverly et al. (2003), on the other hand, involved several management systems and hundreds of unique geographic areas. To support this project, the GrassGro simulator had to be extracted from its user interface and the research team provided with a different interface that could run large batches of simulations.

While GrassGro has been used in diverse ways by a significant number of researchers outside of the development group, the proportion of researchers trained in GrassGro who have then gone on to apply it is about one in four – noticeably lower than for advisors and educators. Why this should be so is unclear. One possibility is that researchers are relatively more willing to invest time in training, just in case GrassGro might prove to be useful in the future. A contributing factor is that both the advisor and educator user groups have benefited from access to dedicated technical support that continues past the initial training process.

'Success' and 'failure' of grassland DS tools

Do the tools in Tables 1 and 2 (and others omitted due to space constraints) together "fall far short of expectation", as McCown (2002) would have it? The answer depends upon what is expected. If the criterion is a radical transformation of the practice of grasslands agriculture, then it clearly has not been met. Impact is only partly determined by user numbers, but the uptake rates of all the tools that have been assessed are too small to form the basis of any such radical change. The only products that can be identified as having directly reached more than 20% of graziers in a large region are those that have been mass-mailed (e.g. the P fertiliser tool of Cayley & Kearney, 2000). A best estimate suggests that GrazFeed is used (directly or via an advisor) by 10-15% of Australian graziers. Adoption rates for simulator-based tools, although increasing steadily (see Figure 1), are lower still: less than 2% for both GrassGro and FARMAX when the ultimate clients (graziers) are used as the basis.

If on the other hand DS tools are expected to "pay for their keep" (Seligman, 1993), then limited adoption and resounding success can go hand in hand. A formal evaluation of GrazFeed has shown an expected benefit:cost ratio of 79 (Moore et al., 2004b), despite its relatively low adoption rate. Many of the decision calculi in Table 1 must have similarly high returns to investment, given their low cost of implementation. In education, however, it is possible to aim for more than a low but useful adoption rate: as discussed previously, the majority of degree students studying grassland agriculture in Australia are now learning from GrassGro.

Critical factors for the future

What factors will determine the impact of DS efforts for grasslands in developed countries over the next decade? In broad terms, the same factors that have applied in the last decade (Donnelly & Moore, 1999): provision of useful user interfaces and client-specific inputs, commitment by the developing organization, effective evaluation, and above all focussing on the place of a DS tool in the operations of the client. Advances in technology and the accumulation of experience have changed the DS landscape in the intervening years. However, nothing has changed with respect to organizational commitment: only the simplest decision calculus will succeed with a short-term investment.

If McCown (2002) is correct, and much of the future of models and DS tools in agricultural decision-making lies with versatile simulators, then a major question facing developers is how to construct interfaces to such simulators that are efficient for the user. Re-casting of GrassGro to be problem-centred, and the transformation of StockPol into the two-tiered FARMAX interface represent attempts to address this question; only time will tell whether these attempts will succeed.

The set of delivery channels for DS tools has broadened considerably in the last 5-10 years. As shown by examples in the preceding sections, it is now possible to deliver via the internet in a variety of ways, such as download from a Web site, writing Java applets for execution over the Internet or using one of a range of server technologies. Internet delivery has clear advantages, especially where a tool is likely to be updated frequently or carries other significant maintenance or distribution costs (e.g. information bases), but the choice of delivery channel should always follow an analysis of the other fundamentals.

The experience of the GRAZPLAN group supports McCown's (2002) contention that grassland managers demand credibility from models before they will base decisions upon them. Scientific sophistication is thus a key element of management relevance. Those who model grasslands are few, and their efforts are fragmented; there is a need to find better ways to share insights and sub-models. In this context, modular simulation frameworks (David *et al.*, 2002; Hillyer *et al.*, 2003; Donnelly *et al.*, 2005) will be an enabling technology that must be adopted.

The most critical factor determining the success of grassland tools is, and will remain, the ability to identify how new models and information can be situated in the operations of the client. New technologies for delivery, a better appreciation of management theory and participative processes of development and application will all be useful, but each by itself is insufficient. The way forward for tool developers must be to bring scientific opportunities, technological advances and social understanding together.

Acknowledgements

I thank the various people who helped me with information and insights about their DS tools, especially Alan Bell, Jerry Cherney, Richard Eckard, Kathy Lewis, Bob McCown and John Ogilvie. The GRAZPLAN project – and GrassGro in particular – has been a team effort: Terry Bolger, John Donnelly, Hugh Dove, Mike Freer, Neville Herrmann, Libby Salmon and Richard Simpson are stimulating colleagues and this paper has benefited from their insights.

References

Beverly, C.R., A.L. Avery, A.M. Ridley & M. Littleboy (2003). Linking farm management with catchment response in a modelling framework. In: M.J. Unkovich & G.J. O'Leary (eds.) Solutions for a better environment: Proceedings of the 11th Australian Agronomy Conference, Geelong, Victoria, Australia, (Feb. 2-6). <www.regional.org.au/au/asa/2003/i/5/beverly.htm>
Cayley, J.W.D., M.C. Hannah, G.A. Kearney & S.G. Clark (1998). Effects of phosphorus fertiliser and rate of stocking on the seasonal pasture production of perennial ryegrass-subterranean clover pasture. *Australian Journal of Agricultural Research*, 49, 233-248.
Cayley, J.W.D. & G.A. Kearney (2000). Profitable use of phosphorus fertiliser for temperate pastoral Australia. *Asian-Australasian Journal of Animal Sciences*, 13 (Suppl.), 191-194.
Clark, S.G., E.A. Austen, T. Prance & P.D. Ball (2003). Climate variability effects on simulated pasture and animal production in the perennial pasture zone of south-eastern Australia. 1. Between year variability in pasture and animal production. *Australian Journal of Experimental Agriculture*, 43, 1211-1219.

Cohen R.D.H., J.P. Stevens, A.D. Moore, J.R. Donnelly & M. Freer (2004). Predicted methane emissions and metabolizable energy intakes of steers grazing a grass/alfalfa pasture and finished in a feedlot or at pasture using the GrassGro decision support tool. *Canadian Journal of Animal Science*, 84, 125-132.

Cresswell, H.P., W.J. Bond, R.J. Simpson, Z. Paydar, S.G. Clark, A.D. Moore, D.J. Alcock, J.R. Donnelly, M. Freer, B.A. Keating, N.I. Huth & V.O. Snow (2002). Soil water balance of three temperate pasture systems in southern Australia. In: N.J. McKenzie, K.J. Coughlan & H.P. Cresswell (eds.) Soil physical measurement and interpretation for land evaluation. CSIRO Publishing, Melbourne, Australia, 332-343.

Crichton, J. (2001). Simulating pasture production to better manage for climate variability in NSW. First Australian Geospatial Information and Agriculture Conference. Sydney, Australia, (July 17-19), 640-654.

Daily, H.G., G.N. Hinch, J.M. Scott & J.V. Nolan (2000). The use of a decision support program to facilitate the teaching of biological principles in the context of agricultural systems. In: Effective teaching and learning at university. Teaching and educational development institute. University of Queensland, Brisbane, Australia. <www.tedi.uq.edu.au/conferences/teach_conference00/papers/daily-hinch-etal.html>

Daily, H.G., J.M. Scott & J.M. Reid (2005). Enhancing grasslands education with decision support tools. *Cork satellite workshop of the XX International Grassland Congress: In association with the European Grassland Federation,* offered paper (in press).

David, O., S.L. Markstrom, K.W. Rojas, L.R. Ahuja & I.W. Schneider (2002). The object modeling system. In: L.R. Ahuja, L. Ma & T.A. Howell (eds.) Agricultural system models in field research and technology transfer. Lewis Publishers, Boca Raton, 317-330.

Donnelly, J.R., M. Freer & A.D. Moore (1998). Using the GrassGro decision support tool to evaluate some objective criteria for the definition of exceptional drought. *Agricultural Systems*, 57, 301-313.

Donnelly J.R., M. Freer, E.M. Salmon, A.D. Moore, R.J. Simpson, H. Dove & T.P. Bolger (2002). Evolution of the GRAZPLAN decision support tools and adoption by the grazing industry in temperate Australia. *Agricultural Systems*, 74, 115-139.

Donnelly, J.R. & A.D. Moore (1999). Decision support: delivering the benefits of grazing systems research. In: J.G. Buchanan-Smith, L.D. Bailey & W.P. McCaughey (eds.) *Proceedings of the XVIII International Grassland Congress*, Winnipeg & Saskatoon, Canada, 469-478.

Donnelly, J.R., E.M. Salmon, R.D.H. Cohen, X.P. Xin & Z.L. Liu (2005). Decision support for temperate grasslands – challenges and pitfalls. *Cork satellite workshop of the XX International Grassland Congress: In association with the European Grassland Federation,* offered paper (in press).

Eckard R.J., G.W. Thomas & D.F. Chapman (2001). Nitrogen fertiliser decision support tools - making cents with best farm practices. *Proceedings of the Grassland Society of Victoria 42nd Annual Conference,* Mt Gambier, Australia, 159pp.

Freer M., A.D. Moore & J.R. Donnelly (1997). GRAZPLAN: decision support systems for Australian grazing enterprises. II. The animal biology model for feed intake, production and reproduction and the GrazFeed DSS. *Agricultural Systems*, 54, 77-126.

Fox, D.G., L.O. Tedeschi, T.P. Tylutki, J.B. Russell, M.E. van Amburgh, L.E. Chase, A.N. Pell & T.R. Overton (2004). The Cornell Net Carbohydrate and Protein System model for evaluating herd nutrition and nutrient excretion. *Animal Feed Science and Technology*, 112, 29-78.

Heard, J.W., D.C. Cohen, P.T. Doyle, W.J. Wales & C.R. Stockdale (2004). Diet Check - a tactical decision support tool for feeding decisions with grazing dairy cows. *Animal Feed Science and Technology,* 112, 177-194.

Hillyer, C, J. Bolte, F. van Evert & A. Lamaker (2003). The ModCom modular simulation system. *European Journal of Agronomy*, 18, 333-34.

Lambers, H. (2003). Dryland salinity: a key environmental issue in southern Australia. *Plant and Soil*, 257, 5-7.

Lewis K.A. & J. Tzilivakis (2000). The role of the EMA software in integrated crop management and its commercial uptake. *Pest Management Science*, 56, 969-973.

Marshall P.R., D.G. McCall & K.L. Johns (1991). Stockpol: a decision support model for livestock farms. *Proceedings of the New Zealand Grassland Association,* 53, 137-140.

McCown, R.L. (2002). Changing systems for supporting farmers' decisions: problems, paradigms, and prospects. *Agricultural Systems,* 74, 179-220.

McPhee, M.J., A.K. Bell, P. Graham, G.R. Griffith & G.P. Meaker (2000). PRO Plus: a whole-farm fodder budgeting decision support system. *Australian Journal of Experimental Agriculture*, 40, 621-630.

McPhee, M.J., G.P. Meaker, P. Graham, P.M. Carberry, B.L. Davies, M.B. Whelan, A.K. Bell & B. Clements (2003). StockPlan: decision tools for exploring management options for drought. In: J. Wilkins (ed.) Southern Beef Update 2003. NSW Agriculture Agricultural Institute, Wagga Wagga, Australia, 25-31.

Moore A.D., J.F. Angus, M.P. Bange, C.J. Crispin, J.R. Donnelly, M. Freer, N.I. Herrmann, H.E. Ottey, D. Richards, E.M. Salmon, M. Stapper & A. Suladze (2004a). Decision support tools for Australian farmers. In: R.A. Fischer, N.C. Turner, J.F. Angus, C.L. McIntyre, M.J. Robertson, A.K. Borrell & D.L. Lloyd (eds.) *Proceedings of the 4th International Crop Science Congress*, Brisbane, Australia. <www.cropscience.org.au>

Moore, A.D., J.R. Donnelly & M. Freer (1997). GRAZPLAN: decision support systems for Australian grazing enterprises. III. Growth and soil moisture submodels, and the GrassGro DSS. *Agricultural Systems*, 55, 535-582.

Moore, A.D., L. Salmon & H. Dove (2004b). The whole-farm impact of including dual-purpose winter wheat and forage brassica crops in a grazing system: a simulation analysis. In: R.A. Fischer, N.C. Turner, J.F. Angus, C.L. McIntyre, M.J. Robertson, A.K. Borrell & D.L. Lloyd (eds.) *Proceedings of the 4th International Crop Science Congress*, Brisbane, Australia. <www.cropscience.org.au>

Rickert, K.G. (1998). Experiences with FEEDMAN, a decision support package for beef cattle producers in south eastern Queensland. *Acta Horticulturae*, No. 476, 227-234.

Seligman, N.G. (1993). Modelling as a tool for grassland science progress. In: M.J. Baker, J.R. Crush & L.R. Humphreys (eds.) *Proceedings of the XVII International Grassland Congress*, Palmerstone North, Hamilton and Lincoln, New Zealand; Rockhampton, Australia, 743-748.

Shaffer, M.J., P.N.S. Bartling & J.C. Ascough (2000). Object-oriented simulation of integrated whole farms: GPFARM framework. *Computers and Electronics in Agriculture*, 28, 29-49.

Stuth J.W., W.T. Hamilton, J.C. Connor & D.P. Sheehy (1993). Decision support systems in the transfer of grassland technology. In: M.J. Baker, J.R. Crush & L.R. Humphreys (eds.) *Proceedings of the XVII International Grassland Congress*, Palmerstone North, Hamilton and Lincoln, New Zealand; Rockhampton, Australia, 749-757.

Stuth J.W., W.T. Hamilton & R. Conner (2002). Insights in development and deployment of the GLA and NUTBAL decision support systems for grazing lands. *Agricultural Systems*, 74, 99-113.

Wheeler D., S.F. Ledgard, C.A.M. deKlein, R.M. Monaghan, P.L. Carey, R.W. McDowell & K.L. Johns (2003). OVERSEER® – moving towards on-farm resource accounting. *Proceedings of the New Zealand Grassland Association*, 65, 191-194.

Decision support for grassland systems in developing countries

P.K. Thornton
International Livestock Research Institute (ILRI), PO Box 30709, Nairobi 00100, Kenya, and Institute of Atmospheric and Environmental Sciences, University of Edinburgh, West Mains Road, Edinburgh, EH9 3JG, UK
Email: p.thornton@cgiar.org

Key points

1. Information flows in complex systems are often themselves highly complex, and decision support approaches based on linear input-output processes may have only limited impact.
2. How decisions are made, and how they can be appropriately supported, is often incompletely understood, in part because of inadequate understanding of the objectives and attitudes of all the decision makers involved.
3. Much of the developing world faces daunting problems in the coming 30 years, and appropriate information could play a critical role in dealing with these.
4. System complexity, household variability, and institutional intricacies have to be embraced rather than avoided, and so decision support might best be orientated towards identifying 'hotspot' areas of the highly disadvantaged and targeting appropriate activities in pursuit of the Millennium Development Goals.
5. Effective decision support could be served by more coordinated efforts to develop and maintain key baseline databases in developing countries, and by innovative, participatory approaches to the processing, adaptation and use of information.

Keywords: models, information, decision-making, intervention mapping, policy

Introduction

Human population in the developing world is expected to increase to 7.6 billion people by 2050, from its current level of 4.7 billion. In sub-Saharan Africa, the numbers of people will more than double to 1.5 billion (FAO, 2004). This is one element in the increase in consumption of animal products projected to 2020, the others being higher incomes, increased urbanisation, and changing dietary preferences (Delgado *et al.*, 1999). Livestock production already accounts for about 40% of global agricultural production, and this figure is increasing (FAO, 2002). By 2020, developing countries will be producing some 60% of the global meat supply and 52% of the milk (Steinfeld, 2001). Given the importance of livestock to the diets and incomes of the rural poor (Thomas & Rangnekar, 2004), understanding how livestock fit into these systems, and how these systems may evolve in the future, are issues of critical importance. Clearly, there may be significant opportunities for resource-poor smallholders to benefit from this likely increase in demand for livestock products. At the same time, there may be substantial environmental, social and public health risks for smallholders; some have called attention to possible environmental problems (Barrett, 2001; Gerber *et al.*, 2004), and Slingenbergh *et al.* (2002) note that severe imbalances may occur in global livestock development; the promotion of commercial and industrial systems, coupled with stringent sanitary regulations, tends to exclude household-based livestock production systems, in which many poor families are engaged. The concentration of livestock production in urban and peri-urban areas also has direct consequences on public health. If benefits are to accrue to the rural and urban poor, through plentiful supplies of livestock products for consumers and increased production marketing opportunities for producers, then many things have to happen.

Food safety issues will have to be addressed, as will potential deleterious impacts on the environment. Policies will be needed that can remove critical market distortions, and promote institutional change in property rights in commercialising smallholder areas (Delgado et al., 2001). In all this, the implications may be enormous, but their local impacts are difficult to foresee. The situation is complicated by the fact that predicted increases in demand for livestock products over the next few decades will occur in concert with climate change and other drivers of change. These drivers will undoubtedly lead to the intensification of agricultural systems in many places (Staal et al., 2001). Working out the implications of increased demand for livestock products for poverty in developing countries will demand a taxing research agenda.

Given the dynamism of the situation in developing countries, no great foresight is needed to predict an increasing importance in the role of information to help decision makers at all levels in the agricultural sector make appropriate decisions, from international trade policy makers to community organisations and livestock farmers, in the quest for sustainable development. In this paper, a short background section highlights the changing nature of information needs and decision support, and touches on the current status of decision support in developing countries. Some recent examples of decision support tools are then described in the areas of research planning and targeting, policymaking, and livestock management. The final section lists some areas that need attention in the future. 'Decision support' is taken here to refer to a process, involving digital methods at some stage, designed to generate or interpret information to assist decision-making. 'Grassland systems' are also interpreted widely, to include pastoral, agro-pastoral, and mixed crop-livestock (grassland-based) systems. The focus of the paper is on the poor; of the 1.3 billion people globally who live on less than $1 per day, some 600 million of these people keep livestock (Thornton et al., 2002).

Decision support: background and status

McCown, (2002a) traces the history of Decision Support Systems (DSSs) back to the early days of operations research in the late 1940s. The special issue of the journal in which that paper appears is devoted to case studies of DSSs, and McCown (2002b) attempts to explain the fairly dismal record of computer-based DSSs in actually impacting on farmers' lives. He is not alone; Matthews et al. (2002) similarly attempt to explain the poor record of crop-soil models outside of the research domain. There are various reasons, but many arise from an inadequate understanding of the context and nature of decision-making. In trying to understand the scope of the issue of information provision in general (of which decision support can be seen as a part), and how it is changing, it is useful to consider the process of innovation itself. A traditional view is that Advanced Research Institutes (ARIs) do upstream research that produces outputs that are then taken on by National Agricultural Research Systems (NARSs) and adapted in some way, to produce something that can then be passed on to the extension services, who then disseminate the innovation to grateful farmers. This simple, linear model has been the norm in the mindsets of many people associated with agricultural research and development for a long time. In many ways, this is not surprising; after all, the Green Revolution can be described as a triumph of this 'Transfer of Technology' approach (Douthwaite et al., 2004). A more modern approach to the innovation process is shown in Figure 1, built around Integrated Natural Resource Management (INRM - Campbell & Hagmann, 2003). Here, the process is anything but linear. Sayer & Campbell (2001) identify three key elements for implementing INRM: management needs to be adaptive; it must move further along the research-management continuum; and the approach must provide for, and be based upon, negotiation among all stakeholders.

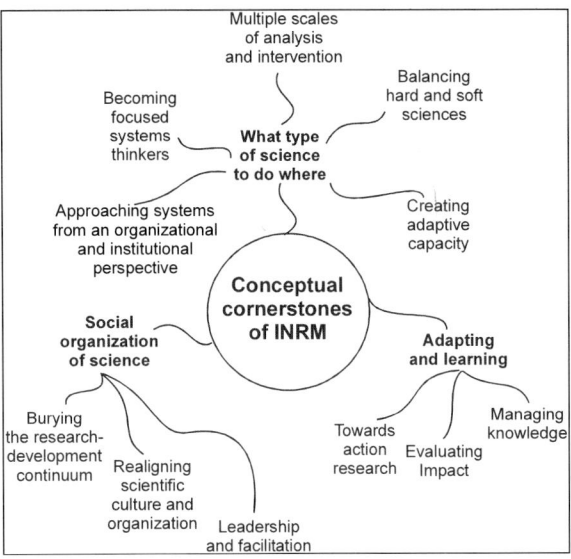

Figure 1 The conceptual pillars of INRM (Campbell & Hagmann, 2003)

It follows that there are key requirements for process facilitation and institutional adaptability if INRM is to be implemented effectively. Decision support tools are seen as being critical to the success of INRM approaches, however van Noordwijk *et al.* (2003), see these more as 'negotiation support systems', in the sense that in complex INRM situations, there are bound to be conflicts between different stakeholders with very different outlooks and objectives. Negotiating resolutions to such conflicts then becomes a key part of the process of INRM research. In any event, however innovation is seen as occurring, there may be niches for decision support at all stages of the process, and the notion of 'a decision maker' has to be expanded to include all stakeholders, not simply the farmer and other direct users of natural resources. The apparently increasing complexity of effective innovation systems has important ramifications for the design and implementation of effective decision support tools; the world is an increasingly complicated place.

Various 'domains' of decision support are shown in Table 1, together with the type of questions that each domain is concerned with, and the tools and information that may be appropriate for helping to answer such questions (NB. Table 1 is by no means exhaustive). 'Decision support' tools may be applied in all domains, involving a wide range of combination of the tools shown in Table 1. For most of these tools, there are numerous recent examples in developing countries, associated with advances in technical power or spatial coverage.

Table 1 Decision support domains, questions to be answered, and tools and information that could be used to address them

Domain	Examples of relevant questions to be answered	Relevant tools	Information sources
Research	Are there adequate frameworks for describing well-understood processes, and can these be easily applied? Do replicable methods and tools for investigating phenomena exist, and can they be easily applied?	Models (conceptual, theoretical, mathematical, etc)	Field data Agricultural statistics
Impact assessment & research planning	What has worked in the past, where, and why? Given limited resources, what should we work on in the future, and where, to maximise research impact? What constitutes a coherent portfolio of research activities that fully addresses an institute's goals?	Models GIS analysis Information systems	Poverty maps Technology intervention maps Agricultural statistics Spatial data layers
Policy formulation	What are the effects of existing policies on different beneficiary groups? What would be the impact of changing institutional and market environments on existing policy outcomes? What would be the impact of policy changes on different groups of beneficiaries?	Policy analysis & sector models Knowledge banks Toolkits Information systems	Agricultural statistics Household surveys
Management & conflict resolution	Which crop and livestock management options are most appropriate for particular conditions, now and in the future? When, where and what type of action should be taken against a crop or livestock disease? Can income be maximised while minimising nutrient losses and maximising capital asset values in livestock?	Models GIS analysis Expert opinion Participatory modelling approaches Integrated assessment	Field data Household surveys Agricultural statistics

For example; Benson (1998) produced maps of spatially- and temporally-variable nitrogen fertilizer recommendations for resource-poor, maize-based crop-livestock systems in Malawi; Thornton *et al.* (2004) provided an assessment of the value of de-stocking decisions in semi-arid grazing lands in southern Africa in response to medium-range weather forecasts of drought; Burnsilver *et al.* (2004) studied the key interactions between livestock, wildlife and pastoralists in Kajiado, Kenya, to help gauge the likely ecological impacts of subdivision of group ranches; and LPP (2004) is one in a series of CD-ROM-based knowledge banks on resource-poor livestock keepers and the animals they keep, in the context of development and sustainable livelihoods, to be used as an information source for informing policy debate.

While the development continues apace of tools and databases that should form strong foundations for different DSSs, the evidence for their use and for them impacting in the four domains shown in Table 1 is more patchy. In the domains of research and impact assessment and research planning, the use of tools and databases is becoming more widespread. For policy formulation and management, however, there is generally less evidence of use and impact. As McCown, (2000b) noted, there may be many reasons for this, but in addition to the increasing complexity of the innovation process associated with research for development, other plausible reasons include the enormous variability in resource endowments and household objectives (making it very difficult to generalise, particularly in the management and conflict resolution domain), and the complexity of information flows between the various stakeholders in any situation. Given a complex and dynamic decision-making context, and the seemingly ever-present problems associated with reliable data acquisition in many developing countries (Minae *et al.*, 2003), it is perhaps not surprising that DSSs have yet to demonstrate substantial impact in some domains. The next section presents examples of recent and on-going work in the domains of impact assessment and research planning (PRIMAS), policy formulation (EXTRAPOLATE), and livestock management (Talking Pictures), that show how some of these difficult problems might be partially addressed.

Examples of DSS

Decision support for research planning and targeting: PRIMAS

PRIMAS (Poverty Reduction Intervention Mapping in Agricultural Systems) is a CD-ROM-based tool that is designed to generate an integrated series of maps on the location of resource-poor livestock keepers and associated natural resource, climatological, communication and marketing data layers for different systems. From an understanding of the problems and needs of resource-poor livestock keepers, there may be several technology options appropriate for the production systems under study. Each of these options may be expected to go someway to solving a perceived problem, or opening up new opportunities for livestock keepers. The PRIMAS tool enables options to be filtered by attempting to match the characteristics of particular options with the characteristics of particular target groups in the landscape (as far as this can be done sensibly in terms of spatial data). A prototype has been developed for Kenya, and this is being refined and its coverage extended. This involves the collation of a great deal of spatial information related to climate, weather, soils, forages and forage availability, roads, markets, cities and towns, predominant livestock species, human and animal population densities, and pest and disease risk. PRIMAS includes a small database of interventions for pastoral and dairy systems. These are summarised according to a common format that describes the intervention, who was responsible for developing it, and where it has been (or is being) applied. PRIMAS allows the user to browse all the available data layers, using a 'map explorer'. The user can carry out a set of overlays, which are basically simple intersections of different spatial data layers. The user can also do simple weighted overlays, where scores or weights can be assigned to the probability of a particular value being associated with a particular spatial variable. Weighting may be useful in assessing variables such as the degree of market integration for target groups, where it decreases with distance from markets and all-weather roads, for example.

PRIMAS uses third-party software, and so can be used and distributed free of any royalty payments. In addition, the user can add interventions and spatial data layers to the system. As an example, consider the distribution of *Pennisetum purpureum* (Napier grass). To map this distribution in PRIMAS, some description of the areas in Kenya where it thrives is needed. Using the information in KARI (1992), *P. purpureum* intervention was described in terms of three constraints: altitudes between 1500 and 2000 m above sea level; rainfall in excess of 750 mm per year, and soil pH greater than 4 (domains may be determined in much more sophisticated ways based on multivariate analysis using discriminant analysis or logistic regression analysis). Figure 2 shows the results in PRIMAS of running (overlaying or intersecting) these three constraint layers.

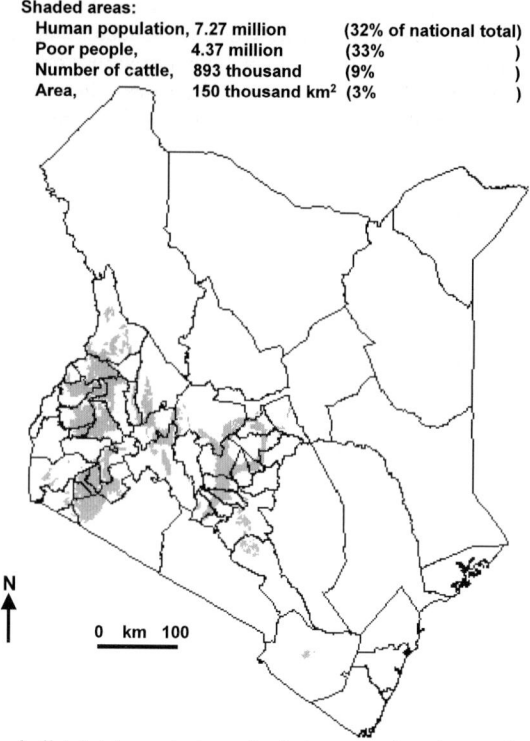

Shaded areas:
Human population, 7.27 million (32% of national total)
Poor people, 4.37 million (33%)
Number of cattle, 893 thousand (9%)
Area, 150 thousand km² (3%)

Figure 2 Kenya and district boundaries, shaded areas showing a domain for *Pennisetum purpureum* (Napier grass)

The shaded areas are those that satisfy the constraints relating to altitude, soil pH and annual rainfall, together with tabular output that summarises the area of the domain, the human population, the livestock population, and the number of resource-poor livestock keepers (Thornton *et al.*, 2002). In this example, the *P. purpureum* domain contains about 33% of Kenya's human population, but less than 3% of the land area of the country. The prototype of PRIMAS is being tested in a project to assess options, from the point of view of farmers' objectives and attitudes, to improve household well-being. An output of this work is to take the experiences gained from study sites in western, central and coastal Kenya, and extrapolate their relevance to East Africa as a whole. A process of up-scaling is underway, whereby potential domains are identified that are similar to the study sites in terms of agro-ecology, distance to markets, population densities, and farming systems. Field visits are made to randomly selected sites within these domains, to 'ground-truth' the domain identification and, if possible, assess the relevance and potential of involving farmers in trying some of the interventions tested earlier in the project at the original study sites. PRIMAS is being used to identify domains related to the key technology interventions being tested within the project, including methods of manure and cattle urine management, and the growing of small areas of vegetables for the local market (Waithaka *et al.*, 2005).

EXTRAPOLATE (EX-ante Tool for RAnking POLicy AlTErnatives) arose out of the need for a decision support tool to assess the impact, both visually and numerically/verbally, of different policy measures. It is a tool that can serve as a filter (in a similar way to PRIMAS) to sift through (in an *ex-ante* fashion) a range of policy measures, to identify those that could be applied in a specific situation to achieve particular outcomes that further policy objectives of specific decision makers. It was envisaged that this would be the first step in assessing potential impact before looking at a situation in more detail. The framework (Figure 3) is circular, and an analysis might start by identifying the opportunity (number 1 in Figure 3), such as the existence of a rapidly expanding market for eggs around a large town. Who are the potential beneficiaries of this expansion (number 2)? Potentially the peri-urban and rural farmers in the vicinity. Who is currently benefiting from this expansion (number 3)? Only a limited number of these peri-urban and rural farmers. What are the constraints to more farmers benefiting from this expanding market (number 4)? There may be several: for example, many farmers have limited physical access to feeds and markets; commercial producers have a competitive edge; many farmers lack the required husbandry skills; and there are poultry diseases that are constraining smallholder producers. Several policies can be envisaged that might have an impact both on the constraints that currently prevent smallholders from taking more advantage of the expanding egg market, and on other opportunities for increasing incomes (number 5). For example, policies that influence rural infrastructure (roads for access, etc), policies that favour commercial producers at the expense of smallholder producers, and existing sanitation rules. If policies are changed, what are the possible impacts on the constraints, and what are the trade-offs involved (number 6)? For example, commercial producers may suffer; it may result in lower food standards; it may increase peri-urban pollution. So if for example, access to markets is improved, sanitation laws are relaxed, or animal health services improved, what are the impacts likely to be, and who will gain and who lose out?

Mainstream economics has a wide array of tools to assess these sorts of policy changes in a rigorous and quantitative fashion (policy analysis matrices, computable general equilibrium models, etc). EXTRAPOLATE is a rapid screening device, to allow the user to carry out quick assessments of likely candidate policy changes that may have particularly beneficial impacts on the poor in particular situations, the most promising of which can then be analysed further using much more rigorous (and time- and data-intensive) methods. EXTRAPOLATE is still being developed and tested; the framework currently implemented is slightly different from that shown in Figure 3, although the underlying ideas are the same. There are various data requirements for setting up a new case study. First, a set of beneficiaries is defined, together with their livelihood statuses. These may include landless labourers, large commercial producers, smallholder producers, and urban consumers. Next, one or more constraints facing each beneficiary group are identified, such as access to household cash, animal disease, and animal feed supplies. These constraints are then quantified on a scale of 1 to 10 in terms of their importance to each beneficiary group. For example, animal disease may be of no direct importance to urban milk consumers, while access to cash is a significant constraint for smallholders. The next step involves identifying a set of outcomes that are affected by the constraints, such as increased employment, increased milk consumption (with associated health and nutrition benefits within the smallholder household and for local (rural) consumers), increased milk sales (with associated income effects), and increased crop and livestock production (with associated consumer price effects). The strength of the impacts that relaxing the various constraints could have on these outcomes (i.e., the marginal strength of impact) is then estimated by the user. So, for example, if households could gain access to more cash, this might allow smallholders to hire more labour and purchase other inputs to raise production. These different marginal changes in outcomes will have different impacts for the different beneficiary groups; e.g. a change that results in increased employment opportunities may benefit a beneficiary group made up predominantly of hired labourers. Increased milk sales, on the other hand, will benefit milk producers in general, but large commercial producers relatively more than smallholders.

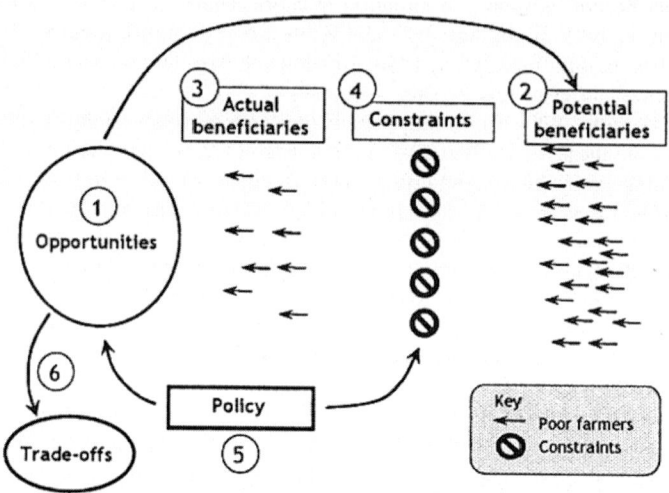

Figure 3 Conceptual model behind EXTAPOLATE

The final data-input step is to identify policies that will have an impact on one or more of the constraints previously identified. These may relate to state provision of veterinary services or to land tenure reform. The potential marginal impacts of each policy on each constraint are then estimated. A policy of subsidised veterinary service provision may have an impact on a disease constraint, but no direct effect on increasing the supply of labour to smallholder households, for example. Once the input data have been specified, the example can be input into the software, and various outputs generated. The livelihood status of the beneficiary groups can be contrasted and compared both before the application of the policy and after, using simple bar charts that reflect any resultant changes in (arbitrary) livelihood scores. Some beneficiary groups may benefit directly from the new policy, while others may be adversely affected. The tool can also summarise the trade-offs amongst beneficiaries that arise from the application of the proposed policy.

EXTRAPOLATE is still highly experimental, but is being tested in Senegal, Kenya and Uganda as part of FAO's Pro-Poor Livestock Policy Initiative. Spatial components are being added to EXTRAPOLATE, so that beneficiaries and constraints can be mapped, where this is appropriate. The usefulness of EXTRAPOLATE for detailed policy analysis remains to be demonstrated, but initial indications are that it has much to offer as a framework for promoting focussed thinking about a case-study and as a rapid screening tool.

This example highlights the work of Thorne & Dijkman, (2001) and Dijkman & Thorne, (2003). The lack of effective linkages among research, extension and farmers is a global problem, and there is a real need to generate information in a way that can simplify interactions between extension services and farmers, and that allows farmers to be actively involved in the evaluation of options. As a response, they (Thorne and Dijkman) developed a software package called Talking Pictures, a dynamic, pictorial system used to represent the nutritional management of dairy cows in smallholder farming systems. Talking Pictures makes use of the Dairy Rationing System for the Tropics (DRASTIC) - itself a decision support tool for planning dairy feeding under tropical conditions (Thorne, 1999). DRASTIC is built around a detailed biological model of protein and energy nutrition, and predicts milk production levels when animals are fed a particular mix of feeds. The nutritional variables needed in the model are assessed from simple indicators of feed quality, so that animal performance predictions can be made for variable feed compositions in the absence of highly quantitative data.

Talking Pictures generates pictorial guides that can be printed out and used in the field. These guides consist of several separate pictorial layers that incorporate information on the genotype, animal condition, the stage of lactation and physiological status, the calf rearing system being used, and the quantity and quality of basal and supplemental feeds. The different layers are dynamically linked, and provide pictorial information concerning expected production outputs, costs, and income. Each hard-copy guide produced in Talking Pictures is generated in such a way that the user can choose one of three options that are appropriate to the user's own conditions (stage of lactation, physiological status, animal condition, etc). Each option is either colour- or pattern-coded, depending on whether colour or black-and-white printers are used to generate the hard-copy guides. The pattern or colour for each of the chosen input layer options is transferred with dry-wipe markers to a reusable laminated 'credit card', leading to a unique sequence of five colours or patterns. In an example developed for Kiambu, Kenya, there are option layers for three stages of lactation, three forage quality levels, three calf rearing systems, and three levels of body weight change (in the software, each of these levels has an appropriate, local picture associated with it). The sequence chosen by the user is then matched to the appropriate sequence out of all the available possibilities, which are listed on three pages and linked to a pictorial representation of the expected production level for the animal in question. The user can then turn to the appropriate supplementation page - indicated by a picture of the expected production level. Different pictorial representations can then be selected of supplementary feeds and different levels of supplementation, which are connected to a picture of the total milk production expected. Each of the supplementation choices also supplies pictorial data on the ratio between milk and concentrate prices at which supplementation of the chosen quantity becomes profitable. In preparing a new guide for a specific area and situation, pictures of the appropriate cow genotype, calf rearing systems, recognizable quantities of feed and milk, and types of concentrates, need to be assembled. These can then be entered into a picture library within Talking Pictures and can then be linked to appropriate biological input data for running DRASTIC. Assembling appropriate photos requires the determination and testing of unit sizes and weights for basal diets, supplements and milk, as well as locally-recognizable pictorial representations of calf feeding systems, stage of lactation, animal conditions, and farmers' perceptions of fodder quality.

A prototype of Talking Pictures was developed for dairy systems (TP-D) in Tanzania, Kenya and India, using participatory methods. Dijkman and Thorne (2003) note that TP-D represents a real innovation in effectively enhancing farmers' personal, dynamic and science-based decision-making capacity that has not been offered by previous static extension materials. Farmers using TP-D increased milk off-take compared with control groupss, augmenting milk-derived income often by as much as 25%. Many of the farmers exposed to TP-D in Tanzania were still using the methodology two years later. The ease with which TP-D guides, based on a format originally developed in Tanzania, have been produced for another country (Kenya) and another sub-continent (India), and the ease with which smallholder farmers in these areas effectively use the guides to address their most significant dairy feeding problems, has shown that TP-D is both valuable and generic. TP-D is currently being applied in further pilot sites in India.

The future

The way in which research for development is done continues to change enormously, and working out the implications for providing effective decision support in an increasingly complex and dynamic environment is an on-going and difficult task. Clearly, all the relevant stakeholders have to be fully involved, and the dimensions of smallholders' problems have to be adequately understood; this includes understanding the nature of household decision making, and what it takes to make it more effective; understanding the flow of information between all the various stakeholders; understanding how new or different knowledge can help; and understanding better how researchers can build on indigenous knowledge that has been accumulated by livestock keepers over many years.

Research for development and poverty alleviation is faced with considerable challenges. On the one hand, enormous impacts are needed if the Millennium Development Goals (Morton, 2001) are to be even partially achieved – the need for up-scaling of local successes is critical. On the other hand, the more that is understood about the importance of local conditions, the variability associated with householders' attitudes and objectives and the socio-cultural milieu within which householders operate, the more challenging it becomes to identify the huge gains that may be widely applicable to vast areas and millions of people. One way to deal with this tension may be to acknowledge that impacts (of policy, technology, information provision, etc.) may often be relatively localised, and that appropriate targeting then becomes critical if the poor are to be reached. Appropriate targeting, however, is highly dependent on data availability, and many areas of the developing world are still hampered by chronic lack of up-to-date data for policy makers and development agencies as well as at farm level (Minae et al., 2003). One problem is that the most important variables for decision support are often those for which we have the least accurate and most uncertain information, such as livestock disease risk and the prevalence and depth of poverty (Robinson, 2002). On-going efforts such as that led by FAO in cooperation with partners are attempting to assemble production data at the district level for developing countries, but even this (relatively aggregated) level of data collection is problematic. The data problem can only adequately be addressed through integrated, collaborative and wide-scale efforts. Targeted decision support efforts in developing countries will have to continue to rely on widespread use of 'expert opinion' and on using proxies of key variables that have not been quantified reliably, in efforts to assemble the information that is required. There is also a great deal to do on information interpretation, and the idea of converting the outputs of complex tools and complex analyses into quick messages in innovative formats that are useful for decision makers. Tools such as Talking Pictures, PRIMAS and EXTRAPOLATE may show the way forward in some respects. Given the uncertainties of the future, and the continuing intensification of agricultural systems in many places, such work on targeting and decision support is crucial if research is to hit the right developments targets in the coming decades.

Acknowledgements

Resources for the development of; PRIMAS came from DFID's Livestock Production Programme, EXTRAPOLATE from the Pro-poor Livestock Policy Initiative, FAO. Without implicating them in any way, I am very grateful to Peter Thorne, Tim Robinson, Jeroen Dijkman, Wyn Richards, Mario Herrero, Carlos Quiros, Maren Radeny, and Nelson Mango.

References

Barrett, J.R. (2001). Livestock farming: eating up the environment? *Environmental Health Perspectives*, 109 (7). <http://ehpnet1.niehs.nih.gov/ docs/2001/109-7/focus1.html>

Benson, T. (1998). Developing flexible fertilizer recommendations for smallholder maize production in Malawi. In: S.R. Waddington (ed.) Soil fertility research for maize-based farming systems in Malawi and Zimbabwe. SoilFertNet, CIMMYT-Zimbabwe, Harare, Zimbabwe, 275-286.

Burnsilver, S., R.B. Boone & K.A. Galvin (2004). Linking pastoralists to a heterogeneous landscape: the case of four Maasai group ranches in Kajiado District, Kenya. In: J. Fox, V. Mishra, R. Rindfuss & S. Walsh (eds.) Linking household and remotely sensed data: methodological and practical problems, Kluwer Academic, Boston, 173-199.

Campbell, B.M. & J. Hagmann (2003). Rising to the challenge of poverty and environmental sustainability: towards a conceptual and operational framework for INRM. In: Building sustainable livelihoods through integrated agricultural research for development, Vol. 2, (Reference materials from the consultative programme development process. FARA, Accra Ghana, 1-20.

Delgado, C., M.W. Rosegrant & S. Meijer (2001). Livestock to 2020: the revolution continues. Proceedings of the International Agricultural Trade Research Consortium (IATRC), Auckland, New Zealand. <http://www.iatrcweb.org/oldiatrc/Papers/Delgado.pdf>

Delgado, C., M.W. Rosegrant, H. Steinfeld, S. Ehui & C. Courbois (1999). Livestock to 2020: the next food revolution. Food, Agriculture and the Environment Discussion Paper 28. IFPRI/FAO/ILRI, Washington, DC, USA, 72pp.

Dijkman, J.T. & P.J. Thorne (2003). Analysis, management and decision support for farmers' feeding strategies: Talking Pictures Phase II. Final Technical Report, DfID-LPP Project R7855. Stirling Thorne Associates, Llangefni, UK, 15pp.

Douthwaite, B., N. de Haan, V.M. Manyong & J.D.H. Keatinge (2004). Blending "hard" and "soft" science: the "follow-the-technology" approach to catalyzing and evaluating technology change. In: B.M. Campbell & J.A. Sayer (eds.) Resource management: linking productivity, the environment and development. CABI Publishing, Wallingford, 15-32.

FAO (2002). World agriculture: towards 2015/30. Food and Agriculture Organisation (FAO), Rome, Italy, 97pp.

FAO (2004). Food and Agriculture Organisation of the United Nations Statistical Databases. <http://apps.fao.org>

Gerber, P., P. Chilonda, G. Franceschini & H. Menzi (2004). Geographical determinants and environmental implications of livestock production intensification in Asia. *Bioresource Technology*, (in press).

KARI (1992). 'Tumbukiza'. A better way to grow Napier Grass for Milk. KARI Information Leaflet, Kenya Agricultural Research Institute-Kakamega and Kitale, 4pp.

LPP (Livestock Production Programme) (2004). Smallstock in development, knowledge bank. CD-ROM, June 2004. DFID and NR International, Aylesford, Kent.

Matthews, R., W. Stephens & T. Hess (2002). Impacts of crop-soil models. In R.B. Matthews, & W. Stephens (eds.) Crop-soil simulation models: applications in developing countries. CABI Publishing, Wallingford, UK, 195-205.

McCown, R.L. (2002a). Locating agricultural decision support systems in the troubled past and socio-technical complexity of 'models for management'. *Agricultural Systems*, 74, 11-25.

McCown, R.L. (2002b). Changing systems for supporting farmers' decisions: problems, paradigms, and prospects. *Agricultural Systems*, 74, 179-220.

Minae, S., D. Baker & J. Dixon (2003). Status of farm data systems and farmer decision support in sub-Saharan Africa. Agrippa. <ftp://ftp.fao.org/UPLOAD/AGRIPPA/628_EN.PDF>

Morton, J.F. (2001). International development goals and livestock research: towards meeting the development assistance criteria targets. In: S.D. Hainsworth, S.H. Godfrey, R.W. Matthewman & J.L. Richards (eds.) Linkages, livestock and livelihoods: promoting coordination in livestock research for poor people. NRI International Ltd., Pembroke, Chatham Maritime, Kent, UK, 19-21.

Robinson, T.P. (2002). Decision support for trypanosomiasis control in Uganda. Report to the Hon. Mary Mugyenyi, Minister of State for Animal Industry. People Livestock and Environment Working Paper number 3. International Livestock Research Institute, Nairobi, Kenya, 25pp.

Sayer, J.A. & B.M. Campbell (2001). Research to integrate productivity enhancement, environmental protection, and human development. *Conservation Ecology*, 5 (2), 32. <URL: http://www.consecol.org/vol5/iss2/art32>

Slingenbergh, J., G. Hendrickx & W. Wint (2002). Will the livestock revolution succeed? *AgriWorld Vision*, 2 (4), 31–33.

Staal, S.J., S. Ehui & J.C. Tanner (2001). Livestock-environment interactions under intensifying production. In: D.R. Lee & C.B. Barrett (eds.) Tradeoffs or synergies? Agricultural intensification, economic development and the environment, CAB International, Wallingford, UK, 345-364.

Steinfeld, H. (2001). Livestock production towards 2020. In: S.D. Hainsworth, S.H. Godfrey, R.W. Matthewman & J.I. Richards (eds.) Proceedings of the first interagency meeting on livestock production and animal health, Natural Resources International Ltd, Chatham Maritime, UK, 7-13.

Thomas, D. & D. Rangnekar (2004). Responding to the increasing global demand for animal products: implications for the livelihoods of livestock producers in developing countries. In: E. Owen, T. Smith, M.A. Steele, S. Anderson, A.J. Duncan, M. Herrero, J.D. Leaver, C.K. Reynolds, J.I. Richards & J.C. Ku Vera (eds.) Responding to the livestock revolution: the role of globalisation and implications for poverty alleviation. British Society of Animal Science Publication 33, Nottingham University Press, 1-35.

Thorne, P.J. (1999). DRASTIC - A Dairy Rationing System for the Tropics. Version 1.01c for Windows 3.1. Natural Resources International Ltd, Chatham Maritime, UK, 48pp.

Thorne, P.J. & J.T. Dijkman (2001). Simple decision support for feed planning. Agrippa. <http://www.fao.org/DOCREP/ARTICLE/AGRIPPA/X9500E06.HTM>

Thornton, P.K., R.H. Fawcett, K.A. Galvin, R.B. Boone, J.W. Hudson & C.H. Vogel (2004). Evaluating management options that use climate forecasts: modelling livestock production systems in the semi-arid zone of South Africa. *Climate Research*, 26, 33-42.

Thornton, P.K., R.L. Kruska, N. Henninger, P.M. Kristjanson, R.S. Reid, F. Atieno, A. Odero & T. Ndegwa (2002). Mapping poverty and livestock in the developing world. International Livestock Research Institute, Nairobi, Kenya, 124pp.

van Noordwijk, M., T.P. Tomich & B. Verbist (2003). Negotiation support models for integrated natural resource management in tropical forest margins. In: B.M. Campbell & J.A. Sayer (eds.) Resource management: linking productivity, the environment and development. CAB International, Wallingford, UK, 87-108.

Waithaka, M.M., P.K. Thornton, M. Herrero & K.D. Shepherd (2005). Bio-economic evaluation of farmers' perceptions of sustainable farms in western Kenya. *Agricultural Systems,* (in press).

Keyword index

adoption	343
agri-environment schemes	305
animal performance	81
animal production	111, 123, 149
animal-plant interactions	97
apomixis	69
biodiversity	13, 193
biopharming	167
climate	265
climate change	179, 251, 279
clover	57
communal range	339
conserved forage	123
consumer	41
dairy	13
decision-making	415
decomposition	209
developing countries	29
development	323
DSS	389
ecosystem functioning	295
ecosystem services	219
ecosystems	193
elevated atmospheric CO_2	251
endophyte	167
extension work	375
farmer participation	359
fatty acids	41
fertilisers	111
forage	167
forage adoption	347
forage and pasture crops	137
forage technology	347
genetic improvement	69
genetic mapping	57
geospatial	389
GIS	389
grass	41
grass breeding	57
GrassGro	403
grassland	179
grassland mitigation	279
grassland systems	251
grazing	295
grazing management	81
greenhouse gas	279
habitat	219
hay	123
herders	339
heterogeneity	97
information	415
intensive	13
intervention mapping	415
irrigation	227
landscape	193

ley farming	137
market access	347
microorganisms	209
milk and beef production	343
models	389, 415
molecular breeding	69
molecular markers	57
multifunctional	13
nitrogen	239
nitrogen fixation	149
nutrients	193
on farm experimentation	359
pastoralists	323
pasture	265
pathogens	239
phosphorus	239
planning tools	403
plant choice	111
policy	227, 265, 323, 415
pollution	13
price projections	29
production	305
quality	41
quantitative eco-physiology	389
radio frequency identification	375
rangeland	265
remote sensing	389
resource integration	111
RFID	375
rotational grazing	339
ruminants	29, 279
salinisation	239
sediment	239
seedling recruitment	305
selection	57, 69
sheep	375
silage	123
smallholders	323
socio-economic factors	347
soil	265
soil protection	219
soil-plant systems	209
species	251
species richness	305
stocking rate	97
sward manipulation	111
systems development	375
technology transfer	149
transformation	167
tropical forages	343
tropical grasslands	81
versatile simulators	403
water	193
yield	251

Author index

Abberton, M.T.	57	Kibblewhite, M.G.	219
Angadi, S.V.	137	Kong, W.D.	209
Argel, P.J.	343	Larbi, A.	339
Askew, M.F.	179	Lascano, C.E.	343
Baas, S.	323	Li, W.	227
Bai, Y.	97	Loreau, M.	295
Baker, B.B.	389	Lüscher, A.	251
Baltenweck, I.	347	Mack, S.	323
Batello, C.	323	Maltoni, S.	111
Battikha, N.	339	McIvor, J.G.	111
Bellotti, W.D.	137	McNabb, W.	167
Boelt, B.	137	Michalk, D.L.	193
Bryan, G.	167	Moloney, A.P.	41
Chadwick, S.A.	375	Moore, A.D.	403
Chapman, D.F.	389	Murphy, J.J.	41
Chen, B.D.	209	Mwangi, D.M.	347
Chen, W.	137	Mwendia, S.W.	347
Christey, M.	167	Nan, Z.B.	209
Christie, P.	209	Nash, D.M.	239
Clark, H.	279	Newton, P.C.D.	251
Cline, S.A.	227	O'Kiely, P.	123
Conner, T.	167	Ominski, K.H.	137
Cooper, A.S.	389	Peel, S.	13
Da Silva, S.C.	81	Peters, M.	149
Daly, C.P.Q.	389	Pinares-Patiño, C.	279
Daniels, N.J.	375	Pollock, C.J.	57
de F. Carvalho, P.C.	81	Porqueddu, C.	111
deKlein, C.	279	Powell, J.M.	137
Delgado, C.L.	29	Rath, M.	13
Dewhurst, R.J.	41	Resende, R.M.S.	69
do Valle, C.B.	69	Reynolds, S.G.	323
Entz, M.H.	137	Richardson, K.	167
Fennessy, P.F.	375	Ringler, C.	227
Follett, R.F.	265	Roberts, N.	167
Franzel, S.	149	Romney, D.	347
Fuhrer, J.	251	Rosegrant, M.W.	227
Ghassali, F.	339	Schuman, G.E.	265
Han, G..	97	Scollan, N.D.	41
Hannaway, D.H.	389	Shelton, H.M.	149
Haygarth, P.M.	239	Speck, P.A.	375
Hector, A.	295	Staal, S.	347
Holmann, F.	343	Stür, W.W.	359
Horne, P.M.	359	Thornton, P.K.	415
Humphreys, M.O.	57	Tiedeman, J.A.	339
Isselstein, J.	305	Valmonte-Santos, R.A.	227
Jank, L.	69	Voisey, C.	167
Johnson, R.	167	Wang, D.	97
Kaiser, A.G.	123	Zhu, Y.G.	209
Kemp, D.R.	193		